AF518284

PLANT VIRUSES
Evolution and Management

PLANT VIRUSES
Evolution and Management

Rihana Bibi
Mridul Pawar
Bhaskar Biswas

2025
Agri Horti Press
New Delhi-110 084

ISBN: 978-93-92969-99-7 (Hardbound)

Published By

Agri Horti Press

New Delhi

E-mail: agrihortipress@gmail.com

Digitally Printed at : Replika Press Pvt. Ltd.

Preface

In the intricate web of plant biology, viruses emerge as formidable adversaries, capable of wreaking havoc on crops, threatening food security, and destabilizing agricultural economies worldwide. Yet, amidst the challenges posed by plant viruses, lies a fascinating story of evolution, adaptation, and innovation. By unraveling the mysteries of plant virus biology and exploring novel management strategies, we can mitigate the impact of these pathogens and safeguard our agricultural systems for future generations.

This book, titled "Plant Viruses: Evolution and Management," embarks on a comprehensive journey through the dynamic world of plant virus biology, evolution, and control. Drawing upon the collective expertise of virologists, plant pathologists, agronomists, and biotechnologists, this volume offers a holistic perspective on the multifaceted challenges and opportunities presented by plant viruses.

From the origins of plant viruses and their coevolutionary dynamics with host plants to the mechanisms of viral replication, transmission, and pathogenesis, "Plant Viruses: Evolution and Management" provides readers with a thorough understanding of the intricate interplay between viruses and their plant hosts. Through a synthesis of cutting-edge research, case studies, and practical insights, we explore the genetic diversity of plant viruses, the emergence of novel strains, and the factors driving viral evolution in agricultural and natural ecosystems.

Moreover, this book delves into innovative strategies for plant virus management, including classical breeding approaches, genetic engineering techniques, and sustainable agronomic practices. By examining the latest advancements in virus detection, diagnosis, and crop protection strategies, we aim to empower researchers, farmers, and policymakers with the knowledge and tools needed to combat plant viruses effectively while minimizing environmental impact.

As we navigate the complexities of plant virus biology and management, it is our hope that "Plant Viruses: Evolution and Management" will serve as a valuable resource for researchers, students, and practitioners engaged in plant pathology, agriculture, and biotechnology. Whether you are exploring the frontiers of virus-host interactions, developing novel control strategies, or advocating for sustainable agricultural practices, this book invites you to join us in the quest to understand, manage, and ultimately conquer plant viruses for the benefit of global food security and agricultural sustainability.

Rihana Bibi

Mridul Pawar

Bhaskar Biswas

Contents

Chapter-1

Special Issue "Plant Viruses: From Ecology To Control"

Plant viruses cause many of the most important diseases threatening crops worldwide. Over the last quarter of a century, an increasing number of plant viruses have emerged in various parts of the world, especially in the tropics and subtropics. As is generally observed for plant viruses, most of the emerging viruses are transmitted horizontally by biological vectors, mainly insects. Reverse genetics using infectious clones—available for many plant viruses—have been used for the identification of viral determinants involved in virus–host and virus–vector interactions. Although many studies have identified a number of factors involved in disease development and transmission, the precise mechanisms are unknown for most of the virus–plant–vector combinations. In most cases, the diverse outcomes resulting from virus–virus interactions are poorly understood. Although significant advances have been made towards understanding the mechanisms involved in plant resistance to viruses, we are far from being able to apply this knowledge to protect cultivated plants from all viral threats.

The aim of this Special Issue was to provide a platform for researchers interested in plant viruses to share their recent results related to the various aspects of plant virology: ecology, virus–plant host interactions, virus–vector interactions, virus–virus interactions, and control strategies. A total of 15 papers have been contributed by 96 authors from 18 countries to the issue, comprising ten research articles, one short communication, and four reviews (**Figure 1**).

Plant virus ecology looks at virus–host-environment interactions and, in a broad sense, includes studies on biodiversity and evolution. Several papers in this Special Issue focused on this topic. Martínez-Turiño et al. [**1**] described how virus host jumping, from *Nicotiana clevelandii* to *Chenopodium foetidum*, can be boosted by adaptation to a bridge plant species, *Arabidopsis thaliana*, using mutants of the potyvirus plum pox virus. Kim et al. [**2**] also analyzed the host adaptation process by characterizing field isolates of the capillovirus apple stem grooving virus sampled from two plant hosts, apple and pear trees, revealing that host adaptation was influenced by the host's codon-usage. Ferro et al. [**3**] studied the complexity of sweepovirus-deltasatellite-plant host interactions by looking at the diversity of satellite and helper virus natural populations, as well as by performing co-inoculation experiments to assess the ability of a number of geminivirids to transreplicate sweet potato leaf curl deltasatellite 1. Montes et al. [**4**], using a genome-wide association strategy, identified a number of genes associated with virulence in *Arabidopsis* *thaliana* genotypes infected by the cucumovirus cucumber mosaic virus that were linked to virus seed transmission. On a more applied aspect, Nancarrow et al. [**5**] estimated the yield losses caused by the luteovirus barley yellow dwarf virus-PAV in wheat and barley by conducting a three-year field study in Australia, highlighting the importance of performing this type of study under varying conditions for specific cultivar–vector–virus combinations.

Figure 1. A word cloud created from the titles of every article published in this Special Issue.

High-throughput sequencing (HTS) technologies have become indispensable tools to characterize plant virus diversity thanks to their ability to detect virtually any virus without prior sequence knowledge. Three papers published in this Special Issue deal with HTS. Kutnjak et al. [**6**] present a critical overview, useful for both beginners and expert scientists interested on plant virome projects, of the steps involved in HTS, including available bioinformatic tools and the steps to be used to implement a structured data analysis. Maachi et al. [**7**] used HTS to identify the viruses present in a number of symptomatic tomato plant samples using two RNA purification methods, illustrating its usefulness with the first identification of three viruses infecting this crop in Spain. Ben Chehida et al. [**8**] went further by showing that nanopore sequencing is a reliable alternative for obtaining complete genome sequences of geminivirids, which is also applicable to other circular ssDNA viruses.

The study of the precise interactions established between viruses and plant hosts to cause a productive infection is a fundamental subject that has received the attention of three research groups contributing to this Special Issue. Teixeira et al. [**9**] reviewed the different levels of geminivirid perception and defense mechanisms acting in the plant host, as well as the counter-defense strategies evolved by geminivirids to overcome such mechanisms of resistance. Kumar and Dasgupta [**10**] reviewed another interesting topic within the scope of this Special Issue, the movement proteins encoded by virus genomes, focusing on structural diversity, multifunctionality, and interaction with plant host proteins. Leastro et al. [**11**] analyzed the association of the movement and capsid proteins of the cilevirus citrus leprosis virus C2 with plant cell membranes by using computer predictions and molecular assays, discussing the implications of their findings for the intracellular movement of the virus.

Deciphering virus–vector interactions is an emerging research subject in plant virology. Thus far, studies concerning begomoviruses, an important group of emerging pathogens transmitted by the whiteflies of the *Bemisiatabaci* complex, have been mostly conducted with tomato yellow leaf curl virus. In this Special Issue, Chi et al. [**12**] showed that a vacuolar protein sorting-associated protein from the Asia II 1 whitefly interacts with the coat protein of the begomovirus cotton leaf curl Multan virus, suggesting that this protein may play an important role in transmission.

Mixed-infections with two or more plant viruses are frequent in the field, with viruses being able to interact in multiple and intricate ways. These interactions are generally categorized as synergistic, antagonistic, or neutral. In this Special Issue, Elvira González et al. [**13**] experimentally analyzed the interactions between an asymptomatic persistent virus, the amalgavirus southern tomato virus, and two well-known acute viruses infecting tomato, the cucumovirus cucumber mosaic virus and the potexvirus pepino mosaic virus, showing that the persistent virus caused a synergistic effect.

Although a number of biotechnological approaches have been developed to produce virus-resistant plants in recent years, the use of classical genetic resistance remains the strategy of choice for practical control of plant viruses. In this Special Issue, Yan et al. [**14**] reviewed the genetic resources used in breeding for resistance to one of the most harmful viral disease affecting tomato globally, tomato yellow leaf curl disease and the begomoviruses that cause it. The authors also summarized some of the future studies aimed at increasing the success and durability of genetic resistance to these and related viruses in a scenario of globalization, climate change, and viral disease emergence. Finally, Sáez et al. [**15**] reported the search for resistance to a cucurbit-adapted strain of tomato leaf curl New Delhi virus, a severe threat for cucurbit production worldwide, in cucumber. They identified, by the first-time, cucumber accessions highly resistant to the virus, and were successful in characterizing the mode of inheritance and location in the cucumber genome of the resistance gene.

Overall, the papers in this Special Issue reveal different perspectives of current research on plant viruses, from applied field studies to investigations into the intricate mechanisms involved in the tripartite interactions between viruses, plants, and vectors.

References

1. Martínez-Turiño, S.; Calvo, M.; Bedoya, L.C.; Zhao, M.; García, J.A. Virus host jumping can be boosted by adaptation to a bridge plant species. *Microorganisms* **2021**, *9*, 805. [**Google Scholar**] [**CrossRef**] [**PubMed**]

2. Kim, J.; Lal, A.; Kil, E.J.; Kwak, H.R.; Yoon, H.S.; Choi, H.S.; Kim, M.; Ali, M.; Lee, S. Adaptation and Codon-Usage Preference of Apple and Pear-Infecting Apple Stem Grooving Viruses. *Microorganisms* **2021**, *9*, 1111. [**Google Scholar**] [**CrossRef**]

3. Ferro, C.G.; Zerbini, F.M.; Navas-Castillo, J.; Fiallo-Olivé, E. Revealing the complexity of sweepovirus-deltasatellite-plant host interactions: Expanded natural and experimental helper virus range and effect dependence on virus-host combination. *Microorganisms* **2021**, *9*, 1018. [**Google Scholar**] [**CrossRef**]

4. Montes, N.; Cobos, A.; Gil-Valle, M.; Caro, E.; Pagán, I. *Arabidopsis thaliana* genes associated with *Cucumber mosaic virus* virulence and their link to virus seed transmission. *Microorganisms* **2021**, *9*, 692. [**Google Scholar**] [**CrossRef**] [**PubMed**]

5. Nancarrow, N.; Aftab, M.; Hollaway, G.; Rodoni, B.; Trębicki, P. Yield losses caused by barley yellow dwarf virus-PAV infection in wheat and barley: A three-year field study in South-Eastern Australia. *Microorganisms* **2021**, *9*, 645. [**Google Scholar**] [**CrossRef**] [**PubMed**]

6. Kutnjak, D.; Tamisier, L.; Adams, I.; Boonham, N.; Candresse, T.; Chiumenti, M.; De Jonghe, K.; Kreuze, J.F.; Lefebvre, M.; Silva, G.; et al. A primer on the analysis of high-throughput sequencing data for detection of plant viruses. *Microorganisms* **2021**, *9*, 841. [**Google Scholar**] [**CrossRef**] [**PubMed**]

7. Maachi, A.; Torre, C.; Sempere, R.N.; Hernando, Y.; Aranda, M.A.; Donaire, L. Use of high-throughput sequencing and two RNA input methods to identify viruses infecting tomato crops. *Microorganisms* **2021**, *9*, 1043. [**Google Scholar**] [**CrossRef**]

8. Ben Chehida, S.; Filloux, D.; Fernandez, E.; Moubset, O.; Hoareau, M.; Julian, C.; Blondin, L.; Lett, J.M.; Roumagnac, P.; Lefeuvre, P. Nanopore sequencing is a credible alternative to recover complete genomes of geminiviruses. *Microorganisms* **2021**, *9*, 903. [**Google Scholar**] [**CrossRef**] [**PubMed**]

9. Teixeira, R.M.; Ferreira, M.A.; Raimundo, G.A.S.; Fontes, E.P.B. Geminiviral triggers and suppressors of plant antiviral immunity. *Microorganisms* **2021**, *9*, 775. [**Google Scholar**] [**CrossRef**] [**PubMed**]

10. Kumar, G.; Dasgupta, I. Variability, functions and interactions of plant virus movement proteins: What do we know so far? *Microorganisms* **2021**, *9*, 695. [**Google Scholar**] [**CrossRef**] [**PubMed**]

11. Leastro, M.O.; Freitas-Astúa, J.; Kitajima, E.W.; Pallás, V.; Sánchez-Navarro, J.A. Membrane association and topology of citrus leprosis virus C2 movement and capsid proteins. *Microorganisms* **2021**, *9*, 418. [**Google Scholar**] [**CrossRef**] [**PubMed**]

12. Chi, Y.; Pan, L.L.; Liu, S.S.; Mansoor, S.; Wang, X.W. Implication of the whitefly protein Vps twenty associated 1 (Vta1) in the transmission of cotton leaf curl Multan virus. *Microorganisms* **2021**, *9*, 304. [**Google Scholar**] [**CrossRef**] [**PubMed**]

13. Elvira González, L.; Peiró, R.; Rubio, L.; Galipienso, L. Persistent Southern tomato virus (STV) interacts with Cucumber mosaic and/or Pepino mosaic virus in mixed-infections modifying plant symptoms, viral titer and small RNA accumulation. *Microorganisms* **2021**, *9*, 689. [**Google Scholar**] [**CrossRef**] [**PubMed**]

14. Yan, Z.; Wolters, A.M.A.; Navas-Castillo, J.; Bai, Y. The global dimension of tomato yellow

leaf curl disease: Current status and breeding perspectives. *Microorganisms* **2021**, *9*, 740. [**Google Scholar**] [**CrossRef**] [**PubMed**]

15. Sáez, C.; Ambrosio, L.G.M.; Miguel, S.M.; Valcárcel, J.V.; Díez, M.J.; Picó, B.; López, C. Resistant sources and genetic control of resistance to ToLCNDV in cucumber. *Microorganisms* **2021**, *9*, 913. [**Google Scholar**] [**CrossRef**] [**PubMed**]

Chapter-2

Virus Host Jumping can be Boosted by Adaptation to a Bridge Plant Species

1. Introduction

Emerging viral diseases are frequently the result of host jumps, when a pathogen gains the ability to infect a new species [**1**,**2**]. Host jumping has received particular attention in the case of animal and human diseases, with the host range breadth being a major determinant of bacterial but also viral emerging outbreaks [**3**,**4**,**5**]. Interspecies jumping is not uncommon among plant viruses, as evidenced by the fairly diverse host ranges and large host range width disparities in viruses derived from a recent radiative evolution [**6**], as well as by the frequent inconsistencies observed between the pathogen and host phylogenies [7]. Indeed, the host range expansion is considered pivotal for emergence of plant pathogens, especially plant viruses [**8**,**9**]—a phenomenon frequently linked to epidemic outbreaks in crops causing substantial yield losses [**10**,**11**]. Viral host jumping goes hand in hand with the concept of adaptive trade-off, according to which a pathogen cannot simultaneously maximize its fitness in all hosts. Thus, viral adaptation to a particular species normally implies a fitness cost in alternative species, and generalist viruses infecting numerous hosts evolve to reach fitness values maximized among hosts, but lower than the optimum they would have reached if had adapted to a single host [**12**].

The genus *Potyvirus* (family *Potyviridae*), is one of the most important groups of plant viruses [**13**]. The genome of potyviruses is a positive-sense single-stranded RNA of ~10 kb, whose 5′ end remains attached to the viral genome-linked protein (VPg) [**14**]. The potyviral genome is translated into a large polyprotein that is proteolytically processed to render at least 9 final products [**15**]. Moreover, frameshifts resulting from RNA polymerase slippage allow production of additional transframe products [**16**,**17**,**18**].

The intrinsic disorder that characterizes VPg protein [**19**,**20**,**21**] enables it to play multiple functions during viral infection [**22**,**23**,**24**]. VPg is involved in viral RNA translation probably both by recruiting host factors to promote translation initiation and by nucleating a protein complex around the viral RNA that protects it from the RNA silencing mechanism and facilitates its traffic to polysomes [**25**]. The interaction of VPg with the eukaryotic translation initiation factor eIF4E and/or its isoform eIF(iso)4E plays an important role in these functions. Compatibility/incompatibility of this interaction is a typical example of host-pathogen coevolution, according to which the virus evolves to match the host factor through adapting VPg, favoured by the large mutational robustness conferred by the intrinsic disorder of this protein [**26**], and the plant evolves to avoid such interaction [**27,28**]. Indeed, the lack of a functional interaction between VPg and eIF4E or eIF(iso)4E causes most cases of recessive resistance to potyviruses [**29,30**]. However, VPg-eIF4E/(iso)4E appears to be only one component of a more complex network involving multiple potyviral components and host factors [**31**]. This assumption is supported by the identification of HCPro as another interaction partner of eIF4E and eIF(iso)4E [**32**] and by the fact that sometimes breakdown of eIF4E-mediated anti-potyvirus resistance has been found associated with mutations in proteins P1 [**33**], P3 [**34**] and CI [**35,36**].

The potyvirus *Plum pox virus* (PPV) is the most serious viral agent of stone fruits, which causes sharka, a devastating disease affecting *Prunus* species [**37**]. In nature, PPV infects most of *Prunus* species, but it can also infect a wide range of herbaceous plants under experimental conditions [**38,39**]. It has been described up to ten different PPV strains, quite distinct in terms of their host range [**40,41**]. Among them, PPV-Dideron (D) is the most widely distributed PPV strain, infecting numerous *Prunus* species, while PPV-Cherry (C) is characterised by a restricted natural host range, limited to cherry trees [**42,43**]. Isolates of these strains also have remarkable differences on infectivity in experimental herbaceous hosts. SwCMp and Rankovic (R) isolates, respectively belonging to PPV-C and PPV-D strains, are highly infectious in *Nicotiana clevelandii* and *Nicotiana benthamiana*; however, while the isolate PPV-R (D strain) causes local lesions in *Chenopodium foetidum* and systemic infections in *Arabidopsis thaliana*, PPV-SwCMp (C strain) is not able to infect these two plant species [**39,44**]. Using chimeric viral cDNA clones, nuclear inclusion a (NIa), which includes VPg and a protease domain, was identified as the major pathogenicity determinant preventing PPV-SwCMp infection in *A. thaliana* and *C. foetidum* [**44**]. In the course of the same work, Calvo et al. identified specific mutations at the VPg protein of PPV-SwCMp, arisen as result of adaptation in *A. thaliana* of a R/SwCMp PPV chimera, which were suggested would contribute to gain infectivity in this host by enabling compatible interactions between VPg and eIF(iso)4E [**44**].

In this work, we have confirmed that the VPg mutations detected in the viral progeny of *A. thaliana* facilitate the efficient infection of this host. Likewise, we have shown that these mutations also provide an infectivity gain in a second restrictive host, *C. foetidum*, without any trade-off in the original host *N. clevelandii*. Furthermore, we discuss the possible relevance of the mutated residues in the viral protein for VPg/eIF(iso)4E compatibility and virus host range definition. The possibility that wild plants serve as bridges that facilitate "host jumps" and emergence of new diseases is also addressed.

2. Materials and Methods

2.1. Viral cDNA Clones

Three previously obtained PPV full-length cDNA clones were employed. pICPPV-NK-lGFP [**45**] and pICPPV-SwCM [**46**], respectively derive from the Nicotiana-adapted PPV isolates, PPV-R, belonging to strain D, and PPV-SwCMp from strain C. The chimeric clone pICPPV-VPgSwCM-R carries the VPg sequence from PPV-SwCMp into the PPV-R backbone [**44**].

Effect of point mutations in the VPg sequence was assayed using three constructs ad-hoc obtained. Amino acids substitutions P114S or F163L in the VPg protein sequence were engineered into the chimeric clone pICPPV-VPgSwCM-R, by replacing the nucleotide triplet CCA by TCA at position 1968–1970 (giving rise to the construct P114S) or substituting the nucleotide triplet TTC by CTC at position 2017–2019 (construct F163L). The double mutant carrying both P114S and F163L substitutions (P114S-F163L) was also generated (**Supplementary Figure S1**). These point mutations were introduced by using the three-step PCR-based mutagenesis method [**47**], using the mutators and flanking primers listed in **Supplementary Table S1**. First mutagenic PCR reactions used the plasmid pICPPV-VPgSwCM-R as template; then, products of these reactions were mixed and employed as templates for a second round PCR. Final overlapping amplicons, digested with *Xho*I (partially in the case of F163L mutagenesis) and *Nru*I, served to replace the corresponding fragment from pICPPV-VPgSwCM-R. Double mutant P114S-F163L was obtained following the same strategy, but by using as template the previously obtained P114S construct (**Supplementary Figure S1 and Supplementary Table S1**).

2.2. Viral Inoculation and Plants Growth Conditions

For mechanical hand inoculation, approximately 15 μL of leaf extracts or plasmid DNAs, (1.0–1.5 μg/μL), were distributed on three leaves of young plants of *N. clevelandii, C. foetidum* or *A. thaliana* (four- to six-leaf stage), previously dusted with Carborundum powder. Leaf extracts used as inocula were obtained as previously described [**44**] from *N. clevelandii* or *A. thaliana* leaves already infected. Primary inoculation of *A. thaliana* plants was done by bombardment with microgold particles coated with DNA using a Helios gene gun (Bio-Rad, Hercules, CA, USA) [**48**]. Microcarrier cartridges were prepared with 1.0 μm diameter gold particles coated at a DNA loading ratio of 2 μg/mg gold and a microcarrier loading amount of 0.5 mg/shot. One cartridge, shot twice onto two leaves of each plant, under a helium pressure of 7.0 bar, was employed.

All plants were grown under glasshouse conditions with 16 h of light photoperiod using natural and supplementary illumination, at a temperature range of 19–23 °C and 67–70% relative humidity. *A. thaliana* ecotype Columbia (Col0-) seeds were vernalized at 4 °C and in vitro grown on MS medium (Sigma Aldrich, St. Louis, MO, USA) containing 0.5 % (*w/v*) of sucrose and 1% (*w/v*) of agar. Once germinated, seedlings were kept for two weeks in a phytotron (Neurtek, Eibar, Spain) under a 14 h photoperiod, at 22 °C and 50–60% relative humidity. After planted out to soil-vermiculite (3:1), plants were grown in controlled environment chambers as mentioned above.

2.3. Assessment of Viral Infection

Viral infection was monitored by visual inspection of PPV-induced symptoms and by immunoblot analysis, as described by Calvo et al. [**44**]. Infection in *C. foetidum* plants was evaluated by registering over time the total number and type of local lesions, discriminating between doubtful, chlorotic or necrotic lesions of variable intensity.

2.4. Viral Progeny Characterization and Sequence and Structure Analyses

For characterization of the viral progeny, appropriate viral DNA fragments covering the entire VPg sequence were amplified from systemically infected tissue of *A. thaliana* and *N.*

clevelandii, by immune-capture-RT-PCR (IC-RT-PCR), as previously described [**44**]. Alternatively, viral progeny in *C. foetidum* was analyzed by direct RT-PCR from total RNA obtained from individual lesions or entire leaves, employing the FavorPrep Plant Total RNA Purification Mini-Kit (Favorgen Biotech, Ping-Tung, Taiwan). Amplification of a region containing the VPg sequence was done using oligos 2295 and 2277, after inoculation with pICPPV-VPgSwCM-R-derived plasmids or with their viral progenies, or with oligos SM-16F and SM-17R after inoculation with pICPPV-NK-lGFP or its viral progeny (**Supplementary Table S1**). Sanger sequencing of amplified fragments was performed by Macrogen Europe (Amsterdam, The Netherlands) using primers SM18-F and/or SM19-R.

For the identification of the *C. foetidum* eIF(iso)4E [Cf-eIF(iso)4E] sequence, total RNA was extracted from *C. foetidum* leaves, as mentioned above, and cDNA was synthetized from it by using the Invitrogen SuperScript III Reverse Transcriptase and Invitrogen hexameric random primers (both from Thermo Fisher Scientific, Rockford, IL, USA), following manufacturer instructions. From the cDNA product, treated with RNase H, a Cf-eIF(iso)4E gene fragment was amplified by using the Thermo Scientific Phusion High-Fidelity DNA Polymerase (Thermo Fisher Scientific, Rockford, IL, USA) and a pair of degenerate oligonucleotides, SM-110F-deg and SM-111R-deg (**Supplementary Table S1**), designed on the basis of the known-sequences of *Chenopodium quinoa* eIF(iso)4E (LOC110697254 and LOC110692931). Sanger sequencing of the amplified gene fragment was performed by Macrogen Europe (Amsterdam, The Netherlands) using the same primers employed for amplification (**Supplementary Table S1**).

Alignment of multiple protein sequences was carried out with the Clustal Omega web server [**49**] (**www.ebi.ac.uk/Tools/msa/clustalo**, accessed on 1 April 2021) using default parameters. Sequences of *Potato virus Y* (PVY) VPg (VPg PVY), human eIF4E (h-eIF4E) and other plant eIF(iso)4E proteins, were retrieved from the Protein Data Bank (PDB, **www.rcsb.org/pdb/** accessed on 1 April **50**] (2021].

Putative spatial localization of specific residues relevant for this study, were mapped over a HADDOCK-derived model, complexing h-eIF4E and PVY VPg, previously generated by Countinho de Oliveira et al. [**51**]. Equivalences between residues of the model and eIF(iso)4Es/PPV VPg were obtained on the basis of corresponding protein alignments. 3D protein structures were visualized by using PyMOL Molecular Graphics System, version 2.1.1 (Schrödinger).

3. Results

3.1. Point Mutations at the VPg Protein in an Avirulent Chimeric Construct of PPV Promote Infection of Arabidopsis thaliana

Previous works had shown that, while the PPV-R isolate belonging to the D strain, efficiently infects *A. thaliana*, a chimeric construct bearing the VPg sequence from a PPV isolate belonging to the C strain in the backbone of PPV-R (PPV-VPgSwCM-R) rarely infects this host [**44**]. Infection of *Arabidopsis* by PPV-VPgSwCM-R appeared to be promoted by emergence of point mutations at the SwCMpVPg coding sequence [**44**]. In order to examine whether the detected changes at VPg sequence are solely responsible for the gain of infectivity, we assayed the effect of these mutations, proline to serine at position 114

(P114S) and phenylalanine to serine at position 163 (F163L), separately engineered into the cDNA clone of the PPV-VPgSwCM-R chimera. Besides, although these modifications had not been concomitantly detected in *A. thaliana*, both changes were also introduced together into the chimeric clone (**Supplementary Figure S1**). The mutated constructs and appropriate controls were biolistically inoculated in *A. thaliana* plants (**Supplementary Figure S2A**). Viral accumulation was checked by an immunoblot assay at 15 dpi (**Figure 1**). VPg mutants P114S and F163L showed similar viral CP accumulation as the PPV-R positive control. RT-PCR amplification and sequencing of the complete VPg gene showed no sequence changes in the viral progeny of the three plants infected with each tested mutant. These results confirm previous assumption launched by Calvo et al. [**44**] that both P114S and F163L mutations were able, by themselves, to facilitate adaptation of the PPV-VPgSwCM-R chimera to *Arabidopsis*. As expected, the wild type PPV-VPgSwCM-R chimera hardly accumulates in inoculated plants, being undetectable in two of the three plants tested. A weak CP signal was detected in the third analyzed plant; however, the analysis of its viral progeny revealed that a P114S mutation had been introduced in the VPg coding sequence, further confirming the relevance of this change for adaptation to *Arabidopsis*.

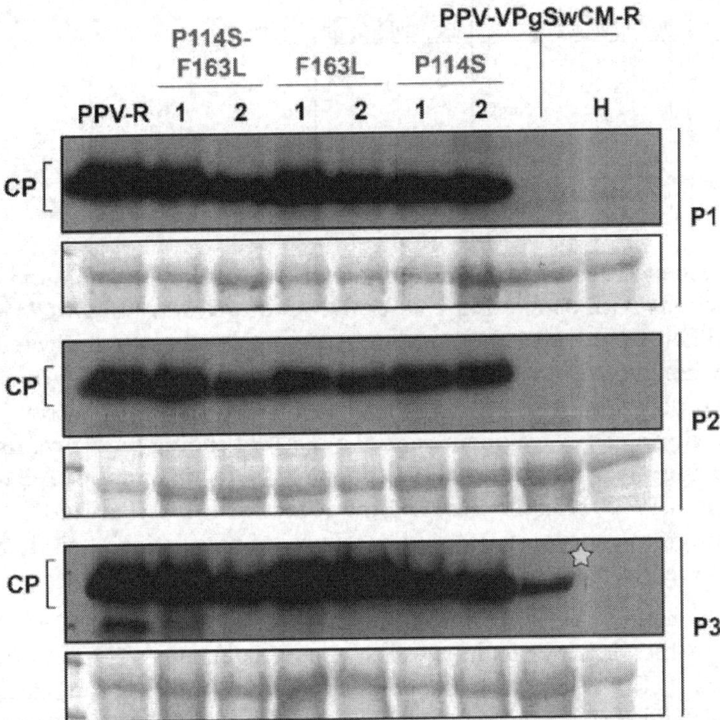

Figure 1. Effect of VPg mutations on *A. thaliana* infection by *Plum pox virus* (PPV). *A. thaliana* plants were inoculated by biolistic with the chimeric clone pICPPV-VPgSwCM-R, its indicated mutant variants (two independent clones, 1 and 2), or the PPV-R clone pICPPV-NK-lGFP. Extracts from upper non-inoculated leaves collected at 15 days after inoculation were subjected to CP-specific immunoblot analysis.

Three individual plants (P1, P2 and P3), inoculated with the specified viruses, were analyzed. An extract of healthy plants (H) was used as a negative control. Blots stained with *Ponceau* red showing the large subunit of the ribulose-1,5-bisphosphate carboxylase-oxygenase (RuBisCO) are included as loading controls. The yellow star indicates that the progeny virus of this plant had incorporated a mutation in the VPg sequence (P114S).

The double mutant P114S-F163L was also infectious and genetically stable in *A. thaliana* plants (**Figure 1**). Although the existence of certain differences at early times of infection cannot be excluded, all VPg mutations seem to allow similar viral accumulation, comparable with that of the positive control PPV-R (**Figure 1**).

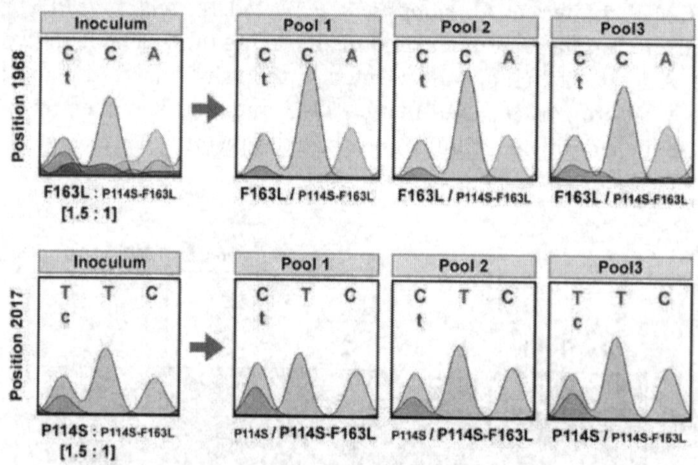

Figure 2. Sequence analysis of viral progeny from *Arabidopsis thaliana* exposed to mixed infections with competing viruses. DNAs of pICPPV-VPgSwCM-R chimeric clones modified by the specified mutations were mixed at the indicated ratio and biolistically inoculated into six *A. thaliana* plants. Viral progenies were analyzed in pools of two plants by reverse transcription-polymerase chain reaction (RT-PCR), then sequencing a cDNA fragment covering the VPg coding sequence. Images show the chromatograms of VPg codons 114 (position 1970–1968 in the viral genome) or 163 (position 2019–2017 in the viral genome). Identified viruses are indicated beneath the chromatograms; smaller letters indicate lower accumulation. A similar result was obtained in a replicate assay.

To find out whether the two specific mutations in the VPg could have a synergist contribution to the adaptation to the new host, appropriate competition assays were carried out. Mixtures of cDNAs, each corresponding to the double mutant and one of the single mutants, were biolistically inoculated in *A. thaliana* plants, at concentrations adjusted to achieve a 1.5:1 ratio, thus conferring some advantage in favour to individual mutations (**Supplementary Figure S2B**). Systemic infection was monitored by immunodetection of CP in upper non-inoculated leaves at 21 dpi. Genotyping of viral progenies was carried out by IC-RT-PCR amplification and sequencing of the VPg coding sequence from three pools of two plants infected with each mutant combination. Examination of viral progeny after the competition F163L vs P114S-F163L did not reveal important differences in the fitness of any of the two types of viruses, which coexisted in all three analyzed pools of plants, maintaining the differences between them already existing in the inoculum (**Figure 2**). Thus, the change P114S does not seem to provide any competitive advantage when it is together with the mutation F163L. Similarly, when mutants P114S-F163L and P114S competed, both viruses coexisted in the three pools of analyzed plants. However, the ratio of the DNA inoculum was reversed in two of the analyzed pools, and the double mutant, despite its lesser initial representation, became the majority virus (**Figure 2**). These results suggest that the enhancement of viral fitness in *A. thaliana* conferred by the F163L mutation in the VPg of PPV-SwCMp might be greater than that provided by the P114S mutation.

3.2. PPV Adaptation to A. thaliana, via Specific Mutations in VPg, Does Not Have a Fitness Cost in Nicotiana clevelandii

Changes introduced in the chimeric virus PPV-VPgSwCM-R that facilitate its amplification in *A. thaliana* are expected to be associated with a loss of fitness in *N. clevelandii*, a host in which the non-mutated virus is completely adapted. To evaluate this possibility, four *N. clevelandii* plants were manually inoculated by hand rubbing with DNAs of the two single mutants, P114S or F163L, as well as with that of the double mutant P114S-F163L. Both, PPV-R and the non-mutated chimera, PPV-VPgSwCM-R, were included as positive controls of infection (**Supplementary Figure S2A**). The results showed that the three viruses with mutations in VPg systemically infected *N. clevelandii* with comparable efficiency. Overall, the onset and severity of disease symptoms (not shown), as well as levels of viral CP accumulation, were similar for all the three viruses and indistinguishable from those induced by the positive controls (**Supplementary Figure S3**).

Next, competition assays were carried out using mixtures containing DNAs from the pICPPV-VPgSwCM-R plasmid and each of the mutants derived from it (P114S, F163L and P114S-F163L), at ratios in which the non-mutated chimera was over-represented (ratio 1.5:1) (**Supplementary Figure S2C**). *N. clevelandii* plants were inoculated by hand-rubbing and systemic infection was confirmed by visual inspection of symptoms and an anti-CP immunoblot assay (data not shown). Four pools, one per type of inoculum, were prepared by joining systemically infected tissue from two plants, collected at 21 dpi. A cDNA fragment covering the complete VPg sequence was amplified by IC-RT-PCR and sequenced (**Figure 3**). The results showed that, despite its lower representation in the inoculum, the mutant F163L was able to completely impose to the non-mutated chimera in all analyzed plants (**Figure 3**A). A similar result was observed after confrontation between the double mutant and the non-mutated chimera. In this case, the double mutant entirely prevailed in the progeny from three of pools, and it was in progress to do it in the fourth sample (**Figure 3**B). The result derived from competition between P114S and the non-mutated chimera was more even. The virus with the original VPg sequence was imposed in one of the four analyzed pools, while the P114S mutant entirely prevailed in two samples and coexisted with advantage in a third case (**Figure 3**C). The result of this competition, in which the mutant was underrepresented in the inoculum, ruled out that the P114S mutation reduces the virus fitness in *N. clevelandii*, and suggested that, in fact, it increases fitness. To further support this assumption, a second competition using comparable amounts of both viruses (ratio 1:1) was done (**Supplementary Figure S2C**), and the viral progeny was analyzed in ten plants, distributed in five pools. The results were in line with previous findings. This time, the mutant P114S was completely dominant in one pool and the majority in the remaining four, largely prevailing the non-mutated virus in two of them (**Figure** 3C).

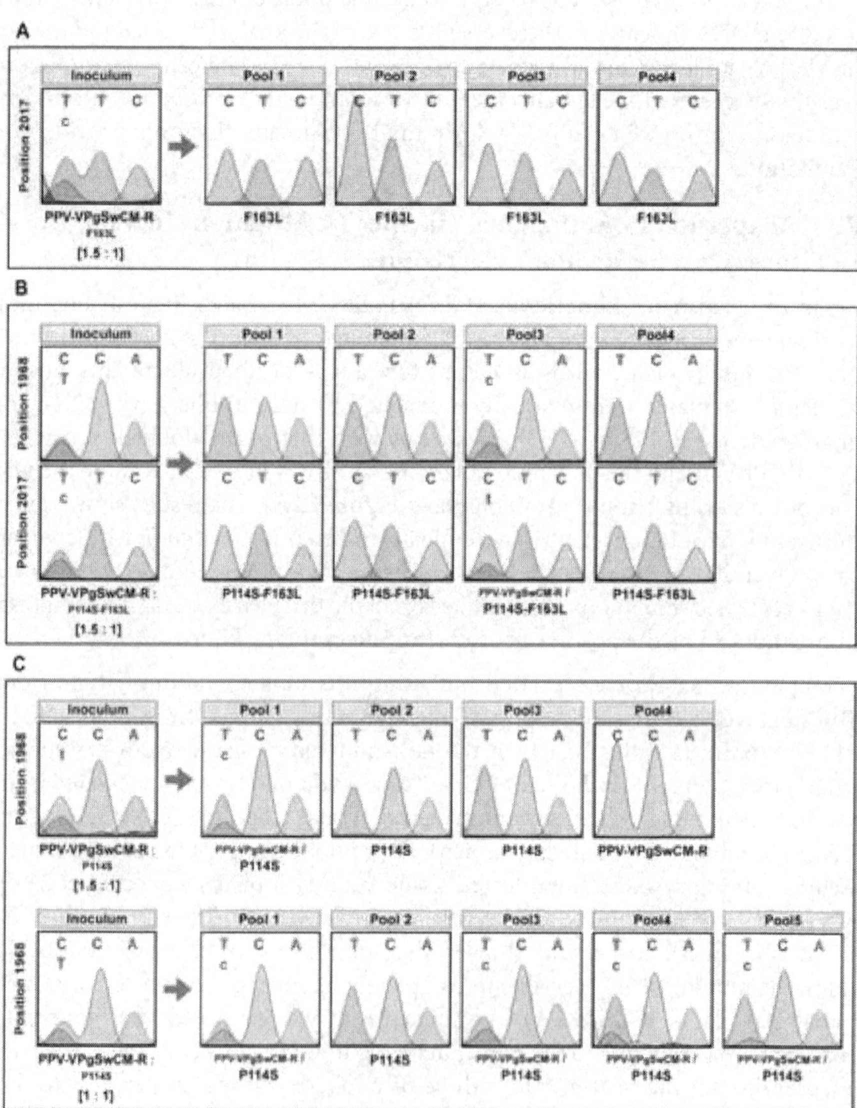

Figure 3. Sequence analysis of viral progeny from *Nicotiana clevelandii* exposed to mixed infections with competing viruses. *N. clevelandii* plants, 8 or 10, were inoculated by hand rubbing with mixtures containing DNAs of pICPPV-VPgSwCM-R DNA and of one version of this chimera modified with the mutation F163L (A), P114S plus F163L (B) or P114S (C). Non-mutated/mutated chimera mixtures at 1.5:1 ratio, were employed for the three competitions. An additional 1:1 ratio mixture was used

for the PPV-VPgSwCM-R vs P114S competition. Viral progenies were analyzed in pools of two plants by reverse transcription-polymerase chain reaction (RT-PCR) amplification and sequencing of a DNA fragment covering the VPg coding sequence. Images show the chromatograms of VPg codons 163 (position 2019–2017 in the viral genome) and/or 114 (position 1970–1968 in the viral genome). Viruses identified are indicated beneath the chromatograms; smaller letters indicate lower accumulation. A similar result was obtained in a replicate assay.

Overall, these results indicate that specific mutations of VPg, in principle needed for adaptation in *A. thaliana*, do not provoke an adverse trade-off in *N. clevelandii*. On the contrary, these VPg changes appear to boost PPV fitness in two unrelated hosts.

The competition experiments in *A. thaliana* reported above did not reveal significant synergistic or additive effects of the P114S and F163L mutations in that host. Additional competition tests were conducted to assess possible accumulative effects of these mutations in *N. clevelandii*. Following an identical procedure to the aforementioned, viral progenies of eight *N. clevelandii* plants, distributed in four pools, were analyzed (**Supplementary Figure S4**). In the competition between F163L and P114S-F163L, the double mutant, underrepresented in the inoculum, did not outcompete the single mutant in any of the samples analyzed. The P114S-F163L mutant was able to reverse its underrepresentation when competing with the P114S single mutant, reflecting that a greater fitness gain could be associated to F163L mutation. However, such a difference would be very subtle because the P114S mutant was completely or almost completely imposed in two samples (**Supplementary Figure S4**). Thus, in *N. clevelandii*, the joint presence of the two mutations in the P114S-F163L mutant does not appear to entail a relevant fitness increase with respect to any of individual F163L or P114S mutations.

3.3. Changes in VPg Protein Resulting from Adaptation to Arabidopsis thaliana Prompt PPV-VPgSwCM-R infection in Chenopodium foetidum

Most PPV isolates, including PPV-R (D strain) induce necrotic local lesions in *C. foetidum*, compatible with a hypersensitive-like response. In contrast, isolates from strain C, in particular the isolate PPV-SwCMp, cannot infect this host, thus emulating what happens in *A. thaliana* [**44**]. Similarly, the defect of chimera PPV-VPgSwCM-R (VPg from PPV-SwCMp in the backbone of PPV-R, **Supplementary Figure S1**) in *A. thaliana* is extensible to *C. foetidum* [**44**]. Having demonstrated that mutations P114S or F163L at the VPg protein of PPV-SwCMp enable to rescue infectivity of PPV-VPgSwCM-R in *A. thaliana*, we decided to check the effect of these mutations on the infectivity of this chimera in *C. foetidum*.

First, we tested whether the adaptation of PPV-VPgSwCM-R to *A. thaliana* facilitated infection of *C. foetidum*. For this purpose, *C. foetidum* leaves were inoculated by hand-rubbing with leaf extracts from *N. clevelandii* plants infected with viral progenies of the PPV-VPgSwCM-R chimera adapted to *A. thaliana* upon introducing VPg mutations P114S or F163L. As controls of infection in this host, *C. foetidum* leaves were also inoculated with extracts of *N. clevelandii* plants infected with PPV-R, PPV-SwCMp or the non-mutated chimera PPV-VPgSwCM-R (**Supplementary Figure S5A**). To warrant delivery of same virus amounts, inoculum concentrations were adjusted by dilution with extracts from healthy *N. clevelandii* (**Supplementary Figure S5B**). PPV-R caused a large number of lesions (more than 60 per leaf) that rapidly necrotized, causing death and dropping of the leaves sometimes before 9 dpi. As expected, no lesions were observed in leaves inoculated with PPV-SwCMp. *C. foetidum* leaves inoculated with the extracts containing the evolved PPV-VPgSwCM-R populations, displayed abundant lesions, although in lesser amount than in those inoculated with PPV-R: at 9 dpi, approximately 6.5 and 17 per leaf for the chimeras with the P114S and F163L mutations, respectively. Moreover, although some of the lesions caused by the evolved chimeras necrotized, in general they were less severe than those caused by PPV-R, remaining alive the inoculated leaves even after 15 dpi (**Supplementary Figure S5C and Supplementary Table S2**). To assess whether the virus further evolved in *C. foetidum*, viral progenies of individual lesions were amplified in *N. clevelandii*, and their VPg sequences were determined after IC-RT-PCR amplification. No changes beyond original modifications introduced during adaptation in *A. thaliana* were detected in any of 4 *N. clevelandii* plants analyzed. A few spots suspected of being viral lesions were observed in leaves inoculated with the non-mutated PPV-VPgSwCM-R, however, we were not able to infect *N. clevelandii* with them, suggesting that either they had not been correctly identified, or their viral load was very low.

To verify that the apparent adaptation of the PPV-VPgSwCM-R to *C. foetidum* was in fact due to the VPg mutations P114S and F163L, the exclusive contribution of these substitutions was assessed making use of corresponding mutated cDNA clones (**Supplementary Figure S1**). pICPPV-VPgSwCM-R and its mutated forms (P114S, F163L and P114S-F163L), as well as the PPV-R cDNA clone pICPPV-NK-lGFP were manually inoculated in *N. clevelandii* plants by hand-rubbing. Leaf extracts from these infected plants were in turn inoculated into *C. foetidum* leaves by hand-rubbing, after adjusting its concentration with extract of healthy *N. clevelandii* to warrant delivery of same virus amounts (**Supplementary Figure S6**). Both mutants, P114S and, in a greater extent, F163L caused abundant local lesions (**Table 1** and **Figure 4**). As in the above experiment using the non-cloned mutant viruses, lesions caused by both mutants were similar and milder than those triggered by PPV-R. Interestingly, simultaneous presence of both mutations fostered a qualitative change in viral symptoms, as reflected in the PPV-R-like lesions induced by the P114S-F163L mutant (**Table 1** and **Figure 4**). Some potential lesions were also detected in a few leaves inoculated with the non-mutated pICPPV-VPgSwCM-R chimera, mainly at 17 dpi, when all leaves inoculated with the rest of viruses were dead (**Table 1**). Viral cDNA from some suspicious lesions could be amplified by IC-RT-PCR, and subsequent sequencing showed that the wild type VPg sequence had been maintained in five analyzed viral progenies. These results confirmed that, although the PPV-VPgSwCM-R chimera was not completely unable to infect *C. foetidum*, its ability to infect this host is greatly enhanced by VPg mutations selected during adaptation to *A. thaliana*.

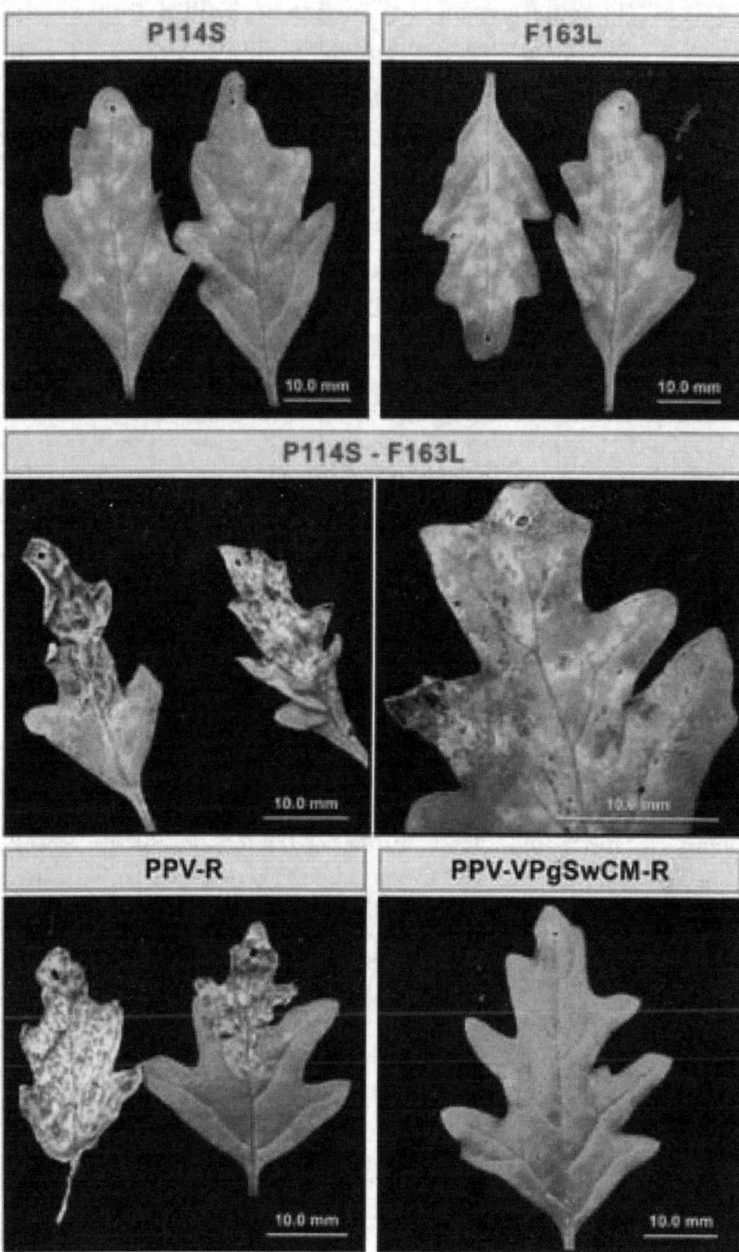

Figure 4. Effect of VPg mutations on *Chenopodium foetidum* infection by *Plum pox virus* (PPV). *C. foetidum* leaves were inoculated by hand rubbing with leaf extracts of *Nicotiana clevelandii* plants, previously infected with two independent clones of pICPPV-VPgSwCM-R, variants of this chimeric clone mutated as indicated, or pICPPV-NK-lGFP (PPV-R isolate). Eight plants (three leaves per plant) per construct (four per clone) were inoculated. Representative images taken at 12 dpi under visible light are shown. Bar, 10.0 mm.

Table 1. Effect of VPg mutations on the infection of the PPV-VPgSwCM-R chimera in Chenopodium foetidum.

Inoculum	Total Inoculated	Number of Lesions [a]	
	Leaves	10 dpi	17 dpi
PPV-VPgSwCM-R	24	1 chl (1) 4 ? (3)	4 chl (3) 1 nec (1) 10 ? (6)
P114S	24	330 chl (24)	DL (24)
F163L	24	400 chl (24)	DL (24)
P114S-F163L	24	255 chl/nec (15) 145 nec (8) DL (1)	DL (24)
PPV-R	12	170 nec (10) DL (2	DL (24)

3.4. Sequence Heterogeneities Between Nicotiana clevelandii, Arabidopsis thaliana and Chenopodium foetidumeIF(iso)4Es map to the eIF4E/VPg Interface

The dysfunction of SwCMpVPg-containing PPV chimera in *A. thaliana* was suggested to be caused by a defect in VPg/eIF(iso)4E interaction, because mutations promoting infection in this host resembled those detected in potyviruses escaping eIF4E/(iso)4E-based resistance [**44**]. Since the mutations selected in *A. thaliana* also boosted infection in *C. foetidum*, we speculated that common characteristics of eIF(iso)4E factors of these two species, not shared by *N. clevelandii* eIF(iso)4E, prevented interaction with SwCMpVPg, and, thus, the infection by SwCMpVPg-containing PPV.

Recently, it has been reported the first high-resolution structure of the VPg from a potyvirus, PVY [**51**]. This study not only revealed interesting structural data concerning the VPg folding, but also identified residues implicated in its interaction with an eIF4E factor. In light of this information, we decided to extrapolate available interaction data of PVY VPg and eIF4E to the proteins subject of interest for our work, by examining heterogeneities in the primary sequence of PPV VPg variants and those of eIF(iso)4Es from the three herbaceous hosts here studied.

The eIF(iso)4E sequences of *N. clevelandii* [Nc-eIF(iso)4E] and *A. thaliana* [At-eIF(iso)4E] were retrieved from NCBI protein database, but that of *C. foetidum* [Cf-eIF(iso)4E] was not available in public databases and had to be specifically obtained for this analysis. Total RNA of *C. foetidum* was retrotranscribed to prepare cDNA, from which to amplify a fragment encoding a segment of the Cf-eIF(iso)4E, by using a pair of degenerated primers designed based on the eIF(iso)4E sequence of *C. quinoa*. The amplicon covered the regions identified as relevant for the interaction with VPg by Coutinho de Oliveira et al. [**51**]. Alignment of the three eIF(iso)4E sequences (**Figure 5**A) revealed that Cf-eIF(iso)4E was not more similar to At-eIF(iso)4E than to Nc-eIF(iso)4E; in fact, the level of identity with the second protein was slightly higher than with the first one (77.3% vs 73.8%).

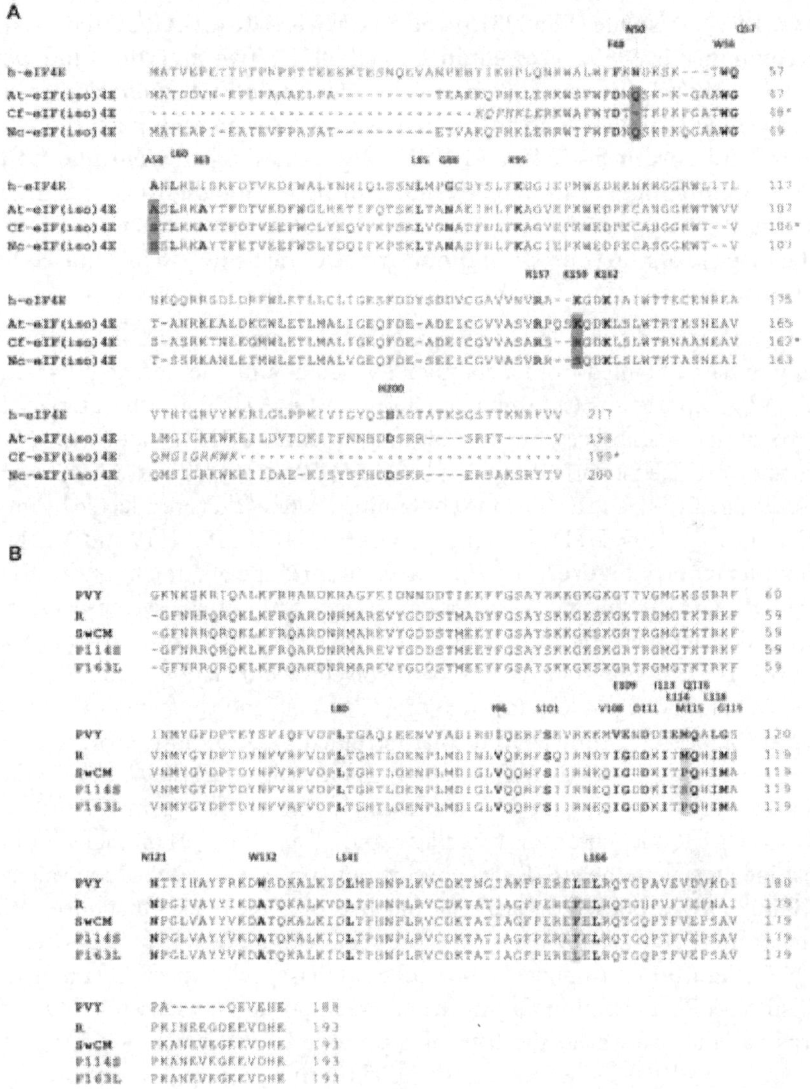

Figure 5. (A) Alignment of the translation initiation factors 4E from *Homo sapiens* (h-eIF4E) (P06730.2) along with eIF(iso)4Es from *Arabidopsis thaliana* [At-eIF(iso)4E] (O04663.2), *Chenopodium foetidum* [Cf-eIF(iso)4E] (partial sequence specifically obtained for this study) and *Nicotiana clevelandii* [Nc-eIF(iso)4E] (KC625579.1). Residues of eIF(iso)4E plant factors aligning with those interacting residues of h-eIF4E appear in bold and, in case of no conservation among the three plant species, are orange highlighted. Amino acids encoded by primers used for PCR amplification of the Cf-eIF(iso)4E sequence are italicized. Asterisks in the numbering of the partial Cf-eIF(iso)4E sequence indicate that for the count, missing amino acids were replaced by the equivalent ones from the full-length sequence of the *Chenopodium quinoa* eIF(iso)4E protein (LOC110697254 and LOC110692931).

(B) Alignment of VPg sequences from *Potato virus Y* (PVY) (QED90173.1) and *Plum pox virus* (PPV), R isolate (EF569215) and SwCMp isolate (SHARCO database, http://w3.pierroton.inra.fr:8060, accessed on 1 April 2021). Two mutations independently arisen at SwCMpVPg, as consequence of the adaptation in *A. thaliana*, are shown in red over a yellow-shaded box. In both panels, specific amino acids perturbed as a result of interaction between h-eIF4E and PVY VPg, according to Coutinho de Oliveira et al. [51], are highlighted in blue. Residues of PPV VPg aligning with the h-eIF4E -interacting PVY VPg residues appear in bold. Protein sequences were aligned using Clustal Omega program (European Bioinformatics Institute), then adjusted by minor manual corrections.

Following, we focussed more closely on regions that are involved in eIF4E/VPg interactions according to data obtained from a complex formed among the human eIF4E (h-eIF4E) and PVY VPg, reported by Coutinho de Oliveira et al. [**51**]. The main interface between these two proteins embraces the cap binding pocket of eIF4E, and includes several residues shown to be perturbed by VPg binding (F48, N50, W56, Q57, A58, L60, G88, R157, K159 and K162), and a VPg loop containing residues that are affected by interaction with eIF4E (V108, E109, D111, I113, E114, M115, Q116, L118, G119 and N121) (**Figure 5** and **Supplementary Figure S7**). When we scrutinize the alignment of the eIF(iso)4Es of *A. thaliana*, *C. foetidum* and *N. clevelandii* along with h-eIF4E, we observed sequence heterogeneities between the three plant eIF(iso)4Es at three positions that align with some of the h-eIF4E amino acids directly involved in interaction with PVY VPg: N50, A58 and K159. However, eIF(iso)4Es of *A. thaliana* and *C. foetidum* only clustered together at the position equivalent to that occupied by K159 in h-eIF4E (see K and R in At-eIF(iso)4E and Cf-eIF(iso)4E, respectively) leaving apart the *N. clevelandii* protein [see S in Nc-eIF(iso)4E] (**Figure 5**A). Interestingly, one of the VPg mutations facilitating PPV-VPgSwCM-R adaptation to *A. thaliana* and *C. foetidum*, P114S, falls into the VPg loop that interacts with the eIF4E cap-binding domain. This residue is equivalent to PVY M115 (**Figure 5**B), proposed to form part of a hydrophobic pocket that buries W56 [**51**], a residue of h-eIF4E close to A58, in turn aligned with a polymorphic position of plant eIF(iso)4Es (**Figure 5**A). This observation suggests that species-specific features governing eIF4E/(iso)4E-VPg interactions at this region are important for susceptibility to different potyvirus variants. However, the lack of a positive correlation between susceptibility to non-mutated PPV-VPgSwCM-R and the clustering of such a polymorphic residue (A in *A. thaliana* and S in both *N. clevelandii* and *C. foetidum* (**Figure 5**A), precludes drawing straightforward conclusions.

The other VPg mutation associated to PPV-VPgSwCM-R adaptation to *A. thaliana* and *C. foetidum*, F163L, appears to be far from the protein-protein interface defined for PVY VPg and h-eIF4E (**Supplementary Figure S7**). However, it is very close to PVY L166 (**Figure 5**B), a neighboring residue to a flexible loop between two ß strands, that has been identified to be perturbed by eIF4E binding [**51**]. Thus, it is tempting to speculate that this target, outside the cap-binding site/VPg interface, is involved in species-specific interactions still to be characterized.

4. Discussion

Viral cross-species jump takes place once a virus develops the ability to infect, replicate and disseminate among individuals of a new host species [**3**,**52**]. This phenomenon has been more frequently described for RNA viruses, mainly among *Rhabdoviridae* and *Picornaviridae* family members, and it has been associated with its huge adaptive plasticity [**53**,**54**]. From the human health perspective, some of the more remarkable examples of zoonotic RNA viruses able to break interspecies barrier by the assistance of an intermediate host are *influenza A virus* (IAV), *human immunodeficiency virus* (HIV) and several respiratory coronaviruses [**55**,**56**,**57**]. Spillover events are also quite often in plant viruses, linked to multiple factors, including ecological conditions, genetic plasticity of virus components, and host factor requirements [**8**,**58**]. Genome nature appears to be one of the elements that determine the capacity of the virus to successfully infect a host variety; and among plant viruses, single-stranded RNA genome viruses are those with largest host-range breadth [**9**].

In this study, we have delved into the contribution of two adaptive changes that affect the VPg protein sequence by promoting infectivity gain of a potyvirus in two non-permissive hosts. Potyviral VPg has a high content of intrinsically disordered regions, a characteristic associated with larger mutational robustness which favourably impacts on viral adaptive plasticity [**21**,**26**]. Here, we took advantage of the PPV chimeric clone pICPPV-VPgSwCM-R that bears the VPg sequence of PPV-SwCMp, an isolate unable to infect *A. thaliana,* in the backbone of the infectious PPV-R isolate [**44**]. We specifically studied infections in different hosts triggered by mutations at this chimeric clone that were engineered on the basis of VPg changes known to emerge during adaptation of PPV-VPgSwCM-R to *A. thaliana* [**44**]. Our results confirm that any of these two mutations, P114S or F163L, is sufficient to prompt the break of resistance to PPV-VPgSwCM-R in *A. thaliana* (**Figure 1**).

More important, the mutations selected to adapt the chimera to *A. thaliana* were also able to boost the infection in a second resistant species, *C. foetidum*. Although PPV-VPgSwCM-R is very poorly infectious in both *A. thaliana* and *C. foetidum*, it is still able to carry out a basal replication in both restrictive hosts as evidenced by the emergence of adaptive mutations in *A. thaliana* and late-onset sporadic lesions in *C. foetidum*. VPg mutations selected in *A. thaliana* promote viral infection in both hosts, but, while in *A. thaliana* the mutant viruses reach amplification levels similar to those of well-fitted PPV-R isolate, the infection they cause in *C. foetidum* is considerably milder than that induced by PPV-R. This indicates that mutations facilitating functional interactions of VPg with *A. thaliana* host factors, also improve matching with the homologous factors of *C. foetidum*, but to a lesser extent (**Figure 1**, **Figure 4**, **Table 1**, **Supplementary Figure S5**, **Supplementary Table S2**).

Although the double mutation P114S-F163L was not detected in the natural adaptation of PPV-VPgSwCM-R to *A. thaliana*, we also engineered it in pICPPV-VPgSwCM-R. No additive or synergistic effects were observed in *A. thaliana*, where the double mutation seemed to confer a little better fitness than the single mutation P114S, but did not improve the F163L performance. In contrast to the effect of the single mutations, concurrence of the two mutations had a differential impact over the typology of lesions caused by the viral chimera in *C. foetidum*, making them similar to those produced by PPV-R. This result further supports the assumption that the effect of P114S and F163L mutations on the coupling between PPV VPg and host-specific plant cofactors is different in *A. thaliana and C. foetidum* (**Figure 4, Table 1**).

The nature of the mutations favouring the adaptation of PPV-VPgSwCM-R to *A. thaliana* led Calvo et al. [**44**] to conclude that resistance to PPV-SwCMp in *A. thaliana* and *C. foetidum* is due to incompatible interactions between PPV VPg and plant eIF(iso)4Es. The establishment of a productive interaction between VPg and eIF4E factors or its isoforms is critical for potyviral infection, as demonstrated by many studies connecting specific mutations in VPg with resistance breakdown events, and numerous examples of eIF4E-mediated plant resistance against potyviruses [**27**,**28**,**30**,**59**]. In this sense, the recently solved potyviral VPg structure and the characterization of a VPg-eIF4E complex have shed light on this issue [**51**]. By using the HADDOCK eIF4E-VPg model generated in that work, we got positional information about the two mutation targets linked to SwCMpVPg adaptation (**Supplementary Figure S7**). We observed that one of these targets (P114 in PPV-SwCMp, S114 in the adapted mutant and M114 in PPV-R) is equivalent to the residue M115 of PVY VPg, which maps to the interface connecting both interacting molecules and whose involvement in such interaction had been experimentally validated by Coutinho de Oliveira et al. [**51**]. Indeed, PVY M115 is proposed to form part of a hydrophobic pocket in which a specific residue of the cap binding domain of h-eIF4E (W56) is buried. This residue is spatially close to another involved in the interaction, A58 of h-eIF4E, equivalent to an amino acid varying among the three analyzed plant eIF(iso)4E factors, *A. thaliana* (A48), *C. foetidum* (S49) and *N. clevelandii* (S50) (**Figure** 5A,B). Interestingly, also positions equivalent to 159 of h-eIF4E, included in the positive patch R157-K159-K162 that interacts with VPg negative amino acids, also at the interface [**51**], exhibit variability among the three herbaceous eIF(iso)4E sequences (**Figure** 5A). The lack of conservation suggests these regions would confer host-specific interaction performances on the protein. Thus, it is plausible that P114S mutation aims to achieve a more optimal fit of VPg, meeting particular requirements for the formation of a complex containing *A. thaliana* eIF(iso)4E. The apparent incongruence found when analyzing sequences of the eIF(iso)4E factors from permissive *N. clevelandii* (S50 and S147) and non-permissive *A. thaliana* (A48 and K149) and *C. foetidum* (S49 and R146) species (**Figure** 5A) could be explained by the fact that it is the functional capacity, rather than the primary sequence of the eIF4E/(iso)4E plant factors, what decides whether a successful infection takes place. Something similar was reported by Estevan et al. [**60**], who observed that, although *Tobacco etch virus* (TEV) uses *A. thaliana* eIF(iso)4E as cofactor, a convenient *trans*-complementation occurs by supplying *Capsicum annuum* eIF4E instead of *C. annuum* eIF(iso)4E.

The second mutation allowing adaptation to *A. thaliana*, F163L, affects a residue that does not match any of the PVY VPg amino acids located in the interface with h-eIF4E, as determined by Coutinho de Oliveira et al. [**51**]. However, experimental data obtained from that work showed that the PVY VPg amino acid L166, equivalent to a close neighbor of the mutated target of PPV VPg, was perturbed by h-eIF4E binding (**Supplementary Figure S7**).

These observations suggest that PPV-SwCMpVPg can adapt to At-eIF(iso)4E by two, probably independent, mechanisms. First, through P114S mutation, by improving the interaction of a VPg eIF4E-binding domain with the cap-binding domain of eIF(iso)4E. The second substitution, mediated by F163L, is less obvious. Although we cannot discard the occurrence of long-distance allosteric interactions between the residue 163 and VPg/eIF4(iso)4E cap binding domain interface, it seems more likely that this amino acid might participate in interactions not identified by Coutinho de Oliveira et al. [**51**] due to intrinsic limitations of their model. In this respect, it is important to remark that in the work of these authors, the VPg structure was obtained from a bacterial-expressed protein lacking the first 37 amino acids. Besides, the VPg-eIF4E complex was generated using the human factor eIF4E, and in absence of other suggested partners, both from the plant [eIF(iso)4G] and from the virus (P1, HCPro, CI) (see **Section 1**). It is clear that, in spite of the high value of the model reported by Coutinho de Oliveira et al. [**51**], more sophisticated studies are required to ascertain the structural details determining the compatibility spectrum between potyviral VPgs and host eIF4E/(iso)4E cofactors.

An important conclusion derived from our work points out to the necessary intervention of an intermediate host to prompt the adaptation to a second non-related host. As mentioned before, sporadic and mild infections by the non-mutated chimera PPV-VPgSwCM-R were detected in *C. foetidum*; but it is only through mutations affecting specific amino acids of SwCMpVPg after adaptation in *A. thaliana* that a robust infection in *C. foetidum* is triggered (**Figure 4**, **Table 1**, **Supplementary Figure S5**, **Supplementary Table S2**). The dynamizing role of intermediate bridge species in virus host range expansion has been described in animal systems, being especially relevant in cases of global pandemics involving animal-to-humans virus jumping [**61,62,63**]. In plants, adaptation of a virus to a particular host can have expanding ecological consequences once enabling adaptation to related plants. This has been described for interactions of PVY and plants of the *Solanaceae* family, in which PVY mutations that break the resistance generated by a particular eIF4E allele concurrently confer adaptation to additional plant genotypes with different eIF4E alleles [**64**]. Our results indicate that a bridge host can also help to break interfamily barriers, which are assumed to frequently restrict the viral host range expansion [**9**]. The reason why PPV-VPgSwCM-R cannot adapt by itself to *C. foetidum* and needs pre-adaptation in *A. thaliana* is probably because local lesions cause bottlenecks preventing fitness gain via natural selection [**65,66**]. However other obstacles, mainly genetic, but also ecological, can make bridge species especially necessary for some host jumpings.

Jumping to new hosts usually brings an adaptive cost in the initial host [**12**]. However, adaptation of PPV-VPgSwCM-R to *A. thaliana* does not seem to imply an adverse trade-off in the previously-adapted host *N. clevelandii* (**Figure 3**). There are previous reports showing fitness losses driven by PPV mutations that are associated to woody-to-herbaceous host jumpings [**46,67,68**]. And there are also examples in which the adaptation of PPV isolates to herbaceous plants did not seem to affect its ability to infect the *Prunus* species from which they came [**68,69**]. However, in these studies a limited trade-off linked to the jump cannot be rule out, as no competition experiments or fitness quantifications have been conducted in them. In our case, the adaptive mutations at SwCMpVPg, selected in *A. thaliana*, not only do not impose a trade-off in *N. clevelandii*, but even confer some better fitness in this host. This observation suggests that adaptation of PPV-R and PPV-SwCMp to *N. clevelandii* is not optimal; probably because these isolates have been replicating in *N. clevelandii* for a long time in human terms, but too short on an evolutionary scale, so they have not been able to fix mutations that provide small fitness gains.

Overall, this study aims to highlight the importance of bridge hosts, exposing the possibility that, as in our case, certain adaptive changes not only contribute to the expansion of the host range as a consequence of the initial jump, but additionally they allow distant species, directly inaccessible, to become regular hosts. In short, these "encounters" with one or more "appropriate intermediaries" could act as shortcuts, radically facilitating the way in which a virus maximizes its host range.

Supplementary Materials

Figure S1: Sequence of VPg in the polyproteins of the chimeric construct pICPPV-VPgSwCM-R [**44**] and its mutated versions. Polyprotein regions derived from *Plum pox virus* (PPV) isolates R and SwCMp are shown in grey and orange, respectively. Underlined amino acids in the VPg protein of SwCMp were mutated to those depicted in red. Figure S2: Schematic representation of the experimental approach followed to assess the effect of VPg mutations on *Plum pox virus* (PPV) infection in different hosts. **(A)** Full-length PPV cDNA clones were mechanically inoculated into *Arabidopsis thaliana* plants by biolistic (represented by a Gene Gun device), or into *Nicotiana clevelandii* and *Chenopodium foetidum* plants by hand-rubbing (represented by a Carborundum bottle). **(B,C)** Competitions assays between single and double mutants in *A. thaliana* and *N. clevelandii* **(B)** or between non-mutated chimera and VPg mutants in *N. clevelandii* **(C)** were carried out using eight plants (P1–P8) and mixtures of DNAs at the specified ratios. Amounts of DNAs used as inocula were adjusted to deliver each virus at specified doses, as the enzymatic pattern rendered after *Eco*RI digestion of each construct shows. Fragments yielded by the *Hind*III-digested ø29 fago DNA used as molecular-weight size marker (M) are also indicated. Figure S3: Effect of VPg mutations on *Plum pox virus* (PPV) infection of *Nicotiana clevelandii. N. clevelandii* plants were inoculated by hand-rubbing with DNAs of the chimeric clone pICPPV-VPgSwCM-R, the indicated chimera-derived mutants (two independent clones, 1 and 2), or the PPV-R clone pICPPV-NK-lGFP. Extracts from upper non-inoculated leaves collected at 15 days after inoculation were subjected to CP-specific immunoblot analysis. Two individual plants (P1 and P2) inoculated with the specified viruses, were analysed. Blots stained with *Ponceau* red showing the large subunit of the ribulose-1,5-bisphosphate carboxylase/oxygenase (RuBisCO) are included as loading controls. Figure S4: Sequence analysis of viral progeny from *Nicotiana clevelandii* exposed to mixed infections with competing viruses. Eight *N. clevelandii* plants were inoculated by hand-rubbing with DNA mixtures containing the indicated pICPPV-VPgSwCM-R-derived mutant clones. In the two competitions, the single mutant was overrepresented in the inoculum with respect to the double mutant (ratio 1.5:1). Viral progenies were analysed in pools of two plants by reverse transcription-polymerase chain reaction (RT-PCR) and sequencing of a cDNA fragment covering the VPg coding sequence. Images show the chromatograms of VPg codons 114 (position 1970–1968 in the viral genome) **(A)** or 163 (position 2019–2017 in the viral genome) **(B)**. Viruses identified are indicated beneath the chromatograms;

smaller letters indicate lower accumulation. Figure S5: Effect of VPg mutations on *Nicotiana clevelandii* infection by *Plum pox virus* (PPV). **(A)** Schematic representation of the experimental approach. **(B)** Prior to *C. foetidum* inoculation, extracts from upper non-inoculated leaves of *N. clevelandii* plants infected with the different viruses, collected at 10 days after inoculation, were subjected to coat protein (CP)-specific immunoblot analysis to estimate virus accumulation, thus allowing to adjust the inoculum doses. Increasing dilutions of the extracts from plants infected with the mutant viruses are indicated. The lower signal observed for the PPV-SwCMp sample is a consequence of the significantly poorer recognition of PPV-SwCMp CP compared to PPV-R CP by the choice antibody. Blots stained with *Ponceau* red showing the large subunit of the ribulose--1,5bisphosphate carboxylase-oxygenase (RuBisCO) are included as loading controls in both cases. **(C)** Images of *C. foetidum* leaves taken under visible light at 15 days post infection, after being inoculated with extracts of infected *N. clevelandii* plants. Bar, 5 mm. Figure S6: Assessment of viral titers in *Nicotiana clevelandii* extracts employed to inoculate *Chenopodium foetidum* plants. *N. clevelandii* plants were inoculated by hand-rubbing with DNA of the chimeric clone pICPPV-VPgSwCM-R, the indicated chimera-derived mutants (two independent clones, 1 and 2), or the PPV-R clone pICPPV-NK-lGFP. Extracts were prepared from upper non-inoculated leaves collected at 7 days post inoculation and their virus titers were adjusted with extracts from healthy leaves on the basis of a previous quantitative anti-CP immunoblot assay. The adjusted extracts were inoculated by hand-rubbing in *C. foetidum* leaves. Equalization of viral titers in the inocula was verified in the CP-specific immunoblot shown in the figure. Blots stained with *Ponceau* red showing the large subunit of the ribulose-1,5-bisphosphate carboxylase-oxygenase (RuBisCO) are included as loading controls. Figure S7: HADDOCK-derived structure of the human eIF4E (h-eIF4E) (blue molecule) in complex with the VPg of *Potato virus Y* (PVY) (PVY VPg) (red molecule), obtained by Countinho de Oliveira et al. [**51**], in which the following residues are highlighted: (i) PVY VPg residues perturbed by h-eIF4E binding (in salmon); (ii) residues in PVY VPg equivalent to those of *Plum pox virus* VPg mediating adaptation of the chimeric virus PPV-VPgSwCM-R to *Arabidopsis thaliana* and *Chenopodium foetidum*, P114 [that match one amino acid specified in (i)] and F163 (both in red); (iii) h-eIF4E residues perturbed by PVY VPg binding, whose equivalents are conserved among *A. thaliana*, *Nicotiana clevelandii* and *C. foetidum* (in cyan); and iv) residues in h-eIF4E perturbed by PVY VPg binding, whose equivalents in eIF(iso)4E from *A. thaliana*, *N. clevelandii* and *C. foetidum* are not fully conserved (in bright blue). Structural details were visualized using PyMOL. On the left, h-eIF4E:PVYVPg complex, pointing key residues addressed in this study. On the right, both molecules, separately disposed, indicating all residues reported to be perturbed after protein:protein interaction, according to Counthino de Oliveira et al. [**51**]. Table S1: Primer list. Table S2: Effect of specific mutations at the VPg of the *Plum pox virus*, isolate SwCMp (strain C), on the infection of *Chenopodium foetidum*.

References

1. Longdon, B.; Brockhurst, M.A.; Russell, C.A.; Welch, J.J.; Jiggins, F.M. The evolution and genetics of virus host shifts. *PLoSPathog.* **2014**, *10*, e1004395. [**Google Scholar**] [**CrossRef**] [**PubMed**] [**Green Version**]

2. Dennehy, J.J. Evolutionary ecology of virus emergence. *Ann. N. Y. Acad. Sci.* **2017**, *1389*, 124–146. [**Google Scholar**] [**CrossRef**] [**PubMed**]

3. Pulliam, J.R. Viral host jumps: Moving toward a predictive framework. *EcoHealth* **2008**, *5*, 80–91. [**Google Scholar**] [**CrossRef**]

4. Shi, Y.; Wu, Y.; Zhang, W.; Qi, J.; Gao, G.F. Enabling the 'host jump': Structural determinants of receptor-binding specificity in influenza A viruses. *Nat. Rev. Microbiol.* **2014**, *12*, 822–831. [**Google Scholar**] [**CrossRef**]

5. dos Santos Bezerra, R.; Valença, I.N.; de Cassia Ruy, P.; Ximenez, J.P.B.; da Silva Junior, W.A.; Covas, D.T.; Kashima, S.; Slavov, S.N. The novel coronavirus SARS-CoV-2: From a zoonotic infection to coronavirus disease 2019. *J. Med. Virol.* **2020**, *92*, 2607–2615. [**Google Scholar**] [**CrossRef**] [**PubMed**]

6. Moury, B.; Desbiez, C. Host range evolution of potyviruses: A global phylogenetic analysis. *Viruses* **2020**, *12*, 111. [**Google Scholar**] [**CrossRef**] [**Green Version**]

7. Wu, B.; Melcher, U.; Guo, X.; Wang, X.; Fan, L.; Zhou, G. Assessment of codivergence of Mastreviruses with their plant hosts. *BMC Evol. Biol.* **2008**, *8*, 335. [**Google Scholar**] [**CrossRef**] [**Green Version**]

8. McLeish, M.J.; Fraile, A.; Garcia-Arenal, F. Evolution of plant-virus interactions: Host range and virus emergence. *Curr. Opin. Virol.* **2019**, *34*, 50–55. [**Google Scholar**] [**CrossRef**]

9. Moury, B.; Fabre, F.; Hebrard, E.; Froissart, R. Determinants of host species range in plant viruses. *J. Gen. Virol.* **2017**, *98*, 862–873. [**Google Scholar**] [**CrossRef**]

10. Jones, R.A.C. Disease pandemics and major epidemics arising from new encounters between indigenous viruses and introduced crops. *Viruses* **2020**, *12*, 1388. [**Google Scholar**] [**CrossRef**]

11. Jones, R.A.C.; Naidu, R.A. Global dimensions of plant virus diseases: Current status and future perspectives. *Ann. Rev. Virol.* **2019**, *6*, 387–409. [**Google Scholar**] [**CrossRef**] [**PubMed**]

12. Elena, S.F.; Fraile, A.; Garcia-Arenal, F. Evolution and emergence of plant viruses. *Adv. Virus Res.* **2014**, *88*, 161–191. [**Google Scholar**] [**PubMed**] [**Green Version**]

13. Valli, A.; García, J.A.; López-Moya, J.J. Potyviruses (*Potyviridae*). In *Encyclopedia of Virology*, 4th ed.; Bamford, D., Zuckerman, M., Eds.; Elsevier: Oxford, UK, 2021; pp. 631–641. [**Google Scholar**]

14. Hari, V. The RNA of tobacco etch virus: Further characterization and detection of protein linked to the RNA. *Virology* **1981**, *112*, 391–399. [**Google Scholar**] [**CrossRef**]

15. Revers, F.; García, J.A. Molecular biology of potyviruses. *Adv. Virus Res.* **2015**, *92*, 101–199. [**Google Scholar**]

16. Rodamilans, B.; Valli, A.; Mingot, A.; San León, D.; Baulcombe, D.; López-Moya, J.J.; García, J.A. RNA polymerase slippage as a mechanism for the production of frameshift gene products in plant viruses of the *Potyviridae* family. *J. Virol.* **2015**, *89*, 6965–6967. [**Google Scholar**] [**CrossRef**] [**PubMed**] [**Green Version**]

17. Olspert, A.; Chung, B.Y.; Atkins, J.F.; Carr, J.P.; Firth, A.E. Transcriptional slippage in the positive-sense RNA virus family *Potyviridae*. *EMBO Rep.* **2015**, *16*, 995–1004. [**Google Scholar**] [**CrossRef**]

18. Hagiwara-Komoda, Y.; Choi, S.H.; Sato, M.; Atsumi, G.; Abe, J.; Fukuda, J.; Honjo, M.N.; Nagano, A.J.; Komoda, K.; Nakahara, K.S.; et al. Truncated yet functional viral protein produced via RNA polymerase slippage implies underestimated coding capacity of RNA viruses. *Sci. Rep.* **2016**, *6*, 21411. [**Google Scholar**] [**CrossRef**]

19. Grzela, R.; Szolajska, E.; Ebel, C.; Madern, D.; Favier, A.; Wojtal, I.; Zagorski, W.; Chroboczek, J. Virulence factor of potato virus Y, genome-attached terminal protein VPg, is a highly disordered protein. *J. Biol. Chem.* **2008**, *283*, 213–221. [**Google Scholar**] [**CrossRef**] [**Green Version**]

20. Rantalainen, K.I.; Uversky, V.N.; Permi, P.; Kalkkinen, N.; Dunker, A.K.; Makinen, K. Potato virus A genome-linked protein VPg is an intrinsically disordered molten globule-like protein with a hydrophobic core. *Virology* **2008**, *377*, 280–288. [**Google Scholar**] [**CrossRef**] [**Green Version**]

21. Charon, J.; Theil, S.; Nicaise, V.; Michon, T. Protein intrinsic disorder within the *Potyvirus* genus: From proteome-wide analysis to functional annotation. *Mol. Biosyst.* **2016**, *12*, 634–652. [**Google Scholar**] [**CrossRef**]

22. Rantalainen, K.I.; Eskelin, K.; Tompa, P.; Mäkinen, K. Structural flexibility allows the functional diversity of potyvirus genome-linked protein VPg. *J. Virol.* **2011**, *85*, 2449–2457. [**Google Scholar**] [**CrossRef**] [**Green Version**]

23. Jiang, J.; Laliberté, J.F. The genome-linked protein VPg of plant viruses-a protein with many partners. *Curr. Opin. Virol.* **2011**, *1*, 347–354. [**Google Scholar**] [**CrossRef**]

24. Martínez, F.; Rodrigo, G.; Aragonés, V.; Ruiz, M.; Lodewijk, I.; Fernández, U.; Elena, S.F.; Daròs, J.A. Interaction network of tobacco etch potyvirus NIa protein with the host proteome during infection. *BMC Genomics* **2016**, *17*, 87. [**Google Scholar**] [**CrossRef**] [**PubMed**]

25. Saha, S.; Mäkinen, K. Insights into the functions of eIF4E-biding motif of VPg in potato virus a infection. *Viruses* **2020**, *12*, 197. [**Google Scholar**] [**CrossRef**] [**PubMed**] [**Green Version**]

26. Charon, J.; Barra, A.; Walter, J.; Millot, P.; Hebrard, E.; Moury, B.; Michon, T. First experimental assessment of protein intrinsic disorder involvement in an RNA virus natural adaptive process. *Mol. Biol. Evol.* **2018**, *35*, 38–49. [**Google Scholar**] [**CrossRef**] [**PubMed**] [**Green Version**]

27. Robaglia, C.; Caranta, C. Translation initiation factors: A weak link in plant RNA virus infection. *Trends Plant Sci.* **2006**, *11*, 40–45. [**Google Scholar**] [**CrossRef**]

28. Charron, C.; Nicolai, M.; Gallois, J.L.; Robaglia, C.; Moury, B.; Palloix, A.; Caranta, C. Natural variation and functional analyses provide evidence for co-evolution between plant eIF4E and potyviral VPg. *Plant J.* **2008**, *54*, 56–68. [**Google Scholar**] [**CrossRef**]

29. Truniger, V.; Miras, M.; Aranda, M.A. Structural and functional diversity of plant virus 3'-Cap-independent translation enhancers (3'-CITEs). *Front. Plant Sci.* **2017**, *8*, 2047. [**Google Scholar**] [**CrossRef**] [**Green Version**]

30. Wang, A.; Krishnaswamy, S. Eukaryotic translation initiation factor 4E-mediated recessive resistance to plant viruses and its utility in crop improvement. *Mol. Plant Pathol.* **2012**, *13*, 795–803. [**Google Scholar**] [**CrossRef**]

31. Ala-Poikela, M.; Rajamaki, M.-L.; Valkonen, J.P.T. A novel interaction network used by potyviruses in virus-host interactions at the protein level. *Viruses* **2019**, *11*, 1158. [**Google Scholar**] [**CrossRef**] [**Green Version**]

32. Ala-Poikela, M.; Goytia, E.; Haikonen, T.; Rajamaki, M.-L.; Valkonen, J.P.T. Helper component proteinase of genus *Potyvirus* is an interaction partner of translation initiation factors eIF(iso)4E and eIF4E that contains a 4E binding motif. *J. Virol.* **2011**, *85*, 6784–6794. [**Google Scholar**] [**CrossRef**] [**Green Version**]

33. Nakahara, K.S.; Shimada, R.; Choi, S.-H.; Yamamoto, H.; Shao, J.; Uyeda, I. Involvement of the P1 cistron in overcoming eIF4E-mediated recessive resistance against *Clover yellow vein virus* in pea. *Mol. Plant Microbe Interact.* **2010**, *23*, 1460–1469. [**Google Scholar**] [**CrossRef**] [**Green Version**]

34. Hjulsager, C.K.; Olsen, B.S.; Jensen, D.M.; Cordea, M.I.; Krath, B.N.; Johansen, I.E.; Lund, O.S. Multiple determinants in the coding region of Pea seed-borne mosaic virus P3 are involved in virulence against sbm-2 resistance. *Virology* **2006**, *355*, 52–61. [**Google Scholar**] [**CrossRef**] [**Green Version**]

35. Abdul-Razzak, A.; Guiraud, T.; Peypelut, M.; Walter, J.; Houvenaghel, M.C.; Candresse, T.; Gall, O.; German-Retana, S. Involvement of the cylindrical inclusion (CI) protein in the overcoming of an eIF4E-mediated resistance against *Lettuce mosaic potyvirus*. *Mol. Plant Pathol.* **2009**, *10*, 109–113. [**Google Scholar**] [**CrossRef**]

36. Sorel, M.; Svanella-Dumas, L.; Candresse, T.; Acelin, G.; Pitarch, A.; Houvenaghel, M.C.; German-Retana, S. Key mutations in the cylindrical inclusion involved in *Lettuce mosaic virus* adaptation to eIF4E-mediated resistance in lettuce. *Mol. Plant Microbe Interact.* **2014**, *27*, 1014–1024. [**Google Scholar**] [**CrossRef**] [**PubMed**] [**Green Version**]

37. García, J.A.; Glasa, M.; Cambra, M.; Candresse, T. *Plum pox virus* and sharka: A model potyvirus and a major disease. *Mol. Plant Pathol.* **2014**, *15*, 226–241. [**Google Scholar**] [**CrossRef**]

38. Llácer, G. Hosts and symptoms of *Plum pox virus*: Herbaceous hosts. *EPPO Bulletin* **2006**, *36*, 227–228. [**Google Scholar**] [**CrossRef**]

39. Decroocq, V.; Sicard, O.; Alamillo, J.M.; Lansac, M.; Eyquard, J.P.; García, J.A.; Candresse, T.; Le Gall, O.; Revers, F. Multiple resistance traits control *Plum pox virus* infection in *Arabidopsis thaliana*. *Mol. Plant Microbe Interact.* **2006**, *19*, 541–549. [**Google Scholar**] [**CrossRef**] [**Green Version**]

40. Hajizadeh, M.; Gibbs, A.J.; Amirnia, F.; Glasa, M. The global phylogeny of *Plum pox virus* is emerging. *J. Gen. Virol.* **2019**, *100*, 1457–1468. [**Google Scholar**] [**CrossRef**]

41. Rodamilans, B.; Valli, A.; García, J.A. Molecular plant-plum pox virus interactions. *Mol. Plant Microbe Interact.* **2020**, *33*, 6–17. [**Google Scholar**] [**CrossRef**]

42. Sihelská, N.; Glasa, M.; Šubr, Z.W. Host preference of the major strains of *Plum pox virus* —Opinions based on regional and world-wide sequence data. *J. Integr. Agr.* **2017**, *16*, 510–515. [**Google Scholar**] [**CrossRef**]

43. Sheveleva, A.; Ivanov, P.; Gasanova, T.; Osipov, G.; Chirkov, S. Sequence analysis of *Plum pox virus* strain C isolates from Russia revealed prevalence of the D96E mutation in the universal epitope and interstrain recombination events. *Viruses* **2018**, *10*, 450. [**Google Scholar**] [**CrossRef**] [**PubMed**] [**Green Version**]

44. Calvo, M.; Martínez-Turiño, S.; García, J.A. Resistance to *Plum pox virus* strain C in *Arabidopsis thaliana* and *Chenopodium foetidum* involves genome-linked viral protein and other viral determinants and might depend on compatibility with host translation initiation factors. *Mol. Plant Microbe Interact.* **2014**, *27*, 1291–1301. [**Google Scholar**] [**CrossRef**] [**PubMed**] [**Green Version**]

45. Pérez, J.J.; Udeshi, N.D.; Shabanowitz, J.; Ciordia, S.; Juárez, S.; Scott, C.L.; Olszewski, N.E.; Hunt, D.F.; García, J.A. *O*-GlcNAc modification of the coat protein of the potyvirus *Plum pox virus* enhances viral infection. *Virology* **2013**, *442*, 122–131. [**Google Scholar**] [**CrossRef**] [**PubMed**] [**Green Version**]

46. Calvo, M.; Malinowski, T.; García, J.A. Single amino acid changes in the 6K1-CI region can promote the alternative adaptation of *Prunus*- and *Nicotiana*-propagated *Plum pox virus* C isolates to either host. *Mol. Plant Microbe Interact.* **2014**, *27*, 136–149. [**Google Scholar**] [**CrossRef**] [**Green Version**]

47. Ho, S.N.; Hunt, H.D.; Horton, R.M.; Pullen, J.K.; Pease, L.R. Site-directed mutagenesis by overlap extension using the polymerase chain reaction. *Gene* **1989**, *77*, 51–59. [**Google Scholar**] [**CrossRef**]

48. López-Moya, J.J.; García, J.A. Construction of a stable and highly infectious intron-containing cDNA clone of plum pox potyvirus and its use to infect plants by particle bombardment. *Virus Res.* **2000**, *68*, 99–107. [**Google Scholar**] [**CrossRef**] [**Green Version**]

49. Sievers, F.; Wilm, A.; Dineen, D.; Gibson, T.J.; Karplus, K.; Li, W.; Lopez, R.; McWilliam, H.; Remmert, M.; Söding, J.; et al. Fast, scalable generation of high-quality protein multiple sequence alignments using Clustal Omega. *Mol. Syst. Biol.* **2011**, *7*, 539. [**Google Scholar**] [**CrossRef**]

50. Berman, H.M.; Bhat, T.N.; Bourne, P.E.; Feng, Z.; Gilliland, G.; Weissig, H.; Westbrook, J. The Protein Data Bank and the challenge of structural genomics. *Nat. Struct. Biol.* **2000**, *7*, 957–959. [**Google Scholar**] [**CrossRef**]

51. de Oliveira, L.C.; Volpon, L.; Rahardjo, A.K.; Osborne, M.J.; Culjkovic-Kraljacic, B.; Trahan, C.; Oeffinger, M.; Kwok, B.H.; Borden, K.L.B. Structural studies of the eIF4E-VPg complex reveal a direct competition for capped RNA: Implications for translation. *Proc. Natl. Acad. Sci. USA* **2019**, *116*, 24056–24065. [**Google Scholar**] [**CrossRef**]

52. Parrish, C.R.; Holmes, E.C.; Morens, D.M.; Park, E.C.; Burke, D.S.; Calisher, C.H.; Laughlin, C.A.; Saif, L.J.; Daszak, P. Cross-species virus transmission and the emergence of new epidemic diseases. *Microbiol. Mol. Biol. Rev.* **2008**, *72*, 457–470. [**Google Scholar**] [**CrossRef**] [**Green Version**]

53. Woolhouse, M.E.J.; Gowtage-Sequeria, S. Host range and emerging and reemerging pathogens. *Emerg. Infect. Dis.* **2005**, *11*, 1842–1847. [**Google Scholar**] [**CrossRef**]

54. Geoghegan, J.L.; Duchene, S.; Holmes, E.C. Comparative analysis estimates the relative frequencies of co-divergence and cross-species transmission within viral families. *PLoSPathog.* **2017**, *13*, e1006215. [**Google Scholar**] [**CrossRef**] [**PubMed**]

55. Seale, J. Crossing the species barrier--viruses and the origins of AIDS in perspective. *J. R. Soc. Med.* **1989**, *82*, 519–523. [**Google Scholar**] [**CrossRef**] [**PubMed**]

56. Taubenberger, J.K.; Morens, D.M. Influenza: The once and future pandemic. *Public Health Rep.* **2010**, *125* (Suppl. 3), 16–26. [**Google Scholar**] [**CrossRef**] [**Green Version**]

57. Zhao, X.; Ding, Y.; Du, J.; Fan, Y. 2020 update on human coronaviruses: One health, one world. *Med. Nov. Technol. Devices* **2020**, *8*, 100043. [**Google Scholar**] [**CrossRef**]

58. Elena, S.F.; Bedhomme, S.; Carrasco, P.; Cuevas, J.M.; de la Iglesia, F.; Lafforgue, G.; Lalić, J.; Pròsper, À.; Tromas, N.; Zwart, M.P. The evolutionary genetics of emerging plant RNA viruses. *Mol. Plant Microbe Interact.* **2011**, *24*, 287–293. [**Google Scholar**] [**CrossRef**] [**PubMed**] [**Green Version**]

59. Truniger, V.; Aranda, M.A. Recessive resistance to plant viruses. *Adv. Virus Res.* **2009**, *75*, 119–159. [**Google Scholar**]

60. Estevan, J.; Marena, A.; Callot, C.; Lacombe, S.; Moretti, A.; Caranta, C.; Gallois, J.L. Specific requirement for translation initiation factor 4E or its isoform drives plant host susceptibility to *Tobacco etch virus*. *BMC Plant Biol.* **2014**, *14*, 67. [**Google Scholar**] [**CrossRef**] [**Green Version**]

61. Sutton, T.C. The pandemic threat of emerging H5 and H7 avian influenza viruses. *Viruses* **2018**, *10*, 461. [**Google Scholar**] [**CrossRef**] [**Green Version**]

62. Olival, K.J.; Hosseini, P.R.; Zambrana-Torrelio, C.; Ross, N.; Bogich, T.L.; Daszak, P. Host and viral traits predict zoonotic spillover from mammals. *Nature* **2017**, *546*, 646–650. [**Google Scholar**] [**CrossRef**]

63. Morens, D.M.; Fauci, A.S. Emerging pandemic diseases: How we got to COVID-19. *Cell* **2020**, *182*, 1077–1092. [**Google Scholar**] [**CrossRef**]

64. Moury, B.; Janzac, B.; Ruellan, Y.; Simon, V.; Ben Khalifa, M.; Fakhfakh, H.; Fabre, F.; Palloix, A. Interaction patterns between *Potato virus Y* and eIF4E-mediated recessive resistance in the *Solanaceae*. *J. Virol.* **2014**, *88*, 9799–9807. [**Google Scholar**] [**CrossRef**] [**Green Version**]

65. Domingo, E.; Holland, J.J. RNA virus mutations and fitness for survival. *Annu. Rev. Microbiol.* **1997**, *51*, 151–178. [**Google Scholar**] [**CrossRef**]

66. García-Arenal, F.; Fraile, A.; Malpica, J.M. Variation and evolution of plant virus populations. *Int. Microbiol.* **2003**, *6*, 225–232. [**Google Scholar**] [**CrossRef**] [**PubMed**]

67. Salvador, B.; Delgadillo, M.O.; Saénz, P.; García, J.A.; Simón-Mateo, C. Identification of *Plum pox virus* pathogenicity determinants in herbaceous and woody hosts. *Mol. Plant Microbe Interact.* **2008**, *21*, 20–29. [**Google Scholar**] [**CrossRef**] [**PubMed**] [**Green Version**]

68. Carbonell, A.; Maliogka, V.I.; Pérez, J.D.J.; Salvador, B.; León, D.S.; García, J.A.; Simón-Mateo, C. Diverse amino acid changes at specific positions in the N-terminal region of the coat protein allow *Plum pox virus* to adapt to new hosts. *Mol. Plant Microbe Interact.* **2013**, *26*, 1211–1224. [**Google Scholar**] [**CrossRef**] [**PubMed**] [**Green Version**]

69. Wallis, C.M.; Stone, A.L.; Sherman, D.J.; Damsteegt, V.D.; Gildow, F.E.; Schneider, W.L. Adaptation of plum pox virus to a herbaceous host (*Pisum sativum*) following serial passages. *J. Gen. Virol.* **2007**, *88*, 2839–2845. [**Google Scholar**] [**CrossRef**] [**PubMed**]

Chapter-3

Adaptation and Codon-Usage Preference of Apple and Pear-Infecting Apple Stem Grooving Viruses

1. Introduction

An understanding of the molecular evolution of viruses is essential to design knowledge-based control strategies. Relatively fewer studies are available on plant viruses covering sequence origin, mutation rates, selection pressure, reassortment, and recombination [**1**] Previous studies suggest point-mutations, recombination, and reassortments as major driving forces in evolution for the natural selection. The mechanisms of adaptation and selections contributing a virus fitness are yet to be explored exactly [**2**].

The adaptation of viruses to novel hosts is a general evolutionary phenomenon of invasion and adaptation to a new niche. The new host may present challenges at the level of viral entry into cells, replication, and transmission/exit. Similarly, the new host may encode different/stringent defense mechanisms, which the newly invading virus has to overcome. Only a small minority of the initial pool of viral genotypes may survive these hurdles and that subsequent adaptation will likely lead to improved adaptation [**3**].

Parallel evolution, i.e., independent evolution of similar or identical features in distinct lineages subject to similar selection pressures [**4**], were reported extensively in both natural and experimental populations of microbes—most often viruses [**4,5,6**], bacteria [**7**], yeast [**8**], and protozoa [**9**]. Differences in the founding genotypes may result in divergent evolutionary trajectories [**10,11**]. It is, therefore, important to know the dynamics of evolution that could be predicted through parallel evolution.

Apple stem grooving virus (type species, genus *Capillovirus*) is an economically important virus of pome fruits and citrus, particularly plant species of *Malus* and *Pyrus* [**12**]. A virion of apple stem grooving virus (ASGV) contains monopartite, linear single-strand positive-sense 6.7–7.4 kb RNA genome. As host plants are often asymptomatic, the virus survives long enough, unnoticed, to adapt to woody plant hosts.

Although ASGV is latent in most commercial rootstock–scion combinations, serious disease burden is evident in apples grown on *Malus sieboldii* rootstocks in Japan [**13**]. In pear, ASGV claimed black necrotic leaf spot disease where a decline up to 50% has been estimated [**14**].

Looking into the importance of the virus, we assessed the dynamics of evolution to examine the molecular basis of adaptation in a viral population under host-derived natural selection in apple and pear plants.

2. Materials and Methods

2.1. Sample Collection

Apple and pear plant samples were collected from major agricultural areas in South Korea (**Figure 1**) in the years 2008 and 2009. Petal and leaf tissues were collected from randomly chosen trees in each field. All collected tissues were stored at –80 °C until plant total RNA was extracted using the Easy-Spin™ (Intron Co., Seongnam, Korea) RNA extraction kit following manufacturer's instructions.

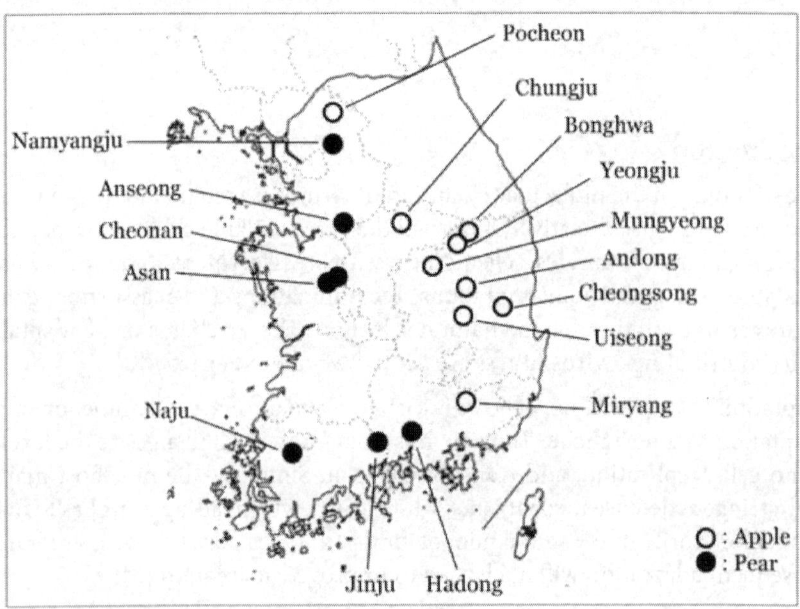

Figure 1. Geography of plant samples collected in this study for apple stem grooving virus amplification.

2.2. cDNA Amplification

Primers for the ASGV coat protein (CP) gene were designed as: *forward primer* -′5AGG CWA TCA GCG GCT GTG′3- and *reverse primer* -′5AGA GTG GAC AAA CTC TAG ACT CTA GAA A›′3. A reference sequence (GenBank AB004063) was used to design the primers. The cDNA syntshesis was carried out in a 5 μL reaction mixture. Briefly, 0.1 ng to 5 μg plant total RNA, 20 pmol reverse primer was taken. The volume was made up to 3 μL with DEPC-treated water. The mixture was incubated for 5 min at 65 °C, followed by chilling on ice. Before incubating reaction mixture for 60 min at 42 °C, 1× reaction buffer (Promega Co., Madison, WI, USA), 1 mM final concentration of dNTP mix and 1 unit of AMV Reverse Transcriptase was added. The product of reverse transcription was directly used in downstream PCR or stored at 20– °C.

A final volume of 25 µL reaction mixture—comprised of 1× Taq buffer, 200 µM dNTPs mix, 1.5 mM $MgCl_2$, 10 pM of each forward and reverse primer, 1 unit of Taq DNA Polymerase (Cat# M7660, Promega Co.), and 5 µL of reverse transcription product—were incubated in a thermal cycler. The incubation program was executed as—denaturation for 3 min at 94 °C, followed by amplification for 35 cycles (94 °C for 30 s, 55 °C for 30 s, and 72 °C accordingly as 1 min per kb amplicon). The amplification was followed with a final extension of 10 min at 72 °C. The PCR product was visualized on 1% agarose gel.

2.3. Phylogenetic Analysis

Amplified PCR amplicons were sequenced in both orientations, assembled and analyzed. The sequences were then submitted to the GenBank (**Table S1**). The ASGV sequences identified in this study were aligned with the previously available sequences in the databases. The evolutionary history was inferred from the nucleotide sequences using the MUSCLE method based on K2+G model (Figure 4). The bootstrap consensus tree was inferred using 100 replicates representing the evolutionary history. The tree is drawn to scale, with branch lengths measured in the number of substitutions per site.

Nucleotide sequences were translated into amino acid sequences. The evolutionary history of the amino acid sequences was inferred using the MUSCLE method, based on the JTT+G best-fit-model (Figure 5). The bootstrap consensus tree was inferred from 100 replicates representing the evolutionary history. All other options followed as described above.

2.4. Protein Structure Prediction

Multi-aligned amino acids sequences were analyzed using Geneious Pro 8 (Biomatters Co., Auckland, New Zealand). Secondary structures were predicted using the original Garnier Osguthorpe Robson algorithm (GOR I) provided by the EMBOSS suite embedded in Geneious Pro 8. Tertiary structures were predicted and modeled using the i-TASSER server [**15**].

2.5. Codon-by-Codon Behavior Analysis for Synonymous and Non-Synonymous Substitutions

Among the sequences identified in this study, only fully characterized ASGV CP sequences were selected. These sequences were analyzed to determine the cumulative behavior of sequence variation in a codon-by-codon manner with the synonymous and non-synonymous analysis program [**16**].

2.6. Codon-by-Codon Maximum Likelihood Analysis of Natural Selection

For each codon, estimates of inferred synonymous (s) and nonsynonymous (n) substitutions were presented along with the numbers of sites that were estimated to be synonymous (S) and nonsynonymous (N). These estimates were produced using joint maximum likelihood reconstructions of ancestral states under a Muse-Gaut model of codon substitution and Tamura-Nei model of nucleotide substitution. For estimating ML values, a tree topology was automatically computed. The test statistic dN-dSwas used to detect codons that had undergone positive selection; where, dS is the number of synonymous substitutions per site (s/S) and dN is the number of nonsynonymous substitutions per site (n/N). The analysis involved 25 Korean ASGV isolates. The codon positions included were 1st + 2nd + 3rd + Noncoding. All positions containing gaps and missing data were eliminated. There were 675 positions in the final dataset. Evolutionary analyses were conducted in MEGA6 [**17**].

2.7. Preferred Codon Usage of ASGV CP Gene

Codon usage of ASGV CP genes was analyzed using Geneious Pro 8 with multi-aligned nucleotide and the corresponding amino acid sequences. Codon usage of ASGV was analyzed in two ways. First, a correspondence analysis for codon usage was performed for all Chinese and Korean isolates. Second, codon usage frequencies among apple and pear isolates were processed on the codon usage database and codon usage analyzer [**18**].

3. Results

A total of 24 amplicons were sequenced in both orientations. Sequencing results confirmed ASGV sequences. The sequences were 710–711 nucleotides long covering complete coat protein gene coding sequence (CDS) except the stop codon (i.e., TAG and, in some cases, a T before the stop codon). The sequences were then submitted to the GenBank (JN792471 to JN792494).

3.1. Phylogenetic Analysis of ASGV Showed Host-Specificity

To determine ASGV host-specificity, 56 CP sequences, including 24 isolates identified in this study (i.e., 13 isolates from apple, 11 isolates from pear) and a former Korean isolate (GenBank AY596172) from pear, identified prior to this study, were used. Phylogenetic analysis based on nucleotide sequences revealed clustering of clear distinct patterns corresponding to their isolation hosts. Additionally, intermediate groups were also observed (Figure 4).

Most ASGV isolates worldwide were categorized in clade 4. Countries from which the isolates were reported are the Latvia, Serbia, and Turkey in Europe; China, India, and Korea in Asia; and Brazil in the Americas. Regardless of the location, ASGV were easily distinguished by their isolation hosts. Similar to the clade 4, ASGV isolates from China in intermediate[2] was also separated to their geography.

Interestingly, clade 3 contains Korean isolates identified on apple trees. Along with this, the clade 3 contains a Japanese isolate identified on apple trees. Most importantly, all isolates collected from pear trees from the distant regions were categorized in intermediate clade 1, referred to as the Korean pear group, and the clade 2. These distant regions included Anseong (GenBank JN792487), Asan (GenBank JN792486), Cheonan (GenBank JN792488), Kimcheon (GenBank JN792484), Naju (GenBank JN792493), Namyangju (GenBank JN792491), and Sangju (GenBank JN792485). To conclude, the distinctly located isolates showed high identities based on their isolation host—e.g., the two pear isolates Cheonan (GenBank JN792492) and Hadong (GenBank JN792494) showed 95.6% nucleotide identities and clustered in the clade 2 (**Figure 2**, Figure 4 and **Figure S1**).

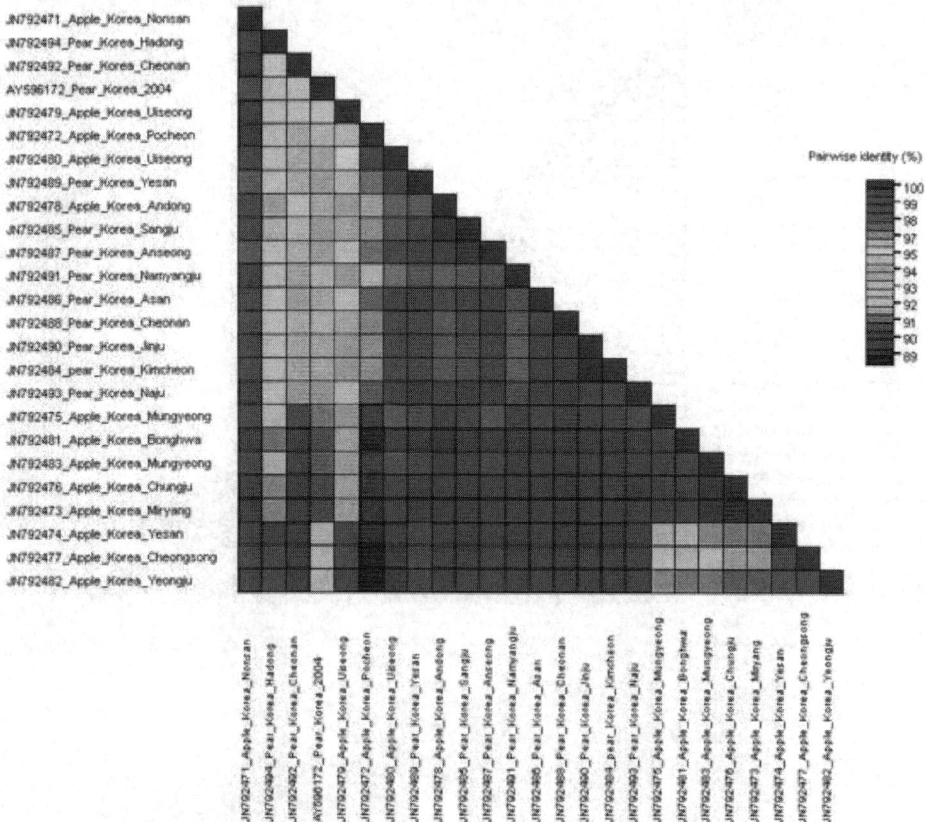

Figure 2. Apple stem grooving virus (ASGV) coat protein gene nucleotide sequence identities. MUSCLE alignment was used implemented in Sequence demarcation tool (SDTV1.2).

The isolates from the pear group exhibited significant differences in nucleotide sequence identity up to 10% against the apple isolates (Figure 6). Four apple isolates cluster into clade 1, GenBank JN792472 (Pochen), JN792479 (Uiseong), and JN792480 (Uiseong) (Figure 4). Although the pear isolates were collected from far flung distant areas, the pear isolates showed close relationship with each other and were distinguished from the apple isolates (**Figure 1** and Figure 4). Isolates from apple and pear identified from the same or nearby regions did not show much relevance. For example, the accession with GenBank JN792492 from Yeongju in pear and JN792478 from Andong in apple were collected from nearby regions were clustered in distinct clades 2 and intermediate[1] clade, respectively (Figures 4 and 5). Similarly, the two accessions from Yesan (GenBank JN792474 (apple) and JN792489 (pear)) shared only 90.0% nucleotide and 91.5% amino acid identities with each other (**Figures S1 and S2**). The distribution of the Yesan isolates into distinct clades, i.e., clade 4 and intermediate[1] clade (Figures 4 and 5) indicates the isolation host specificity (also evident from **Figure 2** and **Figurc 3**).

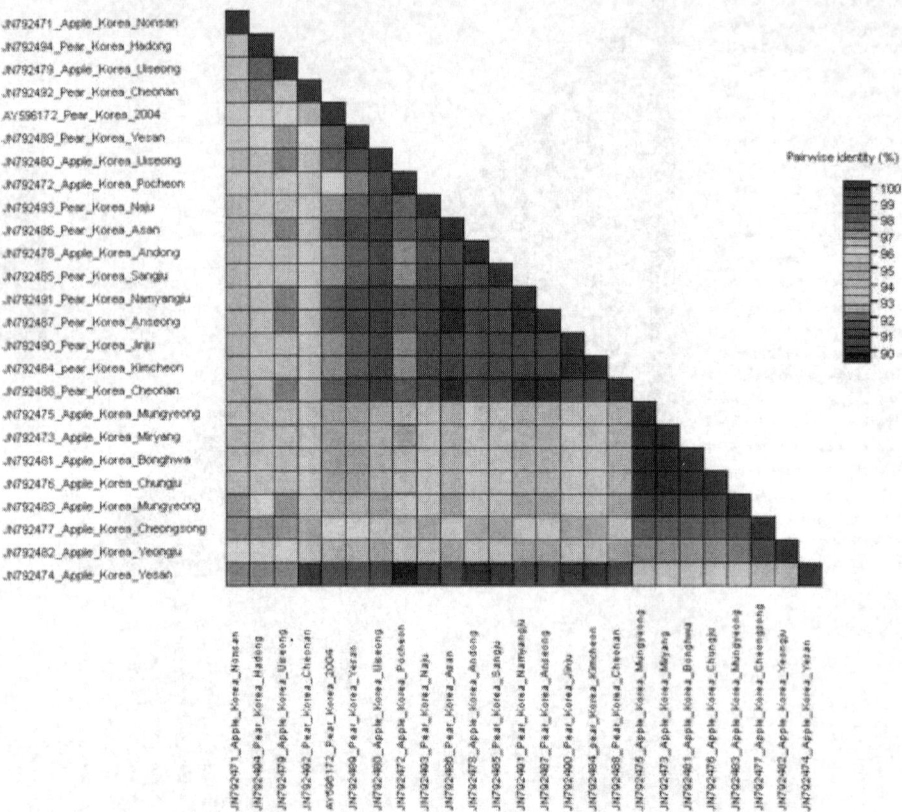

Figure 3. Apple stem grooving virus (ASGV) coat protein amino acid sequence identities. MUSCLE alignment was used implemented in Sequence demarcation tool (SDTV1.2).

The CP sequences were translated to amino acid sequences and aligned using the MUSCLE algorithm. The phylogenetic analysis showed that the amino acids sequences were similar to the nucleotide sequences in context with host relevance (Figure 5). Contrary to the nucleotide sequence tree, the amino acid sequence tree showed low bootstrap support, which may mean that the amino acid sequence variation was lower than the nucleotide sequences.

Both phylogenies followed the same clustering trend (**Figure 4** and **Figure 5**). The Korean apple isolates had Gln^{9}, Ala^{27}, Leu^{30}, Gly^{38}, Lys^{48}, Lys^{96}, Glu^{103}, Arg^{104}, Ala^{108}, Ser^{110}, Ile^{117}, and Arg^{199} residues, whereas the Korean pear isolates had Leu^{9}, Gly^{27}, Ser^{38}, Arg^{48}, Arg^{96}, Glu^{103}, Lys^{104}, Glu^{108}, Met^{110}, Val^{117}, and Lys^{199}. The results of the amino acid sequence analysis showed differences between the apple and pear isolates in terms of amino acids residues. These changes suggest that the differences could be expressed in their structure. To explore these differences, secondary structures of the peptide chains were predicted.

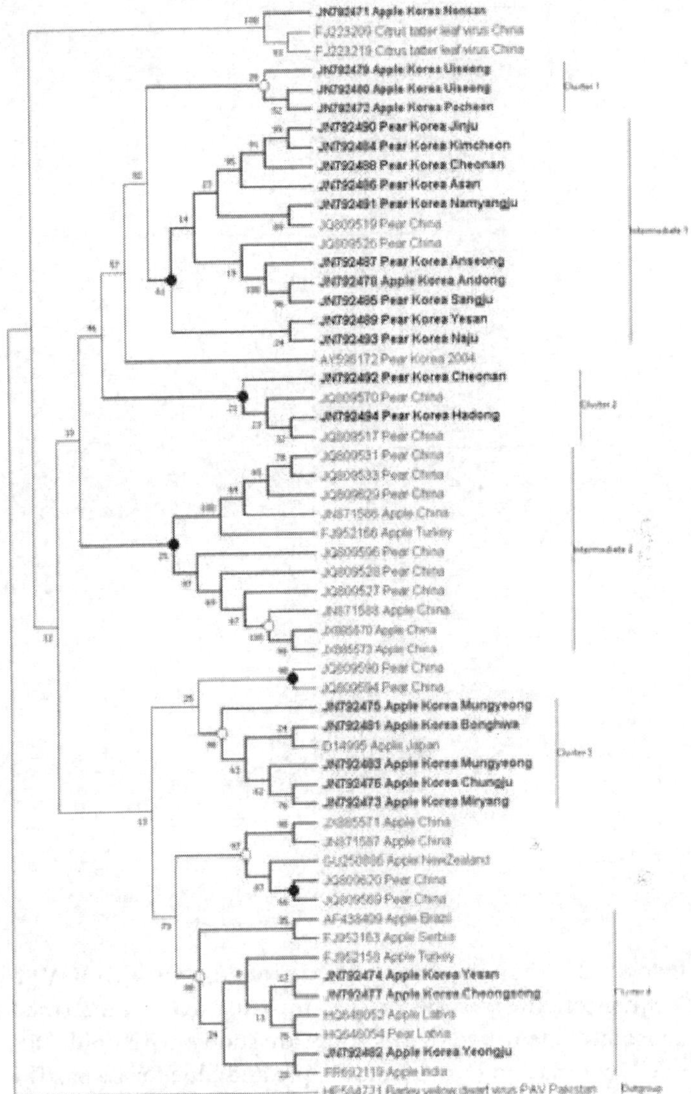

Figure 4. Phylogenetic analysis of apple stem grooving virus (ASGV) coat protein gene nucleotide. The tree is divided into distinct and intermediate clades. The Korean isolates identified in this study are shown with bold. Taxon names are color coded by isolation host (red for apple and blue for pear). The bootstrap support value is displayed on each branch node. The tree was generated using Maximum Likelihood with K2+G best-fit-model implemented in MEGAX.

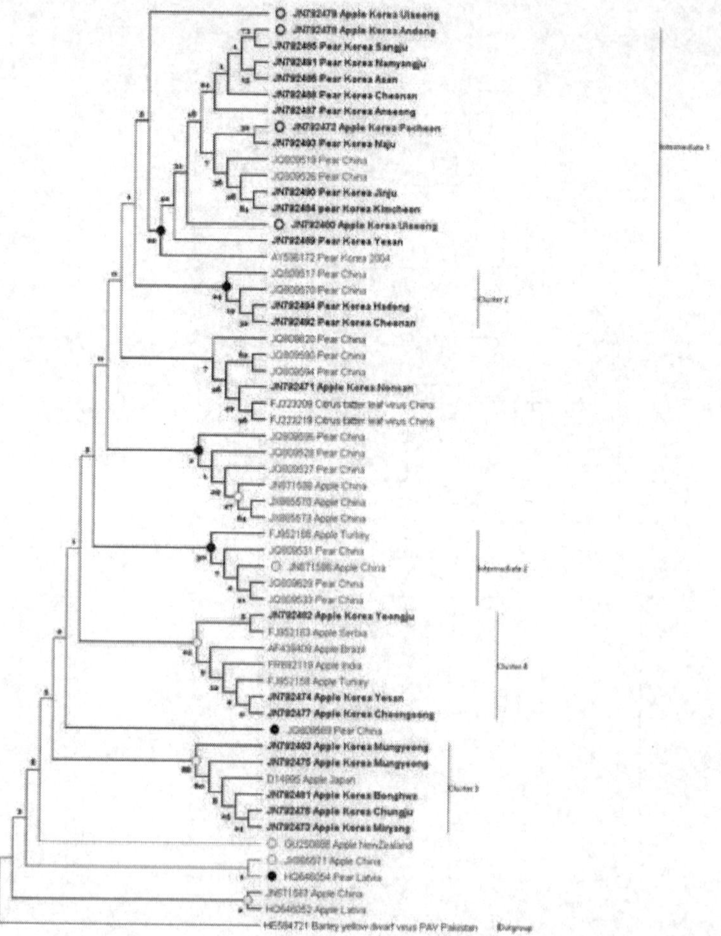

Figure 5. Phylogenetic analysis of apple stem grooving virus (ASGV) coat protein amino acid sequences. The tree was divided into distinct and intermediate clades. The Korean isolates identified in this study are shown with bold. Taxon names are color coded by isolation host (red for apple and blue for pear). The bootstrap support value is displayed on each branch node. The tree was generated using Maximum Likelihood with JTT+G best-fit-model implemented in MEGAX.

The secondary structure predictions showed categorization corresponding to the phylogenetic groupings of nucleotide and amino acid sequences. Representative schematic diagrams of the predicted secondary structures are shown in **Figure S4**. Amino acid residue variations are reflected in the secondary structure prediction. Residue variations in the apple isolate group included Gln^{9}, Ala^{27}, Leu^{30}, Gly^{38}, Lys^{48}, Lys^{96}, Glu^{103}, Arg^{104}, Ala^{108}, Ser^{110}, Ile^{117}, and Arg^{199} compared to the pear isolates and reflected helix length, strand and coil structure, and appearance of the beta-turn.

Among these variations, the most distinguishable locus was located between the 100th and 130th amino acids. At this locus, the apple isolates had shorter helix structure than the pear isolates, due to a change in residues at the 110th position; the pear isolates had a methionine residue instead of serine as in the apple isolates. Similarly, Lys^{198} residue substitution in the pear isolate group for Arg^{198} in the apple isolate group resulted in clear differences in the secondary structure. This residue difference changed the length of the helix structure and appearance of the beta-turn structure (**Figure S4**).

Furthermore, all these variations in the secondary structure are reflected in the tertiary structure prediction model. From the phylogenetic analyses based on the amino acid sequences, a total of eight isolates were selected and the tertiary structures were determined for comparisons between the apple and pear groups. From the pear group, the isolates GenBank JN792486, JN792484, JN79247, JN792485, and JN792472 were selected and from the apple group, the isolates GenBank JN792477, JN 792483, and JN792475 were selected (**Figure S6**). These were representative of the subgroups observed in the phylogenetic tree of the amino acid sequences.

The predicted structures differed significantly from each other. In particular, GenBank JN792486 and JN792484 were the most divergent among the eight isolates. Their structure predictions were quite different from the structure of the apple isolate group; the N-terminal region was folded in a different direction compared with that in the apple isolate group. In contrast, three other sequences had similar structures to the apple isolate group. In addition, all isolates showed structural differences corresponding to their clustering in the amino acid phylogenetic tree (**Figure S2**).

3.2. Comparisons of Nucleotide and Protein Sequences

Few differences were observed when phylogenetic trees based on amino acid and nucleotide sequences were compared. For instance, the maximum nucleotide and amino acid sequence identities of Korean ASGV isolates were 92.8% and 96.4%, respectively (**Figure 6**). As predicted, the amino acid sequence alignment showed fewer differences compared to the nucleotide sequences. The cumulative codon-by-codon behavior analysis for synonymous and non-synonymous substitutions showed a higher rate of synonymous (dS) than non-synonymous (dN) substitutions indicating purifying selection (**Figure 7**).

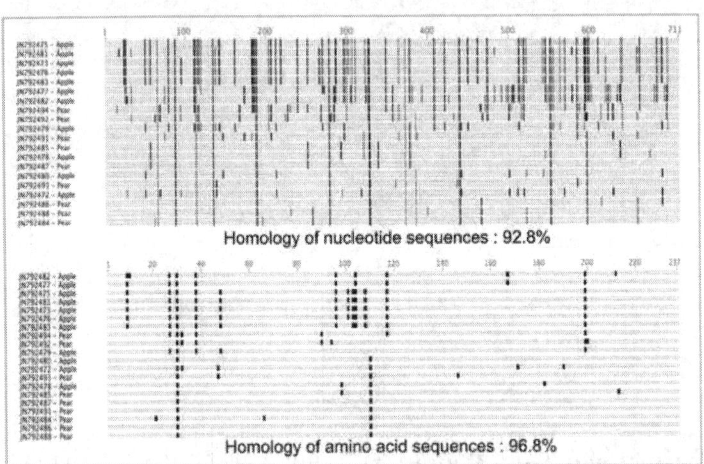

Figure 6. Multi-aligned sequence comparisons between nucleotides and amino acids. Identities of nucleotide and amino acid sequences were different from each

other. Black highlighted positions represent differences between residues and the consensus sequence. Nucleotide sequences showed various residue substitutions coding for the same amino acids.

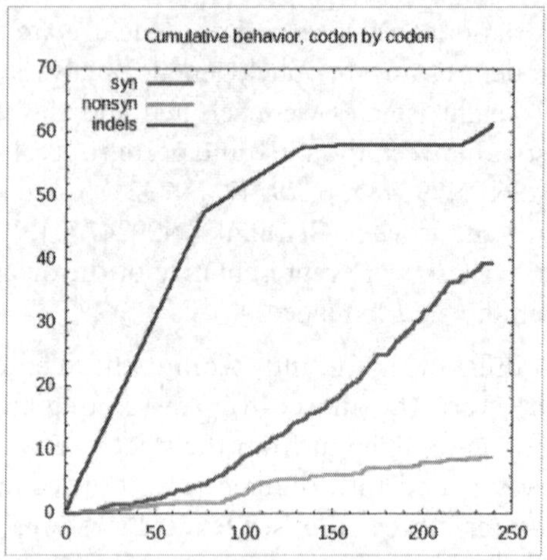

Figure 7. Cumulative behavior (codon-by-codon) of Korean isolates of *Apple stem grooving virus.*

3.3. The CP Gene Corresponds with the Host Codon Usage Frequency

The differences in the nucleotide sequences might better be explained by variation in codon usage. Specifically, Chinese and Korean isolates were clearly separated according to their hosts. More precisely, codon usage of the ASGV CP gene from Korea and China illustrated that isolates could be distinguished by the isolation country as well (**Figure 8**).

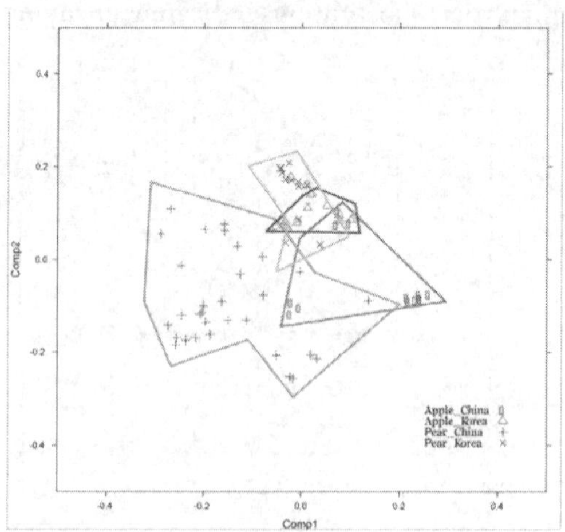

Figure 8. Correspondence analysis with codon usage of apple stem grooving virus (ASGV) coat protein gene from Korea and China. Codon usage of ASGV corresponded with the host and country of collection. Region with red boundary

line indicates in majority the Korean ASGV collected from apple. The area with a blue boundary shows Korean ASGV collected from pear. Region with a pink boundary indicates Chinese ASGV collected from apple. Region with green circumference indicates Chinese ASGV collected from pear.

For the Korean ASGV isolates, the codon usage ratio was examined codon-by-codon via a maximum likelihood analysis of natural selection. Codon usage variations appeared throughout the entire sequence of the CP gene. Specifically, these variations were not relevant to their genomic distance on the CP gene. The dS rate was higher than the dN rate for all codons among the 237 residues (**Figure 9**a). To characterize codon usage variation tendencies, the usage of every codon between isolates obtained from apple and pear were compared on the multiple sequence alignment. We analyzed 208 dN among a total of 237 residues for the Korean ASGV CP gene. The codon usage analysis showed variations between the CP gene of apple and pear isolates at 95 loci among 18 amino acids. Moreover, apple isolates showed more variation (showing more diversity) in codon usage than the pear isolates. Apple isolates showed variation at 79 loci as compared to 56 loci in pear isolates. Both the variation and variety of the codon usage was high in the apple isolates when compared with pear isolates. For example, there was more codon usage variety for glutamine residues among apple isolates, whereas the codon usage was conserved in pear isolates. Similarly, codon usage variations for other amino acids were more highly conserved in pear isolates than apple isolates. Except for those variants, most codon use was similar for the entire CP gene (**Figure 9**b).

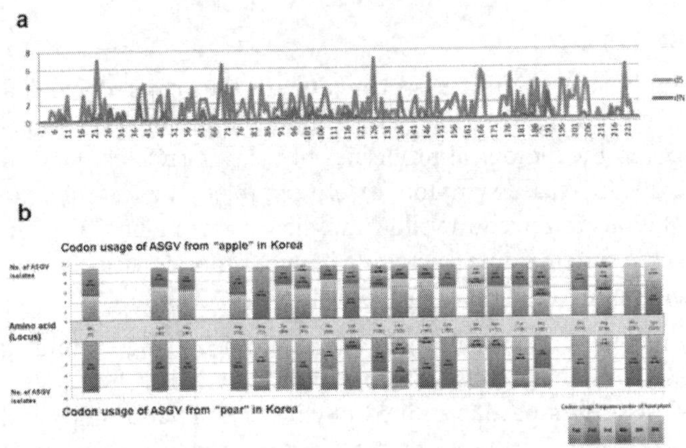

Figure 9. Codon usage analysis of Korean apple stem grooving virus (ASGV) coat protein gene. (a) Maximum likelihood analysis of natural selection codon-by-codon. The analysis involved 25 nucleotide sequences with 675 positions of ASGV codons. Most nucleotide substitutions were synonymous over the entire coat protein gene of Korean ASGV. (b) Analysis of ASGV codon usage at loci of synonymous substitutions compared with host preferred codon usage frequency. Substitutions of virus gene codons tended to follow the preferred codon usage of the isolated host.

Among these variants, codon usages of apple isolates seem to follow the codon usage frequency of the apple host at 21 amino acid residues loci (**Figure 9**b). In pear isolates, the codon for Cys^{150} was UGU only, whereas among the 13 apple isolates, six were UGC and five were UGU. In apple, the UGC codon was %3.4 more frequent than UGU. In pear isolates, for Asp^{73}, the codon was GAC, whereas among the 13 apple isolates five were GAU and six were GAC. The frequency of GAG was 6% higher than that of GAC in apple isolates. Codon usages for Glu were tended to follow the apple gene codon usage. Compared to pear isolates, apple isolates were more likely to follow host codon usage frequency. Codon usage frequency for Glu was similar across pear isolates, with only a 0.2% difference. However, GAG was 6.1% more frequent than GAA in apple isolates. In apple isolates, Glu^{46} was represented by the GAG and GAA codons six and five times, respectively, whereas in pear isolates, only the GAA codon was observed. Similarly, in apple isolates, for Glu^{99}, GAG and GAA appeared eight and three times, respectively, whereas, in pear isolates, they appeared one and eight times, respectively. At the 205th locus, in apple isolates, GAG and GAA were observed five and six times, respectively, whereas among pear isolates, only the GAA codon was observed. At the 226th locus, GAG was observed only in apple isolates, whereas GAA was exclusively observed in pear isolates.

Codon usage variations in apple isolates for Gly^{109}, Gly^{228}, Ile^{171}, Lys^{40}, Leu^{91}, Leu^{129}, Leu^{141}, Pro^{185}, Arg^{214}, Val^{126}, Tyr^{89}, and Tyr^{174} were similar to the apple host codon usage. The codon usage frequency was similar for the similar amino acid, for instance, the codon frequency of apple for glutamine represented a difference of 1 percent only. Furthermore, analysis of Chinese ASGV isolates showed similar trend. These results indicate that the codon usage follows host codon usage frequency (**Figure S3**).

4. Discussion

Mutations, recombinations, and reassortments are the major driving forces explored extensively in plant virus evolution [**2**]. Even single amino acid mutations result in dramatic changes in the biological properties of viruses, such as symptom severity and systemic infectivity. Increased symptom severity in tolerant zucchini cultivars attributed to the point mutations in Zucchini yellow mosaic virus sequence [**19**]. Similarly, the loss and recovery of the systemic infectivity to cucumber mosaic virus in squash linked to a single amino acid mutation in the CP (Thompson et al., 2006).

In a similar way, recombination of viral genomes could also change the biological properties of viruses. RNA-RNA homologous recombination and virus evolution studies have a long history [**20**]. Most studies have examined phylogenetic relationships and viral genome recombination [**1,21,22**]. Furthermore, the effects of viral genome recombination have been studied using artificial recombinant genomes to demonstrate the loss and recovery of viral infectivity [**23**]. Genome reassortment is yet another driving force of evolution in multi-partite viruses. Studies on plant virus genome reassortment have mostly been conducted with the cucumber mosaic virus mainly due its split genome organization, i.e., three-segments [**24,25**].

In this study, 24 ASGV CP isolates (i.e., 13 isolates from apple and 11 isolates from pear) were amplified and analyzed along with 01 former Korean isolates from pear. The amplicons were sequenced in both orientations. Mix infection of a sample with different isolates remains a major possibility; however, was not studied. The sequences were then analyzed to identify the forces driving plant virus evolution. Phylogenetic analyses showed interesting categorizations of the ASGV CP gene. All the analyzed isolates could be categorized in two groups through their corresponding hosts, pear or apple. First, recombinations among virus CP gene were analyzed. All apple and pear isolates were infected with different viruses, Apple chlorotic leaf spot virus (ACLSV) and Apple stem pitting virus, in different combinations. These viruses were classified in the same virus family, *Flexiviridae* [**26**,**27**]. To determine the relationship between recombination and phylogenetic categorization, we analyzed recombination in ASGV CP genes with co-infected virus genes. According to the viral infection status, most apple isolates were co-infected with ACLSV, whereas pear isolates were not. However, we did not detect any recombination event reliable/acceptable among viral species (data not shown).

Second, to explain the categories detected, we took clues from studies regarding host adaptation of viruses. Previous host adaptation studies have examined different viral species with various experimental hosts. Ohshima et al. (2002) analyzed the molecular variability of 66 isolates of turnip mosaic virus (TuMV) isolated from several hosts. In TuMV, phylogenetic categorization showed host and geographical differences. Sacristán et al. (2005) assayed the infectivity of cucumber mosaic virus during the 10th passage of virus inoculation to examine host adaptation. After 10 passages of inoculation, viral accumulation significantly increased on newly introduced hosts. Similarly, viruses on woody plants showed similar phenomena. Symptom severity changed when the viruses were introduced to different hosts [**28**,**29**].

Molecular analysis has been conducted to characterize the biological properties of viruses that have adapted to the introduced hosts. RNA genome sequence variations were analyzed to determine the evolutionary history of viruses through the host passages and examine changes in amino acid sequence substitutions [**30**,**31**].

All sequences were categorized into two major groups accordingly as per their hosts. The bifurcation is evident both in nucleotides and amino acids sequence analysis. These groups showed significant differences in predicted CP structures. Torres et al. (2005) and Pinel-Galzi et al. (2007) previously suggested selective forces and pathways of virus evolution. The present genome variation analysis elucidated the direction of the virus evolution pathway. Together, these studies suggested that genomic evolution is still occurring, and that the direction can be determined. Similar to that in previous studies, we found evidence of evolution in the ASGV CP, with phylogenetic differences corresponding to the hosts in both nucleotide and amino acid levels (**Figure 4** and **Figure 5**). The synonymous microevolution of viral and host genomes suggests that viral genomic evolution is related to adaptation to newly introduced hosts. As Taq DNA polymerase (Cat# M7660, Promega Co.) was used for amplification, there might be some un-intended errors/mutations introduced in the amplicons due to lack of the polymerase proof reading activity. However, a trend of host-related adaption while amplifying multiple samples indicate a minimal impact of the PCR induced mutations.

The phylogenies based on nucleotide and amino acid sequences showed interesting variations. The variations were mostly caused by silent mutations in the genome as an index of wobble hypothesis. All codon usage differences followed the phylogenetic categorization corresponding to their isolated hosts. To explain the relationship between phylogenetic categorization and hosts, viral gene codon usages were compared with host codon usage frequencies. We found that the codon usage variations followed the host codon usage frequency. Most variations that agreed with host codon usage appeared in isolates from apple. For 20 loci of the ASGV CP, more than 5 isolates from apple had higher frequencies of codon similarity to the hosts, whereas pear isolates had lower frequencies of codon similarity. Pear isolates also followed the higher frequency host codons at least at two loci.

In plant evolution studies, codon usage has not been considered an important selective force [**32**]. However, the dicots- and monocots-infecting sobemoviruses were categorized in the same groups [**10**]. Nevertheless, studies of plant virus evolution that examine codon usage were not considered useful to better understand host specificity. Unlike studies of plant viruses, in animal viruses, codon usage is considered an important factor in phylogenetic relationship along with base composition [**33**]. Additional studies of animal viruses suggested that relationships in codon usage between the host and the virus are driving forces of virus evolution [**34**]. Moreover, viral protein codon adaptation to the host, as observed in viral proteins, does not require host-specific recognition [**35**].

We detected host adaptation with synonymous changes in the CP gene among apple isolates of ASGV. In addition, codon usage for the ASGV CP gene was conserved in isolates from pear compared with those from apple. A study on the evolutionary trajectory of *Turnip mosaic virus* reported higher nucleotide diversity among viral genomes that had been introduced to new hosts [**3**]. Codon usage among pear isolates was conserved, unlike the high frequency of codon variations of pear. Conserved codon usage followed tendencies for low frequency codon use of hosts. Our results are similar to the analyses of codon usage bias of Antoniw and Adams (2003). Codon usage variation among apple isolates may, therefore, constitute evidence of ASGV adaptation to hosts. As far as protein structure analysis is concerned, the structural variability was evident in the 3D conformations of the two groups of CP (**Supplementary Figure S3**).

Based on the ASGV CP gene, host preference/adaptation is evident from the nucleotide and protein sequence analysis (**Figure 4** and **Figure 5**). More importantly, non-synonymous nucleotide substitutions were common showing significant separation in accordance with their hosts. Synonymous substitutions of the CP gene reflect parallel evolution. Variations among synonymous nucleotide substitutions follow host codon usage frequency. This indicated that viral genomes evolve alongside host molecular factors. Mutational selection, therefore, may drive microevolution within translational selection forces, especially host codon usage frequency.

Supplementary Materials

Figure S1. Percentage nucleotide identities of Korean Apple stem grooving virus (ASGV) coat protein gene using MUSCLE alignment implemented in SDTv1.2. Figure S2. Percentage amino acid identities of Korean apple stem grooving virus (ASGV) coat protein gene using MUSCLE alignment implemented in SDTv1.2. Figure S3. Synonymous (dS) and Non-synonymous (dN) substitution of the apple stem grooving virus (ASGV) coat protein gene (a). Codon usage of ASGV from apple and pear reported sequences from China (b). Figure S4. Secondary structure prediction of Korean apple stem grooving virus (ASGV) coat protein. Amino acid se-quences were rearranged corresponding to the phylogenetic analysis categorization of nucleotide sequences. Boxes rep-resent the significant differences between the two clades from the phylogenetic analysis. Predicted secondary structure results showed similar categorization of nucleotide phylogenetic analysis. Secondary structures were predicted using the original Garnier Osguthorpe Robson algorithm (GOR I) provided by the EMBOSS suite embedded in Geneious Pro 8. Figure S5. Protein model prediction of selected Korean apple stem grooving virus (ASGV) coat protein sequences using i-TASSER. Phylogenetic relationship of the isolates is also shown. The accessions used for model prediction were under-lined. Table S1. List of coat protein gene sequences used in the nucleotide sequence analyses.

References

1. Ohshima, K.; Yamaguchi, Y.; Hirota, R.; Hamamoto, T.; Tomimura, K.; Tan, Z.; Sano, T.; Azuhata, F.; Walsh, J.A.; Fletcher, J.; et al. Molecular evolution of Turnip mosaic virus: Evidence of host adaptation, genetic recombination and geographical spread. *J. Gen. Virol.* **2002**, *83*, 1511–1521. [**Google Scholar**] [**CrossRef**] [**PubMed**]
2. Roossinck, M.J. Mechanisms of plant virus evolution. *Annu. Rev. Phytopathol.* **1997**, *35*, 191–209. [**Google Scholar**] [**CrossRef**] [**PubMed**]
3. Ohshima, K.; Akaishi, S.; Kajiyama, H.; Koga, R.; Gibbs, A.J. Evolutionary trajectory of turnip mosaic virus populations adapting to a new host. *J. Gen. Virol.* **2010**, *91*, 788–801. [**Google Scholar**] [**CrossRef**] [**PubMed**]
4. Wichman, H.A. Different Trajectories of Parallel Evolution during Viral Adaptation. *Science* **1999**, *285*, 422–424. [**Google Scholar**] [**CrossRef**] [**PubMed**]
5. Novella, I.S.; Zárate, S.; Metzgar, D.; Ebendick-Corpus, B.E. Positive selection of synonymous mutations in vesicular stomatitis virus. *J. Mol. Biol.* **2004**, *342*, 1415–1421. [**Google Scholar**] [**CrossRef**]
6. Wolf, Y.I.; Viboud, C.; Holmes, E.C.; Koonin, E.V.; Lipman, D.J. Long intervals of stasis punctuated by bursts of positive selection in the seasonal evolution of influenza A virus. *Biol. Direct.* **2006**, *1*, 34. [**Google Scholar**] [**CrossRef**] [**Green Version**]
7. Woods, R.; Schneider, D.; Winkworth, C.L.; Riley, M.A.; Lenski, R.E. Tests of parallel molecular evolution in a long-term experiment with *Escherichia coli*. *Proc. Natl. Acad. Sci. USA* **2006**, *103*, 9107–9112. [**Google Scholar**] [**CrossRef**] [**Green Version**]
8. Bazykin, G.A.; Kondrashov, F.A.; Brudno, M.; Poliakov, A.; Dubchak, I.; Kondrashov, A.S. Extensive parallelism in protein evolution. *Biol. Direct.* **2007**, *2*, 20. [**Google Scholar**] [**CrossRef**] [**PubMed**] [**Green Version**]
9. Musset, L.; Le Bras, J.; Clain, J. Parallel Evolution of Adaptive Mutations in Plasmodium falciparum Mitochondrial DNA during Atovaquone-Proguanil Treatment. *Mol. Biol. Evol.* **2007**, *24*, 1582–1585. [**Google Scholar**] [**CrossRef**] [**Green Version**]
10. Zhou, H.; Wang, H.; Huang, L.F.; Naylor, M.; Clifford, P. Heterogeneity in codon usages of sobemovirus genes. *Arch. Virol.* **2005**, *150*, 1591–1605. [**Google Scholar**] [**CrossRef**] [**PubMed**]
11. Pinel-Galzi, A.; Rakotomalala, M.; Sangu, E.; Sorho, F.; Kanyeka, Z.; Traoré, O.; Séréme, D.; Poulicard, N.; Rabenantoandro, Y.; Séré, Y.; et al. Theme and variations in the evolutionary pathways to virulence of an RNA plant virus species. *PLoSPathog.* **2007**, *3*, e180. [**Google Scholar**] [**CrossRef**] [**Green Version**]

12. Magome, H.; Yoshikawa, N.; Takahashi, T.; Ito, T.; Miyakawa, T. Molecular variability of the genomes of capilloviruses from apple, Japanese pear, European pear, and citrus trees. *Phytopathology* **1997**, *87*, 389–396. [**Google Scholar**] [**CrossRef**] [**PubMed**] [**Green Version**]

13. Yanese, H.; Mink, G.I.; Sawamura, K.; Yamaguchi, A. Apple topworking disease. In *Compendium of Apple and Pear Diseases*; Jones, A.L., Aldwinckle, H.S., Eds.; APS Press: St. Paul, MN, USA, 1990; pp. 74–75. [**Google Scholar**]

14. Shim, H.; Min, Y.; Hong, S.; Kwon, M.; Kim, D.; Kim, H.; Choi, Y.; Lee, S.; Yang, J. Nucleotide sequences of a Korean isolate of apple stem grooving virus associated with black necrotic leaf spot disease on pear (Pyrus pyrifolia). *Mol. Cells* **2004**, *18*, 192–199. [**Google Scholar**] [**PubMed**]

15. Roy, A.; Kucukural, A.; Zhang, Y. I-TASSER: A unified platform for automated protein structure and function prediction. *Nat. Protoc.* **2010**, *5*, 725–738. [**Google Scholar**] [**CrossRef**] [**PubMed**] [**Green Version**]

16. Foley, B.; Leitner, T.; Apetrei, C.; Hahn, B.; Mizrachi, I.; Mullins, J.; Rambaut, A.; Wolinsky, S.; Korber, B. (Eds.) *HIV Sequence Compendium 2018*; LA-UR 18-25673; Theoretical Biology and Biophysics Group, Los Alamos National Laboratory: Santa Fe, NM, USA, 2018. [**Google Scholar**]

17. Tamura, K.; Stecher, G.; Peterson, D.; Filipski, A.; Kumar, S. MEGA6: Molecular evolutionary genetics analysis version 6.0. *Mol. Biol. Evol.* **2013**, *30*, 2725–2729. [**Google Scholar**] [**CrossRef**] [**PubMed**] [**Green Version**]

18. Nakamura, Y.; Gojobori, T.; Ikemura, T. Codon usage tabulated from international DNA sequence databases: Status for the year 2000. *Nucleic Acids Res.* **2000**, *28*, 292. [**Google Scholar**] [**CrossRef**] [**Green Version**]

19. Desbiez, C.; Gal-On, A.; Girard, M.; Wipf-Scheibel, C.; Lecoq, H. Increase in Zucchini yellow mosaic virus Symptom Severity in Tolerant Zucchini Cultivars Is Related to a Point Mutation in P3 Protein and Is Associated with a Loss of Relative Fitness on Susceptible Plants. *Phytopathology* **2003**, *93*, 1478–1484. [**Google Scholar**] [**CrossRef**] [**Green Version**]

20. Simon, A.E.; Bujarski, J.J. RNA-RNA Recombination and Evolution in Virus-Infected Plants. *Annu. Rev. Phytopathol.* **1994**, *32*, 337–362. [**Google Scholar**] [**CrossRef**]

21. Revers, F.; Le Gall, O.; Candresse, T.; Le Romancer, M.; Dunez, J. Frequent occurrence of recombinant potyvirus isolates. *J. Gen. Virol.* **1996**, *77 Pt 8*, 1953–1965. [**Google Scholar**] [**CrossRef**]

22. Ali, M.; Tahir, M.; Hameed, S. Phylogenetic and genome-wide pairwise distance analysis of the genus *Luteovirus*. *Pak. J. Agric. Sci.* **2017**, *54*, 363–371. [**Google Scholar**] [**CrossRef**]

23. Hajimorad, M.R.; Eggenberger, A.L.; Hill, J.H. Adaptation of Soybean mosaic virus Avirulent Chimeras Containing P3 Sequences from Virulent Strains to Rsv1-Genotype Soybeans Is Mediated by Mutations in HC-Pro. *Mol. Plant-Microbe Interact.* **2008**, *21*, 937–946. [**Google Scholar**] [**CrossRef**] [**PubMed**]

24. Lin, H.X.; Rubio, L.; Smythe, A.B.; Falk, B.W. Molecular Population Genetics of Cucumber Mosaic Virus in California: Evidence for Founder Effects and Reassortment. *J. Virol.* **2004**, *78*, 6666–6675. [**Google Scholar**] [**CrossRef**] [**PubMed**] [**Green Version**]

25. Chen, Y.; Chen, J.; Zhang, H.; Tang, X.; Du, Z. Molecular evidence and sequence analysis of a natural reassortant between Cucumber mosaic virus subgroup IA and II strains. *Virus Genes* **2007**, *35*, 405–413. [**Google Scholar**] [**CrossRef**] [**PubMed**]

26. Adams, M.J.; Antoniw, J.F.; Bar-Joseph, M.; Brunt, A.A.; Candresse, T.; Foster, G.D.; Martelli, G.P.; Milne, R.G.; Fauquet, C.M. Virology Division News: The new plant virus family Flexiviridae and assessment of molecular criteria for species demarcation. *Arch. Virol.* **2004**, *149*, 1045–1060. [**Google Scholar**] [**CrossRef**] [**PubMed**]

27. Martelli, G.P.; Adams, M.J.; Kreuze, J.F.; Dolja, V.V. Family Flexiviridae: A Case Study in Virion and Genome Plasticity. *Annu. Rev. Phytopathol.* **2007**, *45*, 73–100. [**Google Scholar**] [**CrossRef**] [**PubMed**] [**Green Version**]

28. Moury, B.; Cardin, L.; Onesto, J.P.; Candresse, T.; Poupet, A. Survey of Prunus necrotic ringspot virus in Rose and Its Variability in Rose and *Prunus* spp. *Phytopathology* **2001**, *91*, 84–91. [**Google Scholar**] [**CrossRef**] [**PubMed**] [**Green Version**]

29. Sentandreu, V.; Castro, J.A.; Ayllón, M.A.; Rubio, L.; Guerri, J.; González-Candelas, F.; Moreno, P.; Moya, A. Evolutionary analysis of genetic variation observed in citrus tristeza virus (CTV) after host passage. *Arch. Virol.* **2005**, *151*, 875–894. [**Google Scholar**] [**CrossRef**]

30. Rico, P.; Ivars, P.; Elena, S.F.; Hernández, C. Insights into the selective pressures restricting Pelargonium flower break virus genome variability: Evidence for host adaptation. *J. Virol.* **2006**, *80*, 8124–8132. [**Google Scholar**] [**CrossRef**] [**PubMed**] [**Green Version**]

31. Moury, B. A new lineage sheds light on the evolutionary history of Potato virus Y. *Mol. Plant Pathol.* **2010**, *11*, 161–168. [**Google Scholar**] [**CrossRef**]

32. Antoniw, J.F.; Adams, M.J. Codon usage bias amongst plant viruses. *Arch. Virol.* **2003**, *149*, 113–135. [**Google Scholar**] [**CrossRef**]

33. Jenkins, G.M.; Pagel, M.; Gould, E.A.; de A Zanotto, P.M.; Holmes, E.C. Evolution of base composition and codon usage bias in the genus Flavivirus. *J. Mol. Evol.* **2001**, *52*, 383–390. [**Google Scholar**] [**CrossRef**] [**PubMed**]

34. van Hemert, F.J.; Berkhout, B.; Lukashov, V.V. Host-related nucleotide composition and codon usage as driving forces in the recent evolution of the Astroviridae. *Virology* **2007**, *361*, 447–454. [**Google Scholar**] [**CrossRef**] [**PubMed**] [**Green Version**]

35. Bahir, I.; Fromer, M.; Prat, Y.; Linial, M. Viral adaptation to host: A proteome-based analysis of codon usage and amino acid preferences. *Mol. Syst. Biol.* **2009**, *5*, 1–14. [**Google Scholar**] [**CrossRef**] [**PubMed**]

Chapter-4

Revealing the Complexity of Sweepovirus-Deltasatellite–Plant Host Interactions: Expanded Natural and Experimental Helper Virus Range and Effect Dependence on Virus-Host Combination

1. Introduction

Begomoviruses (genus *Begomovirus*, family *Geminiviridae*) have circular, single-stranded DNA (ssDNA) genomes composed of one or two genomic components. They are encapsidated in twinned quasi-icosahedral (geminate) particles [**1**]. Begomoviruses are responsible for many economically important crop diseases worldwide and are transmitted in nature by whiteflies (Hemiptera: Aleyrodidae) of the *Bemisiatabaci* complex [**2**,**3**]. The sweepoviruses are begomoviruses infecting sweet potato (*Ipomoea batatas*) and other species of the family Convolvulaceae that group in a cluster basal to the main phylogenetic groups in the genus, the Old World (OW) and the New World (NW) begomoviruses [**4**,**5**]. In the last twenty years, a number of sweepoviruses have been identified in various parts of the world, e.g., [**5**,**6**,**7**,**8**,**9**,**10**,**11**,**12**,**13**,**14**,**15**]. The sweepoviruses have the typical genomic organization of the monopartite begomoviruses originating from the OW [**1**,**16**]. The virion-sense strand encodes the coat protein (CP) with function in particle formation and is essential for viral transmission by *B. tabaci* and the V2 protein that is involved in viral movement. The complementary-sense strand encodes the replication-associated protein (Rep), the replication enhancer protein (REn), the transcriptional activator protein (TrAP), and the C4 protein with diverse functions including virus movement and symptom development. In addition to their specific functions, V2, Rep, TrAP, and C4 proteins have been shown to suppress gene silencing. An intergenic region (IR) has a predicted stem-loop structure and contains the nonanucleotide TAATATTAC conserved among geminivirids and iterons, which are repeated short-sequence motifs close to the TATA box of the Rep promoter that are Rep binding sites and, together with the stem-loop structure, form the origin of virion-sense DNA replication.

Three classes of DNA satellites associated with begomoviruses have been identified: betasatellites [**17**], alphasatellites [**18**], and deltasatellites [**19**]. Deltasatellites contain several genome features: small genome size (about a quarter begomovirus DNA component), lack of coding capacity, two stem-loop structures (one containing a conserved nonanucleotide TAATATTAC and another situated close to begomovirus iteron-like sequences), a short region with high sequence identity with the betasatellite conserver region, and an A-rich region [**20**]. Deltasatellites are classified in the genus *Deltasatellite* (family *Tolecusatellitidae*), which includes twelve accepted species [**21,22,23**], three of them include members associated with the sweepovirus sweet potato leaf curl virus (SPLCV) [**19,24**]. To date, the complex sweepovirus-deltasatellite has been found in the OW (continental Spain, the Spanish Canary Islands, and Portugal) [**19,25**] and the NW (Venezuela and Puerto Rico) [**19,24**]. Sweet potato leaf curl deltasatellite 1 (SPLCD1), the first sweepovirus-associated deltasatellite characterized, was found infecting sweet potato in the Canary island of Lanzarote (Spain) and a few blue morning glory (*Ipomoea indica*) plants that were analyzed in a small area of southern continental Spain (Málaga province) [**19**]. *I. indica* is widely grown ornamentally in the Mediterranean basin, including the coastal areas of Spain, where it is frequently naturalized. Considering the vegetative mode of propagation of sweet potato and *I. indica* plants, the report of SPLCV infecting sweet potato and other *Ipomoea* spp. in several countries around the world, and the fact that at least one of the sweepovirus-associated deltasatellites, SPLCD1, is transmitted by *B. tabaci* [**26**], the actual distribution of these deltasatellites could be wider than reported.

Available data about diversity and helper virus range of deltasatellites associated with sweepoviruses, and deltasatellites in general, are very limited. Experimentally, it has been shown that SPLCD1 can be transreplicated by two monopartite OW begomoviruses: tomato yellow leaf curl virus (TYLCV) and tomato yellow leaf curl Sardinia virus (TYLCSV) [**26**]. In some cases, the presence of SPLCD1 reduces the accumulation of the helper begomovirus and symptomatology [**26**].

In this study, the genetic diversity of deltasatellites associated with sweepoviruses infecting *I. indica* plants was analyzed by further sampling the populations where SPLCD1 was initially found, expanding the search to other geographical areas in southern continental Spain and the Canary Islands of Tenerife and Gran Canaria. The sweepoviruses present in the samples coinfected with deltasatellites were also fully characterized by sequencing in order to define the range of sweepoviruses that could act as helper viruses in nature. Additionally, experiments were performed to assess the ability of a number of geminivirids (a monopartite and a bipartite NW begomovirus, a bipartite OW begomovirus, and a curtovirus) to transreplicate SPLCD1 in their natural plant hosts or the experimental host *Nicotiana benthamiana*. The results show that SPLCD1 was transreplicated by all the geminivirids assayed at least in *N. benthamiana*, that SPLCD1did not affect the symptomatology caused by the helper viruses, and that their effect on viral DNA accumulation depended on the helper virus–host plant combination.

2. Materials and Methods

2.1. Plant Samples

Leaf samples from 89 *I. indica* plants were collected in southern continental Spain (Murcia, Granada, Málaga, and Cádiz provinces) and the Spanish Canary Islands (Tenerife and Gran Canaria) in 2015 (**Figure 1** and **Table 1**). Each sample consisted of a few leaves that were transported to the laboratory and held at 4 °C until analysis. Samples from the Canary Islands were dried before being transported to the laboratory. Geographical coordinates and presence of leaf symptoms were recorded (**Table S1**).

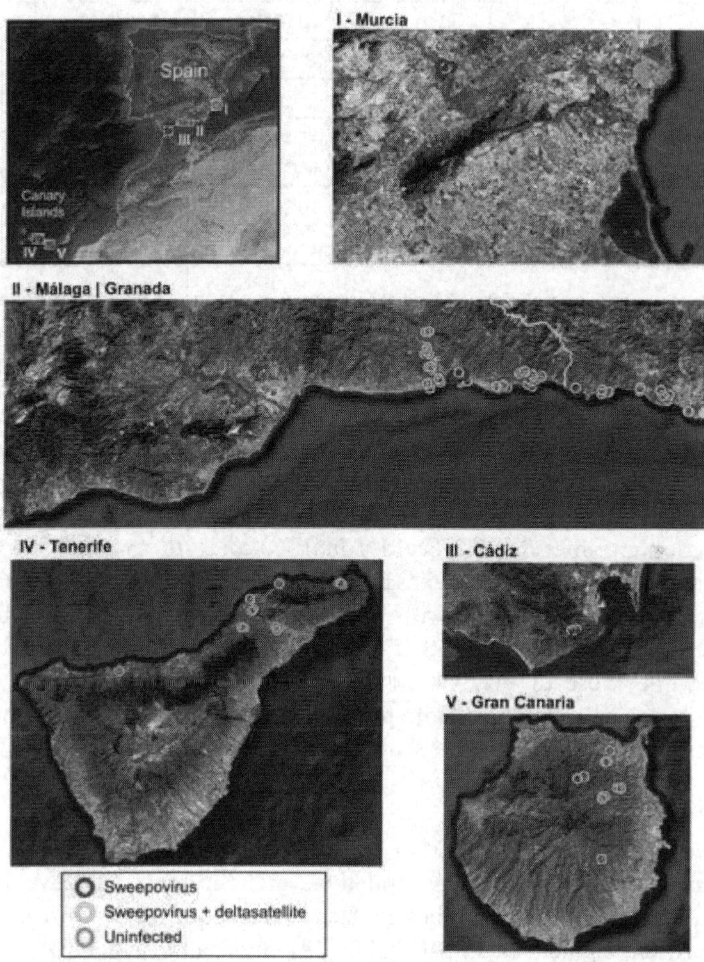

Figure 1. Maps showing the locations of the *Ipomoea indica* plants sampled in southern continental Spain (Murcia (I), Granada (II), Málaga (II), and Cádiz (III) provinces) and the Canary Islands (Tenerife (IV) and Gran Canaria (V)) and analyzed in this work. Samples infected with sweepoviruses (red circles), sweepoviruses plus deltasatellites (yellow circles), and uninfected (blue circles) are indicated in the images.

Table 1. ***Ipomoea indica*** **samples analyzed in this study. Additional details are given in Table S1.**

Province/ Island	Number of Samples	Number of Symptomatic Samples		Number of Infected Samples	
		Yellow Veins (%)	Leaf Curling (%)	Sweepoviruses (%)	Deltasatellites (%)
Murcia	5	1 (20.0)	1 (20.0)	5 (100.0)	1 (20.0)
Granada	5	0 (0.0)	0 (0.0)	5 (100.0)	5 (100.0)
Málaga	52	5 (9.6)	5 (9.6)	46 (88.5	38 (73.1)
Cádiz	3	0 (0.0)	0 (0.0)	3 (100.0)	2 (66.7)
Tenerife	12	3 (25.0)	3 (25.0)	0 (0.0)	0 (0.0)
Gran Canaria	12	8 (66.7)	7 (58.3)	0 (0.0)	0 (0.0)
Continental Spain 1	65	6 (9.2)	6 (9.2)	59 (90.8)	46 (70.8)
Canary Islands 2	24	11 (45.8)	10 (41.7)	0 (0.0)	0 (0.0)
Total	89	17 (19.1)	16 (18.0)	59 (66.3)	46 (51.7)

2.2. DNA Extraction and Cloning

Total DNA was extracted from about 2 cm^2 leaf tissue using a CTAB-based purification method [**27**]. Circular ssDNA was amplified by rolling circle amplification (RCA) with ϕ29 DNA polymerase using the TempliPhi DNA Amplification Kit (GE Healthcare, Little Chalfont, UK). Amplified products were initially digested with the restriction enzyme *Hpa*II, a four-base cutter enzyme, to screen for the putative begomovirus-infected samples. Then, RCA products of the selected samples were digested with the six-base cutter restriction enzymes *Bam*HI, *Eco*RI, *Hind*III, *Nco*I, *Pst*I, and *Sac*I to identify those that cleave the begomoviral and deltasatellite genomes at a single site. RCA products of ~2.7 kbp obtained by digestion with *Nco*I, putatively corresponding to begomovirus genomes, were cloned into a covalently closed pGEM-T-Easy Vector (Promega, Madison, WI, USA), while those of ~0.7 kbp digested with *Pst*I, putatively corresponding to deltasatellite genomes, were cloned into pBluescript II SK(+) (Stratagene, San Diego, CA, USA). Inserts of selected clones were sequenced at Macrogen Inc. (Seoul, Korea).

2.3. Sequence Analysis

Sequences were assembled with SeqMan, part of the Lasergene sequence analysis package (DNAStar Inc., Madison, WI, USA) and then analyzed with the BLASTn algorithm [**28**] for sequence similarity searches in GenBank. Sequences of sweepoviruses and deltasatellites were aligned using MUSCLE [**29**], and pairwise comparisons of all the sequences obtained in this work and selected sequences retrieved from GenBank (**Tables S2 and S3**) were carried out with the program Sequence Demarcation Tool (SDT) v. 1.2 [**30**]. For phylogenetic inference, the maximum likelihood method was used with sequence alignments performed using MUSCLE in MEGA7 [**31**]. The best-fit model of nucleotide substitution was determined based on corrected Akaike information criterion and Bayesian information criterion as implemented in MEGA7 [**31**]. The coefficient of evolutionary differentiation of the SPLCV and SPLCD1 genomes obtained in this work and other isolates previously reported from Spain was estimated using the maximum composite likelihood model with the MEGA 7 program [**31**].

The identification of potential recombinant fragments within sweepovirus and deltasatellites genomes was performed using the seven methods (RDP, GENECONV, BOOTSCAN, MAXIMUM CHI SQUARE, CHIMAERA, SISTER SCAN, and 3SEQ) included in the RDP4 package [**32**] with default settings from the alignment generated by CLUSTAL V algorithm implemented in MEGA 7 [**31**]. Only recombination events detected using at least four methods with *p*-values lower than 10^{-2} were considered.

2.4. Plant Agroinoculation

For agroinoculation assays, *Agrobacterium tumefaciens* cultures harboring each construct were added at 1:1000 dilution to YEP liquid media containing kanamycin (50 μg/mL) and rifampicin (50 μg/mL) and grown for 2 days at 28 °C. Cultures were centrifuged at 3100 g for 20 min at room temperature and then resuspended in 10 mM MES (pH ,(5.6 10 mM $MgCl_2$, and 150 μM acetosyringone, adjusting optical density at 600 nm to 1. Infectious clones of SPLCV and SPLCD1 [**26**], tomato leaf curl New Delhi virus-Spain (ToLCNDV-ES) (a cucurbit-adapted strain hereinafter called simply "tomato leaf curl New Delhi virus" or "ToLCNDV") DNA-A and DNA-B [**33**], Sida golden yellow vein virus (SiGYVV) DNA-A and DNA-B [**34**], tomato leaf deformation virus (ToLDeV) [**35**], and beet curly top virus (BCTV) [**36**] have been described previously. Plants inoculated with *A. tumefaciens* C58C1 cultures containing empty vector (mock) served as negative controls.

N. benthamiana at the four-leaf stage and tomato cv. Moneymaker, *Malvastrumcoromandelianum*, and zucchini cv. Milenio plants at the two-leaf stage were inoculated with *A. tumefaciens* cultures containing clones of viral DNA components and SPLCD1 by stem puncture inoculation. For that, 0.2 mL of *A. tumefaciens* culture was expelled from a 1 mL syringe fitted with a 27G × "2/1 needle into three puncture wounds made in the stem. Inoculated plants were maintained in an insect-free growth chamber (25 °C during the day and 18 °C at night, %70 relative humidity, with a 16 h photoperiod at 250 μmoL s^{-1} m^{-2} of photosynthetically active radiation) until analyzed. At least two independent experiments were performed for each virus-deltasatellite combination.

2.5. Virus and Deltasatellite Detection and Quantification

For molecular hybridization assays, apical leaves of agroinoculated plants were used for tissue blot of petiole cross-sections (tissue printing) performed on positively charged nylon membranes at 28 days post-inoculation. Hybridization was carried out as previously described [**34**] using digoxigenin-labelled DNA probes specific to SPLCD1 [**26**] and each genomic component of SiGYVV [**34**], ToLDeV [**35**], ToLCNDV [**33**], and BCTV [**34**]. The probes were prepared by PCR according to the DIG-labelling detection kit (Roche Diagnostics, Mannheim, Germany). Plants were visually evaluated periodically for symptoms.

For relative quantitative real-time PCR, total DNA was extracted from the leaves used for tissue printing using the DNeasy Plant Mini Kit (Qiagen, Madison, WI, USA). Several pairs of both forward and reverse PCR primers were designed using the PrimerQuest Tool (Integrated DNA Technology, Coralville, IA, USA) and tested for specificity using a standard curve obtained by serial dilution of known quantities of plasmids containing one copy of each viral genome component or the deltasatellite. Additionally, efficiencies of PCR amplification were tested to be close to 100% to select the primers finally used for the assays (**Table S4**). Reactions were conducted in a QuantStudio 5 Real-Time PCR System (Applied Biosystems, Foster City, CA, USA). For that, 1 μL of total DNA was analyzed using the PowerUp SYBR Green Master Mix (Thermo Fisher Scientific, Waltham, MA, USA). PCR reactions were performed as follows: 50 °C for 2 min, 95 °C for 2 min, and 40 cycles of 95 °C for 15 s, 56.5 °C for 15 s and 72 °C for 1 min. Each sample was analyzed in triplicate and virus and deltasatellite genomes were quantified by the $2^{-\Delta\Delta Ct}$ method [**37**], normalizing the amount of target DNA to the amount of plant reference gene DNA (protein phosphatase 2A gene for *N. benthamiana* [**38**] and elongation factor-1α for zucchini [**39**]).

Statistical analyses and graphing to compare the effect of the deltasatellite on geminivirid accumulation were performed using Graphpad Prism 6.0 software (GraphPad Software Inc., San Diego, CA, USA). Unpaired *t* test with Welch's correction or Mann–Whitney test were used, respectively, depending on normal or not normal data distribution (Kolmogorov–Smirnov test). Outlier values were identified by the ROUT method. Differences between means were considered significant when $p \leq 0.05$.

3. Results

3.1. Widespread Presence of Sweepoviruses and Associated Deltasatellites Infecting Ipomoea indica in Spain

Fifty-nine out of sixty-five *I. indica* samples collected from four provinces in southern continental Spain were putatively infected by geminivirids based on the detection of a ~2800 bp DNA fragment after digestion of the RCA products with *Nco*I. In addition, 46 out of these 59 samples were also putatively infected by deltasatellites based on the detection of a ~700-bp DNA fragment after digestion of the RCA products with *Pst*I (**Table 1**). From these samples, 35 full-length sweepovirus genomes (GenBank Accession numbers MW574018-MW574052) and 92 deltasatellite molecules (MW587160-MW587196 and MW587198-MW587252) were obtained (**Table S1**). The 24 *I. indica* samples collected in the Canary Islands of Tenerife and Gran Canaria tested negative both for sweepoviruses and deltasatellites.

A number of the *I. indica* plants sampled showed yellow vein (%19.1) and/or leaf curling (%18.0) symptoms (**Table 1** and **Table S1**). No relation was established between these symptoms and infection by sweepoviruses and/or deltasatellites.

Pairwise comparisons using the SDT program showed that the sweepovirus genome sequences obtained in this work could be divided in two major groups (**Figure S1**). One of the groups that contained 29 sequences with 87.6–99.6% nucleotide identity between them showed the highest identities (94.0–98.9%) with sequences of SPLCV isolates previously described in Spain (EF456741, EF456743, EU839576, EU839578, and FJ151200 [**5**]). Three subgroups, I to III, can be differentiated in this group, showing 96.5–99.6%, 96.4–99.3%, and 92.8–98.4% identity within them, respectively. Thus, in accordance with the current taxonomic guidelines for the genus *Begomovirus* (an isolate having ≥%91 nucleotide identity in full-length genome or DNA-A component to an isolate assigned to a recognized species should be considered to belong to that species) [**40**], these 29 isolates belong to the species *Sweet potato leaf curl virus*. Similarly, the three subgroups could be considered as different strains (≥94% threshold) within that species. The other group consisted of six sequences (MW574021, MW574031, MW574041, MW574043, MW574045, and MW574050) with 94.2–97.7% identity between them that showed the highest identities (92.8–95.8%) with an isolate of sweet potato mosaic virus (SPMV) from Brazil (FJ969831) [**13**] or with the FJ151200 SPLCV isolate, but all of them showed a ≥91% identity with both SPMV and SPLCV isolates. The fact that these six isolates have ≥91% identity with isolates previously assigned to two different species, *Sweet potato leaf curl virus* and *Sweet potato mosaic virus*, and following the species demarcation criteria in the genus [**40**], both species should be merged, the species *Sweet potato mosaic virus* being abolished. Further analysis including all available sweepovirus sequences will determine whether the six abovementioned sequences could be considered to belong to a new strain also containing the isolates previously classified in the species *Sweet potato mosaic virus*.

A recombination analysis performed on the sweepovirus genomes obtained in this work, also including closely related SPLCV isolates previously reported from Spain and a SPMV isolate from Brazil using the seven methods included in the RDP4 package [**32**], revealed a complex recombination pattern for all the sequences (**Figure S2**A). Thus, 25 different recombination events were identified and statistically supported by at least three methods, each present in 1–14 sequences, making a total of 66 recombinant fragments in the set of sequences analyzed (**Figure S2**B). The identified recombinant sequences included five out of the six sweepovirus genomes occupying an intermediate phylogenetic position between SPLCV and SPMV isolates.

Sequencing of the 93 *Pst*I clones confirmed that they corresponded to full-length deltasatellite sequences, with one exception that resulted in being a sweepovirus-deltasatellite chimera (see below). Pairwise comparisons using the SDT program showed that the deltasatellite genome sequences obtained in this work were closely related, showing 89.1–100.0% identity between and 90.8–99.9% within sequences of SPLCD1 isolates previously described from Spain and Portugal [**19,25**] (**Figure S3**). In accordance with the proposed <91% species demarcation threshold for the genus *Deltasatellite* [**22**], these 92 deltasatellite isolates should be classified in the species *Sweet potato leaf curl deltasatellite 1*.

A phylogenetic analysis of all sweepovirus genomes obtained in this work (highlighted in blue in **Figure 2**) showed them grouped in four clades. Clades I, II, and III also contained SPLCV isolates previously characterized from Spain [**5**], thus supporting the pairwise sequence identity results described above. A fourth clade (marked with a yellow star in **Figure 2**) included the six sequences showing a ≥91% identity with both SPLCV and SPMV isolates plus the Brazilian isolate of SPMV (FJ969831) [**13**]. Thus, the mentioned six sequences somehow occupied an intermediate position between SPLCV and SPMV isolates. This is in agreement with the pairwise comparison results that strongly suggested that the species *Sweet potato mosaic virus* should merge with *Sweet potato leaf curl virus* (**Figure S1**). No obvious geographical structure was observed for any of the four clades, with isolates from Málaga province present in all of them. Although the low number of sequences from other regions precluded drawing definitive conclusions, the estimate of the coefficient of evolutionary differentiation for each population pair and all populations together showed that in most cases the genetic diversity within populations was higher than among populations (**Table S5**).

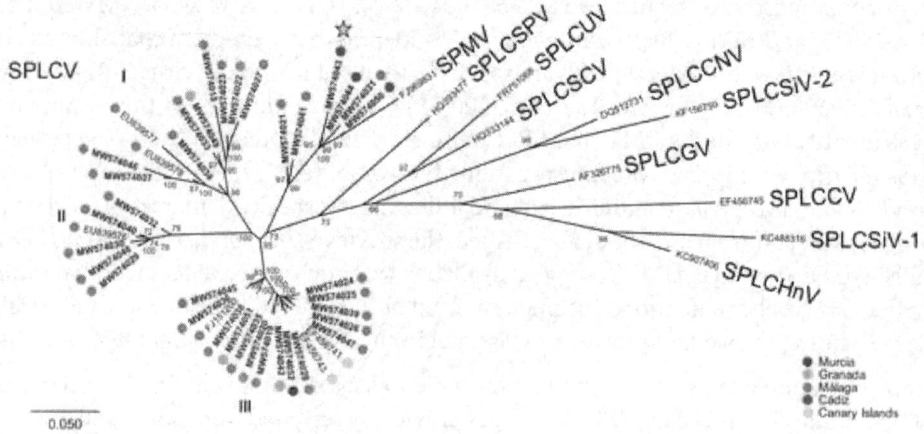

Figure 2. Phylogenetic tree illustrating the relationships of the sweepovirus genomes obtained in this work (highlighted in blue) with closely related sweet potato leaf curl virus (SPLCV) isolates previously reported from Spain and one representative isolate of all other sweepovirus species. Samples were obtained from Murcia (blue dots), Granada (green dots), Málaga (gray dots), and Cádiz (red dots) provinces in southern continental Spain and the Canary Islands (yellow dots). I to III represent major SPLCV clades, and the yellow star corresponds to the clade including sweepovirus described in this work with an intermediate position between SPLCV and sweet potato mosaic virus isolates. The tree was constructed by the maximum likelihood method with the MEGA 7 program using the best fit model, GTR + G+I [31], and bootstrap values (1000 replicates) are shown for supported branches (>70%). The bar below the tree indicates 0.050 nucleotide substitutions per site. Additional details on the sequences and sweepovirus names are included in Tables S1 and S2.

A phylogenetic analysis of all deltasatellite sequences available in GenBank including the genomes obtained in this work (highlighted in blue in **Figure 3**) showed that the SPLCD1 sequences grouped in a single major clade. Numerous minor short-branched clusters, many of them with no bootstrap support, were shown within that major clade. This is in agreement with the pairwise comparison results that showed that the deltasatellites characterized in this work were closely related between them and with the SPLCD1 isolates previously described from Spain and Portugal. As it was shown for SPLCV, no obvious geographical clustering was evident for SPLCD1 isolates, and the estimate of the coefficient of evolutionary differentiation also showed that in most cases the genetic diversity within populations was higher than among populations (**Table S5**). Moreover, no significant recombination events were identified with RDP4.

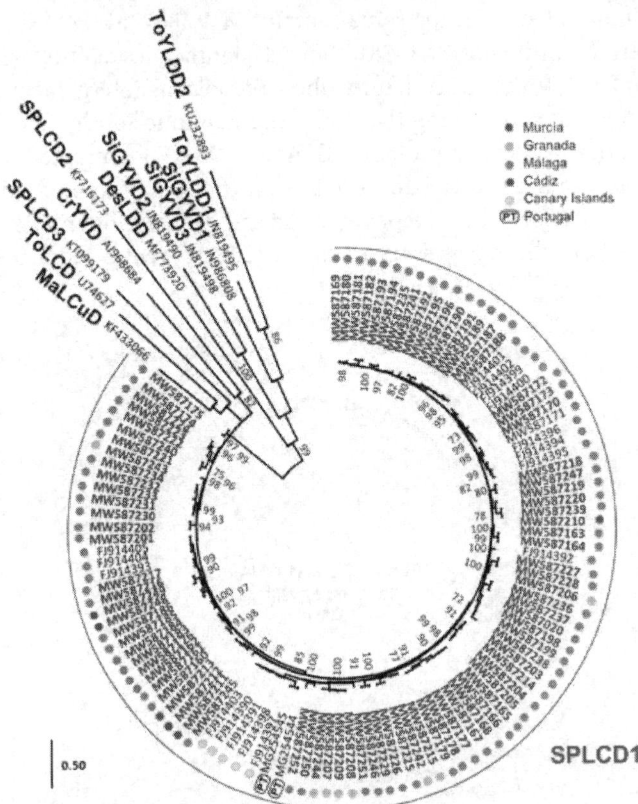

Figure 3. Phylogenetic tree illustrating the relationship of isolates of the deltasatellites obtained in this work (highlighted in blue) with sweet potato leaf curl deltasatellite 1 (SPLCD1) isolates previously reported from Spain and Portugal and one representative isolate of all other deltasatellite species. Samples were obtained from Murcia (blue dots), Granada (green dots), Málaga (gray dots), and Cádiz (red dots) provinces in southern continental Spain and the Canary Islands (yellow dots). The tree was constructed by the maximum likelihood method with the MEGA 7 program using the best fit model T92 + G [31], and bootstrap values (500 replicates) are shown for supported branches (>70%). The bar below the tree indicates 0.50 nucleotide substitutions per site. Additional details on the sequences and deltasatellite names are included in Tables S1 and S3.

3.2. Detection of a Sweepovirus-Deltasatellite Chimera

Sequencing of one of the clones with the characteristic size of deltasatellites resulted to be a sweepovirus-deltasatellite chimera (**Figure 4**). The chimera was found in sample ii16 collected in Málaga province. The insert of this clone (MW587197) was determined to be 699 bp in length, and a BLASTn search showed significant identity with available sequences of deltasatellites for only about half of the length (coordinates 99–443). This DNA fragment included the right half of the conserved stem-loop structure of the deltasatellite, and the A-rich region and showed 97.4% identity to the equivalent region of a deltasatellite isolated from the same sample (MW587238). BLASTn analysis of the remaining insert (coordinates 448–98) showed significant identity with sweepovirus sequences. This DNA fragment contained the sweepovirus intergenic region (including the stem-loop structure containing the conserved nanonucleotide TAATATTAC) and a truncated replication-associated protein and showed 92.0% identity to the corresponding genome region of the SPLCV isolate cloned from the same sample (MW574045). One of the boundaries between the sweepovirus and the deltasatellitemoities of the chimera included four nucleotides not present in the putative parental sequences (CCGAA, in black in **Figure 4**).

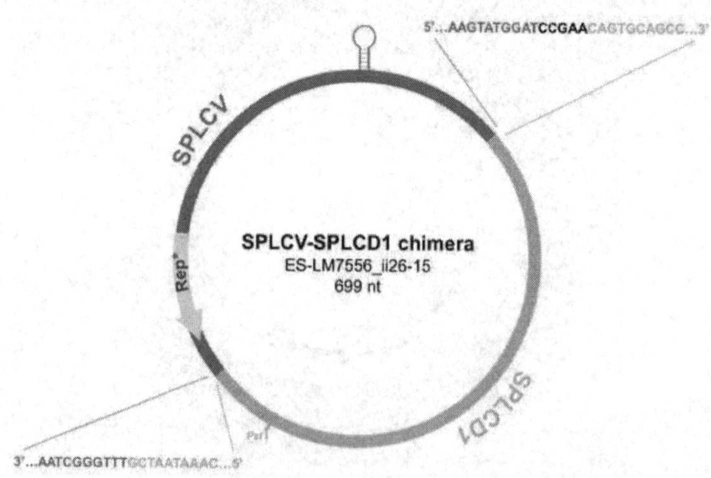

Figure 4. Schematic representation of the chimeric sweepovirus-deltasatellite molecule identified in an *Ipomoea indica* plant from Málaga province (sample ii26). The viral moiety (partial sequence of sweet potato leaf curl virus including a truncated replication-associated protein, Rep *) is represented in orange, and the deltasatellite moiety (partial sequence of sweet potato leaf curl deltasatellite 1) is represented in blue. Nucleotides in black (CCGAA) are not present in the putative parental sequences. The stem-loop structure containing the conserved nanonucleotide TAATATTAC is shown at the top of the graph. The restriction site used for cloning is indicated (*Pst*I).

3.3. Transreplication of Sweet Potato Leaf Curl Deltasatellite 1 by Old World and New World Begomoviruses as Well as by a Curtovirus in Nicotiana benthamiana Plants

N. benthamiana plants agroinoculated with the geminiviridsToLCNDV (a bipartite OW begomovirus), SiGYVV (a bipartite NW begomovirus), ToLDeV (a monopartite NW begomovirus), or BCTV (a curtovirus) or combinations thereof with SPLCD1 were assessed for virus and deltasatellite accumulation (**Table 2, Figure S4**) and symptom development (**Figure 5**).

Figure 5. Symptoms caused by (A) tomato leaf curl New Delhi virus (ToLCNDV), (B) Sida golden yellow vein virus (SiGYVV), (C) tomato leaf deformation virus (ToLDeV), and (D) beet curly top virus (BCTV) alone or in combination with sweet potato leaf curl deltasatellite 1 (δsat) on agroinoculated *Nicotiana benthamiana* plants. Mock-inoculated plants (M) are shown at the left of each panel. Photographs of representative plants were taken at 30 days post-inoculation.

Table 2. Infectivity of sweet potato leaf curl deltasatellite 1 (SPLCD1) in the presence of various geminivirids in Nicotiana benthamiana plants.

Virus (+Deltasatellite)	Number of Infected Plants/Number of Agroinoculated Plants				Total (%)	
	Experiment 1		Experiment 2			
	Virus	SPLCD1	Virus	SPLCD1	Virus	SPLCD1
ToLCNDV	14/15	0/15	15/15	0/15	96.7	0.0
ToLCNDV + SPLCD1	15/15	15/15	15/15	15/15	100.0	100.0
SiGYVV	15/15	0/15	15/15	0/15	100.0	0.0
SiGYVV + SPLCD1	15/15	15/15	15/15	15/15	100.0	100.0
ToLDeV	15/15	0/15	15/15	0/15	100.0	0.0
ToLDeV + SPLCD1	15/15	15/15	15/15	15/15	100.0	100.0
BCTV	15/15	0/15	15/15	0/15	100.0	0.0
BCTV + SPLCD1	15/15	7/15	15/15	8/15	100.0	50.0
SPLCV (positive control)	12/12	0/12	12/12	0/12	100.0	0.0
SPLCV + SPLCD1 (positive control)	12/12	12/12	12/15	12/12	100.0	100.0
Mock (negative control)	0/15	0/15	0/15	0/15	0.0	0.0

ToLCNDV, tomato leaf curl New Delhi virus; SiGYVV, Sida golden yellow vein virus; ToLDeV, tomato leaf deformation virus; BCTV, beet curly top virus; SPLCV, sweet potato leaf curl virus.

Agroinoculation with all viruses and virus-deltasatellite combinations resulted in virtually all *N. benthamiana* plants becoming systemically infected by the viruses, as shown by tissue print hybridization of apical leaves with probes specific for each geminivirid genome or for DNA-A and DNA-B components in the case of bipartite begomoviruses (**Table 2**, **Figure S4**).

In the case of agroinoculation with ToLCNDV, SiGYVV, or ToLDeV in combination with SPLCD1, all the plants infected by the virus were also infected by the deltasatellite, thus showing that these begomoviruses were able to transreplicate SPLCD1 in *N. benthamiana* plants. In the case of plants inoculated with BCTV plus SPLCD1, the deltasatellite was detected only in %50 of the plants infected with the virus (7 out of 15 in Experiment 1 and 8 out of 15 in Experiment 2), thus showing that the curtovirus BCTV is also able to act as a helper virus for SPLCD1 in this host, although not as efficiently as the begomoviruses tested. Positive (SPLCV, SPLCV + SPLCD1) and negative (mock) control plants became infected and remained healthy, respectively.

Plants infected by ToLCNDV showed leaf yellowing and mild curling as well as a severe reduction in plant growth (**Figure** 5A), those by SiGYVV showed mild leaf curling (**Figure** 5B), those by ToLDeV showed leaf deformation (**Figure** 5C) and those by BCTV showed leaf yellowing and curling as well as a severe reduction in plant growth (**Figure** 5D). Co-infection with the deltasatellite did not alter the symptoms caused by each geminivirid.

3.4. Transreplication of Sweet Potato Leaf Curl Deltasatellite 1 in the Natural Plant Hosts of Helper Geminivirids

The geminivirids shown to act as helper viruses of SPLCD1 in *N. benthamiana* were also tested for their ability to transreplicate the deltasatellite in some of their natural plant hosts, i.e., zucchini for ToLCNDV, *Malvastrumcoromandelianum* for SiGYVV and tomato for ToLDeV and BCTV.

Agroinoculation of tomato plants with ToLDeV or BCTV alone or in combination with SPLCD1 resulted in most of the plants being infected by the virus (**Table 3**). The plants infected with ToLDeV or BCTV showed leaf deformation and leaf curling, respectively. None of the virus-infected plants became infected by the deltasatellite assessed by tissue print hybridization.

Table 3. Infectivity of sweet potato leaf curl deltasatellite 1 (SPLCD1) in the presence of various geminivirids in zucchini, *Malvastrumcoromandelianum*, and tomato plants.

Plant Host	Virus (+Deltasatellite)	Number of Infected Plants/Number of Agroinoculated Plants						Total (%)	
		Experiment 1		Experiment 2		Experiment 3			
		Virus	SPLCD1	Virus	SPLCD1	Virus	SPLCD1	Virus	SPLCD1
Zucchini	ToLCNDV	15/15	0/15	15/15	0/15	–	–	100.0	0.0
	ToLCNDV + SPLCD1	15/15	5/15	14/15	9/15	–	–	96.6	46.6
	Mock (negative control)	0/15	0/15	0/15	0/15	–	–	0.0	0.0
M. c.	SiGYVV	17/32	0/32	16/32	0/32	–	–	51.5	0.0
	SiGYVV + SPLCD1	16/32	0/32	15/32	0/32	–	–	48.4	0.0
	Mock (negative control)	0/15	0/15	0/15	0/15	–	–	0.0	0.0
Tomato	ToLDeV	15/15	0/15	60/60	0/60	48/48	0/48	100.0	0.0
	ToLDeV + SPLCD1	14/15	0/15	55/60	0/60	103/108	0/108	93.9	0.0
	BCTV	15/15	0/15	66/66	0/66	–	–	100.0	0.0
	BCTV + SPLCD1	15/15	0/15	57/66	0/66	–	–	88.8	0.0
	Mock (negative control)	0/15	0/15	0/15	0/15	0/15	0/15	0.0	0.0

Agroinoculation of *M. coromandelianum* with SiGYVV alone or in combination with SPLCD1 resulted in approximately %50 of the plants being infected by the virus (**Table 3**), which showed yellow mosaic leaf symptoms. None of the virus-infected plants became infected by the deltasatellite as assessed by tissue print hybridization.

Agroinoculation of zucchini with ToLCNDV alone or in combination with SPLCD1 resulted in almost 100% of the plants being infected by the virus (**Table 3**, **Figure S5**), which showed leaf mosaic and curling symptoms (**Figure 6**). The deltasatellite was detected in approximately 50% of the virus-infected plants (5 out of 15 in Experiment 1 and 9 out of 14 in Experiment 2) (**Table 3**). The presence of the deltasatellite did not modify the symptoms caused by ToLCNDV (**Figure 6**).

Figure 6. Symptoms caused by tomato leaf curl New Delhi virus (ToLCNDV) alone or in combination with sweet potato leaf curl deltasatellite 1 (δsat) on agroinoculated zucchini plants. Mock-inoculated plants (M) are shown at the left of each panel. Photographs of representative plants were taken at 30 days post-inoculation.

3.5. Effect of Sweet Potato Leaf Curl Deltasatellite 1 on Accumulation of Helper Geminivirids: Dependence on the Virus-Host Combination

The accumulation of geminivirids acting as helper for the replication of SPLCD1 was determined by relative quantification with real-time PCR in agroinoculated plants in the presence or absence of the deltasatellite at 30 days post-inoculation (**Table S6, Figure** 7). Genome quantification for ToLCNDV, SiGYVV, ToLDeV, BCTV, and SPLCV (used as a control) was determined in *N. benthamiana* plants and for ToLCNDV also in zucchini plants. For ToLCNDV and SiGYVV, the accumulation of both DNA-A and DNA-B genome components was determined separately.

In *N. benthamiana* plants, accumulation of both DNA-A and DNA-B genome components of ToLCNDV increased in the presence of SPLCD1 in the two experiments performed (**Figure** 7A). On the contrary, in the case of SiGYVV, accumulation of both DNA-A and DNA-B decreased in the presence of SPLCD1 in both experiments (**Figure** 7C). Similarly, the accumulation of the ToLDeV genome also decreased in the presence of SPLCD1 in both experiments (**Figure** 7D). In the case of the curtovirus BCTV, no effect of the presence of SPLCD1 on viral genome accumulation was observed in either of the two experiments (**Figure** 7E). In the only experiment analyzed for SPLCV, used as a control, the negative effect of the deltasatellite on viral genome accumulation previously described [**26**] was confirmed (**Figure** 7F). Interestingly, in contrast to what was observed in *N. benthamiana* plants, the presence of SPLCD1 did not influence the accumulation of ToLCNDV DNA-A or DNA-B in zucchini plants in any of the two experiments performed (**Figure** 7B). A summary of the statistical analysis results is presented in **Table S7**.

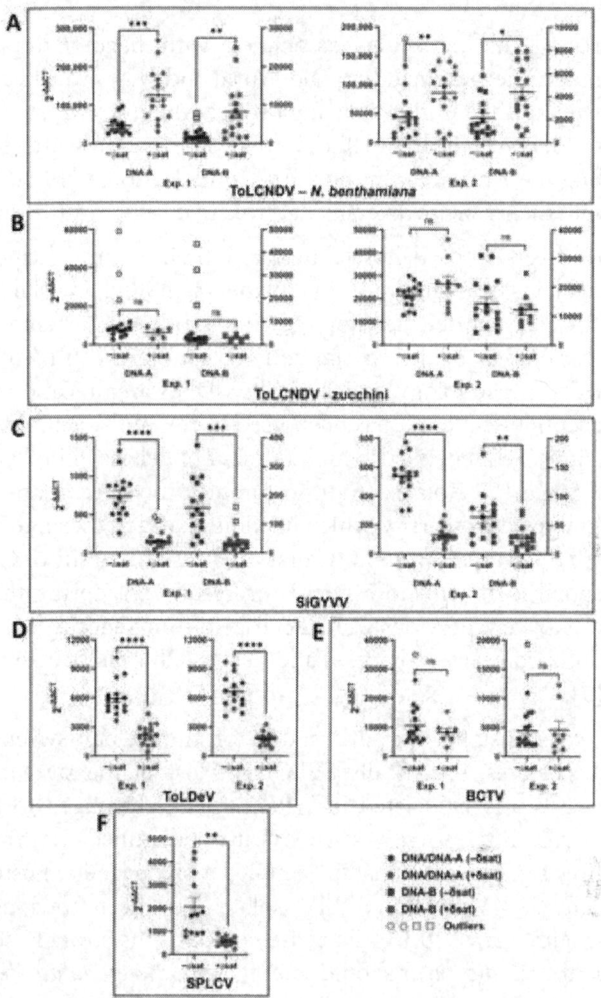

Figure 7. Relative quantification by real-time PCR of helper geminivirids alone (–δsat) or in combination with sweet potato leaf curl deltasatellite 1 (+δsat). (A,B) Tomato leaf curl New Delhi virus (ToLCNDV) DNA-A and DNA-B in *Nicotiana benthamiana* and zucchini, respectively. (C) Sida golden yellow vein virus (SiGYVV) DNA-A and DNA-B, (D) tomato leaf deformation virus (ToLDeV) DNA, (E) beet curly top virus (BCTV) DNA, and (F) sweet potato leaf curl virus (SPLCV) DNA in *N. benthamiana*. Data ($2^{-\Delta\Delta CT}$, Table S6) correspond to plants agroinoculated in two independent experiments (Exp.1, Exp.2) analyzed at 30 days post-inoculation. For SPLCV, included as a control, only Experiment 1 was analyzed. Each circle (DNA/DNA-A) and square (DNA-B) represents one infected plant. Open circles and squares correspond to outlier values. Mean and standard error values are indicated in each graph. Significant differences are labelled with asterisks (n.s., $p > 0.05$; *, $p \leq 0.05$; **, $p \leq 0.01$; ***, $p \leq 0.001$; ****, $p \leq 0.0001$). Additional details of the statistical analysis are provided in Table S7.

4. Discussion

Deltasatellites are ssDNA molecules associated with begomoviruses belonging to different phylogenetic lineages including Old World and New World begomoviruses and sweepoviruses, being unique in that they are non-coding in contrast with alphasatellites and betasatellites [**20**,**41**,**42**]. Deltasatellites have been found in scattered regions around the world including the Americas, Europe, Asia, and Australia [**19**,**20**,**23**,**24**,**25**,**41**], but little is known about their genetic diversity and role in diseases caused by begomoviruses.

In this work the diversity of deltasatellites and their natural helper sweepoviruses infecting *I. indica* plants was analyzed by further sampling in Málaga province, the region where SPLCD1 was detected for the first time [**19**], expanding the sampling to other geographical areas of the coastal zone of southern continental Spain and the Canary Islands. RCA methodology, which allows ssDNA amplification without previous knowledge of nucleotide sequence, has been used to reveal the actual variability present in natural populations. Sequencing of a high number of deltasatellite full-length genomes from continental Spain (92 isolates from 46 samples) revealed a rather homogeneous population with low genetic diversity, which in addition did not seem to be geographically structured. None of the samples from the Canary Islands of Tenerife or Gran Canaria were infected by sweepovirus, a somehow surprising result considering that sweepoviruses infecting sweet potato were previously characterized from Tenerife [**5**]. In another of the Canary Islands, Lanzarote, sweepoviruses and deltasatellites have also been identified in sweet potato [**5**,**19**].

This study also gave insight into the nature of the helper sweepoviruses able to transreplicatedeltasatellites, specifically SPLCD1. Most of the sweepovirus isolates to which SPLCD1 was found associated in *I. indica* plants belonged to the species *Sweet potato leaf curl virus*. Interestingly, sweepoviruses identified in three samples from Málaga and all three samples from Cádiz occupied a phylogenetic position intermediate between SPLCV and SPMV isolates. SPMV isolates have been previously reported only from Brazil and South Africa [**8**,**13**]. Pairwise comparisons strongly suggested that the species *Sweet potato mosaic virus* should merge with *Sweet potato leaf curl virus* but defining a distinct SPLCV strain should include isolates previously identified as SPMV and the six abovementioned Spanish isolates. Five out of these six sweepovirus isolates were found in co-infection with SPLCD1, thus expanding the sweepovirus range to which this deltasatellite is associated in nature.

Recombination between begomoviruses is frequent and contributes significantly to viral diversity, speciation, and evolution (e.g., [**43**,**44**]). The significance of this phenomenon has been well illustrated for sweepoviruses [**5**,**45**]. Recombination events have been identified in all sweepovirus genomes described in this work, including those recombinants that revealed that isolates previously described as SPMV should be considered members of a novel strain of *Sweet potato leaf curl virus*. Deltasatellites described in this work, in their turn, did now show any evidence of recombination between them or with deltasatellites previously described, including representatives of all accepted deltasatellite species.

What has been identified in this work is a recombination event that must have led to the formation of a deltasatellite-sweepovirus chimera with the typical size of a deltasatellite. The putative parentals involved in the generation of this chimeric molecule by recombination were isolates of SPLCV and SPLCD1 closely related to isolates present in the same *I. indica* sample (MW574045 and MW587238). A chimera also containing sweepovirus and deltasatellite sequences was found previously in a sweet potato plant sampled in the Canary Islands [**19**]. This chimera had the typical size of a sweepovirus, and about 70% of its length had high nucleotide identity with an isolate of sweet potato leaf curl Canary virus (V2, CP, and truncated REn genes plus complete IR) cloned from the same sample. The remaining chimera corresponded to almost the full-length sequence of a SPLCD1 genome. Interestingly, in both chimeras one of the recombination points contained an incomplete stem-loop derived from the deltasatellite. In geminivirids, the conserved stem-loop has been identified as a recombination hotspot [**46,47**]. Vegetative propagation is the method of choice for sweet potato, *I. indica*, and other ornamental species of the genus *Ipomoea*. This favors virus accumulation and perpetuation, mixed infections, and occurrence of recombination [**5,45,48**]. This phenomenon may result in the rapid generation of new genomes with adaptive advantages, which could accelerate their evolution and favor the expansion of the host range and, therefore, the emergence of novel diseases (e.g., [**44,49,50**]). The same mechanism that originates viral recombinants could generate the virus-deltasatellite chimera found in this study.

Phylogenetic analysis of a number of deltasatellites associated with NW begomoviruses in Cuba [**20**] or sweepoviruses in Spain [**19**] have revealed some clustering related to geographical origin and plant host. However, in the present study, no clear grouping of deltasatellites was observed related to the geographical origin of the *I. indica* samples.

In order to deepen the understanding of the role that deltasatellites may have on begomovirus diseases and epidemiology, the helper virus range of SPLCD1 was studied experimentally. Previous to this work, the experimental helper virus range of SPLCD1, in addition to SPLCV, was limited to two monopartite OW begomoviruses, TYLCV and TYLCSV [**26**]. In this work, using *N. benthamiana* as a plant host, the helper virus range was successfully extended to a bipartite OW begomovirus, ToLCNDV; a bipartite NW begomovirus, SiGYVV; a monopartite NW begomovirus, ToLDeV; and a curtovirus, BCTV. These compelling results indicate that SPLCD1 has a broad range of helper viruses including members of all major groups of begomoviruses and extending to members of a different virus genus in the family, the genus *Curtovirus*. This helper virus range is wider than that of other deltasatellites for which this has been studied in *N. benthamiana*: Sida golden yellow vein deltasatellite 1 (SiGYVD1) and tomato yellow leaf distortion deltasatellite 2 (ToYLDD2), deltasatellites naturally associated with bipartite NW begomoviruses. SiGYVD1 and ToYLDD2 were maintained by the monopartite NW begomovirusToLDeV, in addition to their respective natural helper begomoviruses, but not by the monopartite OW begomoviruses ACMV, TYLCV, and TYLCSV or the curtovirus BCTV [**34**]. On the other hand, the first deltasatellite to be described, ToLCD, naturally associated with the monopartite OW begomovirus tomato leaf curl virus, was reported to be experimentally transreplicated in *Datura stramonium* plants by TYLCV, the bipartite OW begomovirus African cassava mosaic virus, and BCTV [**41**]. There is no information available about maintenance of ToLCD by NW begomoviruses or sweepoviruses.

In this work, the transreplication of SPLCD1 by ToLCNDV, SiGYVV, ToLDeV, and BCTV was also assessed in their natural host plants. The only positive result was obtained with ToLCNDV in zucchini, with about half of the virus-infected plants also infected by the deltasatellite. The fact that SPLCD1 can be transreplicated by ToLCNDV in zucchini may have significant epidemiological importance. This virus has a wide host range, infecting more than 40 plant species [**51**], and although the primary host is tomato, it also infects other economically important crops such as potato, pepper, and cucurbits. In fact, the isolate used in this work belongs to a strain introduced in the Mediterranean basin, very probably from the Indian subcontinent, adapted to cucurbits [**33,52**]. Considering the wide host range of ToLCNDV and its ability to transreplicate SPLCD1 at least in zucchini, as well as the transmissibility of the deltasatellite by whiteflies [**26**], co-infections involving ToLCNDV and SPLCD1 could provide an opportunity for this complex to expand to other crops and geographical regions where it could have unpredictable consequences.

Previous to this work, the influence of deltasatellites on helper virus accumulation had not been thoroughly addressed, with only a few analyses done by quantifying densitometry of Southern blots for a number of deltasatellite/begomovirus combinations. Summarizing, in most of the deltasatellite/begomovirus combinations agroinoculated in *N. benthamiana* plants (SPLCD1/SPLCV, SPLCD1/TYLCV, SPLCD1/TYLCSV, SiGYVD1/SiGYVV, and ToYLDD2/ToYLDV), the begomovirus accumulation decreased in the presence of the deltasatellite [**26,34**]. In contrast, no effect on virus accumulation was observed when other plant hosts were agroinoculated, including the combinations SPLCD1/SPLCV/*I. setosa*, SPLCD/TYLCV/tomato, SiGYVD1/SiGYVV/*M. coromandelianum*, and ToYLDD2/ToYLDV/*Sidastrummicranthum* [**26,34**]. In the case of the combinations SiGYVD1/ToLDeV and ToYLDD2/ToLDeV in *N. benthamiana*, no effect of the deltasatellite on viral accumulation was observed either [**34**]. In the present work, real-time PCR was used as a more accurate proxy for viral genome quantification. In the case of SPLCD1/SiGYVV and SPLCD1/ToLDeV in *N. benthamiana*, the presence of the deltasatellite decreased the begomovirus accumulation as it has been shown for most of the abovementioned cases. On the contrary, accumulation of the curtovirus BCTV was not affected by the presence of the deltasatellite.

A particular and interesting case is the response of ToLCNDV to the presence of SPLCD1. This is the first time that a deltasatellite has been shown to increase the accumulation of a helper geminivirid, in this case in *N. benthamiana* plants. On the other hand, this effect was not observed in zucchini plants, a natural host for the cucurbit-adapted ToLCNDV isolate used in this study [**33**], where SPLCD1 did not affect virus accumulation. The results of virus quantification obtained in this study, including those for ToLCNDV, were consistent in the two independent experiments performed for each deltasatellite/geminivirid combination and for both genome components (DNA-A, DNA-B) in the case of bipartite begomoviruses. This adds robustness to the otherwise surprising results obtained for ToLCNDV that would reveal the complexity of the deltasatellite/geminivirid/plant host interactions.

The contrasting effect of deltasatellites on helper virus accumulation depending on the virus–host combinations found in this study and in previous research [**26,34**], exemplified in the case of SPLCD1/ToLCNDV combination, has been described for other DNA satellites associated with begomoviruses. Thus, the accumulation of Euphorbia yellow mosaic virus (EuYMV) DNA-A increases in the presence of Euphorbia yellow mosaic alphasatellite (EuYMA) in two plant hosts, *Euphorbia heterophylla* and *N. benthamiana* [**53**]. However, the presence of EuYMA causes a reduction in the accumulation of EuYMV DNA-A in *Arabidopsis thaliana*. For EuYMV DNA-B, no differences in its accumulation are observed in the presence or absence of EuYMA in both *N. benthamiana* and *A. thaliana*, but results in *E. heterophylla* show an increase of EuYMV DNA-B accumulation in the presence of EuYMA [**53**].

In most cases where a possible effect of deltasatellites on the symptomatology caused by the helper geminivirid has been assessed, no symptom modifications have been observed, despite the effect on virus accumulation that was observed for some of the deltasatellite/virus combinations [**26**,**34**,**41**]. The only exception is the effect of SPLCD1 on the symptoms caused by TYLCV and TYLCSV in *N. benthamiana* or tomato [**26**]. In this case, although the symptoms were qualitatively identical in the absence or presence of the deltasatellite, in the latter case milder leaf yellowing and curling was observed.

In nature, known deltasatellites seem to have a narrow helper virus range because closely related isolates of a single begomovirus species have been reported per deltasatellite species [**19**,**20**,**23**,**34**,**41**]. However, in the present study SPLCD1 was found associated with two distinct SPLCV strains, suggesting that a somewhat broader helper virus range could occur naturally. This situation is different from what is observed for betasatellites that are frequently found associated with begomovirus isolates belonging to different species [**42**].

A narrow helper virus range of deltasatellites in turn would restrict the plant host range that deltasatellites could potentially infect. However, the promiscuous replicative nature of deltasatellites that is being revealed in this and other studies [**26**,**34**,**41**], coupled with global trade and whitefly transmission, could facilitate dissemination of deltasatellites to diverse agrosystems with unforeseeable outcomes.

Although deltasatellites and betasatellites are clearly related [**19**], there are fundamental differences between them, including the non-coding nature of the formers. This feature hinders trying to address the identification of deltasatellite motifs/sequences involved in the interaction with their helper viruses and pathogenesis. Anyhow, mutation studies similar to those successfully performed with betasatellites [**54**,**55**] could help to decipher how deltasatellites are able to be transreplicated by different helper geminivirids and whether any of the hypotheses proposed for betasatellite-begomovirus recognition, the "universal Rep" hypothesis or the "universal iteron" hypothesis" [**56**], could also be applied to these small non-coding DNA molecules.

Supplementary Materials

Figure S1: Color-coded matrix of pairwise sequence identity scores generated by alignment of the full-length genomes of the sweepoviruses obtained in this work, Figure S2: Recombination analysis using RDP4 package of the sweepovirus genomes obtained in this work, Figure S3: Color-coded matrix of pairwise sequence identity scores generated by alignment of the full-length genomes of the deltasatellites obtained in this work, Figure S4: DNA hybridization of petiole cross section blots of newly emerged young leaves of *Nicotiana benthamiana* plants agroinoculated with ToLCNDV, SiGYVV, ToLDeV, or BCTV alone or in combination with SPLCD1, Figure S5: DNA hybridization of petiole cross section blots of newly emerged young leaves of zucchini plants agroinoculated with ToLCNDV and SPLCD1, Table S1: Information on the *Ipomoea indica* samples used in this study, Table S2: Sweepovirus sequences retrived from GenBank used for pairwise sequence identity, phylogenetic, and recombination analyses, Table S3: Deltasatellite sequences retrieved from GenBank used for pairwise sequence identity and phylogenetic and recombination analyses, Table S4: List of primers used by real-time PCR to amplify the deltasatellite and viral DNA or plant reference genes, Table S5: Estimate of the coefficient of evolutionary differentiation for SPLCV and SPLCD1 genomes obtained in this work and isolates previously reported from Spain, Table S6: Data of virus quantification by real-time PCR ($^{-}2^{\Delta\Delta Ct}$) used to generate **Figure 7** with Graphpad Prism 6.0 software, Table S7: Summary of statistical analyses performed to evaluate the effect of SPLCD1 on accumulation of helper geminivirids.

References

1. Navas-Castillo, J.; Fiallo-Olivé, E. Geminiviruses (*Geminiviridae*). In *Encyclopedia of Virology*, 4th ed.; Bamford, D.H., Zuckerman, M., Eds.; Academic Press: Oxford, UK, 2021; Volume 3, pp. 411–419. [**Google Scholar**]

2. Navas-Castillo, J.; Fiallo-Olivé, E.; Sánchez-Campos, S. Emerging virus diseases transmitted by whiteflies. *Annu. Rev. Phytopathol.* **2011**, *49*, 219–248. [**Google Scholar**] [**CrossRef**]

3. Fiallo-Olivé, E.; Pan, L.L.; Liu, S.S.; Navas-Castillo, J. Transmission of begomoviruses and other whitefly-borne viruses: Dependence on the vector species. *Phytopathology* **2020**, *110*, 10–17. [**Google Scholar**] [**CrossRef**]

4. Fauquet, C.M.; Stanley, J. Geminivirus classification and nomenclature: Progress and problems. *Ann. Appl. Biol.* **2005**, *142*, 165–189. [**Google Scholar**] [**CrossRef**]

5. Lozano, G.; Trenado, H.P.; Valverde, R.A.; Navas-Castillo, J. Novel begomovirus species of recombinant nature in sweet potato (*Ipomoea batatas*) and *Ipomoea indica*: Taxonomic and phylogenetic implications. *J. Gen. Virol.* **2009**, *90*, 2550–2562. [**Google Scholar**] [**CrossRef**]

6. Albuquerque, L.C.; Inoue-Nagata, A.K.; Pinheiro, B.; Ribeiro, S.D.; Resende, R.O.; Moriones, E.; Navas-Castillo, J. A novel monopartite begomovirus infecting sweet potato in Brazil. *Arch. Virol.* **2011**, *156*, 1291–1294. [**Google Scholar**] [**CrossRef**] [**Green Version**]

7. Banks, G.; Bedford, I.; Beitia, F.; Rodriguez-Cerezo, E.; Markham, P. A novel geminivirus of *Ipomoea indica* (*Convolvulaceae*) from Southern Spain. *Plant Dis.* **1999**, *83*, 486. [**Google Scholar**] [**CrossRef**]

8. Esterhuizen, L.L.; van Heerden, S.W.; Rey, M.E.C.; van Heerden, H. Genetic identification of two sweet-potato-infecting begomoviruses in South Africa. *Arch. Virol.* **2012**, *157*, 2241–2245. [**Google Scholar**] [**CrossRef**]

9. Fiallo-Olivé, E.; Katis, N.I.; Navas-Castillo, J. First report of *Sweet potato leaf curl virus* on blue morning glory in Greece. *Plant Dis.* **2014**, *98*, 700. [**Google Scholar**] [**CrossRef**]

10. Lotrakul, P.; Valverde, R.; Clark, C.; Sim, J.; De La Torre, R. Detection of a geminivirus infecting sweet potato in the United States. *Plant Dis.* **1998**, *82*, 1253–1257. [**Google Scholar**] [**CrossRef**] [**PubMed**] [**Green Version**]

11. Luan, Y.S.; Zhang, J.; Liu, D.M.; Li, W.L. Molecular characterization of sweet potato leaf curl virus isolate from China (SPLCV-CN) and its phylogenetic relationship with other members of the Geminiviridae. *Virus Genes* **2007**, *35*, 379–385. [**Google Scholar**] [**CrossRef**]

12. Mohammed, H.S.; El Hussein, A.A.; El Siddig, M.A.; Ibrahim, F.A.; Navas-Castillo, J.; Fiallo-Olivé, E. First report of *Sweet potato leaf curl virus* infecting sweet potato in Sudan. *Plant Dis.* **2017**, *101*, 849. [**Google Scholar**] [**CrossRef**]

13. Paprotka, T.; Boiteux, L.S.; Fonseca, M.E.N.; Resende, R.O.; Jeske, H.; Faria, J.C.; Ribeiro, S.G. Genomic diversity of sweet potato geminiviruses in a Brazilian germplasm bank. *Virus Res.* **2010**, *149*, 224–233. [**Google Scholar**] [**CrossRef**]

14. Prasanth, G.; Hegde, V. Occurrence of *Sweet potato feathery mottle virus* and *Sweet potato leaf curl Georgia virus* on sweet potato in India. *Plant Dis.* **2008**, *92*, 311. [**Google Scholar**] [**CrossRef**]

15. Wasswa, P.; Otto, B.; Maruthi, M.; Mukasa, S.; Monger, W.; Gibson, R. First identification of a sweet potato begomovirus (sweepovirus) in Uganda: Characterization, detection and distribution. *Plant Pathol.* **2011**, *60*, 1030–1039. [**Google Scholar**] [**CrossRef**]

16. Rojas, M.R.; Hagen, C.; Lucas, W.J.; Gilbertson, R.L. Exploiting chinks in the plant's armor: Evolution and emergence of geminiviruses. *Annu. Rev. Phytopathol.* **2005**, *43*, 361–394. [**Google Scholar**] [**CrossRef**]

17. Briddon, R.W.; Bull, S.E.; Amin, I.; Idris, A.M.; Mansoor, S.; Bedford, I.A.; Dhawan, P.; Rishi, N.; Siwatch, S.S.; Abdel-Salam, A.M. Diversity of DNA β, a satellite molecule associated with some monopartite begomoviruses. *Virology* **2003**, *312*, 106–121. [**Google Scholar**] [**CrossRef**] [**Green Version**]

18. Briddon, R.W.; Bull, S.E.; Amin, I.; Mansoor, S.; Bedford, I.D.; Rishi, N.; Siwatch, S.S.; Zafar, Y.; Abdel-Salam, A.M.; Markham, P.G. Diversity of DNA 1: A satellite-like molecule associated with monopartite begomovirus-DNA β complexes. *Virology* **2004**, *324*, 462–474. [**Google Scholar**] [**CrossRef**] [**PubMed**] [**Green Version**]

19. Lozano, G.; Trenado, H.P.; Fiallo-Olivé, E.; Chirinos, D.; Geraud-Pouey, F.; Briddon, R.W.; Navas-Castillo, J. Characterization of non-coding DNA satellites associated with sweepoviruses (genus *Begomovirus*, *Geminiviridae*)-definition of a distinct class of begomovirus-associated satellites. *Front. Microbiol.* **2016**, *7*, 162. [**Google Scholar**] [**CrossRef**] [**Green Version**]

20. Fiallo-Olivé, E.; Martínez-Zubiaur, Y.; Moriones, E.; Navas-Castillo, J. A novel class of DNA satellites associated with New World begomoviruses. *Virology* **2012**, *426*, 1–6. [**Google Scholar**] [**CrossRef**] [**Green Version**]

21. Adams, M.J.; Kefkowitz, E.J.; King, A.M.Q.; Harrach, B.; Harrison, R.L.; Knowles, N.J.; Kropinski, A.M.; Krupovic, M.; Kuhn, J.H.; Mushegian, A.R.; et al. Changes to taxonomy and the International Code of Virus Classification and Nomenclature ratified by the International Committee on Taxonomy of Viruses (2017). *Arch. Virol.* **2017**, *162*, 2505–2538. [**Google Scholar**] [**CrossRef**]

22. Briddon, R.W.; Navas-Castillo, J.; Fiallo-Olivé, E. ICTV Taxonomic Proposal 2016.021a-kP.A.v2.Tolecusatellitidae. Create the *Tolecusatellitidae*, a New Family of

Single-Stranded DNA Satellites with Two Genera. 2016. Available online: **http://www.ictv.global/proposals-16/2016.021a-kP.A.v2.Tolecusatellitidae.pdf** (accessed on 30 March 2021).

23. Fiallo-Olivé, E.; Navas-Castillo, J. Molecular and biological characterization of a New World mono-/bipartite begomovirus/deltasatellite complex infecting *Corchorus siliquosus*. *Front. Microbiol.* **2020**, *11*, 1755. [**Google Scholar**] [**CrossRef**]
24. Rosario, K.; Marr, C.; Varsani, A.; Kraberger, S.; Stainton, D.; Moriones, E.; Polston, J.E.; Breitbart, M. Begomovirus-associated satellite DNA diversity captured through Vector-Enabled Metagenomic (VEM) surveys using whiteflies (Aleyrodidae). *Viruses* **2016**, *8*, 36. [**Google Scholar**] [**CrossRef**] [**PubMed**] [**Green Version**]
25. Fiallo-Olivé, E.; Lapeira, D.; Louro, D.; Navas-Castillo, J. First report of *Sweet potato leaf curl virus* and *Sweet potato leaf curl deltasatellite 1* infecting blue morning glory in Portugal. *Plant Dis.* **2018**, *102*, 1043. [**Google Scholar**] [**CrossRef**] [**Green Version**]
26. Hassan, I.; Orílio, A.F.; Fiallo-Olivé, E.; Briddon, R.W.; Navas-Castillo, J. Infectivity, effects on helper viruses and whitefly transmission of the deltasatellites associated with sweepoviruses (genus *Begomovirus*, family *Geminiviridae*). *Sci. Rep.* **2016**, *6*, 30204. [**Google Scholar**] [**CrossRef**] [**PubMed**] [**Green Version**]
27. Haible, D.; Kober, S.; Jeske, H. Rolling circle amplification revolutionizes diagnosis and genomics of geminiviruses. *J. Virol. Methods* **2006**, *135*, 9–16. [**Google Scholar**] [**CrossRef**]
28. Altschul, S.F.; Gish, W.; Miller, W.; Myers, E.W.; Lipman, D.J. Basic local alignment search tool. *J. Mol. Biol.* **1990**, *215*, 403–410. [**Google Scholar**] [**CrossRef**]
29. Edgar, R.C. MUSCLE: A multiple sequence alignment method with reduced time and space complexity. *BMC Bioinform.* **2004**, *5*, 1–19. [**Google Scholar**] [**CrossRef**] [**PubMed**] [**Green Version**]
30. Muhire, B.M.; Varsani, A.; Martin, D.P. SDT: A virus classification tool based on pairwise sequence alignment and identity calculation. *PLoS ONE* **2014**, *9*, e108277. [**Google Scholar**] [**CrossRef**]
31. Kumar, S.; Stecher, G.; Tamura, K. MEGA7: Molecular Evolutionary Genetics Analysis version 7.0 for bigger datasets. *Mol. Biol. Evol.* **2016**, *33*, 1870–1874. [**Google Scholar**] [**CrossRef**] [**Green Version**]
32. Martin, D.P.; Murrell, B.; Golden, M.; Khoosal, A.; Muhire, B. RDP4: Detection and analysis of recombination patterns in virus genomes. *Virus Evol.* **2015**, *1*, vev003. [**Google Scholar**] [**CrossRef**] [**Green Version**]
33. Fortes, I.M.; Sánchez-Campos, S.; Fiallo-Olivé, E.; Díaz-Pendón, J.A.; Navas-Castillo, J.; Moriones, E. A novel strain of tomato leaf curl New Delhi virus has spread to the Mediterranean basin. *Viruses* **2016**, *8*, 307. [**Google Scholar**] [**CrossRef**] [**Green Version**]
34. Fiallo-Olivé, E.; Tovar, R.; Navas-Castillo, J. Deciphering the biology of deltasatellites from the New World: Maintenance by New World begomoviruses and whitefly transmission. *New Phytol.* **2016**, *212*, 680–692. [**Google Scholar**] [**CrossRef**] [**Green Version**]

35. Sánchez-Campos, S.; Martínez-Ayala, A.; Márquez-Martín, B.; Aragón-Caballero, L.; Navas-Castillo, J.; Moriones, E. Fulfilling Koch's postulates confirms the monopartite nature of tomato leaf deformation virus: A begomovirus native to the New World. *Virus Res.* **2013**, *173*, 286–293. [**Google Scholar**] [**CrossRef**] [**PubMed**] [**Green Version**]

36. Briddon, R.W.; Watts, J.; Markham, P.G.; Stanley, J. The coat protein of beet curly top virus is essential for infectivity. *Virology* **1989**, *172*, 628–633. [**Google Scholar**] [**CrossRef**]

37. Livak, K.; Schmittgen, T. Analysis of relative gene expression data using real-time quantitative PCR and the $2^{-\Delta\Delta C}{}_{T}$ method. *Methods* **2001**, *25*, 402–408. [**Google Scholar**] [**CrossRef**] [**PubMed**]

38. Liu, D.; Shi, L.; Han, C.; Yu, J.; Li, D.; Zhang, Y. Validation of reference genes for gene expression studies in virus-infected *Nicotiana benthamiana* using quantitative real-time PCR. *PLoS ONE* **2012**, *7*, e46451. [**Google Scholar**] [**CrossRef**]

39. Obrero, A.; Die, J.V.; Román, B.; Gómez, P.; Nadal, S.; González-Verdejo, C.I. Selection of reference genes for gene expression studies in zucchini (*Cucurbita pepo*) using qPCR. *J. Agric. Food Chem.* **2011**, *59*, 5402–5411. [**Google Scholar**] [**CrossRef**] [**Green Version**]

40. Brown, J.K.; Zerbini, F.M.; Navas-Castillo, J.; Moriones, E.; Ramos-Sobrinho, R.; Silva, J.C.; Fiallo-Olivé, E.; Briddon, R.W.; Hernandez-Zepeda, C.; Idris, A.; et al. Revision of *Begomovirus* taxonomy based on pairwise sequence comparisons. *Arch. Virol.* **2015**, *160*, 1593–1619. [**Google Scholar**] [**CrossRef**]

41. Dry, I.B.; Krake, L.R.; Rigden, J.E.; Rezaian, M.A. A novel subviral agent associated with a geminivirus: The first report of a DNA satellite. *Proc. Natl. Acad. Sci. USA* **1997**, *94*, 7088–7093. [**Google Scholar**] [**CrossRef**] [**Green Version**]

42. Zhou, X. Advances in understanding begomovirus satellites. *Annu. Rev. Phytopathol.* **2013**, *51*, 357–381. [**Google Scholar**] [**CrossRef**]

43. Lefeuvre, P.; Martin, D.P.; Hoareau, M.; Naze, F.; Delatte, H.; Thierry, M.; Varsani, A.; Becker, N.; Reynaud, B.; Lett, J.M. Begomovirus 'melting pot' in the south-west Indian Ocean islands: Molecular diversity and evolution through recombination. *J. Gen. Virol.* **2007**, *88*, 3458–3468. [**Google Scholar**] [**CrossRef**]

44. Fiallo-Olivé, E.; Trenado, H.P.; Louro, D.; Navas-Castillo, J. Recurrent speciation of a tomato yellow leaf curl geminivirus in Portugal by recombination. *Sci. Rep.* **2019**, *9*, 1332. [**Google Scholar**] [**CrossRef**] [**PubMed**] [**Green Version**]

45. Albuquerque, L.C.; Inoue-Nagata, A.K.; Pinheiro, B.; Resende, R.O.; Moriones, E.; Navas-Castillo, J. Genetic diversity and recombination analysis of sweepoviruses from Brazil. *Virol. J.* **2012**, *9*, 241. [**Google Scholar**] [**CrossRef**] [**Green Version**]

46. Lefeuvre, P.; Lett, J.M.; Varsani, A.; Martin, D.P. Widely conserved recombination patterns among single-stranded DNA viruses. *J. Virol.* **2009**, *83*, 2697–2707. [**Google Scholar**] [**CrossRef**] [**Green Version**]

47. Monjane, A.L.; van der Walt, E.; Varsani, A.; Rybicki, E.P.; Martin, D.P. Recombination hotspots and host susceptibility modulate the adaptive value of recombination during maize streak virus evolution. *BMC Evol. Biol.* **2011**, *11*, 350. [**Google Scholar**]

[**CrossRef**] [**Green Version**]

48. Valverde, R.A.; Clark, C.A.; Valkonen, J.P. Viruses and virus disease complexes of sweetpotato. *Plant Viruses* **2007**, *1*, 116–126. [**Google Scholar**]
49. Berrie, L.C.; Rybicki, E.P.; Rey, M.E.C. Complete nucleotide sequence and host range of *South African cassava mosaic virus*: Further evidence for recombination amongst begomoviruses. *J. Gen. Virol.* **2001**, *82*, 53–58. [**Google Scholar**] [**CrossRef**]
50. Martin, D.P.; Lefeuvre, P.; Varsani, A.; Hoareau, M.; Semegni, J.Y.; Dijoux, B.; Vincent, C.; Reynaud, B.; Lett, J.M. Complex recombination patterns arising during geminivirus coinfections preserve and demarcate biologically important intra-genome interaction networks. *PLoSPathog.* **2011**, *7*, e1002203. [**Google Scholar**] [**CrossRef**] [**Green Version**]
51. Zaidi, S.S.E.A.; Martin, D.P.; Amin, I.; Farooq, M.; Mansoor, S. *Tomato leaf curl New Delhi virus*: A widespread bipartite begomovirus in the territory of monopartite begomoviruses. *Mol. Plant. Pathol.* **2017**, *18*, 901–911. [**Google Scholar**] [**CrossRef**]
52. Juárez, M.; Tovar, R.; Fiallo-Olivé, E.; Aranda, M.A.; Gosálvez, B.; Castillo, P.; Moriones, E.; Navas-Castillo, J. First detection of *Tomato leaf curl New Delhi virus* infecting zucchini squash in Spain. *Plant Dis.* **2014**, 98. [**Google Scholar**] [**CrossRef**]
53. Mar, T.B.; Mendes, I.R.; Lau, D.; Fiallo-Olivé, E.; Navas-Castillo, J.; Alves, M.S.; Zerbini, F.M. Interaction between the New World begomovirus *Euphorbia yellow mosaic virus* and its associated alphasatellite: Effects on infection and transmission by the whitefly *Bemisiatabaci. J. Gen. Virol.* **2017**, *98*, 1552–1562. [**Google Scholar**] [**CrossRef**] [**PubMed**]
54. Saunders, K.; Briddon, R.W.; Stanley, J. Replication promiscuity of DNA-β satellites associated with monopartite begomoviruses; deletion mutagenesis of the *Ageratum yellow vein virus* DNA-β satellite localizes sequences involved in replication. *J. Gen. Virol.* **2008**, *89*, 3165–3172. [**Google Scholar**] [**CrossRef**] [**PubMed**]
55. Iqbal, Z.; Shafiq, M.; Ali, I.; Mansoor, S.; Briddon, R.W. Maintenance of cotton leaf curl Multan betasatellite by *Tomato leaf curl New Delhi virus*–analysis by mutation. *Front. Plant. Sci.* **2017**, *8*, 2208. [**Google Scholar**] [**CrossRef**] [**PubMed**] [**Green Version**]
56. Nawaz-ul-Rehman, M.S.; Mansoor, S.; Briddon, R.W.; Fauquet, C.M. Maintenance of an Old World betasatellite by a New World helper begomovirus and possible rapid adaptation of the betasatellite. *J. Virol.* **2009**, *83*, 9347–9355. [**Google Scholar**] [**CrossRef**] [**Green Version**]

Chapter-5

Arabidopsis Thaliana Genes Associated With *Cucumber Mosaic Virus* Virulence and their Link to Virus Seed Transmission

1. Introduction

Viruses are major plant pathogens due to the detrimental effect of their infections on the host (i.e., to their virulence) [**1**]. Indeed, they have great impact on agronomic production worldwide, being the second most important cause of economic losses in crops only behind fungi [**2**] and accounting for the largest fraction of plant emerging diseases [**3**]. Hence, understanding the genetic basis of virulence in plant–virus interactions is central in minimizing the damage of virus epidemics.

Although virulence can be intuitively viewed as a pathogen-controlled trait, a large body of evidence indicates that in plant–virus interactions it is modulated by both host and pathogen genetic determinants [**4,5,6,7**]. Plant virus genes and even single mutations determining virulence have been extensively characterized, e.g., [**8,9**]. In parallel, plant genetic determinants have been also studied although at a lesser extent [**8,10,11**]. Interestingly, most of the works on the "host side" addressed the question from a plant pathology perspective. In this context, virulence is often defined either as the virus capacity to gain entrance to the plant or as the virus ability to induce symptoms, generally in the vegetative structures [**12**]. On the one hand, the use of the former definition of virulence led mostly to the identification of plant genes conferring resistance/immunity to virus infection, such as those related to the plant hypersensitive response, RNA silencing machinery or systemic acquired resistance [**13,14,15**]. On the other hand, host genes regulating virus cell-to-cell and long-distance movement, and plant photosynthesis and development [**5,11,16**] have been identified as virulence determinants when symptom severity was used as a proxy. Notably, most of these studies quantified virulence as a qualitative trait (susceptibility *vs.* immunity and mild *vs.* severe symptoms), which may limit the power of the approach to identify genetic determinants by reducing the information contained in qualitative variables as compared with analyses based on quantitative ones [**17**]. Moreover, virulence is also manifested in other plant traits, such as their ability to reproduce and survive, which are often not considered in the above-mentioned studies [**18**] but are relevant in agricultural settings and central in other ecological contexts in which virus infections are also commonplace.

Accumulating data indicates that, besides being of agronomic importance, plant viruses are also major ecological agents in wild ecosystems. Viral infections can drastically reduce the number of individuals in wild plant populations by decreasing the competitive or reproductive abilities of infected individuals [**3**,**19**,**20**]. As a consequence, wild plants evolved quantitative resistance and tolerance in response to infection [**21**,**22**,**23**], which suggests that viruses shape the genetic composition of the host population. Indeed, there is evidence that virus infections may act as a selective force for wild plant populations [**23**]. In wild ecosystems, the main impact of virus infections on the host population is through their detrimental effect on plant fitness [**1**]. Thus, analyses of virulence from an ecological or evolutionary perspective generally used the effect of infection on plant progeny as a proxy, but very little is known on the host genes regulating the effect of virus infection on plant fitness [**18**,**24**]. Certainly, the development of plant symptoms in vegetative structures may potentially impact plant fitness and it is likely that both traits have genetic associations [**10**]. However, the effect of infection on plant growth and on plant reproduction are not necessarily linked, e.g., [**20**,**25**,**26**]. Hence, the host genetic determinants of virulence as the effect of virus infection on plant progeny production are still poorly understood.

Under this definition, virulence has a direct impact not only on the host fitness but also on the pathogen's. For instance, for plant viruses that are horizontally transmitted through insect vectors or by contact, it affects the number of available hosts [**1**]. More importantly, seed production is a major fitness component for vertically transmitted plant viruses, which accounts for at least 25% of all known species [**27**]. Indeed, we have recently shown that virulence is one of the main determinants of *Cucumber mosaic virus* (CMV) and *Turnip mosaic virus* (TuMV) seed transmission rate in *Arabidopsis thaliana* [**28**]. These results suggest that these two traits might have at least partially overlapping genetic regulations, although such link has not been explored yet.

To address these questions, we utilized the interaction between *Cucumber mosaic virus* (CMV, *Bromoviridae*) and *Arabidopsis thaliana* (from here on "Arabidopsis", Brassicaceae) as model. CMV is the plant RNA virus with the broadest host range, infecting about 1200 species in more than 100 plant families, including Arabidopsis [**29**]. This virus is horizontally transmitted by aphids and in Arabidopsis is also seed transmitted with an efficiency that depends on the host per virus genotype x genotype interaction, indicating that the host is involved in controlling this trait [**28**]. CMV is commonly found in wild populations of Arabidopsis at up to 80% prevalence [**21**], and therefore the Arabidopsis–CMV interaction is relevant in nature. Indeed, recent work strongly suggests that CMV infection selects for defenses in Arabidopsis populations of the Iberian Peninsula [**23**]. This geographic region has been shown to contain the largest Arabidopsis genetic diversity in Eurasia due to its role as refugia during the last glaciations [**30**,**31**]. Accordingly, substantial genetic variation has been described for relevant adaptive traits including diversity for responses to pathogens [**23**,**32**,**33**,**34**]. For instance, the infection of ten Iberian Arabidopsis genotypes, representing the variation of the species in this region, with different CMV and TuMV strains showed that virulence (quantitatively measured as the effect of infection on viable seed production) is also controlled by both virus and host genetic determinants [**4**,**23**,**26**].

In this work, we aim at characterizing the Arabidopsis genetic determinants associated with the effect of CMV infection on plant progeny production as a measure of virulence, and to explore the link between the generic control of virulence and of CMV seed transmission rate. To do so, we performed a genome-wide association study (GWAS) with 154 Arabidopsis genotypes from the Iberian Peninsula, for which annotated genomes are publicly available [**35**] and using Arabidopsis seed production in infected and non-infected plants as the quantitative relevant trait. We further investigated whether the identified genetic determinants of virulence play also a role in modulating CMV seed transmission rate.

2. Materials and Methods

2.1. Plant Material

We used 165 Arabidopsis genotypes from the Iberian Peninsula (**Figure 1** and **Table S1**). Genotypes were collected from different populations and selected to cover the genetic and environmental diversity of the species in the region [**36,37**]. This collection spanned 800 km × 700 km, populations being spaced 384.9 ± 3.7 km on the average (from 20.2 to 1038.1 km). Altitudes ranged from 123 to 1670 m above sea level. Each sample was genetically different based on previous single nucleotide polymorphism (SNP) genotyping and genome sequences [**38,39**]. Similar sets of Arabidopsis genotypes have been previously shown to be powerful for fine mapping of genomic regions associated with natural variation of quantitative traits, such as plant life history traits, including seed weight [**40**], seed dormancy [**41**], and flowering time [**39**] both under field and greenhouse conditions.

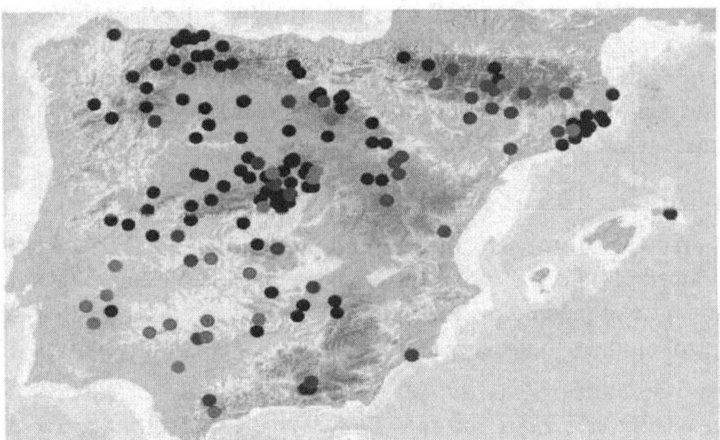

Figure 1. Geographic distribution of *Arabidopsis thaliana* genotypes analyzed in this study. Circles indicate population locations. Colors indicate genetic groups as defined in [39] (group 1: purple, group 2: blue; group 3: red; group 4: green). Grey dots indicate genotypes eliminated from the analysis.

All genotypes used in this study were propagated by selfing during two generations by the single seed descent procedure, in a glasshouse supplemented with lamps to pro- vide a long-day photoperiod. This allowed for the reduction of residual heterozygosity that might contain some wild individuals but also removed any potential maternal and grandmother effects. Seeds were stratified (darkness, 4 °C) for 7 days before germination at 25/20 °C day/night, 16 h light in a greenhouse. Ten-day-old seedlings were transferred to 4 °C, 8 h light, for vernalization for 8 weeks. After vernalization, plants were transplanted to 0.43 L pots containing a 3:1 peat-vermiculite mix and returned to the greenhouse, where they were kept at 25/20 °C day/night and 16 h light (intensity, 120 to 150 mol s/m^2) until the end of the experiment.

2.2. Virus Isolate and Inoculation

Fny-CMV (GenBank accession numbers NC_002034, NC_002035 and NC_001440) was derived from biologically active clones [**42**] by in vitro transcription with T7 RNA polymerase (New England Biolabs, Ipswich, MA, USA), and transcripts were used to infect *Nicotiana benthamiana* plants for virus multiplication. Six Arabidopsis plants per genotype were mechanically inoculated with *N. benthamiana* CMV-infected tissue ground in a solution containing 0.1 M Na_2HPO_4, 0.5 M NaH_2PO_4, and 0.02% DIECA (0.01 M phosphate buffer (pH 7.0), 0.2% sodium diethyldithiocarbamate), and four plants per genotype were mock-inoculated with inoculation buffer. Inoculations were carried out when plants were at developmental stages 1.05 to 1.06 [**43**]. After inoculation, all individuals were randomized in the greenhouse. The efficiency of inoculations was determined by detecting virus presence in all plants. To do so, 15 days post-inoculation three disks with a diameter of 4 mm were collected from different systemically infected rosette leaves. From these plant samples, total RNA extracts were obtained using TRIzol® reagent (Life Technologies, Carlsbad, CA, USA), and 10 ng of total RNA was added to a Brilliant III Ultra-Fast SYBR green qRT-PCR master mix (Agilent Technologies, Santa Clara, CA, USA) according to the manufacturer's recommendations. Specific primers were used to amplify a -154nt fragment of the CMV MP gene [**28**]. Each plant sample was assayed in duplicate on a LightCycler 480 II real-time PCR system (Roche, Indianapolis, IN, USA). The rate of inoculation success was 98.8%. A similar procedure was used to confirm CMV absence in mock-inoculated plants.

2.3. Quantification of Virulence

Seeds were harvested at complete plant senescence, and the total seed weight per plant (*SW*) was obtained for infected and mock-inoculated plants. Virulence (*V*) was estimated as 1 minus the ratio of the total seed weight of each infected plant (SW_i) to the averaged total seed weight of mock inoculated (SW_m) plants from the same genotype.

Seed viability was measured as the germination percentage of 200 seeds per plant. Germination assays were carried out at least 60 days after harvesting to avoid differences on seed dormancy. Relative differences in seed viability between infected and control plants were used to correct *V* values such that only viable seed production was reflected.

2.4. Efficiency of Virus Seed Transmission

The efficiency of CMV seed transmission was estimated as the percentage of infected seeds that gave rise to infected progeny in grow-out tests. One-hundred seeds per replicate were washed in a 10% bleach solution. Then, seeds were placed into Petri dishes containing Murashige–Skoog medium, stratified for 5 days at 4 °C, and germinated in a growth chamber at 22 °C, under 16 h of light (intensity, 120 to 150 mol s/m^2). Following [**28**], seedlings at 15 days poststratification were pooled in groups of 5 for a total of 20 groups per replicate. These groups were tested for the presence of CMV via qRT-PCR as described above. The percentage of virus-infected seeds (*ST*) was estimated using a Poisson distribution as: $p = 1) - 1 - y/n)^{1/k}$, where p is the probability of virus transmission by a single seed, y is the number of positive samples, n is the total number of samples assayed ($n = 20$), and k is the number of seedlings per sample ($k = 5$).

2.5. Data Treatment and Statistical Analyses

In the 165 Arabidopsis genotypes, we analyzed the average variance in *V* across plants of the same genotype. Within-genotype variance in this trait larger than two-fold the average variance in the whole set of genotypes was considered as indicative of an unreliable estimate of *V*. Nine Arabidopsis genotypes showed such inflated variance and were eliminated from the analysis and two more were deleted due to low number of replicates because of inoculation failure, and thus that the final dataset used for the GWAS contained 154 genotypes. We explored the phenotypic variation of *V* across Arabidopsis genotypes using the following general linear mixed model (GLMM): $V = \mu +$ Genotype $+ \varepsilon$, where μ is the overall mean of the phenotypic data, "genotype" corresponds to the genetic differences among the selected Arabidopsis genotypes, and ε is the residual error term. Normality of the residuals was achieved through $(V + 2)^{5.36}$ transformation (Kolmogorov–Smirnov test p-value = 0.143). The factor "genotype" was treated as a random factor.

Broad-sense heritability was estimated as $h^2_b = V_G/(V_G + V_E)$, where V_G is the among-genotypes variance component and V_E is the residual variance. Variance components were determined using GLMMs by the Restricted Maximum Likelihood (REML) method [**44**] as implemented in the R-library lme4 [**45**]. Statistical analyses were conducted using R version 3.6.0 [**46**].

2.6. Genome-Wide Association Study (GWAS)

The 154 Arabidopsis genotypes have been genotyped for 4,932,457 million single nucleotide polymorphisms (SNPs) evenly spaced across the genomes [**35**]. Following [**39**], only SNPs present in at least 55% of the genotypes and with minor allele relative frequency (MAF) > 0.03 were considered in this study, resulting in a total of 2,071,858 SNPs from which 88.6% were present in all accessions. GWAS was run for virulence using the FarmCPU (fixed and random model circulating probability unification) as implemented in the R package GAPIT v3 [**47**]. FarmCPU minimizes false positives by accounting for linkage disequilibrium between SNPs, which reduces model overfitting as compared to other methods [**48**]. To determine the optimal number of principal components to include, forward model selection using the Bayesian information criterion (BIC) was conducted. The Manhattan plot was constructed with R library rMVP [**49**]. Following previous studies [**50,51,52**], a threshold of $-\log_{10}(p) \geq 4$ was considered to identify SNPs associated with natural variation of the trait measured in this study. Within this set of SNPs, we conservatively focused on those below the false discovery rate (FDR) threshold in order to minimize false positives/negatives. For our data set, FDR = 1×10^{-8}. The functional annotation of SNPs was carried out using SnpEff v.4.1 [**53**] and the TAIR database v.10 [**54**]. Gene ontology (GO) annotation enrichment was tested with

PANTHER v16.0 using the binomial test [**55**] and visualized using SimRel for semantic similarity measure with a threshold of $C = 0.7$, as implemented in REVIGO [**56**].

2.7. Bayesian Sparse Linear Mixed Model (BSLMM)

BSLMM as implemented in the GEMMA (genome-wide efficient mixed model association) software was used to infer the genetic architecture of CMV virulence in Arabidopsis [**57**], which allows testing whether the analyzed trait is determined by many loci of small effect (polygenic) or rather determined by a few loci of large effector (oligogenic). Monte Carlo Markov chains (MCMCs) were run for 10 million generations (recording every ten steps), with 10% discarded as burn-in. MAF cut-off was set at 3% and a normalized kinship matrix, more appropriate if SNPs with large effects have low MAF, was included. Normalized kinship matrix was estimated with Tassel 5.2.70 software [**58**]. The proportion of phenotypic variance explained by the available genotypes (PVE) was used as an estimator of the heritability of a given phenotypic trait. PVE is a flexible Bayesian equivalent of the narrow-sense heritability (h^2). BSLMM also estimated the proportion of genetic variance explained by sparse effects (PGE); that is, the proportion of variance explained by loci with large effects.

2.8. Selection of SNPs Linking CMV Virulence and Seed Transmission Rate

A Random Forest (RF) analysis was made to determine the SNPs associated with both CMV virulence and seed transmission rate. In this analysis, we used seed transmission rate estimates of 35 Arabidopsis genotypes with extreme virulence phenotypes as the response variable, and as the input, only the SNPs associated with CMV virulence as determined by GWAS ($-\log_{10}(p) \geq 4$). Arabidopsis genetic group [**39**] and CMV virulence values were included as covariates (randomForest R package [**59**]). The model was run for 2000 trees and mtry = 2, as estimated with the trainControl function from the *caret* R package [**60**]. Significant association of each SNP with the response variable was estimated based on the empirical null distribution of SNPs with no importance in this variable following [**61**], as implemented in the *r2VIM* R package [**62**]. The relative importance of each SNP in CMV seed transmission rate was quantified as % increase in mean square error (MSE), that is, the increase in the error made by the RF in predicting the trait when the SNP is removed from the analysis.

3. Results

3.1. Natural Variation for CMV Virulence in Arabidopsis Genotypes

Overall, CMV-infected plants developed symptoms ranging from mild mosaic to leaf necrosis (**Figure S1**). Virus infection reduced plant progeny production, showing medium–high virulence in Arabidopsis genotypes ($V = 0.035 \pm 0.637$; that is, infected plants produced on average %64 less seeds than mock-inoculated ones). However, virulence greatly varied across Arabidopsis genotypes, from plant sterilization ($V = 1$ in 13 genotypes) to overcompensation of the effect of infection on plant progeny production (negative values of V in 16 genotypes, with minimum values of 1.869– in genotype Lam-0) (**Figure 2** and **Table S1**). This ample variation did not depend on the Arabidopsis genetic group ($F = 1.801$; $p = 0.150$) but did vary according to the plant genotype ($F = 12.619$; $p = {}^{4-}10 \times 1$).

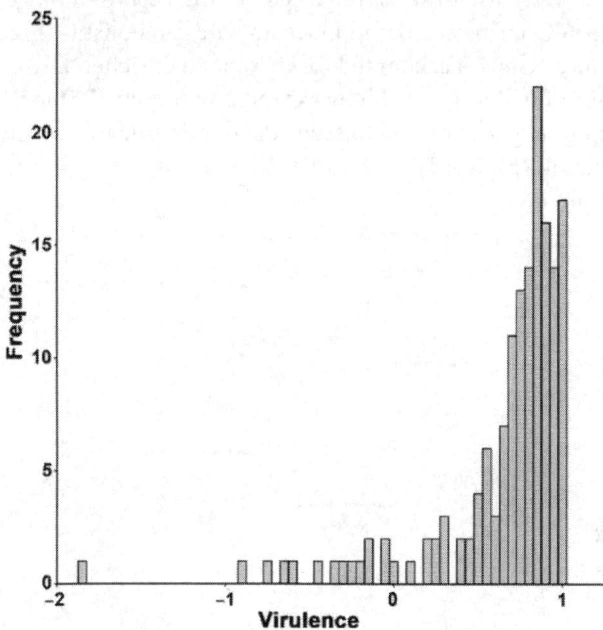

Figure 2. Frequency distributions of *Cucumber mosaic virus* (CMV) virulence in 154 *Arabidopsis thaliana* genotypes.

The broad-sense heritability of CMV virulence in Arabidopsis showed moderate–high values: $h^2_b = 0.68$. Therefore, there was significant genetic variation among the studied Arabidopsis genotypes for the analyzed trait, such that the utilized population allowed a meaningful GWAS.

3.2. Genetic Architecture of Virulence in Arabidopsis

The 154 Arabidopsis genotypes accounted for 2.07 million SNPs, which were included in a BSLMM to evaluate how much of the variance in CMV virulence was explained by these SNPs. A moderate percentage of the phenotypic variance was explained by the genotyped SNPs [PVE: 51%; 95% ETPI (equal-tail probability intervals): 15–76%]. The strength of association of a SNP with CMV virulence above or equal to 0.25 was considered an indication of a large effect SNP that contributed the most to the phenotype. These large effect variants explained 58% (PGE: 58%; 95% ETPI: 32–96%) of the variance in CMV virulence. The number of SNPs with a large effect size was low (mean: 10; 95% ETPI: 4–18). Thus, the BSLMM indicated that Arabidopsis genetic architecture of CMV virulence involved few SNPs with detectable large effects.

In parallel with the BSLMM, we performed a GWAS. The results indicated that 223 SNPs were significantly associated with CMV virulence in Arabidopsis ($-\log_{10}(p) \geq 4$), with effect size varying from 0.08 to 0.42 (**Table S2**). To provide global insights into the biological processes associated with CMV virulence, we performed a GO term enrichment analysis based on the 223 detected SNPs (**Figure 3**). We identified 37 enriched GO terms ($p < {}^{3-}10 \times 1$), which were reduced to 30 after eliminating redundant terms (**Table S3**). Similarity analyses of these 30 GO terms indicated that the largest clusters grouped functions related to response to stress (eight GO terms), cell wall biogenesis and cellular metabolisms (four GO terms each) and glucosinolate catabolism (three GO terms). GO terms in these four categories accounted for 63% of all enriched terms (**Figure 3**). Most other terms corresponded to general housekeeping functions (**Table S3**). One exception was SNP enrichment in genes related to seed coat development, a function highly related to progeny production (**Figure 3**).

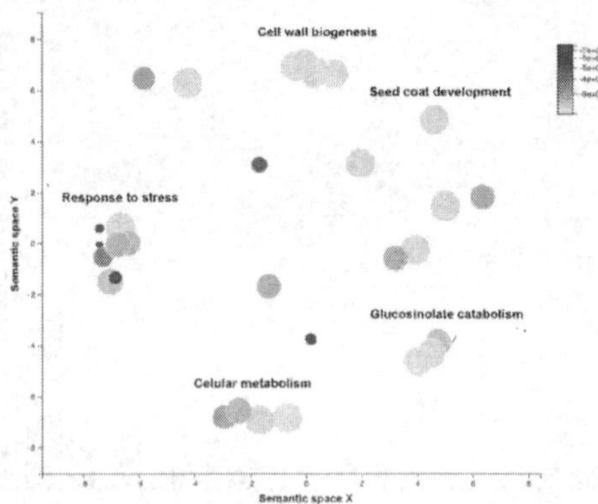

Figure 3. Scatterplot of enriched gene ontology (GO) terms for Arabidopsis determinants of CMV virulence. Color scale indicates significance of the term enrichment (–log$_{10}$(p). Dot diameter indicates fold enrichment.

Only eight of the 223 SPNs passed the FDR threshold (FDR-adjusted p-value < 3×10^{-3} and $-\log_{10}(p) \geq 8$) (**Figure 4**), representing most of the SNPs with the largest effects in virulence (effect size > 0.20) (**Table S2**). With few exceptions (nine SNPs, four of them affecting the same gene), the rest of the 215 SNPs had effects lower than 0.20. Therefore, only a reduced number of SNPs with large effects size were strongly associated with CMV virulence in Arabidopsis, in agreement with BSLMM.

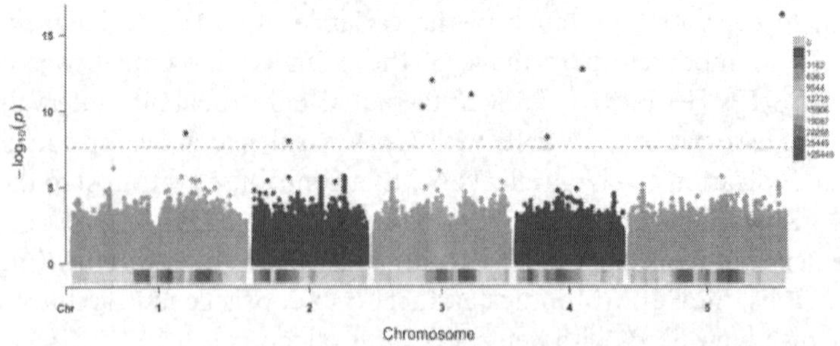

Figure 4. Manhattan plot illustrating the *Arabidopsis thaliana* genomic regions associated with CMV virulence. Dashed line indicates false discovery rate (FDR) threshold. Color scale represents single nucleotide polymorphism (SNP) density.

Altogether, our analyses indicate that Arabidopsis genetic determinants of CMV virulence are mainly involved in functional categories associated with stress response and cell wall biogenesis. Of the identified SNPs, a few have detectable large effects and are strongly associated with the analyzed trait, whereas a larger number of SNPs are less strongly associated with virulence, most of them with smaller effect size.

3.3. Arabidopsis Genetic Determinants of CMV Virulence Identified by GWAS

We focused on the eight SNPs that passed the FDR threshold (**Figure 4**). MAF values for these SNPs varied between 0.08 and 0.11. For all of them, the minor variant had a negative effect on CMV virulence (**Table 1**).

Table 1. SNPs significantly associated with virulence according to FDR threshold.

SNP [a]	Protein	Description	FDR	Effect
C5_26478732	GMD1	Cell wall metabolism	8.92×10^{-11}	−0.42
C3_10139431	PLL18	Cell wall metabolism	6.53×10^{-7}	−0.40
C4_11603821	HSP20-like	Abiotic stress response	1.81×10^{-7}	−0.22
C3_16927277	ZAT8	Biotic and abiotic stress response	3.96×10^{-6}	−0.30
C3_8500254	RTFL13	Flowering time	2.05×10^{-5}	−0.23
C1_20023793	AT1G53635	Unknown	9.81×10^{-4}	−0.23
C4_5465990	ORTHL	DNA methylation	1.56×10^{-3}	−0.21
C2_6211877	LURP1	Biotic stress response	2.51×10^{-3}	−0.22

[a] Each SNP is named by the chromosome (C) in which it is located and its position in the chromosome.

The two Arabidopsis SNPs with the strongest association with CMV virulence were located at genes encoding a GDP-d-mannose 4,6-dehydratase (GMD1) and a peptate lyase-like protein (PLL18) (FDR < 6.53×10^{-7}). Both genes are involved in cell wall metabolism [**63,64**]. Of the remaining six SNPs strongly associated with virulence, three were involved in response to biotic and abiotic stresses: heat shock protein 20-like (HSP20-like), the zinc finger protein ZAT8 and the *late upregulated in response to Hyaloperonosporaparasitica protein 1* (LURP**65**] (1,**66,67,68,69**]. Another two were related to different biological processes such as flowering time (ROTUNDIFOLIA like 13 protein, RTFL13) [**70**] and DNA methylation (ORTHRUS-like protein, ORTHL) [**71**], and onc had unknown function (AT1G53635). To confirm the association of these SNPs with CMV virulence, we performed General Linear Models (GLMs) comparing CMV virulence in Arabidopsis genotypes with different variants in the eight detected SNPs. In all cases, analyses indicated significant differences ($F_{1,142-152} \geq 8.74$; $p < {}^{3-}10 \times 3.6$), supporting the role of these SNPs in the analyzed trait, with minor variants always showing a reduced virulence (**Figure 5**).

In summary, the Arabidopsis SNPs strongly associated with CMV virulence are located in genes mostly involved in cell wall biogenesis and in stress response in agreement with GO term enrichment analyses, and minor variants of the SNPs in these genes generally reduce CMV virulence.

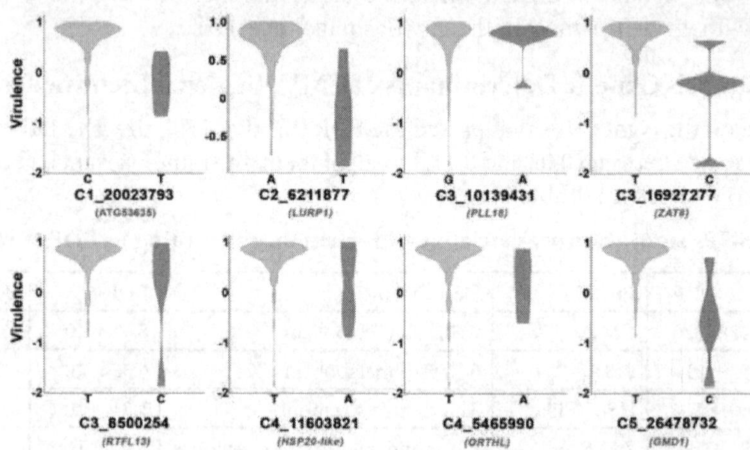

Figure 5. Distribution of CMV virulence across Arabidopsis genotypes harboring major (blue) and minor (red) variants of the eight SNPs strongly associated with this trait. Each SNP is indicated by the chromosome (C) and the position in which it is located, and the corresponding gene is shown in grey.

3.4. Link between Arabidopsis Genetic Determinants of CMV Virulence and Seed Transmission

To analyze the potential genetic links between CMV virulence and seed transmission rate in the Arabidopsis genome, we selected 35 plant genotypes representing extreme virulence phenotypes in which we estimated the per cent of CMV seed transmission (*ST*) (**Table S1**). *ST* varied from absence of infected seeds (genotypes Cdc3-, Iso4- and Lum0-) to %100 of infected seeds (genotypes Alm0-, Amu0-, Ang0- and Aul0-). Then, we performed a RF analysis including the 223 SNPs associated with CMV virulence as predictors of virus seed transmission rate. The RF detected only seven of these SNPs as significantly associated with this trait ($p < 0.041$) (**Table 2**). None of the eight SNPs strongly associated with virulence was so with virus seed transmission rate.

Table 2. SNPs significantly associated with CMV seed transmission rate according to Random Forest (RF). CIPK2: calcineurin B-like (CBL)-interacting protein kinase 2; JAL4: Jacalin-related lectin 4; MAC5C: MOS4-ASSOCIATED COMPLEX SUBUNIT 5C protein. MEG: Maternally expressed gene.

SNP [a]	Protein	Description	*p*-Value	% Inc. MSE [b]
C5_2202898	CIPK2	Response to abiotic stress	0.012	8.11
C5_2198824	MAC5C	Cell wall metabolism	0.012	2.77
C5_11777493	AT5G31963	Unknown	2×10^{-16}	0.83
C2_7038732	MEG	Embryogenesis	2×10^{-16}	0.73
C3_15811438	AT3G44030	Unknown	0.041	0.72
C1_12253373	JAL4	Response to biotic stress	0.041	0.53
C1_16528990	AT1G43750	Unknown	0.033	0.49

Each SNP is named by the chromosome (C) in which it is located and its position in the chromosome. [b] Per cent of increase in the mean square error (MSE).

GLMs comparing CMV seed transmission rate in Arabidopsis genotypes with different variants in the seven SNPs indicated significant differences only for the two with the highest % MSE ($F_{1,34} \geq 8.74$; $p < 0.033$) (**Figure 6**). The two SNPs associated with both CMV virulence and seed transmission rate were located in genes encoding the calcineurin B-like (CBL)-interacting protein kinase 2 (CIPK2) and the MOS4-ASSOCIATED COMPLEX SUBUNIT 5C protein (MAC5C). The former is a calcium regulated protein [**72**], and the latter has been reported as being involved in cell wall metabolism [**73**].

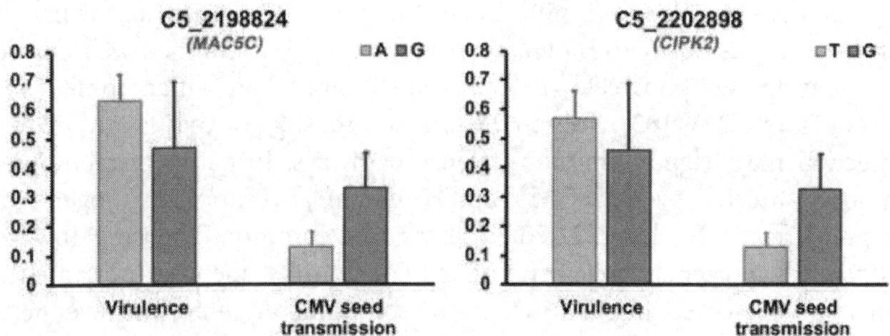

Figure 6. Distribution of CMV virulence and seed transmission rate across Arabidopsis genotypes harboring major (blue) and minor (red) variants of the two SNPs associated with both traits. Letters indicate the nucleotide at each SNP for the major and minor variants. Seed transmission rate is shown as *ST*/100 to fit in the same scale as virulence. Each SNP is indicated by the chromosome (C) and the position in which it is located, and the corresponding gene is shown in grey.

Hence, Arabidopsis genes involved in both CMV virulence and seed transmission rate control stress responses and cell wall metabolism.

4. Discussion

Most analyses of the host genetic determinants of plant virus virulence have identified genes involved in symptom development and/or plant susceptibility [**8,11**]. In contrast, genes associated with the effect of virus infection on plant progeny production are poorly characterized [**74**]. Identifying this genetic control is of great relevance to understand plant–virus interactions from an agronomic and ecological perspective: first, seed production is a key trait for wild plant population dynamics and a trait of agronomic importance [**75,76**]. Second, it modulates the chances for virus horizontal transmission (i.e., determines the number of available susceptible individuals) [**1**], and is central for viruses that disperse through seeds [**27,77**]. Here, through a GWAS including 154 plant genotypes, we identified Arabidopsis genes involved in the effect of CMV infection on seed production and we explored the potential link of these genes with the efficiency of virus seed transmission.

We identified SNPs in eight plant genes strongly associated with the effect of CMV infection on seed production. Three of these genes (*HSP20-like*, *ZAT8* and *LURP1*) were related to the response to biotic and abiotic stresses. The first two genes have been shown to be overexpressed in response to heat and salinity [**78,79**]. However, there is also evidence of their role in plant response to pathogens. For instance, many HSPs have been reported to alter their expression patterns upon infection by multiple pathogens [**80**]. Indeed, in *N. benthamiana* and rice HSP20 expression has been shown to change upon infection by *Impatiens necrotic spot virus* and *Rice stripe virus*, respectively, facilitating virus replication and within-host movement [**68,81**]. Similarly, Arabidopsis transcription factor *ZAT8* knock-out mutants have been reported to have higher susceptibility to *Pseudomonas syringae* [**67**], and in the same host other transcription factors of the ZAT family are responsive to infection by viruses, including CMV [**82**]. The third gene, *LURP1*, is also regulated in response to biotic stresses, being required for basal defense to *Hyaloperonosporaarabidopsidis* [**69**]. Interestingly, our results are not the first evidence of a link between defenses to *H. arabidopsidis* and CMV. Another well-characterized resistance gene to this oomycete (*RPP8*) is allelic to *RCY-1*, which confers resistance to the isolate Y of CMV [**83**]. These results and ours suggest that *H. arabidopsidis* and CMV infection may trigger common defense pathways. In nature, Arabidopsis plants are commonly infected by both CMV and *H. arabidopsidis*, both pathogens imposing selective pressures on the host [**21,23,84**], such that common defense pathways would be selectively advantageous while minimizing energy investment at the same time. The evolutionary processes leading to this common solution is an interesting avenue of future research. It should be noted that the expression of *HSP20-like*, *ZAT8* and *LURP1* has been mostly studied in vegetative plant tissues, and their effect has been generally linked to reduced pathogen load and/or lower effect of infection on plant growth. The association of these three genes with the effect of infection on seed production might be explained if: (i) their differential expression in vegetative tissues affects plant reproduction, or (ii) these genes are also differentially regulated in plant reproductive organs. In any case, our results provide tentative evidence of a novel function for these genes in plant–virus interactions.

Perhaps more interestingly, the SNPs with the strongest association, and the largest effect, with virulence were located in genes related to the cell wall metabolism (*GMD1* and *PLL18*). Current information on changes in the composition, structure and function of plant cell walls in response to infection has been largely restricted to non-viral pathogens [**85,86**]. The scant work on plant–virus interactions focused on the association between cell wall modifications and virus cell-to-cell and, to a lesser extent, systemic movement. For instance, pectin methylesterases and hydroxyproline-rich glycoproteins are needed for efficient *Tobacco mosaic virus* and *Potato virus Y* movement in tobacco and potato, respectively [**87,88,89**]. It could be argued that GMD1 and PLL18 work in a similar way: certain variants of these proteins might facilitate virus cell-to-cell and systemic colonization from entry points at plant leaves and therefore promote the invasion of reproductive structures. However, none of the two proteins encoded by these genes are significantly expressed in Arabidopsis leaves. Rather, their expression is largely restricted to flowers and pollen grains as they are mainly involved in the synthesis and remodeling of cell walls in these organs [**63,64**]. Hence, we hypothesize that CMV infection alters the functioning of these proteins, leading to the malformation of cell walls in flowers and/or pollen, which may affect plant fertility and therefore seed production. Arabidopsis genotypes with variants in these genes less prone to CMV effects would suffer less from infection. Further functional analyses of these GMD1 and PLL18 upon virus infection will allow testing this hypothesis. Regardless the mechanism involved, our results reveal a new role of the cell wall in plant–virus interactions as a regulator of the effect of infection on plant fitness. The identification of four large effect size SNPs that did not pass the FDR threshold but were significantly associated with CMV virulence within a gene encoding the prolyl 4-hydroxylase 11 (P4H11) (**Table S2**), which is also involved in cell wall biogenesis [**90**], and the results of our GO term enrichment, further support this conclusion.

The other two Arabidopsis genes strongly associated with CMV virulence were *RTFL13* and *ORTHL*. RTFL13 has been shown to be a regulator of flowering time that is expressed during cambium formation in Arabidopsis inflorescences [**70**]. There is ample evidence of the link between flowering time and the effect of virus infection on seed production [**25,26,91**]. Although the genetic bases are unknown, flowering genes are thought to control this process [**18**]. In agreement, *RTFL13* would be a candidate gene to be involved in such control. As for ORTHL, an epigenetic regulator of embryogenesis [**92**], any effect of virus infection on embryo development would directly impact plant fitness. Our results suggest that such effects would occur through alterations of the epigenetic regulation of plant reproduction. Indeed, virus infections have been repeatedly shown to alter plant epigenetics (including DNA methylation) [**93,94**]. Hence, the identification of these two genes as potential regulators of CMV virulence support the idea that at least part of the host control of this trait occurs at the plant reproductive stage, and independently of the virus effects on vegetative structures.

Although none of the Arabidopsis SNPs strongly associated with CMV virulence were also modulating virus seed transmission rate, our RF analysis identified two SNPs that were linked to both traits: CIPK2 and MAC5C. CIPK2 is a protein kinase associated to calcineurin B-like protein (CBLs), which is a group of calcium sensors. In plants, calcium fluxes are important for sensing a variety of stimuli, including abiotic conditions and pathogens, and in these processes both CBLs and CIPKs are involved [**72,95**]. For instance, tobacco CIPK2 has been shown to be involved in tolerance to drought [**72**]. Although in Arabidopsis the role of CIPK2 is not known, other CIPKs (such as CIPK6) have been demonstrated to induce plant resistance to pathogens when silenced. Thus, it could be hypothesized that CIPK2 acts in a similar way in response to CMV. Interestingly, CIPK2 is highly expressed in flowers [**96**], which may explain its common role in virulence and seed transmission: the differential expression of CIPK2 may induce local resistance to the virus in flowers, minimizing the effects of infection in seed production and preventing seed invasion. Similarly, MAC5C, part of the secondary cell wall regulation network [**73**], is expressed during seed germination [**97**]. In infected seeds, MAC5C may contribute to minimize the effect of CMV presence on secondary wall metabolism, promoting seed viability and favoring virus seed transmission. As part of the MAC complex, MAC5C also has a role in plant resistance to pathogens [**98**]. The specific mechanism by which MAC5C is involved in plant defenses is not well characterized [**99**]. However, this work was restricted to analysis in plant vegetative structures. Similarly to CIPK2, MAC5C could also influence virulence and seed transmission rate by restricting virus invasion of reproductive organs. It is worth mentioning that here we did not quantify CMV load in the reproductive plant organs, which would have contributed to understand whether certain variants of these two genes prevent virus invasion. On the other hand, we show that variants with lower virulence had higher seed transmission rate, suggesting a negative correlation between both traits, which is in agreement with our previous results [**28**], strengthening the evidence of a role of these genes as multi-trait regulators.

In summary, this work identifies genes related to stress response and cell wall metabolism as potential regulators of the effect of virus infection on plant progeny production and of virus seed transmission rate. Although functional analyses are needed to validate these results, our GWAS provides tentative evidence of novel roles for these plant genes in plant–virus interactions, opening a new avenue of future research.

Supplementary Materials

Figure S1: Symptoms induced by CMV infection in Arabidopsis. Upper line shows mock-inoculated plants and lower line CMV-infected plants of Arabidopsis genotypes Cem-0 (A and E), Gra-0 (B and F), Ses-0 (C and G) and Vas-0 (D and H). Table S1: List of Arabidopsis genotypes utilized in the GWAS and average virulence (*V*) and seed transmission rate (*ST*) values per genotype; Table S2: SNPs in the Arabidopsis genome associated with CMV virulence; Table S3: GO enriched terms in the 223 host SNPs associated with CMV virulence in *Arabidopsis thaliana*.

References

1. Read, A.F. The evolution of virulence. *Trends Microbiol.* **1994**, *2*, 73–76. [**Google Scholar**] [**CrossRef**]
2. Oerke, E.-C. Crop losses to pests. *J. Agric. Sci.* **2006**, *144*, 31–43. [**Google Scholar**] [**CrossRef**]
3. Anderson, P.K.; Cunningham, A.A.; Patel, N.G.; Morales, F.J.; Epstein, P.R.; Daszak, P. Emerging infectious diseases of plants: Pathogen pollution, climate change and agrotechnology drivers. *Trends Ecol. Evol.* **2004**, *19*, 535–544. [**Google Scholar**] [**CrossRef**] [**PubMed**]
4. Pagán, I.; Alonso-Blanco, C.; García-Arenal, F. The relationship of within-host multiplication and virulence in a plant-virus system. *PLoS ONE* **2007**, *2*, e786. [**Google Scholar**] [**CrossRef**] [**Green Version**]
5. Pallás, V.; García, J.A. How do plant viruses induce disease? Interactions and interference with host components. *J. Gen. Virol.* **2011**, *92*, 2691–2705. [**Google Scholar**] [**CrossRef**]
6. Paudel, D.B.; Sanfaçon, H. Exploring the diversity of mechanisms associated with plant tolerance to virus infection. *Front. Plant Sci.* **2018**, *9*, 1575. [**Google Scholar**] [**CrossRef**] [**PubMed**]
7. Shukla, A.; Pagán, I.; García-Arenal, F. Effective tolerance based on resource reallocation is a virus-specific defence in *Arabidopsis thaliana. Mol. Plant Pathol.* **2018**, *19*, 1454–1465. [**Google Scholar**] [**CrossRef**] [**Green Version**]
8. García, J.A.; Pallás, V. Viral factors involved in plant pathogenesis. *Curr. Opt. Virol.* **2015**, *11*, 21–30. [**Google Scholar**] [**CrossRef**] [**PubMed**]
9. Palukaitis, P. Determinants of pathogenesis. In *Cucumber Mosaic Virus*; Palukaitis, P., García-Arenal, F., Eds.; APS Press: Washington, DC, USA, 2019; pp. 145–154. [**Google Scholar**]

10. Ascencio-Ibáñez, J.T.; Sozzani, R.; Lee, T.J.; Chu, T.M.; Wolfinger, R.D.; Cella, R.; Hanley-Bowdoin, L. Global analysis of Arabidopsis gene expression uncovers a complex array of changes impacting pathogen response and cell cycle during geminivirus infection. *Plant Physiol.* **2008**, *148*, 436–454. [**Google Scholar**] [**CrossRef**] [**PubMed**] [**Green Version**]

11. García-Ruíz, H. Host factors against plant viruses. *Mol. Plant Pathol.* **2019**, *20*, 1588–1601. [**Google Scholar**] [**CrossRef**] [**PubMed**]

12. Dieckmann, U.; Metz, J.A.; Sabelis, M.W.; Sigmund, K. Adaptive dynamics of infectious diseases. In *Pursuit of Virulence Management*; Cambridge University Press: Cambridge, UK, 2002. [**Google Scholar**]

13. Kang, B.-C.; Yeam, I.; Jahn, M.M. Genetics of plant virus resistance. *Annu. Rev. Phytopathol.* **2005**, *43*, 581–621. [**Google Scholar**] [**CrossRef**] [**Green Version**]

14. de Ronde, D.; Butterbach, P.; Kormelink, R. Dominant resistance against plant viruses. *Front. Plant Sci.* **2014**, *5*, 307. [**Google Scholar**] [**CrossRef**] [**PubMed**] [**Green Version**]

15. Faoro, F.; Gozzo, F. Is modulating virus virulence by induced systemic resistance realistic? *Plant Sci.* **2015**, *234*, 1–13. [**Google Scholar**] [**CrossRef**] [**PubMed**]

16. Carr, J.P.; Murphy, A.M. Host responses: Susceptibility. In *Cucumber Mosaic Virus*; Palukaitis, P., García-Arenal, F., Eds.; APS Press: Washington, DC, USA, 2019; pp. 47–58. [**Google Scholar**]

17. Drummond, D.A.; Raval, A.; Wilke, C.O. A single determinant dominates the rate of yeast protein evolution. *Mol. Biol. Evol.* **2006**, *23*, 327–3337. [**Google Scholar**] [**CrossRef**] [**PubMed**] [**Green Version**]

18. Pagán, I.; García-Arenal, F. Tolerance to plant pathogens: Theory and experimental evidence. *Int. J. Mol. Sci.* **2018**, *19*, 810. [**Google Scholar**] [**CrossRef**] [**Green Version**]

19. Malmstrom, C.M.; McCullough, A.J.; Johnson, H.A.; Newton, L.A.; Borer, E.T. Invasive annual grasses indirectly increase virus incidence in California native perennial bunchgrasses. *Oecologia* **2005**, *145*, 153–164. [**Google Scholar**] [**CrossRef**]

20. Vijayan, V.; López-González, S.; Sánchez, F.; Ponz, F.; Pagán, I. Virulence evolution of a sterilizing plant virus: Tuning multiplication and resource exploitation. *Virus Evol.* **2017**, *3*, vex033. [**Google Scholar**] [**CrossRef**] [**PubMed**] [**Green Version**]

21. Pagán, I.; Fraile, A.; Fernández-Fueyo, E.; Montes, N.; Alonso-Blanco, C.; García-Arenal, F. *Arabidopsis thaliana* as a model for the study of plant–virus co-evolution. *Philos. Trans. R. Soc. Lond. B* **2010**, *365*, 1983–1995. [**Google Scholar**] [**CrossRef**] [**PubMed**] [**Green Version**]

22. Moreno-Pérez, M.G.; Pagán, I.; Aragón-Caballero, L.; Cáceres, F.; Fraile, A.; García-Arenal, F. Ecological and genetic determinants of *Pepino mosaic virus* emergence. *J. Virol.* **2014**, *88*, 3359–43368. [**Google Scholar**] [**CrossRef**] [**Green Version**]

23. Montes, N.; Alonso-Blanco, C.; García-Arenal, F. *Cucumber mosaic virus* infection as a potential selective pressure on *Arabidopsis thaliana* populations. *PLoSPathog.* **2019**, *15*, e1007810. [**Google Scholar**] [**CrossRef**] [**Green Version**]

24. Pagán, I.; García-Arenal, F. Tolerance of plants to plant pathogens: A unifying view. *Annu Rev. Plant Pathol.* **2020**, *18*, 9.1–9.20. [**Google Scholar**] [**CrossRef**]

25. Pagán, I.; Alonso-Blanco, C.; García-Arenal, F. Host responses in life-history traits and tolerance to virus infection in *Arabidopsis thaliana. PLoSPathog.* **2008**, *4*, e1000124. [**Google Scholar**] [**CrossRef**] [**PubMed**] [**Green Version**]

26. Montes, N.; Vijayan, V.; Pagán, I. Trade-offs between host tolerances to different pathogens in plant-virus interactions. *Virus Evol.* **2020**, *6*, veaa019. [**Google Scholar**] [**CrossRef**] [**PubMed**] [**Green Version**]

27. Sastry, K.S. *Seed-Borne Plant Virus Diseases*; Springer: New Delhi, India, 2013. [**Google Scholar**]

28. Cobos, A.; Montes, N.; López-Herranz, M.; Gil-Valle, M.; Pagán, I. Within-host multiplication and speed of colonization as infection traits associated with plant virus vertical transmission. *J. Virol.* **2019**, *93*, e01078-19. [**Google Scholar**] [**CrossRef**] [**PubMed**] [**Green Version**]

29. Palukaitis, P.; García-Arenal, F. *Cucumber Mosaic Virus*; APS Press: Washington, DC, USA, 2019. [**Google Scholar**]

30. Brennan, A.C.; Méndez-Vigo, B.; Haddioui, A.; Martínez-Zapater, J.M.; Picó, F.X.; Alonso-Blanco, C. The genetic structure of *Arabidopsis thaliana* in the south-western Mediterranean range reveals a shared history between North Africa and southern Europe. *BMC Plant Biol.* **2014**, *14*, 17. [**Google Scholar**] [**CrossRef**] [**Green Version**]

31. Lee, C.R.; Svardal, H.; Farlow, A.; Exposito-Alonso, M.; Ding, W.; Novikova, P.; Alonso-Blanco, C.; Weigel, D.; Nordborg, M. On the post-glacial spread of human commensal *Arabidopsis thaliana. Nat. Commun.* **2017**, *8*, 14458. [**Google Scholar**] [**CrossRef**] [**PubMed**] [**Green Version**]

32. Jorgensen, T.H.; Emerson, B.C. Functional variation in a disease resistance gene in populations of *Arabidopsis thaliana. Mol. Ecol.* **2008**, *17*, 4912–4923. [**Google Scholar**] [**CrossRef**]

33. Huard-Chauveau, C.; Perchepied, L.; Debieu, M.; Rivas, S.; Kroj, T.; Kars, I.; Bergelson, J.; Roux, F.; Roby, D. An atypical kinase under balancing selection confers broad-spectrum disease resistance in Arabidopsis. *PLoS Genet.* **2013**, *9*, e1003766. [**Google Scholar**] [**CrossRef**] [**PubMed**] [**Green Version**]

34. Karasov, T.L.; Kniskern, J.M.; Gao, L.; DeYoung, B.J.; Ding, J.; Dubiella, U.; Lastra, R.O.; Nallu, S.; Roux, F.; Innes, R.W.; et al. The long-term maintenance of a resistance polymorphism through diffuse interactions. *Nature* **2014**, *512*, 436–440. [**Google Scholar**] [**CrossRef**] [**PubMed**] [**Green Version**]

35. The 1001 Genomes Consortium. 1,135 genomes reveal the global pattern of polymorphism in *Arabidopsis thaliana. Cell* **2016**, *166*, 481–491. [**Google Scholar**] [**CrossRef**] [**PubMed**] [**Green Version**]

36. Picó, F.X.; Méndez-Vigo, B.; Martinez-Zapater, J.M.; Alonso-Blanco, C. Natural genetic variation of *Arabidopsis thaliana* is geographically structured in the Iberian Peninsula. *Genetics* **2008**, *180*, 1009–1021. [**Google Scholar**] [**CrossRef**] [**PubMed**] [**Green Version**]

37. Méndez-Vigo, B.; Picó, F.X.; Ramiro, M.; Martinez-Zapater, J.M.; Alonso-Blanco, C. Altitudinal and climatic adaptation is mediated by flowering traits and FRI, FLC, and PHYC genes in Arabidopsis. *Plant Physiol.* **2011**, *157*, 1942–1955. [**Google Scholar**] [**CrossRef**] [**Green Version**]

38. Gomaa, N.H.; Montesinos-Navarro, A.; Alonso-Blanco, C.; Picó, F.X. Temporal variation in genetic diversity and effective population size of Mediterranean and subalpine Arabidopsis thaliana populations. *Mol. Ecol.* **2001**, *20*, 3540–3554. [**Google Scholar**] [**CrossRef**] [**PubMed**]

39. Tabas-Madrid, D.; Méndez-Vigo, B.; Arteaga, N.; Marcer, A.; Pascual-Montano, A.; Weigel, D.; Picó, F.X.; Alonso-Blanco, C. Genome-wide signatures of flowering adaptation to climate temperature: Regional analyses in a highly diverse native range of *Arabidopsis thaliana. Plant Cell Environ.* **2018**, *41*, 1806–1820. [**Google Scholar**] [**CrossRef**] [**PubMed**]

40. Manzano-Piedras, E.; Marcer, A.; Alonso-Blanco, C.; Picó, F.X. Deciphering the adjustment between environment and life history in annuals: Lessons from a geographically-explicit approach in *Arabidopsis thaliana. PLoS ONE* **2014**, *9*, e87836. [**Google Scholar**] [**CrossRef**] [**PubMed**] [**Green Version**]

41. Vidigal, D.S.; Marques, A.C.; Willems, L.A.; Buijs, G.; Méndez-Vigo, B.; Hilhorst, H.W.; Bentsink, L.; Picó, F.X.; Alonso-Blanco, C. Altitudinal and climatic associations of seed dormancy and flowering traits evidence adaptation of annual life cycle timing in *Arabidopsis thaliana. Plant Cell Environ.* **2016**, *39*, 1737–1748. [**Google Scholar**] [**CrossRef**] [**Green Version**]

42. Rizzo, T.M.; Palukaitis, P. Construction of full-length cDNA clones of cucumber mosaic virus RNAs 1, 2 and 3: Generation of infectious RNA transcripts. *Mol. Gen. Genet.* **1990**, *222*, 249–256. [**Google Scholar**] [**CrossRef**]

43. Boyes, D.C.; Zayed, A.M.; Ascenzi, R.; McCaskill, A.J.; Hoffman, N.E.; Davis, K.R.; Görlach, J. Growth stage–based phenotypic analysis of Arabidopsis: A model for high throughput functional genomics in plants. *Plant Cell* **2001**, *13*, 1499–1510. [**Google Scholar**] [**CrossRef**] [**PubMed**] [**Green Version**]

44. Lynch, M.; Walsh, B. *Genetics and Analysis of Quantitative Traits*; Sinauer Associates Inc.: Sunderland, MA, USA, 1998. [**Google Scholar**]

45. Bates, D.; Mächler, M.; Bolker, B.; Walker, S. Fitting Linear Mixed-Effects Models using lme4. *J. Stat. Soft.* **2015**, *67*, 1–48. [**Google Scholar**] [**CrossRef**]

46. R Core Team. *R: A Language and Environment for Statistical Computing*; R Foundation for Statistical Computing: Vienna, Austria, 2018. [**Google Scholar**]

47. Tang, Y.; Liu, X.; Wang, J.; Li, M.; Wang, Q.; Tian, F.; Su, Z.; Pan, Y.; Liu, D.; Lipka, A.E.; et al. GAPIT Version 2: An enhanced integrated tool for genomic association and prediction. *Plant Genome* **2016**, *9*, 2. [**Google Scholar**] [**CrossRef**] [**PubMed**] [**Green Version**]

48. Liu, X.; Huang, M.; Fan, B.; Buckler, E.S.; Zhang, Z. Iterative usage of fixed and random effect models for powerful and efficient genome-wide association studies. *PLoS Genet.* **2016**, *12*, e1005767. [**Google Scholar**] [**CrossRef**]

49. Yin, L.; Zhang, H.; Tang, Z.; Xu, J.; Yin, D.; Zhang, Z.; Yuan, X.; Zhu, M.; Zhao, Z.; Li, X.; et al. rMVP: Memory-Efficient, Visualize-Enhanced, Parallel-Accelerated GWAS Tool. R Package Version 1.0.4. 2020. Available online: **https://CRAN.R-project.org/package=rMVP** (accessed on 26 February 2021).

50. Van Rooijen, R.; Aarts, M.G.M.; Harbinson, J. Natural genetic variation for acclimation of photosynthetic light use efficiency to growth irradiance in Arabidopsis. *Plant Physiol.* **2015**, *167*, 1412–1484. [**Google Scholar**] [**CrossRef**] [**PubMed**] [**Green Version**]

51. Thoen, M.P.M.; Olivas, N.H.D.; Kloth, K.J.; Coolen, S.; Huang, P.-P.; Aarts, M.G.M.; Bac-Molenaar, J.A.; Bakker, J.; Bouwmeester, H.J.; Broekgaarden, C.; et al. Genetic architecture of plant stress resistance: Multi-trait genome-wide association mapping. *New Phytol.* **2017**, *213*, 1346–1362. [**Google Scholar**] [**CrossRef**] [**Green Version**]

52. Rubio, B.; Cosson, P.; Caballero, M.; Revers, F.; Bergelson, J.; Roux, F.; Schurdi-Levraud, V. Genome-wide association study reveals new loci involved in Arabidopsis thaliana and *Turnip mosaic virus* (TuMV) interactions in the field. *New Phytol.* **2019**, *221*, 2026–2038. [**Google Scholar**] [**CrossRef**] [**PubMed**]

53. Cingolani, P.; Platts, A.; Wangle, L.; Coon, M.; Nguyen, T.; Wang, L.; Land, S.J.; Lu, X.; Ruden, D.M. A program for annotating and predicting the effects of single nucleotide polymorphisms, SnpEff. *Fly* **2012**, *6*, 80–92. [**Google Scholar**] [**CrossRef**] [**Green Version**]

54. Berardini, T.Z.; Reiser, L.; Li, D.; Mezheritsky, Y.; Muller, R.; Strait, E.; Huala, E. The arabidopsis information resource: Making and mining the "gold standard" annotated reference plant genome. *Genesis* **2015**, *53*, 474–485. [**Google Scholar**] [**CrossRef**] [**PubMed**] [**Green Version**]

55. Mi, H.; Ebert, D.; Muruganujan, A.; Mills, C.; Albou, L.-P.; Mushayamaha, T.; Thomas, P.D. PANTHER version 16: A revised family classification, tree-based classification tool, enhancer regions and extensive API. *Nucl. Acids Res.* **2021**, *49*, D394–D403. [**Google Scholar**] [**CrossRef**]

56. Supek, F.; Bosnjak, M.; Skunca, N.; Smuc, T. REVIGO summarizes and visualizes long lists of gene ontology terms. *PLoS ONE* **2011**, *6*, e21800. [**Google Scholar**] [**CrossRef**] [**PubMed**] [**Green Version**]

57. Zhou, X.; Carbonetto, P.; Stephens, M. Polygenic modeling with Bayesian Sparse Linear Mixed Models. *PLoS Genet.* **2013**, *9*, e1003264. [**Google Scholar**] [**CrossRef**] [**Green Version**]

58. Bradbury, P.J.; Zhang, Z.; Kroon, D.E.; Casstevens, T.M.; Ramdoss, Y.; Buckler, E.S. TASSEL: Software for association mapping of complex traits in diverse samples. *Bioinformatics* **2007**, *23*, 2633–2635. [**Google Scholar**] [**CrossRef**] [**PubMed**]

59. Liaw, A.; Wiener, M. Classification and Regression by randomForest. *R News* **2002**, *2*, 18–22. [**Google Scholar**]

60. Kuhn, M. Caret: Classification and Regression Training. R Package Version 6.0-86. 2020. Available online: **https://CRAN.R-project.org/package=caret** (accessed on 26 February 2021).
61. Janitza, S.; Celik, E.; Boulesteix, A. A computationally fast variable importance test for random forests for high-dimensional data. *Adv. Data Anal. Classif.* **2018**, *12*, 885–915. [**Google Scholar**] [**CrossRef**]
62. Szymczak, S.; Holzinger, E.; Dasgupta, A.; Malley, J.D.; Molloy, A.M.; Mills, J.L.; Brody, L.C.; Stambolian, D.; Bailey-Wilson, J.E. r2VIM: A new variable selection method for random forests in genome-wide association studies. *BioData Min.* **2016**, *9*, 1–15. [**Google Scholar**] [**CrossRef**] [**PubMed**] [**Green Version**]
63. Bonin, C.P.; Freshour, G.; Hahn, M.G.; Vanzin, G.F.; Reiter, W.-D. The *GMD1* and *GMD2* genes of Arabidopsis encode isoforms of GDP-D-mannose 4,6-dehydratase with cell type-specific expression patterns. *Plant Physiol.* **2003**, *132*, 883–892. [**Google Scholar**] [**CrossRef**] [**Green Version**]
64. Sun, L.; van Nocker, S. Analysis of promoter activity of members of the *PECTATE LYASE-LIKE* (*PLL*) gene family in cell separation in Arabidopsis. *BMC Plant Biol.* **2010**, *10*, 152. [**Google Scholar**] [**CrossRef**] [**Green Version**]
65. Ciftci-Yilmaza, S.; Mittler, R. The zinc finger network of plants. *Cell Mol. Life Sci.* **2008**, *65*, 1150–1160. [**Google Scholar**] [**CrossRef**]
66. Waters, E.R.; Vierling, E. Plant small heat shock proteins—Evolutionary and functional diversity. *New Phytol.* **2020**, *227*, 24–37. [**Google Scholar**] [**CrossRef**] [**PubMed**] [**Green Version**]
67. Yin, M. Genetic Dissection of Nitric Oxide Signalling Network in Plant Defence Response. Ph.D. Thesis, The University of Edinburgh, Edinburgh, UK, 2014. Available online: **http://hdl.handle.net/1842/10462** (accessed on 26 February 2021).
68. Li, J.; Xiang, C.-Y.; Yang, J.; Chen, P.-P.; Zhang, H.-M. Interaction of HSP20 with a viral RdRp changes its sub-cellular localization and distribution pattern in plants. *Sci. Rep.* **2015**, *5*, 14016. [**Google Scholar**] [**CrossRef**] [**PubMed**] [**Green Version**]
69. Knoth, C.; Eulgem, T. The oomycete response gene *LURP1* is required for defense against *Hyaloperonosporaparasitica* in *Arabidopsis thaliana*. *Plant J.* **2008**, *55*, 53–64. [**Google Scholar**] [**CrossRef**] [**PubMed**]
70. Davin, N.; Edger, P.P.; Hefer, C.A.; Mizrachi, E.; Schuetz, M.; Smets, E.; Myburg, A.A.; Douglas, C.J.; Schranz, M.E.; Lens, F. Functional network analysis of genes differentially expressed during xylogenesis in soc1ful woody Arabidopsis plants. *Plant J.* **2016**, *86*, 376–390. [**Google Scholar**] [**CrossRef**] [**PubMed**] [**Green Version**]
71. Kraft, E.; Bostick, M.; Jacobsen, S.E.; Callis, J. ORTH/VIM proteins that regulate DNA methylation are functional ubiquitin E3 ligases. *Plant J.* **2008**, *56*, 704–715. [**Google Scholar**] [**CrossRef**] [**Green Version**]
72. Wang, Y.; Sun, T.; Li, T.; Wang, M.; Yang, G.; He, G. A CBL-interacting protein kinase TaCIPK2 confers drought tolerance in transgenic tobacco plants through regulating the stomatal movement. *PLoS ONE* **2016**, *11*, e0167962. [**Google Scholar**] [**CrossRef**] [**PubMed**]

73. Taylor-Teeples, M.; Lin, L.; de Lucas, M.; Turco, G.; Toal, T.W.; Gaudinier, A.; Young, N.F.; Trabucco, G.M.; Veling, M.T.; Lamothe, R.; et al. An Arabidopsis gene regulatory network for secondary cell wall synthesis. *Nature* **2015**, *517*, 571–575. [**Google Scholar**] [**CrossRef**] [**PubMed**] [**Green Version**]

74. Bartoli, C.; Roux, F. Genome-Wide Association Studies in plant pathosystems: Toward an ecological genomics approach. *Front. Plant Sci.* **2017**, *8*, 763. [**Google Scholar**] [**CrossRef**] [**PubMed**] [**Green Version**]

75. Madden, L.V.; Jeger, M.J.; van den Bosch, F. A theoretical assessment of the effects of vector-virus transmission mechanism on plant virus disease epidemics. *Phytopathology* **2000**, *90*, 576–594. [**Google Scholar**] [**CrossRef**] [**Green Version**]

76. Hily, J.M.; García, A.; Moreno, A.; Plaza, M.; Wilkinson, M.D.; Fereres, A.; Fraile, A.; García-Arenal, F. The relationship between host lifespan and pathogen reservoir potential: An analysis in the system *Arabidopsis thaliana*-Cucumber mosaic virus. *PLoSPathog.* **2014**, *10*, e1004492. [**Google Scholar**] [**CrossRef**]

77. Pagán, I. Movement between plants: Vertical transmission. In *Cucumber Mosaic Virus*; Palukaitis, P., García-Arenal, F., Eds.; APS Press: Washington, DC, USA, 2019; pp. 185–198. [**Google Scholar**]

78. Yin, M.; Wang, Y.; Zhang, L.; Li, J.; Quan, W.; Yang, L.; Wang, Q.; Chan, Z. The Arabidopsis Cys2/His2 zinc finger transcription factor ZAT18 is a positive regulator of plant tolerance to drought stress. *J. Exp. Bot.* **2017**, *68*, 2991–3005. [**Google Scholar**] [**CrossRef**] [**PubMed**]

79. Sewelam, N.; Kazan, K.; Hüdig, M.; Maurino, V.G.; Schenk, P.M. The *AtHSP17.4C1* gene expression is mediated by diverse signals that link biotic and abiotic stress factors with ROS and can be a useful molecular marker for oxidative stress. *Int. J. Mol. Sci.* **2019**, *20*, 3201. [**Google Scholar**] [**CrossRef**] [**PubMed**] [**Green Version**]

80. Haq, S.; Khan, A.; Ali, M.; Khattak, A.M.; Gai, W.-X.; Zhang, H.-X.; Wei, A.-M.; Gong, Z.-H. Heat shock proteins: Dynamic biomolecules to counter plant biotic and abiotic stresses. *Int. J. Mol. Sci.* **2019**, *20*, 5321. [**Google Scholar**] [**CrossRef**] [**Green Version**]

81. Senthil, G.; Liu, H.; Puram, V.G.; Clark, A.; Stromberg, A.; Goodin, M.M. Specific and common changes in *Nicotiana benthamiana* gene expression in response to infection by enveloped viruses. *J. Gen. Virol.* **2005**, *86*, 2615–2625. [**Google Scholar**] [**CrossRef**] [**PubMed**] [**Green Version**]

82. Mitsuya, Y.; Takahashi, Y.; Berberich, T.; Miyazaki, A.; Matsumura, H.; Takahashi, H.; Terauchi, R.; Kusano, T. Spermine signaling plays a significant role in the defense response of *Arabidopsis thaliana* to cucumber mosaic virus. *J. Plant Physiol.* **2009**, *166*, 626–643. [**Google Scholar**] [**CrossRef**] [**PubMed**]

83. Takahashi, H.; Miller, J.; Nozaki, Y.; Takeda, M.; Shah, J.; Hase, S.; Ikegami, M.; Ehara, Y.; Dinesh-Kumar, S.P.; Sukamto. *RCY1*, an *Arabidopsis thaliana RPP8/HRT* family resistance gene, conferring resistance to cucumber mosaic virus requires salicylic acid, ethylene and a novel signal transduction mechanism. *Plant J.* **2002**, *32*, 655–667. [**Google Scholar**] [**CrossRef**] [**PubMed**]

84. Herlihy, J.; Ludwig, N.R.; van den Ackerveken, G.; McDowell, J.M. Oomycetes used in Arabidopsis research. In *The Arabidopsis Book*; BioOne: Washington, DC, USA, 2019; pp. 1–26. [**Google Scholar**]

85. Bellincampi, D.; Cervone, F.; Lionetti, V. Plant cell wall dynamics and wall-related susceptibility in plant–pathogen interactions. *Front. Plant Sci.* **2014**, *5*, 228. [**Google Scholar**] [**CrossRef**] [**PubMed**] [**Green Version**]

86. Bethke, G.; Thao, A.; Xiong, G.; Li, B.; Soltis, N.E.; Hatsugai, N.; Hillmer, R.A.; Katagiri, F.; Kliebenstein, D.J.; Pauly, M.; et al. Pectin biosynthesis is critical for cell wall integrity and immunity in *Arabidopsis thaliana. Plant Cell* **2016**, *28*, 537–556. [**Google Scholar**] [**CrossRef**] [**PubMed**] [**Green Version**]

87. Chen, M.-H.; Sheng, J.; Hind, G.; Handa, A.K.; Citovsky, V. Interaction between the tobacco mosaic virus movement protein and host cell pectin methylesterases is required for viral cell-to-cell movement. *EMBO J.* **2000**, *5*, 913–920. [**Google Scholar**] [**CrossRef**] [**PubMed**]

88. Lionetti, V.; Raiola, A.; Cervone, F.; Bellincampi, D. Transgenic expression of pectin methylesterase inhibitors limits tobamovirus spread in tobacco and Arabidopsis. *Mol. Plant. Pathol.* **2014**, *15*, 265–274. [**Google Scholar**] [**CrossRef**]

89. Otulak-Kozieł, K.; Kozieł, E.; Lockhart, B.E.L. Plant cell wall dynamics incompatible and incompatible potato response to infection caused by *Potato Virus Y* (PVY^{NTN}). *Int. J. Mol. Sci.* **2018**, *19*, 862. [**Google Scholar**] [**CrossRef**] [**Green Version**]

90. Vlad, F.; Spano, T.; Vlad, D.; Bou Daher, F.; Ouelhadj, A.; Kalaitzis, P. Arabidopsis prolyl 4-hydroxylases are differentially expressed in response to hypoxia, anoxia and mechanical wounding. *Physiol. Plant.* **2007**, *130*, 471–483. [**Google Scholar**] [**CrossRef**]

91. Hily, J.M.; Poulicard, N.; Mora, M.A.; Pagán, I.; García-Arenal, F. Environment and host genotype determine the outcome of a plant-virus interaction: From antagonism to mutualism. *New Phytol.* **2016**, *209*, 812–822. [**Google Scholar**] [**CrossRef**] [**PubMed**]

92. Liu, S.; Yu, Y.; Ruan, Y.; Meyer, D.; Wolff, M.; Xu, L.; Wang, N.; Steinmetz, A.; Shen, W.-H. Plant SET- and RING-associated domain proteins in heterochromatinization. *Plant J.* **2007**, *52*, 914–926. [**Google Scholar**] [**CrossRef**] [**PubMed**]

93. Baulcome, D.C.; Dean, C. Epigenetic regulation in plant responses to the environment. *Cold Spring Harb. Perspect. Biol.* **2014**, *6*, a019471. [**Google Scholar**] [**CrossRef**] [**PubMed**]

94. Cao, M.; Du, P.; Wang, X.; Yu, Y.Q.; Qiu, Y.H.; Li, W.; Gal-On, A.; Zhou, C.; Li, Y.; Ding, S.W. Virus infection triggers widespread silencing of host genes by a distinct class of endogenous siRNAs in Arabidopsis. *Proc. Natl. Acad. Sci. USA* **2014**, *111*, 14613. [**Google Scholar**] [**CrossRef**] [**Green Version**]

95. Sardar, A.; Nandi, A.K.; Chattopadhyay, D. CBL-interacting protein kinase 6 negatively regulates immune response to *Pseudomonas syringae* in Arabidopsis. *J. Exp. Bot.* **2017**, *68*, 3573–3584. [**Google Scholar**] [**CrossRef**] [**PubMed**] [**Green Version**]

96. Klepikova, A.V.; Kasianov, A.S.; Gerasimov, E.S.; Logacheva, M.D.; Penin, A.A. A high-resolution map of the *Arabidopsis thaliana* developmental transcriptome based

on RNA-seq profiling. *Plant J.* **2016**, *88*, 1058–1070. [**Google Scholar**] [**CrossRef**] [**PubMed**]

97. Dekkers, B.J.; Pearce, S.; van Bolderen-Veldkamp, R.P.; Marshall, A.; Widera, P.; Gilbert, J.; Drost, H.G.; Bassel, G.W.; Müller, K.; King, J.R.; et al. Transcriptional dynamics of two seed compartments with opposing roles in Arabidopsis seed germination. *Plant. Physiol.* **2013**, *163*, 205–215. [**Google Scholar**] [**CrossRef**] [**PubMed**] [**Green Version**]
98. Jia, T.; Zhang, B.; You, C.; Zhang, Y.; Zeng, L.; Li, S.; Johnson, K.C.M.; Yu, B.; Li, X.; Chen, X. The Arabidopsis MOS4-Associated Complex promotes microRNA biogenesis and precursor messenger RNA splicing. *Plant Cell* **2017**, *29*, 2626–2643. [**Google Scholar**] [**CrossRef**] [**PubMed**] [**Green Version**]
99. Monaghan, J.; Xu, F.; Xu, S.; Zhang, Y.; Li, X. Two putative RNA-binding proteins function with unequal genetic redundancy in the MOS4-associated complex. *Plant. Physiol.* **2010**, *154*, 1783–1793. [**Google Scholar**] [**CrossRef**] [**PubMed**] [**Green Version**]

Chapter-6

Yield Losses Caused by Barley Yellow Dwarf Virus-Pav Infection in Wheat and Barley: a Three-Year Field Study in South-Eastern Australia

1. Introduction

Cereals, a staple food in many parts of the world, are continually threatened by abiotic (e.g., temperature, water, and nutrition stress) and biotic (e.g., weeds, pests, and diseases) factors. It has been estimated that plant diseases cost the global economy approximately USD 220 billion each year [**1**]. Diseases caused by viruses result in significant economic losses worldwide through crop failure and yield and quality losses, as well as increased input costs associated with disease management and prevention [**2**]. Barley, cereal, and maize yellow dwarf viruses (referred to collectively throughout this manuscript as YDVs) belong to the family *Luteoviridae* and are among the most widespread and important viruses affecting cereals worldwide. They commonly infect wheat (*Triticum aestivum*), barley (*Hordeum vulgare*), oat (*Avena sativa*), and other species belonging to the family *Poaceae*. Currently, ten barley yellow dwarf (BYDV), cereal yellow dwarf (CYDV), and maize yellow dwarf virus (MYDV) species are listed on the ICTV master species list: BYDV-kerII, BYDV-kerIII, BYDV-MAV, BYDV-PAS, and BYDV-PAV have been assigned to the *Luteovirus* genus; CYDV-RPS, CYDV-RPV, and MYDV-RMV have been assigned to the *Polerovirus* genus; and BYDV-GPV and BYDV-SGV have not been assigned to a genus [**3**]. BYDV-PAV, BYDV-MAV, CYDV-RPV, and MYDV-RMV have been found in Australia [**4,5**], where BYDV-PAV is considered the most abundant YDV species, particularly in the South-Eastern Australian state of Victoria [**4,6,7,8,9**]. YDVs are phloem-limited viruses which are persistently transmitted from infected to healthy plants by aphids [**10,11**]. The bird cherry-oat aphid (*Rhopalosiphumpadi*) and the corn aphid (*Rhopalosiphummaidis*) are the most common vectors of YDVs in Australia [**12**].

Symptoms of YDV infection include yellowing and/or reddening of leaves, stunted growth, and reduced root biomass [**11**], however infection can also be symptomless. Numerous studies have reported significant yield losses due to YDV infection [**13,14,15,16,17,18,19**], and losses of up to 93% have been reported in field experiments after artificial inoculation

[**18,20**]. In Australia, cereals such as wheat, barley, and oats are widely grown, and yield losses of up to 72 kg/ha have been reported for each 1% increase in virus incidence (percentage of plants infected), resulting in losses of up to 3790 kg/ha [**17,19,21**]. Symptoms and yield losses associated with YDV infection are usually more severe when plants are infected at the early growth stages [**11,22,23,24**]. The impacts of YDV infection can vary depending on factors such as virus species, host cultivar, plant growth stage at the time of infection, aphid vector, and environmental conditions [**13,15,24,25,26**].

Despite the implementation of the latest disease management strategies and an increased use of insecticides to control virus vectors over the past thirty years, YDVs still occur with high incidence in cereal fields in South-Eastern Australia, particularly in higher-rainfall regions [**7,9**]. Furthermore, a recent study found that YDVs were more prevalent, and occurred with higher incidence, in cereals in the region throughout 2014–2017 [**7**] than had previously been reported more than thirty years earlier [**8**]. Therefore, it is likely that yield losses associated with YDVs in South-Eastern Australia have been underestimated in recent years due to a lack of current yield loss data.

Four field experiments were conducted over three years with varying seasonal conditions to quantify the effects of BYDV-PAV, the most prevalent YDV species found in cereals in South-Eastern Australia, on yield and yield components of wheat and barley that are currently grown in the region. This study aimed to provide current yield loss data that can be used to obtain a more accurate and updated estimation of the impact of YDVs in South-Eastern Australia.

2. Materials and Methods

2.1. Field Sites and Experiments

Four field experiments to quantify the effects of BYDV-PAV infection on plant growth, yield, and yield components of field-cultivated wheat and barley under typical conditions were conducted at two sites in the Wimmera region, Victoria in South-Eastern Australia. Experiments 1 (2015), 2 (2015), and 3 (2017) were conducted at the Agriculture Victoria Plant Breeding Centre at Vectis (36°44′ S, 142°6′ E). Experiment 4 (2018) was conducted at the Agriculture Victoria Wimmera Research Station at Longerenong (36°40′ S, 142°18′ E). The Vectis and Longerenong field sites are both located in the Wimmera region and are approximately 18 km apart. The long-term (1990–2018) mean annual maximum temperature and mean annual rainfall of the Wimmera region were 21.6 °C and 404 mm, respectively (**www.bom.gov.au**; **www.longpaddock.qld.gov.au/silo**, accessed on 5 February 2021) [**27**]. Each experiment was designed using a randomized block design with 6 replicates (**Figure 1**A,B) and was direct-seeded using a cone seeder (PJ green, Grovedale, Australia) with 183 mm row spacing and a target establishment density of 150 plants/m^2. The wheat cultivar Yitpi was evaluated in experiment 1, the barley cultivar Hindmarsh was evaluated in experiment 2, while the wheat cultivar Mace was evaluated in experiments 3 and 4. Yitpi, a commonly grown cultivar in the Wimmera region, has recently been replaced by Mace. Field sites were maintained using agronomic practices typical for the region.

Figure 1. (A) The experimental layout in wheat and barley; (B) Field inoculation cages used to cover the virus-treated plots during inoculation; (C) Widespread leaf-yellowing symptoms of barley yellow dwarf virus (BYDV) background infection observed in non-inoculated control plots of wheat in 2017 in South-Eastern Australia.

2.2. Virus Propagation and Aphid Colony

The isolate of BYDV-PAV used to inoculate the virus-treated plots in each field experiment was collected from an infected oat plant in Horsham, Victoria, Australia with the virus identity confirmed by tissue blot immunoassay (TBIA) [**9**] and reverse transcription-polymerase chain reaction (RT-PCR) [**28**]. Viruliferous *R. padi* were maintained on infected wheat plants (cv. Yitpi) contained in cages in plant growth chambers at 20 °C with a 14:10 h light:dark photoperiod. Aphids were allowed a virus acquisition period of at least 7 days before they were used to inoculate plants in the field experiments.

2.3. Inoculation of Virus-Infected Plots with BYDV-PAV

In experiments 1 and 2, the three treatments were: (1) early BYDV inoculation (BYDV 1, inoculated at the seedling stage, Zadoks growth stage Z12 where two leaves had emerged) [**29**]; (2) later BYDV inoculation (BYDV 2, inoculated at the early tillering stage, Z22, where two tillers were visible); and (3) a non-inoculated control. In experiments 3 and 4, the two treatments were early BYDV inoculation (inoculated at the seedling stage, Z12) and a non-inoculated control. In each experiment, treatments were applied to plots 180 cm × 3 rows (1.62 m^2) in size. Plots of wheat and/or barley selected for BYDV-PAV infection were inoculated with the virus by placing plant sections containing viruliferous *R. padi* alongside each row of plants within the plot. All plants within the plot were then covered with a large field cage (**Figure 1**B) for 3–5 days to contain the aphids and prevent virus contamination of control plots. Plants within each inoculated and control plot were sprayed with pyrethrum (Yates, active ingredient: pyrethrins) and Confidor (Bayer, active ingredient: imidacloprid) insecticides immediately after the cages were removed. All plots were then sprayed with insecticide as part of the normal spray program throughout the growing season.

2.4. Assessment of BYDV-PAV Incidence

In each experiment, 45–60 whole tillers were randomly collected from individual plants from each plot before maturity and tested for BYDV-PAV to assess inoculation success and levels of background infection using TBIA [**9**]. The number of positive tillers in the samples collected from each plot was recorded and the percentage of positive tillers was calculated.

2.5. Harvest Assessments

At plant maturity, the height of 12–15 plants in each plot was measured, recorded, and averaged. The above-ground portion of all plants from each plot was then hand harvested, placed into large paper bags, and transported to the laboratory, where plant biomass was measured. Samples were threshed using a Hans-Ulrich Hege 16 laboratory thresher (Wintersteiger, Ried imInnkreis, Austria). Grain was then aspirated with a vacuum separator (Kimseed, Wangara, Australia), counted using a Numigral seed counter (CHOPIN Technologies, Cedex, France), weighed, and 1000-grain weight was calculated. Grain size was assessed by passing each grain sample through 2.8 mm, 2.5 mm, and 2.2 mm sieves using a Sortimat laboratory sorting machine (Pfeuffer GmbH, Kitzingen, Germany) and calculating the percentage of grain in each range. Harvest index was calculated by dividing the grain yield (g) by plant biomass (g). In experiments 3 and 4, grain protein (%) and moisture content (%) were measured using a CropScan 3000B whole-grain analyzer (Next Instruments, Condell Park, Australia). Additionally, the number of heads in each plot was counted in experiments 3 and 4, then the total grain weight per head and the number of grains per head were calculated.

2.6. Weather Data

All rainfall and temperature data used to represent the Wimmera region, and therefore the Vectis and Longerenong field sites, were obtained from the Bureau of Meteorology (BOM) website (**www.bom.gov.au**) and SILO (**www.longpaddock.qld.gov.au/silo**) [**27**] from weather station number 79,028 (accessed on 5 February 2021), which was located in the paddock adjacent to the Longerenong field site. Long-term rainfall and temperature values were calculated using all available data from 1961–2018 while average annual temperature and rainfall data were used to demonstrate the variation in weather conditions in the Wimmera region in each year of the field study (**Table 1**).

Table 1. Three-monthly annual rainfall (mm) and mean maximum temperature (°C) for the years 2014–2018 and three-monthly long-term mean rainfall (mm) and long-term mean maximum temperature (°C) for the years 1961–2018 for the Wimmera region, South-Eastern Australia.

	Year	Jan–Mar	Apr–Jun	Jul–Sep	Oct–Dec
Rainfall (mm)	2014	36	133	64	33
	2015	88	77	60	28
	2016	69	110	222	108
	2017	56	135	128	112
	2018	24	82	79	42
Long-term mean	1961–2018	67	107	132	97
Mean maximum temperature (°C)	2014	30.9	18.4	16.4	27.3
	2015	29.1	17.1	15.4	29.3
	2016	30.2	18.4	14.8	24.0
	2017	30.3	17.9	15.3	27.2
	2018	30.6	19.0	16.0	26.7
Long-term mean	1961–2018	28.9	17.8	15.2	24.5

2.7. Data Analysis

Grain weights from each plot were converted to grain yield (kilograms per hectare, kg/ha) prior to analysis. Yield losses (kg/ha) were calculated by subtracting the mean yield of the inoculated plots from the mean yield of the non-inoculated control plots with percentage yield loss also calculated. Means and standard errors of the means (SEMs) were calculated using GenStat 14th Edition (VSN International, Hemel Hempstead, UK). The normality of residuals was checked using quantile–quantile plots (**Figures S1–S4**). Analysis of variance (ANOVA) and Tukey's honestly significant difference (HSD) test were used to analyze data from experiments 1 and 2. Experiments 3 and 4 consisted of two treatments, therefore paired, two-sided *t*-tests were used to analyze data from experiments 3 and 4 instead of ANOVA. The relationship between the natural BYDV-PAV background infection present in the non-inoculated control plots (x) and grain yield (y) in experiment 3 was tested for normality using the Shapiro–Wilk test and analyzed using polynomial linear regression and Pearson's product moment correlation. Normality, ANOVA, *t*-tests, linear regression analysis, and correlations were performed using R version 4.0.0 (R Foundation for Statistical Computing, Vienna, Austria), and differences were considered statistically significant at $p < 0.05$. Figures were produced in SigmaPlot (Systat Software, San Jose, CA, USA).

3. Results

During the three years of this study, four individual field experiments were conducted to quantify yield losses associated with BYDV-PAV infection under varying seasonal conditions in the Wimmera region, South-Eastern Australia. Weather conditions varied between the 3 years of the study (**Table 1**), resulting in different rainfall, grain yield, green bridge, and background virus infection (**Figure 1**C) each year, and higher levels of BYDV-PAV background infection were observed in non-inoculated control plots of wheat in 2017 (22–60%) compared to 2015 (4–19%) and 2018 (3–17%).

3.1. Experiment 1 (2015); Wheat (cv. Yitpi)

When comparing the early and later BYDV-PAV-inoculated treatments to the non-inoculated control treatment, plant height was significantly reduced by both early (17%) and later (14%) infection (**Figure 2**A,B). Plant biomass was significantly reduced by 50% by early infection and 39% by later infection (**Figure 2**C). Grain yield was significantly reduced by both early and later infection; early infection reduced grain yield by 84% (1358 kg/ha) while later infection reduced grain yield by 75% (1214 kg/ha) (**Figure 2**D). Grain number was also significantly reduced by 84% ($p < 0.001$) by early infection and by %74 ($p < 0.001$) by later infection, while harvest index was reduced by %69 ($p < 0.001$) by early infection and by %60 ($p < 0.001$) by later infection. Thousand-grain weight was not significantly affected by either early ($p = 0.37$) or later ($p = 0.17$) infection; similarly, grain size (measured as the proportion of grain in each of the >2.8 mm, 2.8–2.5 mm, 2.5–2.2 mm, and <2.2 mm size ranges) was not significantly affected by either early ($p \geq 0.41$) or later ($p \geq 0.19$) infection. When comparing the early and later BYDV-PAV infection treatments to each other, there were no significant differences between early and later infection in any of the parameters measured (i.e., plant height ($p = 0.57$), plant biomass ($p = 0.12$), grain yield ($p = 0.61$), grain number ($p = 0.54$), harvest index ($p =$ -1000 ,(0.72grain weight ($p = 0.87$), or grain size ($p \geq 0.72$)) (**Figure S5 and Table S1**). The typical virus symptom of leaf yellowing/reddening was observed in plots inoculated with BYDV-PAV, and the stunted growth of plants in the inoculated plots was obvious at harvest (**Figure 2**A, left). Due to the plants maturing earlier than expected as a result of the hot and dry conditions of 2015, not all of the wheat samples collected to assess BYDV-PAV incidence in each plot contained enough sap for an accurate estimation of virus incidence, however the mean incidence was at least 63% in the inoculated plots and 9% in the non-inoculated control plots, and the majority of plants in the inoculated plots were symptomatic. Additionally, a previous study showed that natural YDV background infection was low in the Wimmera region in 2015 [7]. Rainfall was 37% below average, and the mean maximum temperature was 1.1 °C above the long-term mean (**Table 1**).

3.2. Experiment 2 (2015); Barley (cv. Hindmarsh)

When comparing the early and later BYDV-PAV-inoculated treatments to the non-inoculated control treatment, plant height was not significantly affected by either early or later infection (**Figure 2**A,E), but plant biomass was significantly reduced by 38% by early infection and 31% by later infection (**Figure 2**F). Grain yield was significantly reduced by both early and later infection; early infection reduced grain yield by 60% (1352 kg/ha) while later infection reduced grain yield by 64% (1456 kg/ha) (**Figure 2**G). Grain number was also significantly reduced by 56% ($p = 0.002$) by early infection and %62 ($p <$ 0.001) by later infection, while harvest index was reduced by %35 ($p = 0.051$) by early infection and by %47 ($p = 0.009$) by later infection. While -1000grain weight was not significantly affected by either early ($p = 0.06$) or later ($p = 0.18$) infection, grain size was significantly affected by infection at both times; when compared to the non-inoculated control treatment, significantly more smaller grains (<2.2 mm in size) were obtained after both early (%69, $p = 0.001$) and later (%59, $p = 0.005$) infection, and significantly fewer larger grains (2.5–2.2 mm in size) were obtained after both the early (%47, $p < 0.001$) and later (%38, $p = 0.005$) infection. When comparing the early and later BYDV-PAV inoculation treatments to each other, there were no significant differences between early and later infection in any of the parameters measured (i.e., plant height ($p = 0.82$), plant biomass ($p = 0.70$), grain yield ($p = 0.94$), grain number ($p = 0.92$), harvest index (p = -1000 ,(0.66grain weight ($p = 0.80$), or grain size ($p \geq 0.66$)) (**Figure S6 and Table S2**). The typical virus symptom of leaf yellowing was observed in plots inoculated with BYDV-PAV, and the stunted growth of plants in the inoculated plots was visible at harvest (**Figure 2**A, right). Most of the barley samples collected to assess BYDV-PAV incidence in each plot did not contain enough sap for accurate estimation of virus incidence; however, the mean BYDV-PAV incidence was at least 20% in the inoculated plots and 3% in the non-inoculated control plots, and the majority of plants in the inoculated plots were symptomatic.

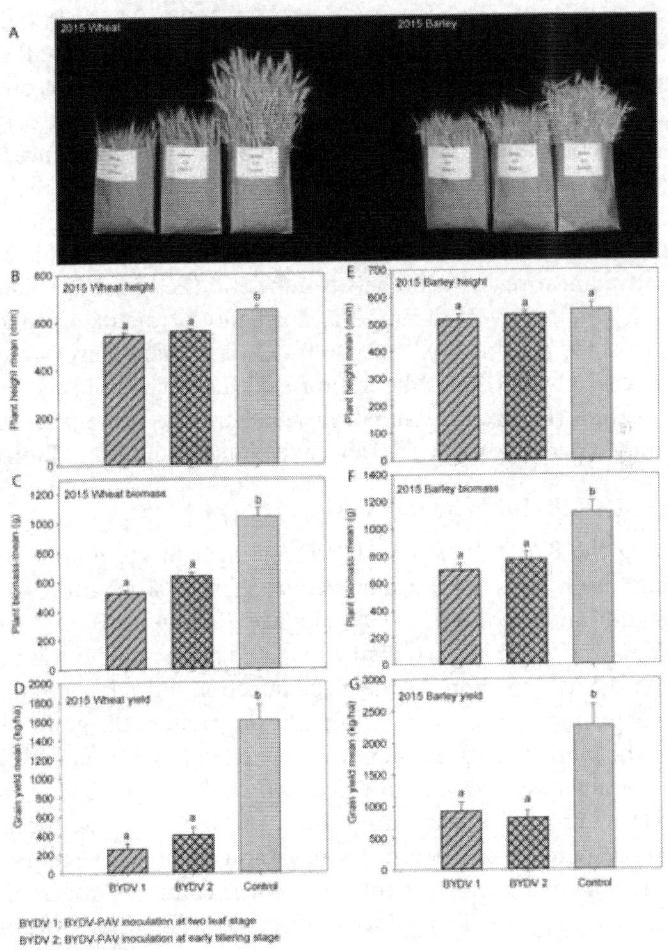

Figure 2. The effect of early (BYDV 1), later (BYDV 2), or no (Control) barley yellow dwarf virus-PAV (BYDV-PAV) inoculation on: (A) plant growth of wheat (left)

and barley (right); (B) plant height; (C) plant biomass; (D) grain yield of wheat in experiment 1 (2015) and (E) plant height; (F) plant biomass; (G) grain yield of barley in experiment 2 (2015) in South-Eastern Australia. Error bars represent standard error; means with the same letter are not significantly different at $p < 0.05$.

3.3. Experiment 3 (2017); Wheat (cv. Mace)

When comparing the BYDV-PAV-inoculated treatment to the non-inoculated control treatment, plant height was significantly reduced by 6% (**Figure 3**A), while plant biomass was significantly reduced by 15% (**Figure 3**B) due to virus infection. Grain yield was reduced by 18% (1038 kg/ha, **Figure 3**C), and grain number was reduced by 15% (p = 0.050) by virus infection, however these differences were not statistically significant at the $p < 0.05$ level. While infection did not significantly affect the number of heads per plot (p = 0.99), it did significantly reduce the grain weight per head (p = 0.04) and the number of grains per head (p = 0.01) by %17 and %15, respectively. There were no significant effects of BYDV-PAV infection on harvest index (p = -1000 ,(0.35grain weight (p = 0.54), grain size ($p \geq 0.08$), grain protein (p = 0.43), or grain moisture (p = 0.20) (**Figure S7 and Table S3**). The typical virus symptom of leaf yellowing/reddening was observed in both inoculated and control plots (**Figure 1**C); however, stunted plant growth in the inoculated plots was not visually obvious at harvest. The mean BYDV-PAV incidence was 89% in the inoculated plots and 37% in the non-inoculated control plots. Linear regression analysis performed to quantify the relationship between grain yield and the unusually high level of natural BYDV-PAV background infection observed in the non-inoculated control plots revealed a negative linear relationship between the two (Pearson's correlation coefficient R = −0.88, p = 0.0197) after confirming the normality of the data using the Shapiro–Wilk normality test (W statistic = 0.9185, p = 0.274, significance level = 0.05) (**Figure 4**). Grain yield decreased by 0.91% (72 kg/ha) for each 1% increase in natural BYDV-PAV background infection (**Figure 4**). Rainfall in Horsham was average in 2017, while the mean maximum temperature was 1.1 °C above the long-term mean (**Table 1**).

3.4. Experiment 4 (2018); Wheat (cv. Mace)

When comparing the BYDV-PAV-inoculated treatment to the non-inoculated control treatment, plant height was significantly reduced by 15% (**Figure 5**A), while plant biomass was significantly reduced by 39% (**Figure 5**B) due to BYDV infection. Grain yield was reduced by 41% (923 kg/ha, **Figure 5**C), and grain number was reduced by 34% (p = 0.01) by BYD-PAV infection. While virus infection did not significantly affect the number of heads per plot (p = 0.15), it did significantly reduce the grain weight per head ($p < 0.001$) and the number of grains per head (p = 0.001) by %32 and %24, respectively. Thousand-grain weight was significantly reduced by %10 ($p < 0.001$) by BYDV-PAV infection. Similarly, grain size was significantly affected by virus infection with %15 more smaller grains (2.8–2.5 mm in size, p = 0.04) and %18 fewer larger grains (>2.8 mm in size, p = 0.04) obtained after virus infection when compared to the non-inoculated control treatment. There were no significant effects of BYDV-PAV infection on harvest index (p = 0.16), grain protein (p = 0.07), or grain moisture (p = 0.81) (**Figure S8 and Table S4**). The typical virus symptom of leaf yellowing/reddening was observed in both inoculated and control plots, however the stunted growth of plants in the inoculated plots was not obvious at harvest. The mean BYDV-PAV incidence was 85% in the inoculated plots and 10% in the non-inoculated control plots. Rainfall in Horsham was 44% below average in 2018 while the mean maximum temperature was 1.5 °C above the long-term mean (**Table 1**).

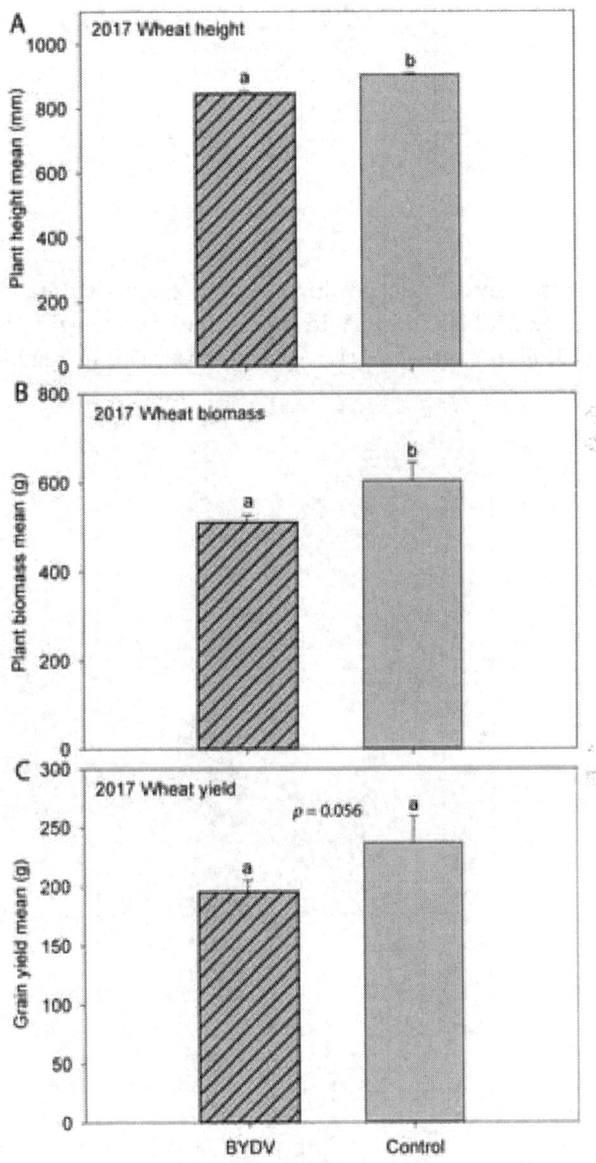

Figure 3. The effect of early (BYDV) or no (Control) BYDV-PAV inoculation on: (A) plant height; (B) plant biomass; (C) grain yield of wheat in experiment 3 (2017) in South-Eastern Australia. Error bars represent standard error; means with the same letter are not significantly different at $p < 0.05$.

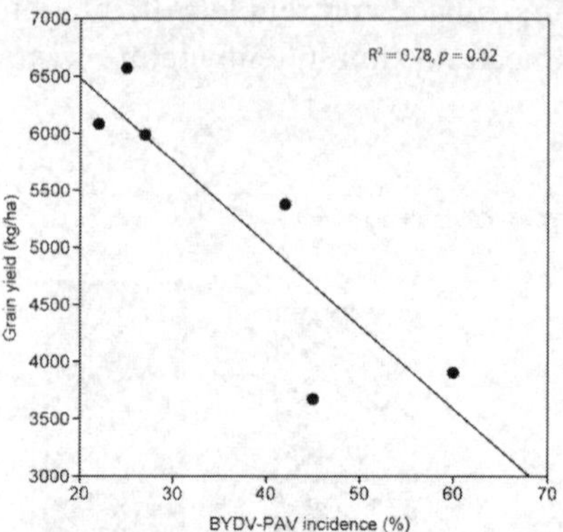

Figure 4. The negative linear relationship between grain yield (kg/ha) and natural background incidence of BYDV-PAV in non-inoculated control plots of wheat in experiment 3 (2017) in South-Eastern Australia revealed by regression analysis.

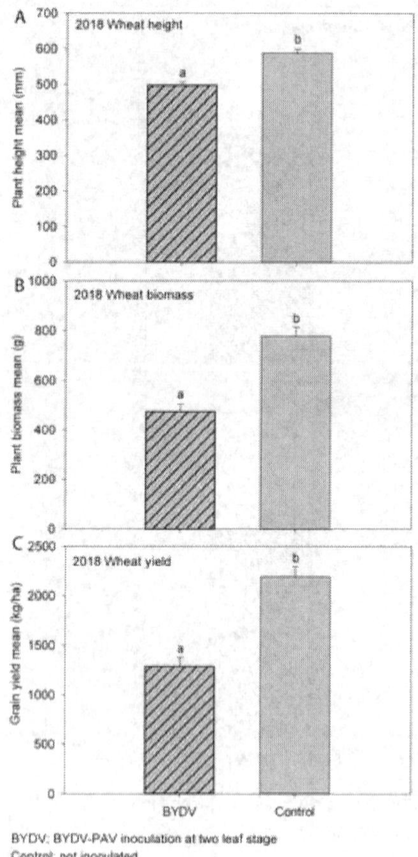

Figure 5. The effect of early (BYDV) or no (Control) BYDV-PAV inoculation on: (A) plant height; (B) plant biomass; (C) grain yield of wheat in experiment 4 (2018) in South-Eastern Australia. Error bars represent standard error; means with the same letter are not significantly different at $p < 0.05$.

4. Discussion

This study quantified yield losses associated with BYDV-PAV infection in cereals grown under field conditions in South-Eastern Australia. In four experiments conducted over three years, we report yield reductions of up to 84% (1358 kg/ha) in wheat and 64% (1456 kg/ha) in barley, along with an estimated yield loss impact factor of 0.91% (72 kg/ha) for each one percent increase in natural BYDV-PAV infection in wheat. Although a previous study of yield losses associated with an isolate of CYDV-RPV in Victoria was published more than 30 years ago [**24**] and showed yield losses of up to 79% as a result of virus infection, our study quantifies yield losses associated with BYDV-PAV, the most prevalent YDV species found in cereal fields in the region [7]. Despite the implementation of the latest control strategies, grain yield and quality losses resulting from virus infection are still common and often devastating [**30**]; however, the impact of viruses on cereal production in South-Eastern Australia is still often underestimated. In some years, YDVs are prevalent and can occur with high incidences in cereals in South-Eastern Australia [7]. Although the losses reported in this study were the result of artificially introduced aphids, they demonstrate the severe yield losses that can result from BYDV-PAV infection in the region, particularly when high numbers of viruliferous aphids are present. Combined with the recently updated YDV incidence data [7], this yield loss information will assist with the provision of a more accurate understanding of the impacts of YDV infection in currently grown cereal cultivars in South-Eastern Australia.

Yield losses due to BYDV-PAV infection were recorded each year of the experiment but were particularly severe and obvious in the hot and dry year of 2015, where virus infection severely reduced already low yields. While severe yield losses such as these have previously been reported in cereals [**13,16,18,20,31,32**], the same severe losses were not observed in our experiments in 2017 or 2018; however, the 41% yield loss recorded in 2018 (experiment 4) was still severe and similar to those reported by others [**14,22,23,24,25**]. The high level of natural BYDV-PAV background infection present in the non-inoculated control plots in wheat in 2017 (experiment 3) is likely to have reduced the difference between the inoculated and non-inoculated treatments, thereby masking the severity of yield losses resulting from BYDV-PAV infection.

Linear regression analysis of grain yield with natural BYDV-PAV background infection detected in the non-inoculated control plots revealed a negative linear relationship between the two, showing a decrease in yield of 0.91% (72 kg/ha) for each 1% increase in natural BYDV-PAV infection; similar negative relationships between yield loss and BYDV incidence were reported by others [**16,17,21,23,24**]. The lowest and highest incidence levels used for regression analysis were 22% and 60%, respectively, so any predictions based on incidence levels outside of this range would be based on extrapolation; also, it is not known if linearity still applies to the relationship when incidences are outside of this range. It is not known when the background infection occurred, and other YDV species may also have been present in the non-inoculated control plots so the incidences used for regression analysis may be a slight underestimation.

The effects of early (inoculated at the seedling stage) and later (inoculated at early tillering) inoculation with BYDV-PAV was assessed in wheat (experiment 1) and barley (experiment 2) in 2015. Significant reductions in yield, grain number, plant biomass, and harvest index were recorded after both early and later infection, in both wheat and barley, when compared to the non-inoculated control plots. Additionally, there was no statistically significant difference between early and later BYDV-PAV infection in grain yield or any other parameter measured (grain number, 1000-grain weight, grain size, plant height, biomass, and harvest index) in either wheat or barley when the two inoculation times were compared to each other, showing that significant yield losses can occur in wheat and barley following both early and later infection. While the majority of BYDV yield loss studies report greater yield losses from early BYDV infection and that yield losses associated with late infection are not significant [**23,24**], other studies have reported lower but significant yield losses from late inoculation, particularly in susceptible cultivars [**11,14,16**], while others have reported no significant differences in yield between early and later inoculation [**15,16,22**]. This variation seems to be influenced by factors such as cultivar, virus isolate, time of inoculation, and environmental conditions, among others. In this study, the early inoculation was done at the seedling stage (Z12–13) and the later inoculation was done at early-tillering (Z21–22), while in a number of other studies, late inoculation was done at or after early stem extension (Z30) [**22,24**]. Therefore, our later inoculation was applied at a relatively early growth stage in comparison to some other studies.

In each experiment, yield losses resulting from BYDV-PAV infection were accompanied by similar-sized reductions in grain number per plot while 1000-grain weight and grain size were affected to a lesser extent and differences were only significant in some experiments, indicating that the yield losses were primarily due to the presence of less grain rather than smaller grain. Furthermore, virus infection did not significantly affect the number of heads per plot but did significantly reduce the number of grains per head and the weight of grain from each head. Thousand-grain weight and grain size were not significantly affected by infection in wheat in experiment 1 (despite the severe yield loss observed) or experiment 3, but were significantly affected in experiment 4. In cereals, the number of tillers per plant [**14,24,33**], heads per meter, seeds per head, along with 1000-grain weight [**15,16,31,34**], number of heads with aborted terminal spikelets [**24**], and grain size/quality [**17,25**] have all been reported to be significantly affected by BYDV infection in some hosts and/or cultivars. Although leaf symptoms typical of BYDV infection were observed in each experiment, stunted plant growth was only obvious at harvest in experiments 1 and 2 (**Figure 2**A,E) and was not noticeable at harvest in experiments 3 or 4, despite the 39% reduction in biomass and 41% yield loss recorded in experiment 4. The reduced visibility of the effects of BYDV infection at harvest in experiment 4 suggests that infection at these levels is unlikely to be noticed in the field, which in turn is likely to have contributed to the continued underestimation of the importance of BYDV noted in previous studies [**21**]. It is not clear why the effects of BYDV infection were so much more obvious in wheat in experiment 1 than experiment 4. Plant height and biomass were reduced by a similar amount in each experiment, however harvest index (the ratio of grain yield to above-ground plant biomass) calculations show that BYDV infection affected yield more than plant biomass in experiment 1 but not in experiments 3 or 4. Differences in factors such as growing conditions and cultivar, among others, are likely to have contributed to the difference in harvest index between 2015 and 2018.

The effects of BYDV infection on parameters such as yield, grain number, grain size, grain weight, harvest index, and symptom expression vary between experiments and studies, and can be influenced by several factors. Environmental factors such as soil moisture, rainfall, and temperature, and their effects on plants, vectors, and viruses, are likely to have contributed to the variation in results between studies and experiments. While it has been suggested that plants infected with BYDV and other viruses may be more drought tolerant than non-infected plants [**35,36,37**], the results of this study do not support these findings. While describing BYDV in 1953, Oswald and Houston [**11**] noted that damage caused by BYDV infection was particularly severe in a drought year, and others have also reported more severe yield losses from BYDV infection in years of lower rainfall [**15,25,38,39**]. Given that the root system of a plant infected with BYDV is also often just as stunted, and sometimes even more so, than the visible, above-ground portion of the plant [**11,40,41,42,43**], it has been suggested that the roots of infected plants may be too shallow to reach or obtain adequate water and nutrients in dry conditions [**11**]. This is one possible explanation for the especially severe yield losses observed in experiment 1, as 2015 was the second year in a row of well below average rainfall in the region (**Table 1**).

Differences in factors such as YDV species/isolate, host cultivar, time of infection, and aphid vector species are also likely to have contributed to the variation in results between studies and experiments. For example, when Monneveux et al. [**39**] reported higher yield losses due to BYDV infection in the drought year of 1988 than the average rainfall year of 1987, they also reported that the same severe yield losses were not observed in lines that were tolerant to BYDV. Others have also reported significant differences in the response of different cultivars to BYDV infection, especially between susceptible and resistant cultivars [**13,14,25**]. Given that different wheat cultivars were used in 2015 and 2018, this is also likely to have contributed to the different results obtained in those years. Furthermore, Baltengberger et al. [**13**] reported more severe symptoms and greater yield losses in plants infected with both BYDV-PAV and CYDV-RPV than they did in plants inoculated with one isolate or the other singly. A high level of *Septoria tritici* was also observed in experiment 3 and may also have played a role, as significant interactions between BYDV and other diseases have been reported [**44,45**]. Thus, some of the yield loss reported here may be attributed to other factors that have not been analyzed or captured here, but some of which have been discussed.

In conclusion, the four randomized, replicated field experiments conducted in this study quantified the yield losses associated with BYDV-PAV infection in cereals grown in South-Eastern Australia. The results of these experiments demonstrate the potential for severe yield losses that can result from infection with BYDV-PAV, the most prevalent species of BYDV in South-Eastern Australia, showing yield losses of up to 84% (1358 kg/ha) in wheat and 64% (1456 kg/ha) in barley due to BYDV infection. Additionally, an estimated yield loss impact factor of 0.91% (72 kg/ha) for each one percent increase in natural BYDV-PAV infection was obtained for wheat. Yield losses varied between experiments and years, demonstrating that losses can be influenced by many factors, such as cereal cultivar and environmental factors, and illustrating the importance of conducting these experiments under varying conditions. The results of this study will assist with the provision of more accurate estimates of current yield losses in cereals due to BYDV infection and a more accurate understanding of the importance of BYDV in Victoria and more broadly in South-Eastern Australia.

Supplementary Materials

Figure S1: Quantile–quantile plots to assess normality in experiment 1 (wheat, 2015); Figure S2: Quantile–quantile plots to assess normality in experiment 2 (barley, 2015); Figure S3: Quantile–quantile plots to assess normality in experiment 3 (wheat, 2017); Figure S4: Quantile–quantile plots to assess normality in experiment 4 (wheat, 2018); Figure S5: Percent change associated with early and later BYDV infection in experiment 1 (wheat, 2015); Table S1: *p*-Values calculated according to Student's *t*-tests or ANOVA for experiment 1 (wheat, 2015); Figure S6: Percent change associated with early and later BYDV infection in experiment 2 (barley, 2015); Table S2: *p*-Values calculated according to Student's *t*-tests or ANOVA for experiment 2 (barley, 2015); Figure S7: Percent change associated with early BYDV infection in experiment 3 (wheat, 2017); Table S3: *p*-Values calculated according to Student's *t*-tests or ANOVA for experiment 3 (wheat, 2017); Figure S8: Percent change associated with early BYDV infection in experiment 4 (wheat, 2018); Table S4: *p*-Values calculated according to Student's *t*-tests or ANOVA for experiment 4 (wheat, 2018).

References

1. Food and Agriculture Organization of the United Nations (FAO). Plant Health and Food Security. Available online: **http://www.fao.org/3/a-i7829e.pdf** (accessed on 3 March 2021).
2. Bos, L. Crop losses caused by viruses. *Crop Prot.* **1982**, *1*, 263–282. [**Google Scholar**] [**CrossRef**]
3. International Committee on Taxonomy of Viruses (ICTV). ICTV Master Species List 2019 v1. Checklist Dataset. Available online: **https://talk.ictvonline.org/files/master-species-lists/m/msl/9601**; **https://doi.org/10.15468/i4jnfv** (accessed on 3 March 2021).
4. Sward, R.J.; Lister, R.M. The identity of barley yellow dwarf virus isolates in cereals and grasses from mainland Australia. *Aust. J. Agric. Res.* **1988**, *39*, 375–384. [**Google Scholar**] [**CrossRef**]
5. McKirdy, S.J.; Jones, R.A.C. Occurrence of barley yellow dwarf virus serotypes MAV and RMV in over-summering grasses. *Aust. J. Agric. Res.* **1993**, *44*, 1195–1209. [**Google Scholar**] [**CrossRef**]
6. Jones, R.A.C.; McKirdy, S.J.; Shivas, R.G. Occurrence of barley yellow dwarf viruses in over-summering grasses and cereal crops in Western Australia. *Australas Plant Pathol.* **1990**, *19*, 90–96. [**Google Scholar**] [**CrossRef**]
7. Nancarrow, N.; Aftab, M.; Freeman, A.; Rodoni, B.; Hollaway, G.; Trębicki, P. Prevalence and incidence of yellow dwarf viruses across a climatic gradient: A four-year field study in south eastern Australia. *Plant Dis.* **2018**, *102*, 2465–2472. [**Google Scholar**] [**CrossRef**] [**Green Version**]
8. Sward, R.J.; Lister, R.M. The incidence of barley yellow dwarf viruses in wheat in Victoria. *Aust. J. Agric. Res.* **1987**, *38*, 821–828. [**Google Scholar**] [**CrossRef**]
9. Trębicki, P.; Nancarrow, N.; Bosque-Pérez, N.A.; Rodoni, B.; Aftab, M.; Freeman, A.; Yen, A.; Fitzgerald, G. Virus incidence in wheat increases under elevated CO2: A

4-year field study of yellow dwarf viruses from a free air carbon dioxide facility. *Virus Res.* **2017**, *241*, 137–144. [**Google Scholar**] [**CrossRef**]

10. Gray, S.; Gildow, F.E. Luteovirus-aphid interactions. *Annu. Rev. Phytopathol.* **2003**, *41*, 539–566. [**Google Scholar**] [**CrossRef**]
11. Oswald, J.W.; Houston, R. The yellow dwarf virus disease of cereal crops. *Phytopathology* **1953**, *43*, 128–136. [**Google Scholar**]
12. Thackray, D.J.; Diggle, A.J.; Jones, R.A.C. BYDV PREDICTOR: A simulation model to predict aphid arrival, epidemics of barley yellow dwarf virus and yield losses in wheat crops in a Mediterranean-type environment. *Plant Pathol.* **2009**, *58*, 186–202. [**Google Scholar**] [**CrossRef**]
13. Baltenberger, D.E.; Ohm, H.W.; Foster, J.E. Reactions of oat, barley and wheat to infection with Barley yellow dwarf virus isolates. *Crop Sci.* **1987**, *27*, 195–198. [**Google Scholar**] [**CrossRef**]
14. Choudhury, S.; Larkin, P.; Meinke, H.; Hasanuzzaman, M.D.; Johnson, P.; Zhou, M. Barley yellow dwarf virus infection affects physiology, morphology, grain yield and flour pasting properties of wheat. *Crop Pasture Sci.* **2019**, *70*, 16–25. [**Google Scholar**] [**CrossRef**]
15. Gildow, F.E.; Frank, J.A. Barley yellow dwarf virus in Pennsylvannia: Effect of the PAV isolate on yield components of Noble spring oats. *Plant Dis.* **1988**, *72*, 254–256. [**Google Scholar**] [**CrossRef**]
16. Gill, C.C. Assessment of losses on spring wheat naturally infected with barley yellow dwarf virus. *Plant Dis.* **1980**, *64*, 197–203. [**Google Scholar**] [**CrossRef**]
17. McKirdy, S.J.; Jones, R.A.C.; Nutter, F.W. Quantification of yield losses caused by barley yellow dwarf virus in wheat and oats. *Plant Dis.* **2002**, *86*, 769–773. [**Google Scholar**] [**CrossRef**] [**PubMed**] [**Green Version**]
18. Pike, K.S. A review of barley yellow dwarf virus grain yield losses. In *World Perspectives on Barley Yellow Dwarf Virus*; Burnett, P.A., Ed.; CIMMYT: Texcoco, Mexico, 1990; pp. 356–361. [**Google Scholar**]
19. Thackray, D.J.; Ward, L.T.; Thomas-Carroll, M.L.; Jones, R.A.C. Role of winter-active aphids spreading barley yellow dwarf virus in decreasing wheat yields in a Mediterranean-type environment. *Aust. J. Agric. Res.* **2005**, *56*, 1089–1099. [**Google Scholar**] [**CrossRef**]
20. Banttari, E.E. Occurrence of aster yellows in barley in the field and its comparison with barley yellow dwarf. *Phytopathology* **1965**, *55*, 838–843. [**Google Scholar**]
21. Banks, P.M.; Davidson, J.L.; Bariana, H.; Larkin, P.J. Effects of barley yellow dwarf virus on the yield of winter wheat. *Aust. J. Agric. Res.* **1995**, *45*, 935–946. [**Google Scholar**] [**CrossRef**]
22. El Yamani, M.; Hill, J.H. Identification and importance of barley yellow dwarf virus in Morocco. *Plant Dis.* **1990**, *74*, 291 294. [**Google Scholar**] [**CrossRef**]

23. Perry, K.L.; Kolb, F.L.; Sammons, B.; Lawson, C.; Cisar, G.; Ohm, H. Yield effects of barley yellow dwarf virus in soft red winter wheat. *Phytopathology* **2000**, *90*, 1043–1048. [**Google Scholar**] [**CrossRef**] [**PubMed**] [**Green Version**]

24. Smith, P.R.; Sward, R.J. Crop loss assessment studies on the effects of barley yellow dwarf virus in wheat in Victoria. *Aust. J. Agric. Res.* **1982**, *33*, 179–185. [**Google Scholar**] [**CrossRef**]

25. Edwards, M.C.; Fetch, T.G.J.; Schwarz, P.B.; Steffenson, B.J. Effect of barley yellow dwarf virus infection on yield and malting quality of barley. *Plant Dis.* **2001**, *85*, 202–207. [**Google Scholar**] [**CrossRef**] [**PubMed**]

26. Rochow, W.F. Biological properties of four isolates of barley yellow dwarf virus. *Phytopathology* **1969**, *59*, 1580–1589. [**Google Scholar**] [**PubMed**]

27. Jeffrey, S.J.; Carter, J.O.; Moodie, K.B.; Beswick, A.R. Using spatial interpolation to construct a comprehensive archive of Australian climate data. *Environ. Model. Softw.* **2001**, *16*, 309–330. [**Google Scholar**] [**CrossRef**]

28. Nancarrow, N.; Constable, F.; Finlay, K.; Freeman, A.; Rodoni, B.; Trębicki, P.; Vassiliades, S.; Yen, A.; Luck, J. The effect of elevated temperature on barley yellow dwarf virus-PAV in wheat. *Virus Res.* **2014**, *186*, 97–103. [**Google Scholar**] [**CrossRef**]

29. Zadoks, J.C.; Chang, T.T.; Konzak, C.F. A decimal code for the growth stages of cereals. *Weed Res.* **1974**, *14*, 415–421. [**Google Scholar**] [**CrossRef**]

30. Scholthof, K.; Adkins, S.; Czosnek, H.; Palukaitis, P.; Jacquot, E.; Hohn, T.; Hohn, B.; Saunders, K.; Candresse, T.; Ahlquist, P.; et al. Top 10 plant viruses in molecular plant pathology. *Mol. Plant Pathol.* **2011**, *12*, 938–954. [**Google Scholar**] [**CrossRef**]

31. Fitzgerald, P.J.; Stoner, W.N. Barley yellow dwarf studies in wheat (*Triticum aestivum* L). *Crop Sci.* **1967**, *7*, 337–341. [**Google Scholar**] [**CrossRef**]

32. Gill, C.C. Epidemiology of barley yellow dwarf virus in Manitoba and effect of the virus on yield of cereals. *Phytopathology* **1970**, *60*, 1826–1830. [**Google Scholar**] [**CrossRef**]

33. Yount, D.J.; Martin, J.M.; Carroll, T.W.; Zaske, S.K. Effects of barley yellow dwarf virus on growth and yield of small grains in Montana. *Plant Dis.* **1985**, *69*, 487–491. [**Google Scholar**] [**CrossRef**]

34. Jensen, S.G.; Fitzgerald, P.J.; Thysell, J.R. Physiology and field performance of wheat infected with barley yellow dwarf virus. *Crop Sci.* **1971**, *11*, 775–780. [**Google Scholar**] [**CrossRef**]

35. Davis, T.S.; Bosque-Pérez, N.A.; Foote, N.E.; Magney, T.; Eigenbrode, S. Environmentally dependent host-pathogen and vector-pathogen interactions in the Barley yellow dwarf virus pathosystem. *J. Appl. Ecol.* **2015**, *52*, 1392–1401. [**Google Scholar**] [**CrossRef**]

36. Roossinck, M.J. Plant Virus Ecology. *PLoSPathog.* **2013**, *9*, e1003304. [**Google Scholar**] [**CrossRef**]

37. Xu, P.; Chen, F.; Mannas, J.P.; Feldman, T.; Sumner, L.W.; Roossinck, M.J. Virus infection improves drought tolerance. *New Phytol.* **2008**, *180*, 911–921. [**Google Scholar**] [**CrossRef**] [**PubMed**]

38. Hoffman, T.K.; Kolb, F.L. Effects of barley yellow dwarf virus on yield and yield components of drilled winter wheat. *Plant Dis.* **1998**, *82*, 620–624. [**Google Scholar**] [**CrossRef**]

39. Monneveux, P.; St-Pierre, C.A.; Comeau, A.; Makkouk, K.M. Barley yellow dwarf virus tolerance in drought situations. In *Barley Yellow Dwarf in West Asia and North Africa*; Comeau, A., Makkouk, K.M., Eds.; ICARDA: Aleppo, Syria, 1992; pp. 209–220. [**Google Scholar**]

40. Erion, G.G.; Riedell, W.E. Barley yellow dwarf virus effects on cereal plant growth and transpiration. *Crop Sci.* **2012**, *52*, 2794–2799. [**Google Scholar**] [**CrossRef**]

41. Hoffman, T.K.; Kolb, F.L. Effects of barley yellow dwarf virus on root and shoot growth of winter wheat seedlings grown in aeroponic culture. *Plant Dis.* **1997**, *81*, 497–500. [**Google Scholar**] [**CrossRef**] [**Green Version**]

42. Liang, X.; Rashidi, M.; Rogers, C.; Marshall, J.; Price, W.J.; Rashed, A. Winter wheat (*Triticum aestivum*) response to Barley yellow dwarf virus at various nitrogen application rates in the presence and absence of its aphid vector, Rhopalosiphumpadi. *Entomol. Exp. Appl.* **2019**, *167*, 98–107. [**Google Scholar**] [**CrossRef**]

43. Malmstrom, C.M.; Bigelow, P.; Trębicki, P.; Busch, A.K.; Friel, C.; Cole, E.; Abdel-Azim, H.; Phillippo, C.; Alexander, H.M. Crop-associated virus reduces the rooting depth of non-crop perennial native grass more than non-crop associated virus with known viral suppressor of RNA silencing (VSR). *Virus Res.* **2017**, *241*, 172–184. [**Google Scholar**] [**CrossRef**] [**PubMed**]

44. Comeau, A.; Jedlinski, H. Successful breeding for barley yellow dwarf virus resistance or tolerance: A systemic approach related to other agronomic characteristics. In *World Perspectives on Barley Yellow Dwarf Virus*; Burnett, P.A., Ed.; CIMMYT: Texcoco, Mexico, 1990; pp. 441–451. [**Google Scholar**]

45. Comeau, A.; Pelletier, G.J. Predisposition to Septoria leaf blotch in oats affected by barley yellow dwarf virus. *Can. J. Plant Sci.* **1976**, *56*, 13–19. [**Google Scholar**] [**CrossRef**] [**Green Version**]

Chapter-7

A Primer on the Analysis of High-Throughput Sequencing Data for Detection of Plant Viruses

1. Introduction

High-throughput sequencing (HTS) technologies have become an integral part of research and diagnostics toolbox in life sciences, including phytopathology and plant virology [**1**]. HTS enables the untargeted acquisition of extremely large amounts of sequence data from diverse sample types and thus represents an ideal and unique solution for the generic detection of highly diverse viruses. In the past decade, sequencing prices have significantly decreased, and the technology has become accessible to many more research and diagnostic labs. From the first uses of HTS for detection of plant viruses in 2009 [**2**,**3**,**4**,**5**], the use of this technology for detection of known and new plant viruses and the characterization of viromes in different plant species has intensified dramatically. Many different bioinformatics tools have been developed and different pipelines have been used to detect and identify plant viruses represented in HTS datasets. The variation in results associated with the use of different pipelines in different labs has highlighted the significance of understanding different approaches [**6**]. Arguably, one of the main challenges for less experienced users of HTS is to understand, select, and properly use tools for the analysis of HTS data intended for detection and identification of plant virus sequences. In this review, we aim to present the different and often complementary approaches used for analysis of HTS data for the detection of plant viruses. We provide a short introduction to the laboratory work required and then describe the possible steps in data processing for the detection of plant viruses, including quality control and trimming of the sequences, *de novo* assembly, sequence similarity searches, and taxonomic classification of the identified viral sequences. By including a short glossary (**Figure 1**), checklists, and comparison tables, we aim to present the topic to the widest possible audience and thus encourage the use of HTS technologies by researchers with limited experience in the field.

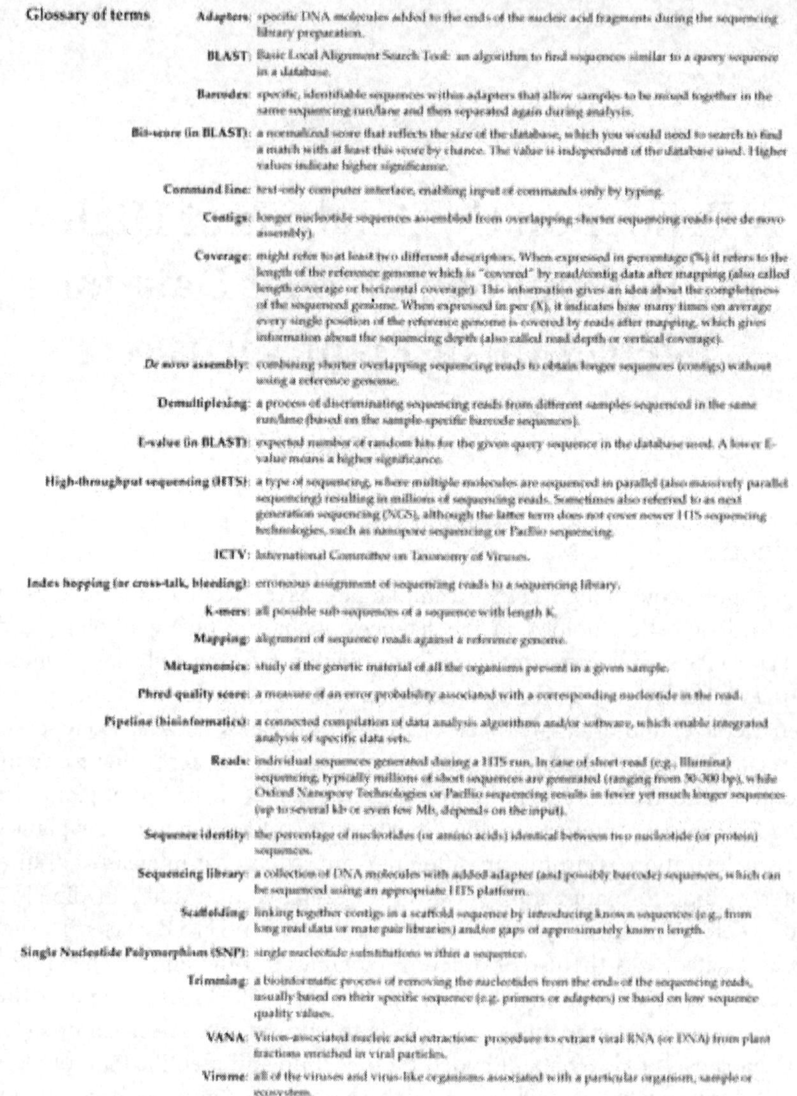

Glossary of terms

Adapters: specific DNA molecules added to the ends of the nucleic acid fragments during the sequencing library preparation.

BLAST: Basic Local Alignment Search Tool: an algorithm to find sequences similar to a query sequence in a database.

Barcodes: specific, identifiable sequences within adapters that allow samples to be mixed together in the same sequencing run/lane and then separated again during analysis.

Bit-score (in BLAST): a normalized score that reflects the size of the database, which you would need to search to find a match with at least this score by chance. The value is independent of the database used. Higher values indicate higher significance.

Command line: text-only computer interface, enabling input of commands only by typing.

Contigs: longer nucleotide sequences assembled from overlapping shorter sequencing reads (see de novo assembly).

Coverage: might refer to at least two different descriptors. When expressed in percentage (%) it refers to the length of the reference genome which is "covered" by read/contig data after mapping (also called length coverage or horizontal coverage). This information gives an idea about the completeness of the sequenced genome. When expressed in per (X), it indicates how many times on average every single position of the reference genome is covered by reads after mapping, which gives information about the sequencing depth (also called read depth or vertical coverage).

***De novo* assembly**: combining shorter overlapping sequencing reads to obtain longer sequences (contigs) without using a reference genome.

Demultiplexing: a process of discriminating sequencing reads from different samples sequenced in the same run/lane (based on the sample-specific barcode sequences).

E-value (in BLAST): expected number of random hits for the given query sequence in the database used. A lower E-value means a higher significance.

High-throughput sequencing (HTS): a type of sequencing, where multiple molecules are sequenced in parallel (also massively parallel sequencing) resulting in millions of sequencing reads. Sometimes also referred to as next generation sequencing (NGS), although the latter term does not cover newer HTS sequencing technologies, such as nanopore sequencing or PacBio sequencing.

ICTV: International Committee on Taxonomy of Viruses.

Index hopping (or cross-talk, bleeding): erroneous assignment of sequencing reads to a sequencing library.

K-mers: all possible sub-sequences of a sequence with length K.

Mapping: alignment of sequence reads against a reference genome.

Metagenomics: study of the genetic material of all the organisms present in a given sample.

Phred quality score: a measure of an error probability associated with a corresponding nucleotide in the read.

Pipeline (bioinformatics): a connected compilation of data analysis algorithms and/or software, which enable integrated analysis of specific data sets.

Reads: individual sequences generated during a HTS run. In case of short-read (e.g., Illumina) sequencing, typically millions of short sequences are generated (ranging from 50-300 bp), while Oxford Nanopore Technologies or PacBio sequencing results in fewer yet much longer sequences (up to several kb or even few Mb, depends on the input).

Sequence identity: the percentage of nucleotides (or amino acids) identical between two nucleotide (or protein) sequences.

Sequencing library: a collection of DNA molecules with added adapter (and possibly barcode) sequences, which can be sequenced using an appropriate HTS platform.

Scaffolding: linking together contigs in a scaffold sequence by introducing known sequences (e.g., from long read data or mate pair libraries) and/or gaps of approximately known length.

Single Nucleotide Polymorphism (SNP): single nucleotide substitutions within a sequence.

Trimming: a bioinformatic process of removing the nucleotides from the ends of the sequencing reads, usually based on their specific sequence (e.g. primers or adapters) or based on low sequence quality values.

VANA: Virion-associated nucleic acid extraction: procedure to extract viral RNA (or DNA) from plant fractions enriched in viral particles.

Virome: all of the viruses and virus-like organisms associated with a particular organism, sample or ecosystem.

Figure 1. Glossary of terms commonly used in bioinformatics analysis of high-throughput sequencing (HTS) data for plant virus detection.

2. What Should I Anticipate and How Should I Prepare?

Modern sequencing platforms can generate massive amounts of data, and not all laboratories wishing to use HTS in their projects have the necessary infrastructure and bioinformatics expertise, which, for example, is one of the main challenges identified for the adoption of these technologies in diagnostic laboratories [7]. The cost of the bioinformatics analysis in a HTS project was estimated to be around 15% of the total cost of a program (an example for whole genome analysis in cancer research), and it includes the salary of the bioinformatician and cost of data storage [8].

Some commercial sequence analysis software is able to handle HTS data (see **Section 4.3.8**), with dedicated modules for common operations (e.g., mapping and assembly). These software solutions are usually easy to use, regardless of the user's bioinformatics skill, but they are also quite expensive and might be limited for some analyses (specific applications). Furthermore, some "all in 1" viral-detection focused pipelines are available (see **Section 4.3.8**), which require only limited bioinformatics knowledge or only the help of a skilled computer scientist at the installation stage.

However, for in-depth analysis of plant virus sequence data that goes beyond detection and species classification, the use of dedicated bioinformatics software, without an easy-to-use graphical user interface, is often needed to optimize time and efforts. These programs have in a large part been developed and optimized for the Linux platform; they can be used in the command line only and so require specific computing skills. Considering the number of steps with the average HTS analysis pipeline and the number of samples studied, automation quickly becomes a priority. This can be achieved by writing scripts as well as grouping and ordering all the steps of the analysis, which also require expertise in programing languages (e.g., shell, Python, R). Finally, for the interpretation of the analysis results, skills beyond pure bioinformatics are needed. A close collaboration between a bioinformatician and a plant virologist (or a plant virologist trained in bioinformatics) is needed to achieve a meaningful interpretation of the results.

Beyond the skills of users, IT resources must also be addressed. The amount of data generated by each project must be anticipated in order to have raw data storage space available beforehand and to ensure that data is safely stored at least for several years after the end of projects. Depending on the sequencing platform, the total size of the raw data can become very large. For example, the Illumina NextSeq platform can generate from 120 to ≈300 Gbases (Gb) per run, leading to file sizes varying between 39 and 170 Gb depending on the read length. A stable and fast internet connection is often needed to facilitate the efficient transfer of large data files. The computing resources also need to be anticipated. For time-efficient analysis, it is often necessary to have a more powerful machine than an average workstation to run the various parts of pipelines, regardless of the software used. An alternative to the acquisition of a powerful computer is making use of online bioinformatics platforms and cloud computing solutions. These platforms generally have a structure adapted to the use of software making high demands on system resources (e.g., computing clusters). Many research centers or universities host a Galaxy instance, which represents a very good alternative to the Linux platforms, in a more "user friendly" interface.

3. Starting the Project: How Do I Prepare Samples and Sequence Nucleic Acids?

Sampling, nucleic acids extraction, viral enrichment, and sequencing library preparation are essential steps before HTS itself. Since these steps can influence the sequencing results, we briefly summarize here the most important considerations for some of these processes. An extensive description of how to control all of these steps is in preparation in forthcoming international guidelines for the use of HTS tests for the diagnostic of plant pests [**9**]. After obtaining the nucleic acids suitable for further analysis using HTS, the approximate amount of sequence data required per each sample should be estimated according to the goals of the study. If an external sequencing provider will perform HTS, this number, together with some general characteristics of the samples, should be communicated with the provider.

3.1. Input Material and Nucleic Acids Preparation

The extraction step separates the nucleic acids (including viral nucleic acids) from other cellular components. There are many methods that can be used to obtain high-quality nucleic acids intended for HTS [**10,11,12,13**]. The efficiency of an extraction method is evaluated by the quantity of nucleic acids obtained, their integrity, and the absence of contaminants that inhibit the enzymatic activities involved in the preparation of sequencing libraries. Irrespective of the chosen nucleic acid extraction procedure and library preparation methodology, it is recommended to collect several samples per plant or that tissue from distributed locations on a plant is combined into a single sample to overcome the uneven distribution of viruses, especially in the case of low titer viruses. Different types of nucleic acids can be used as inputs for HTS, which can be combined with different viral enrichment methods. No method is universal [**11,14**]; each favors certain viral families or certain experimental objectives [**15**]. For example, total RNA or small RNA sequencing might be most straightforward and universal to use for single samples. On the other hand, for sequencing of pools of many samples, or to optimize the detection of viruses with a low titer, methods that allow the enrichment of viral nucleic acids such as Virion-Associated Nucleic Acids extraction (VANA) or the purification of double-stranded RNA might be preferred. The choice for one of the approaches should be based on the research question and study design. The purpose of the following sections is to help make the most appropriate choices for sample preparation.

3.1.1. Total RNA/DNA

Extraction of total RNA and/or, to a lesser extent, DNA is a widely used approach for HTS analysis of plant tissues infected with viruses. Simple and robust, the method can be carried out according to several standard extraction protocols in solid phase or in liquid phase or using commercial kits (mostly based on silica-membrane or magnetic bead purification). The extraction and sequencing of total DNA can be sometimes used specifically for the detection of DNA viruses, while sequencing of total RNA is a very generic approach and can be used for detection of all types of DNA and RNA viruses and viroids [**15**]. The high abundance of nucleic acids from the host plant co-extracted with viral nucleic acids can greatly limit the sequencing sensitivity. The relative proportion of viral sequences in the total extracted RNA can be increased by the depletion of the plant ribosomal RNA [**16,17**] and the proportion of sequences of circular DNA viruses in extracted DNA can be enriched by rolling circle amplification [**18,19,20**].

3.1.2. Small RNA (sRNA)

The plant immune system responds to the presence of viruses by activating a defense response that leads to the cleavage of double-stranded forms of viral RNA into small RNAs (sRNA) of 21 and 22 nucleotides (nt) as well as, more marginally, of 24 nt [**21**]. The analysis of sRNAs facilitates the reconstruction of the complete genomes of infecting RNA and DNA viruses or viroids, as well as those of integrated endogenous viral elements (EVEs) if they are transcribed [**2,15,22,23**]. Since sRNAs are more stable than longer RNA molecules, the method is promising for use in old or even ancient plant samples [**24**], and since only very short reads are needed to sequence sRNAs, the method is relatively cost efficient. On the other hand, *de novo* assembly from short sequences might not work very well for targets present at a very low titer [**15**] and might lead to chimeric sequences in case of multiple infections with different virus strains [**25**]. For the same reason, pooled samples used in metagenomic studies including a large number of plants are not recommended to be analyzed with sRNA sequencing. Due to their short lengths, analyses of recombination events on a read level are also not feasible with sRNA [**22**].

3.1.3. Virion-Associated Nucleic Acids (VANA)

The extraction of Virion-Associated Nucleic Acids (VANA) enriches the samples in nucleic acids of viral origin by semi-purifying the viral particles by ultracentrifugation. Viral particles are separated from most of the organelles and plant debris by one or two differential ultracentrifugation cycles depending on the viral family and the plant material. After purification of the particles and a nuclease treatment to degrade non-protected nucleic acids, the viral nucleic acids are extracted according to a standard extraction protocol also used for the extraction of total RNA/DNA. Initially developed for the biochemical characterization of viral particles in the 1970s, VANA was used in pioneering studies of prospecting for viral diversity in wild asymptomatic plants before the advent of HTS [**26,27**]. Then, the approach was extended to the preparation of nucleic acids intended for HTS [**28,29**]. It achieves balanced enrichment in high-quality viral RNA and DNA and allows the use of up to several hundred grams of starting material. However, it is based on the stability of the viral particles mainly determined by the pH and the concentration of salts in the extraction buffer. Unsuitable for high throughput, and relying on numerous laboratory operations, the approach only identifies the encapsidated viral nucleic acids as well as the viruses of the *Endornaviridae* family, which are devoid of capsids but encapsulated in membranous vesicles [**28,30**]. Moreover, certain viral families are difficult to purify, and VANA is also not the method of choice for the extraction of viruses from plants with high content of phenolic and polysaccharide compounds [**31**].

3.1.4. Double-Stranded RNA

The majority of plant viruses have RNA genomes, accounting for 75% of the total number of viruses reported [**32**]. While plants do not produce large quantities of double-stranded (ds)RNAs, RNA viruses generate high molecular weight dsRNA structures during replication, so their enrichment is a popular strategy used for virus diagnostics [**10,13,33,34**]. The extraction of dsRNA purifies nucleic acids from double-stranded RNA viruses but also from most single-stranded RNA viruses, viroids as well as from some DNA viruses [**35,36,37,38**]. This approach allows the detection of a very wide range of RNA virus species [**30,39**]. Sequencing of dsRNA is likely not the most effective method for the detection of negative sense single-stranded RNA viruses [**37**]. It is also a laborious approach, even if a number of modified protocols have been developed to overcome this limitation [**13,34,40,41,42**].

3.2. Library Preparation and Sequencing

Following nucleic acid extraction, different methods have been developed for library preparation using commercially available kits and automated systems. As inputs, the extracted and possibly virus-enriched nucleic acids described in the previous sections can be used. The type of the library preparation and exact protocol is dependent on the input nucleic acids (e.g., total RNA or DNA, sRNA, dsRNA). Specific libraries are prepared for different HTS platforms. The library preparation step usually consists of fragmenting the nucleic acids and the ligation of short oligonucleotides (adaptors) at one or both extremities of the fragments in order to allow the sequencing. There are two main groups of HTS platforms: (i) short read HTS (also termed next-generation sequencing—NGS), producing reads up to several hundred nucleotides, and (ii) long read HTS (also termed single molecule sequencing—SMS), producing reads up to hundreds of kilobases (kb). Currently, the most commonly used sequencing platform is Illumina (short read HTS), and, for long read HTS, Pacific Biosciences (PacBio) and Oxford Nanopore Technologies. Nanopore sequencing is rapidly developing and is expected to be more widely used in the future [**43**]. Most of the available protocols recommend assessing the quality and quantity of the nucleic acids before library preparation. The integrity and purity of the nucleic acids can be assessed using spectrophotometric and fluorescence-based assays. For some enrichment approaches (e.g., VANA, dsRNA extraction), the concentrations of the obtained nucleic acids can be below the input required for library preparation so that a random amplification step may be required prior to library construction [**13**].

Several samples can be pooled and sequenced in the same sequencing run (multiplexing). In this case, the oligonucleotides ligated to the nucleic acids during library preparation also include unique barcode sequences that are specific for each sample. After sequencing, the reads are allocated to the appropriate sample according to the barcode used. Most commonly, the raw sequencing read data output is converted to a fastq file format. The fastq files represent an input for the bioinformatics analysis described in the following paragraphs.

Important consideration, when preparing samples for sequencing, is also, how many samples to pool in the same sequencing run/lane, i.e., how many reads (or nucleotides) are needed for the sensitive detection of different possible viruses in the plant sample. The answer is not straightforward, and it might depend on the sequencing approach, type of the matrix (host plant species, different parts of the plant), present virus(es), and other variables [**15,17,38**], such as, e.g., season, but also the sensitivity of the bioinformatics pipeline used (e.g., reads vs. contigs analysis) [**6**]. Some starting general recommendations regarding this problem are given in this primer; however, these need to be adjusted after performing a pilot study on a specific system, considering employed sample preparation, sequencing, and analysis approach.

3.3. Contamination

Contamination is common in all sensitive molecular diagnostic methods and has been reported in HTS diagnostics [**44,45**]. Contamination has been shown to enter sequencing systems in diverse ways, from sample cross-contamination [**46**] to external contamination of consumables [**47**]. Whilst some of the most commonly used HTS platforms from Illumina were subjected to significant hardware and procedural changes as a result of within-instrument DNA carry over, contamination can still be a significant issue in sensitive molecular diagnostics applications. The fundamentals of contamination control for diagnostics remain consistent. Key to achieving this is the separation of procedures into different locations, operating a one-way system (from clean reagents to DNA samples) within those locations and using negative controls at various stages to identify contamination. Sample-to-sample and reagent contamination are common in any molecular technique. Physically separating steps involving samples, purified DNA, and clean reagents is the best approach to preserve the integrity of future experiments. Known healthy control samples (not blanks), included from NA-extraction through to sequencing should be included in each run to identify incidences of contamination but are frequently excluded due to cost constraints.

4. How Do I Analyze the Data?

Figure 2 outlines typical steps that can be followed once the fastq file has been obtained. The first is a quality control (QC) check. This is followed by pre-processing steps, including trimming low-quality bases, removing adapter sequences, and discarding very short and low-quality reads, followed by further QC filtering (**Section 4.1**). Then, reads passing QC are ready for analysis either directly or after assembly into contigs (**Section 4.2**). Reads or contigs can optionally be mapped to a host reference genome, and, in this way, host sequences can be removed (**Section 4.3.3**). Then, reads or contigs are used to query a database of known viral sequences or motifs (**Section 4.3.2, Section 4.3.3, Section 4.3.4** and **Section 4.3.5**). Results need to be carefully inspected for correct taxonomic classification (**Section 4.3.7**). The described steps can be performed using the tools indicated in the flow chart (**Figure 2**) or other available tools. Finally, the same analyses can also be performed using user-friendly free software with graphical user interfaces (GUI) available online or using commercial software as described in **Section 4.3.8.**

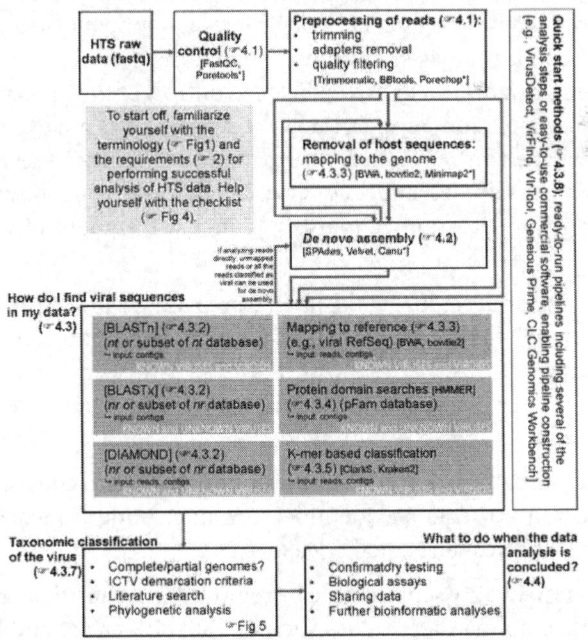

Figure 2. Flowchart representing different approaches for the analysis of HTS data for the detection of plant viruses. Boxes represent different steps in data analysis and interpretation. Arrows connect different possible sequences of the analysis steps. As

an example, a non-exhaustive list of possible analysis tools is added in the square brackets at each of the analysis steps. Tools designated with * are intended for use with long-read or, specifically, nanopore sequencing data. Pointing hands lead to the text sections (or figures) with more detailed description of the corresponding steps.

4.1. Demultiplexing, Quality Control, and Trimming

Each sequencing platform produces a series of quality metrics associated with the data produced from each sequencing run. A discussion of the metrics with the sequencing data provider is important before accepting any sequencing data.

If the run was successful, the first step is the demultiplexing of barcoded samples, which is usually carried out using the sequencing platform software or performed by the sequencing data provider. In the event that data has not been demultiplexed, third-party tools such as Cutadapt [**48**] can be used to demultiplex the Illumina data by looking for specific barcode sequences present in the samples. Alternatively, demultiplexing tools developed by the sequencing platform provider are frequently accessible as stand-alone tools, such as Illumina's bcl2fastq software [**49**], or Oxford Nanopore Technologies' guppy scripts [**50**].

Barcode misassignments, also termed index hopping/cross-talk/bleeding, can occur due to the technical reasons during each sequencing run and result into erroneous assignment of a small fraction of reads from one sample to another one [**51**]. This represents a problem when using HTS for detection purposes, since it might often be difficult to distinguish index hopping from, e.g., very low titer virus infection in the sample. The amount of index hopping differs between different sequencing platforms, but it was, e.g., shown to be higher for newer Illumina sequencing devices using nonpatterned flow cells [**52**]. To mitigate this problem, it is advised to know the identity of all the samples sequenced in the same sequencing run or/and to use dedicated controls of the procedure. For example, including a control sample containing a known virus (which is not expected to be present in other samples in the run) could help estimate the amount of the crosstalk from the control sample to other samples, and vice versa. In addition, using unique double indexes in sequencing library preparation can largely reduce the amount of the index hopping [**53**].

Adapter sequences introduced during the library preparation process need to be removed. Tools such as Cutadapt [**48**], Trimmomatic [**54**], and Porechop [**55**] or NanoFilt [**56**] can be used to carry out this process, with the latter two working specifically for data generated using nanopore sequencers. At this step, contaminant filtering for synthetic molecules and/or spike-in is also recommended.

Sequencing data are usually provided in the fastq format, which consists of four lines per sequence [**57**], including a sequence identifier, raw nucleotide sequence, a separator line (containing + sign), and sequence quality values.

Nucleotides with a low-quality score should be removed to ensure that only high-accuracy bases remain. With Illumina data, values such as Q20 (1% error) and Q30 (0.1% error) are often used when trimming data, but this value depends on the application and the sequencing platform used. If accuracy is of the utmost importance (e.g., for detection of SNPs), selecting a higher quality score will be beneficial. If accuracy is less important (e.g., for detection of virus), then relaxing constraints on quality when trimming will allow more data to be available for downstream applications.

Quality control reports can be generated by tools such as FastQC [**58**], MultiQC [**59**], or, specifically for nanopore sequencing data, Poretools [**60**] or NanoStat [**56**]. This allows for the visual inspection of metrics such as per base sequence quality, sequence length distribution, and GC (guanine–cytosine) content. These reports can be generated both before and after trimming, to assess the impact of trimming on different quality parameters. A number of tools exist to trim sequencing reads based on quality scores, sequence length, or other metrics. These include but are not limited to Sickle [**61**], Trimmomatic [**54**], Cutadapt [**48**], BBDuk (**https://sourceforge.net/projects/bbmap/**, accessed on 13 April 2021) and NanoFilt for nanopore sequencing data [**56**]. Illumina data, particularly longer MiSeq reads, suffer from lower quality toward the 3′ end of the read. Many trimming strategies start at the 3′ end of such reads and determine the position at which the quality (or the average quality in a region) is high enough to keep.

The order in which these processes are carried out can vary, and some tools can be used to carry out multiple steps at the same time. The final output should be a series of demultiplexed samples with reads that have an acceptable sequence quality and no longer contain sequences added during the sequencing process (e.g., adapters, barcodes).

4.2. De Novo Assembly

HTS technologies provide us with shorter (e.g., Illumina) or longer (e.g., Oxford Nanopore Technologies, PacBio) sequence reads, which usually need to be assembled in silico to reconstruct complete or near-complete genomes. Compared to bacteria or eukaryotes, most viral genomes are very small. Nevertheless, high mutation rates and the great diversity of some viral populations [**62**] can represent a challenge for in silico genome reconstruction. Assembling a genome is similar to solving a "Jigsaw puzzle". Similar to a puzzle, there could be pieces fitting together (overlapping reads), missing pieces (regions with low coverage, sequencing bias), and damaged parts (sequencing errors). The process for which individual reads are combined to form longer contiguous sequences is named *de novo* sequence assembly, and the nucleotide fragments obtained through this process are called contigs [**63**].

The intrinsic features of short vs. long read output, from the computational point of view, has led to the development of two major groups of assembly algorithms: (i) de Bruijn graph (DBG) and (ii) the overlap-layout-consensus (OLC) methods. In the first case, DBGs are constructed using k-mers, which are substring of the reads of length k; whereas for OLC, the overlap graphs are constructed directly from reads, eliminating the redundant ones. The use of k-mers is more widely applied for the assembly of short reads, whilst the OLC approach is most appropriate for long read data [**63,64**].

For short HTS reads, many de Bruijn graph assemblers are available, such as SOAPdenovo2 [**65**], ALLPATHS-LG [**66**], ABySS [**67**], Velvet [**68**], IDBA-UI [**69**], and (rna)SPAdes [**70,71,72**]. One of the first and most widely used and cited assemblers [**73**] in viral metagenomics [**6**] is the open-source software Velvet, which is followed by the more user-friendly and commercially-available CLC Genomics Workbench (**https://digitalinsights.qiagen.com**, accessed on 13 April 2021) and Geneious Prime (**https://www.geneious.**

com, accessed on 13 April 2021). The latter has the advantage of providing a graphical interface for command-line assembly programs such as Velvet and Spades.

Different factors can positively influence the quality of the *de novo* assembly, e.g., a preliminary filtering step to eliminate the genomic host plant reads [**23**] or the selection of appropriate k-mer values based on the read length [**6**]. Moreover, approaches in which *de novo* assemblies using different k-mer values are generated and then reassembled can generally improve the completeness of *de novo* genome assemblies, but this can be a laborious and computationally lengthy process. Usually, higher sequencing depth and a higher fraction of viral reads in the dataset will positively affect the completeness of assembled viral genomes; however, extremely high coverage might have a negative effect on the completeness of the assembly when using some assemblers; thus, in such cases, the assembly of subsampled data might give better results [**15**]. Since reads of some viruses can be present in a very low number, it is important not to set too low cut-offs for contig length [**6**], e.g., a number around or slightly above the 2× length of an average read length is recommended. Finally, the use of an additional scaffolding step when using paired-end data can sometimes further increase the length of a contig. Nevertheless, despite improvements in *de novo* assembly algorithms, 3' and 5′ ends of viral genomes usually cannot be obtained in full through *de novo* assembly.

Although long-read HTS platforms can produce reads close to full-length viral genomes, a major issue that could affect the *de novo* assembly step is the higher error rate (%15–5) of these technologies [**74**]. Long-read assemblers can algorithmically correct base errors before/when building contigs. PBcR [**75**], Canu [**76**], Falcon [**77**], and Pomoxis [**78**] are some of the OLC-based *de novo* assemblers available. Long read nanopore sequencing has recently been successfully applied to virus discovery, detection, and reconstruction of virus genomes; in these studies, Canu is the most cited assembler [**79,80,81,82**].

Contigs generated by *de novo* assembly can be used in subsequent similarity searches, and finally, viral contigs can be used for phylogenetic or recombination analysis. If this is so, it is important to check the quality of the contig by mapping the trimmed reads (explained in **Section 4.3**) to the viral contig followed by visual inspection of the mapping and to check the completeness of expected open reading frames contained in such contigs. For contigs generated by *de novo* assembly of nanopore sequencing reads, additional quality checking steps might be needed such as assembly polishing [**81**] or correction of the consensus sequences using quality data of mapping reads [**82**].

When the presence of specific viruses is already known, viral genomes can be reconstructed by mapping the reads (explained in **Section 4.3**) to the closest reference sequences obtained from sequence databases (after initial similarity searches, **Section 4.3**). Then, this is followed by the extraction of new consensus sequence from the mapping, which is an approach known as reference guided assembly. Sometimes, parts of the viral genomes are obtained by *de novo* assembly and other parts are obtained through reference guided assembly; such an approach is also known as combined assembly.

4.3. How Do I Find and Classify Viral Sequences in My Data?

Identification of viral reads/contigs in massive HTS datasets is most frequently performed by comparing sequences against known and annotated sequences in databases. This can be done on the level of reads or contigs *de novo* assembled from the reads. Since longer sequences in almost all cases improve the ability to identify similarities regardless of the method or databases used, an assembly of quality checked raw reads is generally recommended prior to similarity searches. At the same time, a prior assembly will also generally reduce the computing time needed for the similarity search steps, as up to millions of reads can be assembled in a single contig. The annotation of HTS reads, or contigs, on the basis of similarity with known viral sequences can be performed using three main strategies: homology searches with tools such as Basic Local Alignment Search Tool - BLAST [**83**], read/contig mapping against reference viral genomes using tools such as BWA [**84**], and the search for encoded, conserved protein motifs using tools based on Hidden Markov Models (HMMs) such as HMMER [**85**]. Each of these approaches and, in turn, each of the specific programs used to perform them, has advantages and drawbacks. In many cases, they should be seen as complementary rather than mutually exclusive possibilities. Several additional alternatives have also been proposed. For example, the use of e-probes (short unique pathogen-specific reference sequences) [**86**] or the analysis of the frequency of specific k-mer sequences (see **Section 4.3.5**). A summary of tools commonly used for similarity searches is presented in **Table 1**.

Table 1. Summary of the most commonly used similarity search strategies with advantages and limitations for each of the strategies.

Tool Name	Advantages	Limits and Considerations	Important Thresholds
BLASTx or BLASTn	High sensitivity	Slow, intensive use of computing power if a large database is used, BLASTx needed for the detection of divergent novel viruses, BLASTn needed for the detection of viroids and noncoding regions of viral genomes or satellites; performance improved by prior assembly of contigs.	Minimum percentage of identity; length of identified region of similarity; minimal e-value, bit-score.
MegaBLAST	Faster than BLASTn, handles longer sequences	Less sensitive than BLASTn, only useful for detection of nucleotide sequences very similar to the ones in the used database; performance improved by prior assembly of contigs.	Minimum percentage of identity; length of identified region of similarity; minimal e-value, bit-score.
BLASTp	High sensitivity	Slow, need to translate nucleotide sequences to proteins first; performance improved by prior assembly of contigs; not applicable for viroids or noncoding regions of viral genomes or satellites.	Minimum percentage of identity; length of identified region of similarity; minimal e-value, bit-score.
DIAMOND	Faster than BLASTx	Less sensitive, annotation less accurate than BLAST; performance improved by prior assembly of contigs; only available for searches against protein databases; not applicable for viroids or noncoding regions of viral genomes or satellites.	Minimum percentage of identity; length of identified region of similarity; minimal e-value, bit-score; use sensitive mode.

Tool Name	Advantages	Limits and Considerations	Important Thresholds
Burrows-Wheeler transform-based mapping algorithms (e.g., BWA or Bowtie2)	Does not require prior assembly of contigs, high sensitivity for short sequences	Only allows detection of known agents. Difficult to adjust mapping stringency to (1) allow detection of divergent isolates while (2) avoiding cross-mapping between related agents; prior assembly of contigs reduces cross-mapping between related agents.	Mapping stringency (e.g., mismatch penalties, gap open/extension penalties, percent of read length matching reference, minimum percentage of identity)
HMMER or HMMScan	High efficiency for detection of distant homologs	Annotation more complex for protein families shared between cellular organisms and viruses; not applicable for viroids or noncoding regions of viral genomes or satellites.	Minimal e-value.
K-mer based classification algorithms (Kraken or Taxonomer)	Fast	Requires large computer memory; accuracy may be limited for the shorter genomes of plant viruses; the confidence scoring of the results is not straight forward.	C/Q ratio for Kraken (advise the manual).

4.3.1. Databases

The database(s) against which sequences are compared is/are of utmost importance for the efficiency and completeness of the annotation process. The more complete the collection of viral sequences, the greater the likelihood of detecting and identifying the presence of a virus. For BLAST and BLAST-like approaches, the most used databases are the non-redundant nucleotide database (nr/nt, named also just nt) hosted by the NCBI, the non-redundant GenBank protein database (nr) or the viral RefSeq database. The GenBank non-redundant nucleotide and protein databases are the most comprehensive and most frequently updated public databases, limiting the time from discovery of a novel virus to its availability for comparisons (provided the local version of these databases is also regularly updated). However, the size of these databases has the drawback of increasing the computing time/power needed to perform a comparison. The reduced viral RefSeq database has the benefit of a better annotation/curation at the expense of the number of included sequences and of less frequent updates. For read mapping approaches, smaller dedicated databases are generally used, such as a subset of all viral sequences from the NCBI nt database, viral RefSeq, or a smaller, locally developed and curated database (for example, one or several isolates of every virus known to infect the crop of interest). For conserved protein motifs searches, the most common databases are PFAM [**87**] and CDD [**88**]. The identification of viral sequences is critically dependent upon the quality of the database(s) used. For example, some plant-derived proteins might also be misidentified as viral if only a virus sequence database is used for similarity searches, because some viral proteins are related to plant encoded proteins. Typical examples are heat shock proteins (i.e., Hsp70h) found in closteroviruses [**89**] or reverse transcriptase proteins of *Caulimoviridae* that have homologs among retrotransposons. Wrongly annotated sequences in the public databases can also lead to erroneous annotations.

Although this is generally not implemented at the moment, comparing the identified viral sequences with databases of retrotransposons [**90**] or to databases created from the systematic screening of plant genomes for integrated viral sequences [**91,92,93**] may provide an efficient strategy to differentiate transcripts derived from integrated viral elements from autonomously replicating viruses.

4.3.2. BLAST and BLAST-Like Approaches

BLAST programs are the most widely used and among the most accurate in detecting sequence similarity [**94**]. The BLAST suite [**95**] comprises different algorithms, each with its own use:

- ❖ BLASTn can be used to compare a nucleotide sequence with a nucleotide database. It is less computationally intensive than BLASTx, but because of the higher divergence rate of nucleotide sequences, it is less efficient for the annotation of novel viruses not represented in the database used.
- ❖ BLASTp can be used to compare a protein sequence with a database of protein sequences.
- ❖ BLASTx can be used to compare a nucleotide sequence translated in all six reading frames with a database of protein sequences. While computationally intensive, it is the most efficient BLAST program for the annotation of novel viruses.
- ❖ tBLASTn can be used to compare a protein sequence with all six possible reading frames of a nucleotide database and is often used to identify proteins in new, unannotated genomes.
- ❖ tBLASTx can be used to compare all six reading frames of a nucleotide sequence with all six reading frames of a nucleotide database. It is the costliest in computation time.
- ❖ MegaBLAST can be used to compare nucleotide sequences expected to be already present or closely related to those in a nucleotide database. It can be much faster than BLASTn and is able to handle much longer sequences but deals less efficiently with very divergent sequences.

Short sequences may lead to false positives in BLAST searches, and for this reason, other approaches should be preferred for very short reads or contigs. All BLAST programs return a table of results, which contain several parameters, among which some are particularly important to check: the identity threshold (threshold for the percentage of identical nucleotides between the query sequence and a hit in a database), e-value (expected number of random hits in the used database for a given query sequence), and query coverage (percentage of the query sequence covered by the database hit). It is very important to consider that some of these values depend on the size of the database used and that the use of too stringent parameters (e.g., identity threshold >85% and e-value smaller than 10^{-10}) may lead to a failure to detect some divergent viruses [**6**]. BLAST is very widely used, but it remains, in the case of millions/billions of reads analyses, a time-consuming algorithm. Restricting the database used to specific taxa (e.g., viruses) can speed up BLAST searches, but care should be taken, as this frequently leads to the identification of viral reads that on closer examination, using complete databases, are in fact host sequences (e.g., plant sequences). An extremely fast but considerably less sensitive alternative to BLAST is BLAT (BLAST-Like Alignment Tool) [**96**]. Another faster alternative to BLASTx is DIAMOND [**97**], which runs at 500–20,000× the speed of BLAST while maintaining a high level of sensitivity, especially if using the sensitive mode. However, the DIAMOND annotations have been observed to be less optimal in virus species identification than BLAST ones (ML and TC personal observations).

4.3.3. Mapping Reads (or Contigs) to Reference Database

Mapping tools are commonly used as a filtering step to remove host genome sequences or as a complement to similarity searches on short nucleotide sequences. Reads originating from the host genome can be partially removed by mapping the complete dataset to reference genomic sequences of corresponding host (if available) and then using only unmapped reads for further analyses. A reference genome sequence of the host must be chosen carefully, since it can affect the analysis. Choosing divergent variety/genotype of the host might reduce the efficiency of the host reads removal. Furthermore, reference host genomes might contain contaminating or genome-integrated viral sequences; thus, some viral reads can be lost in this step.

Mapping tools can be also used to perform the alignment of reads or contigs against a reference viral database (e.g., NCBI Viral RefSeq database or a custom developed database containing one or more complete or partial viral genomes). In comparison to BLAST programs, most of the mapping tools such as Bowtie2 [**98**] or BWA [**84**] build an index for the reference genome or the reads, increasing the speed of the analysis if used against a limited, virus-specific database. The mapping strategy is potentially more sensitive to detect viruses with low number of reads in analyzed datasets [**6**], in particular when using 21–24 nt sRNA sequences. Consequently, it is also sensitive to cross-sample contamination due to index-hopping, which may require the development of strategies to set a positivity threshold. On the other hand, mapping strategies are inefficient at detecting novel viruses or viroids that are absent from the database used. Mapping stringency parameters (see **Table 1**) critically affect the outcome of the analyses and should be optimized keeping in mind the objective of the experiment. Too stringent parameters may result in the failure to detect divergent viral isolates. Too relaxed parameters may also give rise to erroneous results through the mapping of related host genes on a viral genome or through cross-mapping the reads of a virus on the genome of a related virus. These problems can be minimized by first mapping all HTS reads against the reference viral database. Then, any reads that map to a virus are remapped against the host genome sequence. If the mapping score is higher for the host genome, the read is discarded. Tools such as Pathoscope [**99**] can help with cross-mapping between virus species by weighting reads that map to more than one viral sequence. An efficient strategy, besides counting the number of mapped reads on a particular reference genome, considers the portion of this genome covered by the mapped reads and depth of coverage, the percent similarity between mapped reads, and the reference or other similar indicators to eliminate potential false positive results. Including suitable reference samples as controls during sample preparation and sequencing can help to eliminate such errors [**9**]. Similar to reads, contigs generated by *de novo* assembly, can also be mapped to the reference databases. The longer the contig, the fewer erroneous mapping results are expected. However, the same recommendations for careful inspection of mapping results apply.

4.3.4. Protein Domain Searches

Searching for known viral domains by matching translated amino acid sequences of reads/contigs with Hidden Markov Models (HMMs) of known protein domains using programs such as HMMER [**100**] or HMMScan is a popular alternative to BLASTx. With this method, sequences are first translated in all possible reading frames, and the translated protein sequences are compared to a database of conserved protein motifs such as Protein Families database—PFAM [**87**], viral profile HMMs—vFAM [**101**], and Conserved Domains Database—CDD [**88**]. These approaches are faster than BLAST-based homology searches and more effective than mapping or BLAST searches for the detection of very distant homologs [**102**] and therefore possibly for the detection of novel, very divergent viruses. Similar to BLAST, a significance e-value is calculated, allowing the evaluation of the significance of a match. This e-value can be used to filter results, striking a balance between low values and the reporting of false-positives, and high values and the failure to detect a divergent virus.

4.3.5. K-mer Approaches and Machine Learning-Based Approaches

Nucleotide k-mer-based approaches can be used to annotate sequences based on the presence and frequency of specific k-mers. Comparing these frequencies is computationally less demanding and faster than sequence alignment but requires a lot of computer memory. Even if most of the k-mer-based classification tools, such as Kraken [**103,104**], Kaiju [**105**], or Taxonomer [**106**], are not dedicated toward the detection of plant viruses, they can be used for such purpose. Kodoja [**107**] uses a combination of such tools for the taxonomic classification of plant viruses in metagenomic data. Most of the tools are not very user friendly, and the use of k-mer tools for plant virus detection is fairly new; thus, some questions remain to be answered, e.g., the usability of k-mer tools on small RNA datasets [**107**].

Methods based on machine learning are being developed for the detection of viral sequences in metagenomics datasets. Several tools have already been published, e.g., ViraMiner [**108**], DeepVirFinder [**109**], or Virnet [**110**] for human virus detection purpose. Given a metagenome with known composition, machine learning approaches attempt to find some meaningful patterns that allow differentiating the host from the virus. When the unknown metagenome dataset is provided, the software should be able to discriminate virus sequences from host sequences using the learnt pattern. Machine learning tools are new in this field; thus, we still lack their in-depth comparison with the more known approaches discussed above.

4.3.6. Which Analysis Approach Should I Choose?

The variety in similarity-based search approaches is striking. Choosing the most relevant one will depend on criteria such as the aims of the study (diagnostic, metagenomics) and the time/computational power available. Whichever program/approach is selected, it is important to consider its limitations and to properly set the key parameters to avoid false-positive or false-negative results. Fast programs can be used as a filtering step and then validated by slower approaches, or alternatively, two approaches can be used to validate each other, or multiple approaches can be used in parallel, for example an optimized approach for the detection of known viruses and a separate approach for novel virus discovery. If computational time or power is not a serious limitation, combining several approaches may enhance the ability to obtain an accurate annotation [**111**]. Here, we provide a checklist, identifying the most important considerations, which should be taken into account when analyzing HTS data (**Figure 3**).

Most important considerations to keep in mind during the data processing

I. **Quality control and sequence preprocessing**

   a. What is the average quality of the sequences? [For Illumina, the Phred values histogram have the peak around 37-40]

   b. Is the size distribution of sequences in accordance to library preparation approach? [For example, peak at 21-24 bp for sRNAs]

   c. Do you have a sufficient number of reads for the detection limit you want to achieve? [In general, we recommend 3 - 5 million reads (150-250 nucleotides long) per sample for total RNA-seq or 1-4 million reads for sRNAs. A million reads would suffice for both in most cases. However, in some cases, *e.g.*, for detection of viruses in fruit trees, much more reads will be needed.]

   d. Are there not too many read duplicates? [In case of lots of reads duplicates, for example> 20%, there might have been too many PCR cycles during the library preparation, leading to a low diversity library which lowers the limit of detection.]

   e. Are the adaptor, primer, barcode sequences, spiked sequences etc. removed?

   f. If the end of the reads is of lower quality, did you consider quality trimming?

II. ***De novo* assembly**

   a. Are the parameters set according to the input sequence data? [For example, k-mer length for de Brujin graph assemblers.]

   b. Are the cut-off values set to accommodate detection of widest possible range of viruses? [Coverage, contig length cut-offs: set contig length cut-off at low lengths, e.g., twice the length of the reads to detect also possible low-titer viruses assembled only in short contigs.]

III. **Similarity searches**

   a. Does the method or combination of methods you use allow for detections of known and unknown viruses and viroids? [Perform similarity searches both on level of nucleotide and translated protein sequences]

   b. Is the database used up to date?

   c. How reliable are the viral hits? Are the E-values etc. interpreted correspondingly to the used database? [Use more stringent filtering parameters or expect much more false-positive hits with smaller, e.g., virus only databases; check the relevant results manually and by another analysis approach.]

   d. What portion of the length of the viral genome is covered by the reads / contigs, and how many reads/contigs are assigned to the virus? [If only a very small fraction of genome is covered or very small number of reads is assigned, it might be a false positive.]

   e. What fraction of the reads is assigned to be of viral origin, and does this more or less agree with your expectations based on the literature and your experience? [The expected number of viral reads depends partially on factors you can control such as quality of RNA extraction, addition of rRNA removal step, but it can also be out of your control since this also depends on the host plant and the viral load]

   f. Can any of the hits be a process or index-crosstalk contaminations?

   g. Can any of the viral sequences correspond to inactive viral sequences integrated in the host genome or host sequences with reported similarity to host genes?

   h. What are the % identities between the reads/contigs and the detected virus? Are detected viruses new or known viral species (go to Figure 4)?

Figure 3. Checklist of the most important considerations to keep in mind during HTS data processing for detection of plant viruses.

Moreover, when analyzing the data obtained from long-read technologies, one should pay special attention to using approaches that enable the efficient processing of such data. Mapping algorithms have been developed for the processing of long read data with higher error rates, such as Minimap2 [**112**]. For BLASTx-like similarity searches, algorithms that can handle frame-shift mutations (caused by the relatively higher error rates), such as DIAMOND [**97**], are preferred. Assembly and polishing of long read data can improve further processing [**113**] and improve the chances for the correct identification of viral sequences in the data.

4.3.7. Taxonomic Classification

To assign viruses to taxonomic ranks, demarcation criteria specifically set for different viral genera need to be followed. Often, identities <75% at the nucleotide or protein level are indicative of a new viral species; however, the threshold might be also lower or higher, such as at <91% for begomoviruses. Identities <60% might be indicative of a new viral genus; however, the threshold might be also lower or higher, such as <45% within *Betaflexiviridae* family. As noted, these criteria differ substantially between virus families and genera, but up-to-date information is published by the International Committee on Taxonomy of Viruses (ICTV) in the latest taxonomy reports [**114,115**] that can be found online (**https://talk.ictvonline.org/taxonomy/**, accessed on 13 April 2021). Once a sequence is identified to a family or genus level, a pairwise sequence comparison (PASC) webtool [**116**] to support virus classification, hosted by NCBI (**https://www.ncbi.nlm.nih.gov/sutils/pasc/**, accessed on 13 April 2021), can quickly provide an indication on how a new sequence fits in that genus or family. In cases where virus sequence identity is near the limit of the identity cut-off values for different species, additional information and/or justification may be required for their definite classification. These could include biological information such as host species, vector species, or symptom types, and if enough isolates have been sequenced, population genomics approaches can also be employed [**117**].

Strains of viruses do not fall under official taxonomy. Rather, they are definitions utilized by communities of practice around virus species and would thus require a review of the literature concerning the specific virus species to be able to classify the sequence to a particular strain or phylotype. This is a process that generally includes phylogenetic analysis of the identified sequence with published virus (reference) sequences.

The approach described above can be rather straightforward if complete genomes of viruses with a single genome segment have been assembled. However, things can become more ambiguous in situations where a new virus has multiple genome segments or have been incompletely assembled, resulting in several contigs corresponding to different parts of a viral genome. The individual contigs for a novel virus may be equally distantly related to several known viruses and can then show the highest level of similarity with different viruses, which could lead to the erroneous interpretation that several new viruses are found in the same sample. This issue will often manifest itself in the previous step of similarity searches, and, to resolve this, the first recommended step is to identify the taxonomic position of all the best hits identified for the different viral contigs. If several best hits fall within the same genus or family, one could suspect they may correspond to the same virus. The next step would be to investigate the general viral genome structures in the identified genus or family from the ICTV reports and ascertain if the different best hits correspond to the same or different genomic regions for that type of virus. If they are all different, it is likely that a single new species is present; if the same region is covered by multiple contigs that differ significantly from each other, then the scenario of multiple new viruses belonging to a similar taxonomic group is more probable. A checklist in **Figure**

4 contains the most important points to keep in mind for the taxonomic classification of viral sequences obtained by HTS.

Taxonomic classification

I. **If you obtained one or more single apparently full-length sequences of clearly distinguishable viruses:**
   a. What taxonomic group does the virus correspond to, based on database annotations?
   b. What are the taxonomic demarcation criteria for the identified taxonomic group (https://talk.ictvonline.org/taxonomy/)?
   c. If falling within a known family or genus, how does the sequence fit, based on taxonomic criteria of that group (https://www.ncbi.nlm.nih.gov/sutils/pasc/)?
      i. If clearly falling within or outside of a taxonomic group based on sequence demarcation and genome organization criteria, define species or new species. Perform phylogenetic analysis with other isolates from same and related species for support.
         - Define strains based on literature if relevant.
      ii. If not clearly falling within or outside of the corresponding group, consult disciplinary literature for guidance, or define as unclassified related virus and refer to ICTV.
   d. If falling outside of known taxonomic groupings based on ICTV criteria, perform phylogenetic analysis of conserved proteins with most closely related virus groups to determine evolutionary position. Based on these analyses, suggestions can be made for new taxonomic groupings for consideration by the ICTV.

II. **If you obtained apparently partial sequences or sequences corresponding to multiple genome segments of one or more viruses:**
   a. Do sequences show highest similarity to same or different viruses?
      i. If highest similarity is always the same virus, follow checklist starting from step I.a. using each individual contig to check for consistency in step I.c. If inconsistent, perform phylogenetic analysis of individual contigs for evolutionary consistency.
      ii. If highest similarity is to different viruses, check if sequences correspond to the same taxonomic grouping at family or genus level
         - If yes, check if contigs cover the same or different parts of the viral genome
            - If contigs cover different parts of the genome, they probably correspond to a single virus, follow checklist starting from step I.a. using each individual contig to check for consistency in step I.c. If inconsistent, perform phylogenetic analysis of individual contigs for evolutionary consistency.
            - If contigs cover the same part of the genome, separate contigs covering similar regions and analyze them individually following the checklist from I.a. checking for consistency in step I.c. If inconsistent, perform phylogenetic analysis of individual contigs for evolutionary consistency.

Figure 4. Checklist of the most important considerations during taxonomic classification of plant viruses detected by HTS.

Sequences of new viruses belonging to previously undescribed families and/or genera can often only be reliably aligned by using the translated amino acid sequences of conserved genes such as polymerases and coat proteins. In these cases, phylogenies generated with viruses from related genera or families are needed to determine the exact taxonomic position. Additional criteria, such as number of open reading frames and overall genomic organization, need to be considered when classifying a virus as a member of a new genus or family. When there is uncertainty, viruses can be categorized as unclassified new species until new evidence arises that can support a definite classification.

Irrespective of the situation encountered, to become an officially recognized new species, generally, a near complete genome sequence, including the complete coding sequence information, is required by the ICTV to assign a "sequence only" virus to a species level. If relevant supportive biological data are available, that rule is more relaxed and will be determined by the relevant virus family study groups.

4.3.8. "Quick Start" Methods

Depending on the computational background of the user, there are different ways to approach the analysis. Many software solutions are available for detecting the presence of (plant) viruses in HTS datasets, which have been summarized recently by several reviews [**118,119**]. For beginners or newcomers in the field, all these tools can be overwhelming. The quick-start guide (**Figure 5**) might be handy to select an appropriate tool or pipeline.

Quick-start guide to start analyzing HTS data for virus detection

Where do I get (test) data?

Using well-characterized datasets is crucial to evaluate the classical performance criteria of an analysis pipeline, such as diagnostic sensitivity (depending on false negatives), reproducibility and false discovery rate (depending on false positives).

What?	More information	Links
10 Illumina sRNA datasets used in performance testing study involving 21 labs	https://doi.org/10.1094/PHYTO-02-18-0067-R (accessed on 13. 4. 2021)	https://github.com/plantvirology/COST_Action_PT/releases (accessed on 13. 4. 2021).
7 semi-artificial datasets composed of real Illumina RNA-seq datasets from virus-infected plants spiked with artificial virus reads, 3 real datasets and 8 completely artificial datasets. Each dataset addresses specific challenges that could prevent virus detection.	https://doi.org/10.5281/zenodo.4584718 (accessed on 13. 4. 2021)	https://gitlab.com/ilvo/VIROMOCKchallenge (accessed on 13. 4. 2021).

How do I choose an analysis pipeline?

The choice of a suitable analysis pipeline depends on the type of data, the application, available resources and bioinformatics skills. Regardless of these considerations, each pipeline must roughly contain the different analysis steps as explained in the main text (chapter 4) and in Figure 2. Some suggestions for pipelines for analyzing Illumina RNA-seq data for virus detection are given below (summarized on https://gitlab.com/ilvo/phbn-wp2-training) (accessed on 13. 4. 2021).

Bioinformatics skill level	Available resources	Recommended type of pipeline	Suggested software (more info: Table 2)
Low to moderate	Low	Web- or cloud-based tool	VirFind*, VirusDetect*, IDTaxa, Kaiju **dedicated to plant virus detection*
Low to moderate	Moderate, willing to pay license fee	GUI-based commercial software	CLC Genomics Workbench, Geneious Prime *Pre-built pipelines available at: https://gitlab.com/ilvo/phbn-wp2-training* (accessed on 13. 4. 2021).
Low to moderate	Moderate, limited to open source software	GUI-based open source software	VirTool, Galaxy with Kodoja plug-in installed *Ask your IT department to set up a local instance*
Moderate to high	Moderate to high (Linux-based OS)	Dedicated command-line software packages	VirusDetect, virAnnot, Kodoja[a], Angua[a] *[a]Available as conda package, which eases installation*
High	Moderate to high (Linux-based OS)	Custom-built pipeline combining different command-line software packages	Combination of selected tools for each step mentioned in Figure 2, automated using a shell script or pipeline building software (e.g., Snakemake, Nextflow).

How do I interpret the data?

The interpretation of the results is highly dependent on the pipeline you use. Make yourself familiar with the different steps of the chosen pipeline and possible drawbacks of each step by thoroughly reading the manual(s). A helpful guide to identify the weak points of your pipeline can be the checklist in Figure 3. Also, the taxonomic classification of your sequences should not be taken for granted, and should be considered carefully as explained in Figure 4. Finally, a confirmation of your virus/viroid presence by an independent technique is strongly recommended as discussed in chapter 4.4.1.

Figure 5. Quick-start guide assisting selection of analysis approaches for plant virus detection from HTS data.

Among these options, easy-to-use pipelines that do not require extensive computational expertise might be a good start. These pipelines present a user-friendly interface on-line or directly on the computer. A first group of pipelines can be considered as "all in one": they automatically start on the raw data to deliver the final results as a list of viruses detected. They may or may not allow the adaptation of parameters. A second group corresponds to pipelines for which the different steps of the process have to be done separately and independently. This is the case when using commercial software such as CLC Genomics

Workbench or Geneious Prime, which both also enable the building of customized "all-in-one" workflows. **Table A1** summarizes the pros and cons of the most common "easy-to-use" analysis solutions. Ease of use may generate a false sense of confidence in the results and, as with all pipelines, understanding of the steps and the parameters of the pipelines, as well as critical interpretation of the results is always required.

4.4. What to Do When The Data Analysis Is Concluded?

4.4.1. Identity Confirmation by an Independent Technique

As for many other test methods, HTS may sometimes provide false-positive results. Therefore, if consequential, it is important that HTS results are confirmed.

The need to confirm the identity of a pest depends on the context of the analysis and on the type of organism identified (e.g., identification of a quarantine compared to an endemic pest). The results must be confirmed in cases considered critical to national or international plant protection programs. These are the detection of a pest in an area where it is not known to occur or in a consignment originating from a country where it is declared to be absent; and also, when a pest is identified by a laboratory for the first time (EPPO PM 7/76, 2019). The identity of any uncharacterized pest with potential risks to plant health should also be confirmed by another test. Whilst a virus in its common host is unlikely to require confirmation (if not regulated), it may be useful if associated with different symptoms (e.g., an emerging strain).

When confirmation is needed, it is recommended to use a test or a combination of tests based on different biological principles (e.g., ELISA or targeted PCR instead of resequencing the sample using the same protocol). If available, validated tests should be used and a new sample extract obtained for analysis. The selection of confirmatory tests depends on the performance characteristics required; the general characteristics of methods for plant virology have been reviewed [**120**]. If no other tests are available to confirm the identity of the pest (i.e., poorly characterized and uncharacterized organisms), primers should be designed and tested, based on the HTS sequence data and available sequence information in the sequence databases. Alternatively, generic primers that enable the amplification of viruses within a genus or family, including the targeted one(s), followed by Sanger sequencing of the amplicons could be used to confirm the identity.

4.4.2. Biological Characterization Post HTS Detection

Based on HTS, the list of thus far unknown or poorly characterized viruses for which only genome data are available is rapidly increasing [**121**]. This presents a challenge for the further steps necessary to determine the causative relationship to a disease and guide phytosanitary diagnostic laboratories on data interpretation and recommendations. Viruses for which only genome data are available can indeed be taxonomically assigned, but the real challenge is to attribute biological meaning to their detection. The interpretation of the biological relevance applies mainly to poorly characterized and uncharacterized or newly discovered viruses. For example, the viral sequences detected may correspond to a *bona fide* virus infecting other organisms associated with the sample, including bacteria, fungi, or arthropods [**122,123**] or to viral sequences integrated into the plant genome [**124,125**]. As stated previously [**125**], relevant scientific expertise is essential for sound biological interpretation of HTS results, in particular when identifying a target with a low titer, a poorly characterized species, an uncharacterized organism, or sequences integrated in the host genome [**6**,**126**]. In this latter case, careful phylogenetic analysis, including retrotransposons and viruses reported only from integration events in plant genomes [**91**,**92**,**93**] may provide critical information on whether the sequences identified correspond to an autonomously replicating (episomal) virus or to cellular transcripts from integrated viral elements. This may need to be validated by specific experiments to confirm or disprove an episomal replication scenario.

The extent to which additional biological characterization is performed depends largely on the potential risk the organism(s) would pose to plant health, although the acquisition of such data may take time or may not be possible (e.g., lack of human and/or financial resources). The scaled and progressive scientific framework proposed by Massart et al. [**125**] is a useful tool for guiding the biological characterization and the risk assessment of an uncharacterized or poorly characterized plant virus detected by HTS.

4.4.3. Sharing Data to Leverage Knowledge

After the detection of the virus in the laboratory, the researcher or diagnostician faces an important dilemma: when and how to share data publicly. As shown by recent examples [**127**,**128**,**129**], pre-publication data sharing between laboratories brings valuable information to address the risks raised by a virus. Sharing data will give a more global picture of its geographical repartition, its genetic diversity, its host range and symptomatology, allowing a contextualized risk analysis and avoiding unnecessary regulatory action. When shared, the genome information usefulness is leveraged. Data sharing must also include metadata from the sample (e.g., origin, species, cultivar, time point, organ of sampling). Nevertheless, data sharing is not always easy due to regulatory implications, and for commercial work, laboratories may be bound by confidentiality agreements [**7**]. In addition to sharing sequence data itself, sharing of analysis pipelines, protocols, and experiences between labs can greatly contribute to the harmonization of the field and provide useful resources for newcomers to the field. The recently established Plant Health Bioinformatics Network (PHBN) aims to foster this approach and provide protocols, pipelines (**https://gitlab.com/ilvo/phbn-wp2-training**, accessed on 13 April 2021), and reference datasets (**https://gitlab.com/ilvo/VIROMOCKchallenge**, accessed on 13 April 2021) [**130**] that can be widely employed. It also aims to organize community efforts to advance certain aspects of plant health bioinformatics (**https://gitlab.com/ilvo/PHBN-WP4-RNAseq_Community_Screening**, accessed on 13 April 2021).

4.4.4. Recombination Analysis

Recombination is common in some genera of plant viruses, and the presence of recombination events can have impacts on downstream analysis such as phylogenetics. Thus, identification of recombination is a useful first step, prior to further genome analysis. The most popular software solutions, which detect recombination patterns comparing full or partial viral genomes and run on Windows, are RDP4 [**131**], SimPlot [**132**], and TreeOrder Scan [**133**]. ViReMa (Viral Recombination Mapper) can be used for the detection of recombination junctions, as well as insertion/substitution events and multiple recombinations within single reads [**134**], and it has been successfully applied for the analysis of recombination events in plant virus genomes [**22**,**135**,**136**].

4.4.5. Additional Bioinformatics Analyses

Further analyses, beyond viral detection and taxonomic classification, can be performed on HTS data, depending on the goal of the study. For instance, the large amount of sequence data generated by HTS allows a good resolution of the within-host genetic diversity of the viral populations [**22**]. Assessing the genetic diversity within and among viral populations can provide a better understanding of virus evolution and help to determine population genetic parameters or epidemiological patterns [**137,138**]. This can be done using single nucleotide polymorphism (SNP) calling algorithms, which need to allow the detection of low-frequency variants expected in virus populations. Phylogenetic relationships among the detected and previously known viruses can also be investigated using fast neighbor-joining algorithms [**139**], more precise maximum likelihood approaches [**140,141**], or Bayesian analysis approaches [**142**]. Freeware phylogenetic analysis suites, such as MEGAN [**143**], or phylogenetic analysis algorithms integrated within commercial software, such as CLC Genomics Workbench and Geneious Prime, can be used. Studying the time of emergence of viral species and strains including the distribution of the genetic diversity across geographical sites can be done using software such as BEAST [**144**], TempEst [**145**] and SPAGeDI [**146**].

5. Conclusions and Outlook

In this review, we aimed to provide an informative primer on the generation and analysis of HTS data for the detection of plant viruses. Even though the field of HTS is transforming rapidly and new platforms and analysis tools are being developed constantly, the basic concepts of data analysis reviewed here will remain relevant in the future. In the next few years, we expect a great increase in the use of the long-read HTS platforms. New algorithms and pipelines for analysis of data will continually be developed, building on some of the concepts described above. These developments are likely to focus on two main areas. Firstly, the adoption of deep learning approaches will likely be more and more integrated into the field of virus detection, on different levels, from similarity searches to the estimation of detection confidence levels, to enable the more robust detection of virus sequences that are more distantly related to those we currently recognize. Secondly, with the further development of nanopore sequencing-based platforms, potentially facilitating on-site HTS analysis of samples, we will need faster and more memory-efficient analysis approaches to enable rapid data analysis, potentially away from centralized facilities. Moreover, guidelines are being developed to enable the validation and verification of HTS-based detection of plant pathogens in research and diagnostic settings, which also include bioinformatics steps of the analysis [**9**]. These guidelines will provide detailed information on how to use appropriate controls and which specific results parameters to use to ensure the validity of the results, which is briefly covered in **Figure 3** and **Figure 4** in this text. Finally, we encourage the readers to use this guide as a starting point for the selection of appropriate analysis approaches and to get further informed about the specifics of the algorithms (**Figure 5**). By combining knowledge on the analysis approaches with a sound plant virology background, we can maximize the potential of these technologies and provide sound interpretation of the results.

Appendix A

Table A1. List of selected easy-to-use analysis solutions for detection of plant viruses with their pros and cons.

Pipeline	Brief Description	Web Link/Publication	Pros	Cons
VirusDetect	Virus discovery using sRNA and RNAseq sequences	http://virusdetect.feilab.net, accessed on 13 April 2021 [147]	• Easy to use: single command to run one or multiple datasets simultaneously. • Performs *de novo* assembly and reference mapping in parallel, including optional host genome subtraction and identified contigs through BLASTn and BLASTx. • Automatic results organization and presentation in html table providing key metrics on coverage, sequence depth, virus and genus name, and link to visual map and NCBI GenBank reference sequence. • Options to modify key assembly, mapping, and reporting parameters. • Windows version with visual interface and automatic quality control and trimming to be released in 2021. • Available via user account online.	• Uses complete NCBI GenBank database for viruses (divided along host type) for reference mapping and identity searches. NCBI GenBank sequences are poorly curated and may lead to reports of wrong results. • Creating and formatting new custom or up-to-date NCBI GenBank reference library is not very straightforward and ready formatted updates are not uploaded very regularly to the VirusDetect webpage. • Currently requires Linux environment, which is an impediment for many diagnosticians. • Default reporting cutoff settings are optimized for siRNA to minimize false positives due to index-hopping; however, they may lead to the non-reporting of low concentration viruses.

Pipeline	Brief Description	Web Link/Publication	Pros	Cons
Virtool	HTS sample manager with virus detection, discovery and analysis workflows	www.virtool.ca, accessed on 13 April 2021 https://github.com/virtool/virtool, accessed on 13 April 2021 [36]	• Open source modern graphical optimized for cloud computing. • User and group control with password protection, sample data management, security, and QA features. • Support for multiple workflows and versioned databases for viral and non-viral pathogens. • Can process short and long reads (Illumina). • Result visualization, filtering, and sorting. • HTTP API for automation or integration with other services such as LIMS. • Can also be controlled via the command line for more complex tasks.	• Requires some computational skills for user (or help of informatician) to install as a local server on Linux operating system. • Limited ability to change parameters within a workflow.
virAnnot	Command-line tool for virus detection and viral diversity estimation	[148]	• Wide options to modify assembly, mapping, annotation, and clustering parameters. • Performs parallel analysis of samples from the same dataset. • Estimation of viral diversity through Operational Taxonomic Units (OTUs). • Easy results visualization with Krona and phylogenic trees.	• Requires a Linux environment, which is an impediment for many diagnosticians. • Need a cluster access for the annotation step. • Requires a good knowledge of command-line and Unix packages installation.
VirFind	Online virus discovery tool	http://virfind.org, accessed on 13 April 2021 [149]	• Available via user account online. • Performs reference mapping, *de novo* assembly, and conserved domain searches in parallel or subsequently.	• Analysis by online version can take several days. • Output only in text files: experience needed for further interpretation.

Pipeline	Brief Description	Web Link/Publication	Pros	Cons
Angua	Command-line tool for virus detection	https://fred.fera.co.uk/smcgreig/angua3, accessed on 13 April 2021	• Simple: can be executed with one command but has a number of parameters/tools that can be tweaked • Uses full nt and nr GenBank databases so is sensitive • Manual inspection of results with a local MEGAN installation improves accuracy • Supports single and paired-end analysis • Supports BLASTn/MEGAN parallelization	• Requires a Linux environment, which is an impediment for many diagnosticians. • Dependent on locally stored nt and nr GenBank databases. • BLASTx stage can take a long time. • Manual inspection of results with a local MEGAN installation is required.
Kodoja	k-mer based command-line tool for virus detection	https://github.com/abaizan/kodoja, accessed on 13 April 2021 [107]	• Available as Galaxy plug-in or as command-line tool that can be installed using conda. • k-mer based rather than assembly and mapping, which makes it more sensitive and computationally less intensive.	• Requires a Linux environment for the command-line tool, which is an impediment for many diagnosticians.
Truffle	Targeted virus detection using e-probes based approach	[150]	• Results easy to interpret, good sensitivity. • Requires relatively low computational resources.	• Undescribed virus or viral strain will not be detectable using this pipeline. • Only grapevine and citrus viruses are available; however, e-probes for other viruses can be designed. • Requires a Linux environment, which is an impediment for many diagnosticians.
Kaiju	Online metagenomic analysis tool	http://kaiju.binf.ku.dk/, accessed on 13 April 2021 [105]	• Both standalone and web server available. • Quick analysis not requiring any knowledge in bioinformatics and data analysis. • Prepared downloadable databases available.	• Not specifically made for virus detection. • Protein based, hence blind for non-coding sequences (viroids, satellites).

Pipeline	Brief Description	Web Link/Publication	Pros	Cons
Galaxy	Workflow system for computational analyses	https://usegalaxy.org, accessed on 13 April 2021 [151]	• Web-based platform. • Open source. • Vast choice of computational biology tools.	• Limit in data upload, unless if you establish own local galaxy server. • Not specifically made for virus detection.
ID-Seq	Online metagenomic analysis tool	https://idseq.net/, accessed on 13 April 2021 [152]	• Easy-to-use visual interface of results. • Quick analysis not requiring any knowledge in bioinformatics and data analysis.	• Not possible to change parameters of the workflow. • Complementary software needed for reads alignment. • Not specifically made for virus detection.
Geneious Prime	Software for molecular biology and sequence analysis	https://www.geneious.com, accessed on 13 April 2021	• Graphical interface. • Multiple plugins available, including some frequently used freeware assembly algorithms. • Automated, customizable workflows. • Constant release of updated versions and customer support. • Nice and efficient visualization tools. • Free trial version available.	• Licensed, including license fee; • HTS data analysis requires computational resources.
CLC Genomics Workbench	Comprehensive software solution of molecular biology analysis tools	https://digitalinsights.qiagen.com/products-overview/discovery-insights-portfolio/analysis-and-visualization/qiagen-clc-genomics-workbench/, accessed on 13 April 2021	• Graphical interface. • Automated, customizable workflows. • Constant release of updated versions and customer support. • Nice and efficient visualization tools. • Free trial version available.	• Expensive ongoing licensing fee. • HTS data analysis requires computational resources.

References

1. Villamor, D.E.V.; Ho, T.; Al Rwahnih, M.; Martin, R.R.; Tzanetakis, I.E. High throughput sequencing for plant virus detection and discovery. *Phytopathology* **2019**, *109*, 716–725. [**Google Scholar**] [**CrossRef**]

2. Kreuze, J.F.; Perez, A.; Untiveros, M.; Quispe, D.; Fuentes, S.; Barker, I.; Simon, R. Complete viral genome sequence and discovery of novel viruses by deep sequencing of small RNAs: A generic method for diagnosis, discovery and sequencing of viruses. *Virology* **2009**, *388*, 1–7. [**Google Scholar**] [**CrossRef**] [**PubMed**] [**Green Version**]

3. Adams, I.P.; Glover, R.H.; Monger, W.A.; Mumford, R.; Jackeviciene, E.; Navalinskiene, M.; Samuitiene, M.; Boonham, N. Next-generation sequencing and metagenomic analysis: A universal diagnostic tool in plant virology. *Mol. Plant Pathol.* **2009**, *10*, 537–545. [**Google Scholar**] [**CrossRef**] [**PubMed**]

4. Al Rwahnih, M.; Daubert, S.; Golino, D.; Rowhani, A. Deep sequencing analysis of RNAs from a grapevine showing syrah decline symptoms reveals a multiple virus infection that includes a novel virus. *Virology* **2009**, *387*, 395–401. [**Google Scholar**] [**CrossRef**] [**PubMed**] [**Green Version**]

5. Donaire, L.; Wang, Y.; Gonzalez-Ibeas, D.; Mayer, K.F.; Aranda, M.A.; Llave, C. Deep-sequencing of plant viral small RNAs reveals effective and widespread targeting of viral genomes. *Virology* **2009**, *392*, 203–214. [**Google Scholar**] [**CrossRef**] [**Green Version**]

6. Massart, S.; Chiumenti, M.; De Jonghe, K.; Glover, R.; Haegeman, A.; Koloniuk, I.; Komínek, P.; Kreuze, J.; Kutnjak, D.; Lotos, L.; et al. Virus detection by high-throughput sequencing of small RNAs: Large-scale performance testing of sequence analysis strategies. *Phytopathology* **2019**, *109*, 488–497. [**Google Scholar**] [**CrossRef**] [**PubMed**] [**Green Version**]

7. Olmos, A.; Boonham, N.; Candresse, T.; Gentit, P.; Giovani, B.; Kutnjak, D.; Liefting, L.; Maree, H.J.; Minafra, A.; Moreira, A.; et al. High-throughput sequencing technologies for plant pest diagnosis: Challenges and opportunities. *EPPO Bull.* **2018**, *48*, 219–224. [**Google Scholar**] [**CrossRef**] [**Green Version**]

8. Weymann, D.; Laskin, J.; Roscoe, R.; Schrader, K.A.; Chia, S.; Yip, S.; Cheung, W.Y.; Gelmon, K.A.; Karsan, A.; Renouf, D.J.; et al. The cost and cost trajectory of whole-genome analysis guiding treatment of patients with advanced cancers. *Mol. Genet. Genomic Med.* **2017**, *5*, 251–260. [**Google Scholar**] [**CrossRef**] [**PubMed**]

9. Valitest EU Project Consortium Guidelines for the Selection, Development, Validation and Routine Use of High-Throughput Sequencing Analysis in Plant Health Diagnostic Laboratories: Grant Agreement N. 773139: Deliverable N° 2.2. (Confidential). 2020. Available online: **https://www.valitest.eu/work_packages/** (accessed on 13 April 2021).

10. Maliogka, V.I.; Minafra, A.; Saldarelli, P.; Ruiz-García, A.B.; Glasa, M.; Katis, N.; Olmos, A. Recent advances on detection and characterization of fruit tree viruses using high-throughput sequencing technologies. *Viruses* **2018**, *10*, 436. [**Google Scholar**] [**CrossRef**] [**Green Version**]

11. Roossinck, M.J. Deep sequencing for discovery and evolutionary analysis of plant viruses. *Virus Res.* **2017**, *239*, 82–86. [**Google Scholar**] [**CrossRef**] [**PubMed**]

12. Roossinck, M.J.; Martin, D.P.; Roumagnac, P. Plant virus metagenomics: Advances in virus discovery. *Phytopathology* **2015**, *105*, 716–727. [**Google Scholar**] [**CrossRef**] [**Green Version**]

13. Marais, A.; Faure, C.; Bergey, B.; Candresse, T. Viral double-stranded RNAs (dsRNAs) from plants: Alternative nucleic acid substrates for high-throughput sequencing. In *Viral Metagenomics: Methods and Protocols*; Pantaleo, V., Chiumenti, M., Eds.; Humana Press: New York, NY, USA, 2018; pp. 45–53. ISBN 978-1-4939-7682-9. [**Google Scholar**]

14. Massart, S.; Olmos, A.; Jijakli, H.; Candresse, T. Current impact and future directions of high throughput sequencing in plant virus diagnostics. *Virus Res.* **2014**, *188*, 90–96. [**Google Scholar**] [**CrossRef**]

15. Pecman, A.; Kutnjak, D.; Gutiérrez-Aguirre, I.; Adams, I.; Fox, A.; Boonham, N.; Ravnikar, M. Next generation sequencing for detection and discovery of plant viruses and viroids: Comparison of two approaches. *Front. Microbiol.* **2017**, *8*. [**Google Scholar**] [**CrossRef**] [**Green Version**]

16. Boone, M.; De Koker, A.; Callewaert, N. Survey and summary capturing the "ome": The expanding molecular toolbox for RNA and DNA library construction. *Nucleic Acids Res.* **2018**, *46*, 2701–2721. [**Google Scholar**] [**CrossRef**]

17. Visser, M.; Bester, R.; Burger, J.T.; Maree, H.J. Next-generation sequencing for virus detection: Covering all the bases. *Virol. J.* **2016**, *13*, 4–9. [**Google Scholar**] [**CrossRef**] [**PubMed**] [**Green Version**]

18. Idris, A.; Al-Saleh, M.; Piatek, M.J.; Al-Shahwan, I.; Ali, S.; Brown, J.K. Viral metagenomics: Analysis of begomoviruses by illumina high-throughput sequencing. *Viruses* **2014**, *6*, 1219–1236. [**Google Scholar**] [**CrossRef**] [**Green Version**]

19. Sukal, A.C.; Kidanemariam, D.B.; Dale, J.L.; Harding, R.M.; James, A.P. Assessment and optimization of rolling circle amplification protocols for the detection and characterization of badnaviruses. *Virology* **2019**, *529*, 73–80. [**Google Scholar**] [**CrossRef**] [**PubMed**]

20. Wyant, P.S.; Strohmeier, S.; Schäfer, B.; Krenz, B.; Assunção, I.P.; de Andrade Lima, G.S.; Jeske, H. Circular DNA genomics (circomics) exemplified for geminiviruses in bean crops and weeds of northeastern Brazil. *Virology* **2012**, *427*, 151–157. [**Google Scholar**] [**CrossRef**] [**PubMed**] [**Green Version**]

21. Vivek, A.T.; Zahra, S.; Kumar, S. From current knowledge to best practice: A primer on viral diagnostics using deep sequencing of virus-derived small interfering RNAs (vsiRNAs) in infected plants. *Methods* **2020**, *183*, 30–37. [**Google Scholar**] [**CrossRef**]

22. Kutnjak, D.; Rupar, M.; Gutierrez-Aguirre, I.; Curk, T.; Kreuze, J.F.; Ravnikar, M. Deep sequencing of virus-derived small interfering RNAs and RNA from viral particles shows highly similar mutational landscapes of a plant virus population. *J. Virol.* **2015**, *89*, 4760–4769. [**Google Scholar**] [**CrossRef**] [**PubMed**] [**Green Version**]

23. Seguin, J.; Rajeswaran, R.; Malpica-López, N.; Martin, R.R.; Kasschau, K.; Dolja, V.V.; Otten, P.; Farinelli, L.; Pooggin, M.M. De novo reconstruction of consensus master genomes of plant RNA and DNA viruses from siRNAs. *PLoS ONE* **2014**, *9*, e88513. [**Google Scholar**] [**CrossRef**] [**PubMed**]

24. Smith, O.; Clapham, A.; Rose, P.; Liu, Y.; Wang, J.; Allaby, R.G. A complete ancient RNA genome: Identification, reconstruction and evolutionary history of archaeological Barley Stripe Mosaic Virus. *Sci. Rep.* **2014**, *4*, 4003. [**Google Scholar**] [**CrossRef**] [**Green Version**]

25. Turco, S.; Golyaev, V.; Seguin, J.; Gilli, C.; Farinelli, L.; Boller, T.; Schumpp, O.; Pooggin, M.M. Small RNA-omics for virome reconstruction and antiviral defense characterization in mixed infections of cultivated solanum plants. *Mol. Plant-Microbe Interact.* **2018**, *31*, 707–723. [**Google Scholar**] [**CrossRef**] [**PubMed**] [**Green Version**]

26. Melcher, U.; Muthukumar, V.; Wiley, G.B.; Min, B.E.; Palmer, M.W.; Verchot-Lubicz, J.; Ali, A.; Nelson, R.S.; Roe, B.A.; Thapa, V.; et al. Evidence for novel viruses by analysis of nucleic acids in virus-like particle fractions from Ambrosia psilostachya. *J. Virol. Methods* **2008**, *152*, 49–55. [**Google Scholar**] [**CrossRef**] [**PubMed**]

27. Muthukumar, V.; Melcher, U.; Pierce, M.; Wiley, G.B.; Roe, B.A.; Palmer, M.W.; Thapa, V.; Ali, A.; Ding, T. Non-cultivated plants of the tallgrass prairie preserve of northeastern oklahoma frequently contain virus-like sequences in particulate fractions. *Virus Res.* **2009**, *141*, 169–173. [**Google Scholar**] [**CrossRef**] [**PubMed**]

28. Bernardo, P.; Charles-Dominique, T.; Barakat, M.; Ortet, P.; Fernandez, E.; Filloux, D.; Hartnady, P.; Rebelo, T.A.; Cousins, S.R.; Mesleard, F.; et al. Geometagenomics illuminates the impact of agriculture on the distribution and prevalence of plant viruses at the ecosystem scale. *ISME J.* **2018**, *12*, 173–184. [**Google Scholar**] [**CrossRef**] [**Green Version**]

29. Filloux, D.; Dallot, S.; Delaunay, A.; Galzi, S.; Jacquot, E.; Roumagnac, P. Metagenomics approaches based on virion-associated nucleic acids (VANA): An innovative tool for assessing without a priori viral diversity of plants. *Methods Mol. Biol.* **2015**, *1302*, 249–257. [**Google Scholar**] [**CrossRef**] [**PubMed**]

30. Ma, Y.; Marais, A.; Lefebvre, M.; Theil, S.; Svanella-Dumas, L.; Faure, C.; Candresse, T. Phytovirome analysis of wild plant populations: Comparison of double-stranded rna and virion-associated nucleic acid metagenomic approaches. *J. Virol.* **2019**, *94*. [**Google Scholar**] [**CrossRef**] [**PubMed**]

31. Roossinck, M.J. Plants, viruses and the environment: Ecology and mutualism. *Virology* **2015**, *479–480*, 271–277. [**Google Scholar**] [**CrossRef**] [**PubMed**] [**Green Version**]

32. Hull, R. Origins and evolution of plant viruses. In *Plant Virology*; Elsevier: London, UK, 2014; pp. 423–476. [**Google Scholar**]

33. Al Rwahnih, M.; Daubert, S.; Golino, D.; Islas, C.; Rowhani, A. Comparison of next-generation sequencing versus biological indexing for the optimal detection of viral pathogens in grapevine. *Phytopathology* **2015**, *105*, 758–763. [**Google Scholar**] [**CrossRef**] [**PubMed**] [**Green Version**]

34. Kesanakurti, P.; Belton, M.; Saeed, H.; Rast, H.; Boyes, I.; Rott, M. Screening for plant viruses by next generation sequencing using a modified double strand RNA extraction protocol with an internal amplification control. *J. Virol. Methods* **2016**, *236*, 35–40. [**Google Scholar**] [**CrossRef**] [**PubMed**]

35. Loconsole, G.; Saldarelli, P.; Doddapaneni, H.; Savino, V.; Martelli, G.P.; Saponari, M. Identification of a single-stranded DNA virus associated with citrus chlorotic dwarf disease, a new member in the family geminiviridae. *Virology* **2012**, *432*, 162–172. [**Google Scholar**] [**CrossRef**] [**PubMed**] [**Green Version**]

36. Rott, M.; Xiang, Y.; Boyes, I.; Belton, M.; Saeed, H.; Kesanakurti, P.; Hayes, S.; Lawrence, T.; Birch, C.; Bhagwat, B.; et al. Application of next generation sequencing for diagnostic testing of tree fruit viruses and viroids. *Plant Dis.* **2017**, *101*, 1489–1499. [**Google Scholar**] [**CrossRef**] [**PubMed**] [**Green Version**]

37. Weber, F.; Wagner, V.; Rasmussen, S.B.; Hartmann, R.; Paludan, S.R. Double-stranded RNA is produced by positive-strand RNA viruses and DNA viruses but not in detectable amounts by negative-strand RNA viruses. *J. Virol.* **2006**, *80*, 5059–5064. [**Google Scholar**] [**CrossRef**] [**PubMed**] [**Green Version**]

38. Gaafar, Y.Z.A.; Ziebell, H. Comparative study on three viral enrichment approaches based on RNA extraction for plant virus/viroid detection using high-throughput sequencing. *PLoS ONE* **2020**, *15*, e0237951. [**Google Scholar**] [**CrossRef**] [**PubMed**]

39. Thapa, V.; McGlinn, D.J.; Melcher, U.; Palmer, M.W.; Roossinck, M.J. Determinants of taxonomic composition of plant viruses at the nature conservancy's tallgrass prairie preserve, Oklahoma. *Virus Evol.* **2015**, *1*, vev007. [**Google Scholar**] [**CrossRef**] [**PubMed**]

40. Blouin, A.G.; Ross, H.A.; Hobson-Peters, J.; O'Brien, C.A.; Warren, B.; MacDiarmid, R. A new virus discovered by immunocapture of double-stranded RNA, a rapid method for virus enrichment in metagenomic studies. *Mol. Ecol. Resour.* **2016**, *16*, 1255–1263. [**Google Scholar**] [**CrossRef**] [**PubMed**] [**Green Version**]

41. Kobayashi, K.; Tomita, R.; Sakamoto, M. Recombinant plant dsRNA-binding protein as an effective tool for the isolation of viral replicative form dsRNA and universal detection of RNA viruses. *J. Gen. Plant Pathol.* **2009**, *75*, 87–91. [**Google Scholar**] [**CrossRef**]

42. Roossinck, M.J.; Saha, P.; Wiley, G.B.; Quan, J.; White, J.D.; Lai, H.; Chavarría, F.; Shen, G.; Roe, B.A. Ecogenomics: Using massively parallel pyrosequencing to understand virus ecology. *Mol. Ecol.* **2010**, *19*, 81–88. [**Google Scholar**] [**CrossRef**]

43. Chalupowicz, L.; Dombrovsky, A.; Gaba, V.; Luria, N.; Reuven, M.; Beerman, A.; Lachman, O.; Dror, O.; Nissan, G.; Manulis-Sasson, S. Diagnosis of plant diseases using the nanopore sequencing platform. *Plant Pathol.* **2019**, *68*, 229–238. [**Google Scholar**] [**CrossRef**]

44. Lusk, R.W. Diverse and widespread contamination evident in the unmapped depths of high throughput sequencing data. *PLoS ONE* **2014**, *9*, e110808. [**Google Scholar**] [**CrossRef**] [**Green Version**]

45. Laurence, M.; Hatzis, C.; Brash, D.E. Common contaminants in next-generation sequencing that hinder discovery of low-abundance microbes. *PLoS ONE* **2014**, *9*, e97876. [**Google Scholar**] [**CrossRef**] [**PubMed**]

46. Schmieder, R.; Edwards, R. Fast identification and removal of sequence contamination from genomic and metagenomic datasets. *PLoS ONE* **2011**, *6*, e17288. [**Google Scholar**] [**CrossRef**] [**PubMed**] [**Green Version**]

47. Naccache, S.N.; Greninger, A.L.; Lee, D.; Coffey, L.L.; Phan, T.; Rein-Weston, A.; Aronsohn, A.; Hackett, J.; Delwart, E.L.; Chiu, C.Y. The perils of pathogen discovery: Origin of a novel parvovirus-like hybrid genome traced to nucleic acid extraction spin columns. *J. Virol.* **2013**, *87*, 11966–11977. [**Google Scholar**] [**CrossRef**] [**PubMed**] [**Green Version**]

48. Martin, M. Cutadapt removes adapter sequences from high-throughput sequencing reads. *EMBnet. J.* **2011**, *17*, 10. [**Google Scholar**] [**CrossRef**]

49. *Illumina bcl2fastq and bcl2fastq2 Conversion Software*; v.2.20; Illumina: San Diego, CA, USA, 2019; Available online: **https://emea.support.illumina.com/sequencing/sequencing_software/bcl2fastq-conversion-software.html** (accessed on 13 April 2021).

50. Oxford Nanopore Technologies Guppy: Local Accelerated Basecalling for Nanopore Data. Available online: **https://community.nanoporetech.com/downloads** (accessed on 13 April 2021).

51. Illumina Effects of Index Misassignment on Multiplexing and Downstream Analysis (770-2017-004-D). Available online: **https://www.illumina.com/content/dam/illumina-marketing/documents/products/whitepapers/index-hopping-white-paper-770-2017-004.pdf** (accessed on 13 April 2021).

52. van der Valk, T.; Vezzi, F.; Ormestad, M.; Dalén, L.; Guschanski, K. Index hopping on the Illumina HiseqX platform and its consequences for ancient DNA studies. *Mol. Ecol. Resour.* **2020**, *20*, 1171–1181. [**Google Scholar**] [**CrossRef**]

53. MacConaill, L.E.; Burns, R.T.; Nag, A.; Coleman, H.A.; Slevin, M.K.; Giorda, K.; Light, M.; Lai, K.; Jarosz, M.; McNeill, M.S.; et al. Unique, dual-indexed sequencing adapters with UMIs effectively eliminate index cross-talk and significantly improve sensitivity of massively parallel sequencing. *BMC Genom.* **2018**, *19*, 30. [**Google Scholar**] [**CrossRef**] [**PubMed**] [**Green Version**]

54. Bolger, A.M.; Lohse, M.; Usadel, B. Trimmomatic: A flexible trimmer for Illumina sequence data. *Bioinformatics* **2014**, *30*, 2114–2120. [**Google Scholar**] [**CrossRef**] [**PubMed**] [**Green Version**]

55. Wick, B. Porechop. Available online: **https://github.com/rrwick/Porechop** (accessed on 13 April 2021).

56. De Coster, W.; D'Hert, S.; Schultz, D.T.; Cruts, M.; Van Broeckhoven, C. NanoPack: Visualizing and processing long-read sequencing data. *Bioinformatics* **2018**, *34*, 2666–2669. [**Google Scholar**] [**CrossRef**] [**PubMed**]

57. Cock, P.J.A.; Fields, C.J.; Goto, N.; Heuer, M.L.; Rice, P.M. The Sanger FASTQ file format for sequences with quality scores, and the Solexa/Illumina FASTQ variants. *Nucleic Acids Res.* **2009**, *38*, 1767–1771. [**Google Scholar**] [**CrossRef**] [**PubMed**] [**Green Version**]

58. Andrews, S. FastQC. Available online: **https://www.bioinformatics.babraham.ac.uk/projects/fastqc/** (accessed on 13 April 2021).

59. Ewels, P.; Magnusson, M.; Lundin, S.; Käller, M. MultiQC: Summarize analysis results for multiple tools and samples in a single report. *Bioinformatics* **2016**, *32*, 3047–3048. [**Google Scholar**] [**CrossRef**] [**Green Version**]

60. Loman, N.J.; Quinlan, A.R. Poretools: A toolkit for analyzing nanopore sequence data. *Bioinformatics* **2014**, *30*, 3399–3401. [**Google Scholar**] [**CrossRef**] [**PubMed**]

61. Najoshi Sickle—A Windowed Adaptive Trimming Tool for FASTQ Files Using Quality. Available online: **https://github.com/najoshi/sickle** (accessed on 13 April 2021).

62. Andino, R.; Domingo, E. Viral quasispecies. *Virology* **2015**, *479–480*, 46–51. [**Google Scholar**] [**CrossRef**] [**PubMed**] [**Green Version**]

63. Paszkiewicz, K.; Studholme, D.J. De novo assembly of short sequence reads. *Brief. Bioinform.* **2010**, *11*, 457–472. [**Google Scholar**] [**CrossRef**] [**Green Version**]

64. Sohn, J.-I.; Nam, J.-W. The present and future of de novo whole-genome assembly. *Brief. Bioinform.* **2018**, *19*, 23–40. [**Google Scholar**] [**CrossRef**] [**Green Version**]

65. Luo, R.; Liu, B.; Xie, Y.; Li, Z.; Huang, W.; Yuan, J.; He, G.; Chen, Y.; Pan, Q.; Liu, Y.; et al. SOAPdenovo2: An empirically improved memory-efficient short-read de novo assembler. *Gigascience* **2012**, *1*, 2047-217X-1-18. [**Google Scholar**] [**CrossRef**] [**PubMed**]

66. Gnerre, S.; MacCallum, I.; Przybylski, D.; Ribeiro, F.J.; Burton, J.N.; Walker, B.J.; Sharpe, T.; Hall, G.; Shea, T.P.; Sykes, S.; et al. High-quality draft assemblies of mammalian genomes from massively parallel sequence data. *Proc. Natl. Acad. Sci. USA* **2011**, *108*, 1513–1518. [**Google Scholar**] [**CrossRef**] [**Green Version**]

67. Simpson, J.T.; Wong, K.; Jackman, S.D.; Schein, J.E.; Jones, S.J.M.; Birol, I. ABySS: A parallel assembler for short read sequence data. *Genome Res.* **2009**, *19*, 1117–1123. [**Google Scholar**] [**CrossRef**] [**PubMed**] [**Green Version**]

68. Zerbino, D.R.; Birney, E. Velvet: Algorithms for de novo short read assembly using de Bruijn graphs. *Genome Res.* **2008**, *18*, 821–829. [**Google Scholar**] [**CrossRef**] [**Green Version**]

69. Peng, Y.; Leung, H.C.M.; Yiu, S.M.; Chin, F.Y.L. IDBA-UD: A de novo assembler for single-cell and metagenomic sequencing data with highly uneven depth. *Bioinformatics* **2012**, *28*, 1420–1428. [**Google Scholar**] [**CrossRef**] [**Green Version**]

70. Bankevich, A.; Nurk, S.; Antipov, D.; Gurevich, A.A.; Dvorkin, M.; Kulikov, A.S.; Lesin, V.M.; Nikolenko, S.I.; Pham, S.; Prjibelski, A.D.; et al. SPAdes: A new genome assembly algorithm and its applications to single-cell sequencing. *J. Comput. Biol.* **2012**, *19*, 455–477. [**Google Scholar**] [**CrossRef**] [**PubMed**] [**Green Version**]

71. Nurk, S.; Bankevich, A.; Antipov, D.; Gurevich, A.A.; Korobeynikov, A.; Lapidus, A.; Prjibelski, A.D.; Pyshkin, A.; Sirotkin, A.; Sirotkin, Y.; et al. Assembling single-cell genomes and mini-metagenomes from chimeric MDA products. *J. Comput. Biol.* **2013**, *20*, 714–737. [**Google Scholar**] [**CrossRef**] [**Green Version**]

72. Bushmanova, E.; Antipov, D.; Lapidus, A.; Prjibelski, A.D. RnaSPAdes: A de novo transcriptome assembler and its application to RNA-Seq data. *Gigascience* **2019**, *8*, giz100. [**Google Scholar**] [**CrossRef**] [**PubMed**] [**Green Version**]

73. Edwards, D.J.; Holt, K.E. Beginner's guide to comparative bacterial genome analysis using next-generation sequence data. *Microb. Inform. Exp.* **2013**, *3*, 2. [**Google Scholar**] [**CrossRef**] [**Green Version**]

74. Rang, F.J.; Kloosterman, W.P.; de Ridder, J. From squiggle to basepair: Computational approaches for improving nanopore sequencing read accuracy. *Genome Biol.* **2018**, *19*, 90. [**Google Scholar**] [**CrossRef**] [**PubMed**] [**Green Version**]

75. Koren, S.; Schatz, M.C.; Walenz, B.P.; Martin, J.; Howard, J.T.; Ganapathy, G.; Wang, Z.; Rasko, D.A.; McCombie, W.R.; Jarvis, E.D.; et al. Hybrid error correction and de novo assembly of single-molecule sequencing reads. *Nat. Biotechnol.* **2012**, *30*, 693–700. [**Google Scholar**] [**CrossRef**] [**PubMed**] [**Green Version**]

76. Koren, S.; Walenz, B.P.; Berlin, K.; Miller, J.R.; Bergman, N.H.; Phillippy, A.M. Canu: Scalable and accurate long-read assembly via adaptive κ-mer weighting and repeat separation. *Genome Res.* **2017**, *27*, 722–736. [**Google Scholar**] [**CrossRef**] [**PubMed**] [**Green Version**]

77. Chin, C.S.; Peluso, P.; Sedlazeck, F.J.; Nattestad, M.; Concepcion, G.T.; Clum, A.; Dunn, C.; O'Malley, R.; Figueroa-Balderas, R.; Morales-Cruz, A.; et al. Phased diploid genome assembly with single-molecule real-time sequencing. *Nat. Methods* **2016**, *13*, 1050–1054. [**Google Scholar**] [**CrossRef**] [**PubMed**] [**Green Version**]

78. Oxford Nanopore Technologies Pomoxis—Bioinformatics Tools for Nanopore Research. Available online: **https://github.com/nanoporetech/pomoxis** (accessed on 13 April 2021).

79. Filloux, D.; Fernandez, E.; Loire, E.; Claude, L.; Galzi, S.; Candresse, T.; Winter, S.; Jeeva, M.L.; Makeshkumar, T.; Martin, D.P.; et al. Nanopore-based detection and characterization of yam viruses. *Sci. Rep.* **2018**, *8*, 17879. [**Google Scholar**] [**CrossRef**] [**Green Version**]

80. Boykin, L.M.; Sseruwagi, P.; Alicai, T.; Ateka, E.; Mohammed, I.U.; Stanton, J.A.L.; Kayuki, C.; Mark, D.; Fute, T.; Erasto, J.; et al. Tree lab: Portable genomics for early detection of plant viruses and pests in sub-saharanafrica. *Genes* **2019**, *10*, 632. [**Google Scholar**] [**CrossRef**] [**Green Version**]

81. Naito, F.Y.B.; Melo, F.L.; Fonseca, M.E.N.; Santos, C.A.F.; Chanes, C.R.; Ribeiro, B.M.; Gilbertson, R.L.; Boiteux, L.S.; de Cássia Pereira-Carvalho, R. Nanopore sequencing of a novel bipartite new world begomovirus infecting cowpea. *Arch. Virol.* **2019**, *164*, 1907–1910. [**Google Scholar**] [**CrossRef**]

82. Leiva, A.M.; Siriwan, W.; Lopez-Alvarez, D.; Barrantes, I.; Hemniam, N.; Saokham, K.; Cuellar, W.J. Nanopore-based complete genome sequence of a srilankan cassava

mosaic virus (geminivirus) strain from Thailand. *Microbiol. Resour. Announc.* **2020**, *9*. [**Google Scholar**] [**CrossRef**] [**Green Version**]

83. Altschul, S.F.; Gish, W.; Miller, W.; Myers, E.W.; Lipman, D.J. Basic local alignment search tool. *J. Mol. Biol.* **1990**, *215*, 403–410. [**Google Scholar**] [**CrossRef**]
84. Li, H.; Durbin, R. Fast and accurate short read alignment with burrows-wheeler transform. *Bioinformatics* **2009**, *25*, 1754–1760. [**Google Scholar**] [**CrossRef**] [**Green Version**]
85. Finn, R.D.; Clements, J.; Eddy, S.R. HMMER web server: Interactive sequence similarity searching. *Nucleic Acids Res.* **2011**, *39*, W29–W37. [**Google Scholar**] [**CrossRef**] [**PubMed**] [**Green Version**]
86. Stobbe, A.H.; Daniels, J.; Espindola, A.S.; Verma, R.; Melcher, U.; Ochoa-Corona, F.; Garzon, C.; Fletcher, J.; Schneider, W. E-probe diagnostic nucleic acid analysis (EDNA): A theoretical approach for handling of next generation sequencing data for diagnostics. *J. Microbiol. Methods* **2013**, *94*, 356–366. [**Google Scholar**] [**CrossRef**] [**PubMed**]
87. Punta, M.; Coggill, P.C.; Eberhardt, R.Y.; Mistry, J.; Tate, J.; Boursnell, C.; Pang, N.; Forslund, K.; Ceric, G.; Clements, J.; et al. The Pfam protein families database. *Nucleic Acids Res.* **2012**, *40*, 290–301. [**Google Scholar**] [**CrossRef**] [**PubMed**]
88. Marchler-Bauer, A.; Panchenko, A.R.; Shoemarker, B.A.; Thiessen, P.A.; Geer, L.Y.; Bryant, S.H. CDD: A database of conserved domain alignments with links to domain three-dimensional structure. *Nucleic Acids Res.* **2002**, *30*, 281–283. [**Google Scholar**] [**CrossRef**] [**PubMed**]
89. Agranovsky, A.A.; Boyko, V.P.; Karasev, A.V.; Koonin, E.V.; Dolja, V.V. Putative 65 kDa protein of beet yellows closterovirus is a homologue of HSP70 heat shock proteins. *J. Mol. Biol.* **1991**, *217*, 603–610. [**Google Scholar**] [**CrossRef**]
90. Amselem, J.; Cornut, G.; Choisne, N.; Alaux, M.; Alfama-Depauw, F.; Jamilloux, V.; Maumus, F.; Letellier, T.; Luyten, I.; Pommier, C.; et al. RepetDB: A unified resource for transposable element references. *Mob. DNA* **2019**, *10*, 6. [**Google Scholar**] [**CrossRef**] [**PubMed**] [**Green Version**]
91. Geering, A.D.W.; Maumus, F.; Copetti, D.; Choisne, N.; Zwickl, D.J.; Zytnicki, M.; McTaggart, A.R.; Scalabrin, S.; Vezzulli, S.; Wing, R.A.; et al. Endogenous florendoviruses are major components of plant genomes and hallmarks of virus evolution. *Nat. Commun.* **2014**, *5*, 5269. [**Google Scholar**] [**CrossRef**] [**Green Version**]
92. Diop, S.I.; Geering, A.D.W.; Alfama-Depauw, F.; Loaec, M.; Teycheney, P.-Y.; Maumus, F. Tracheophyte genomes keep track of the deep evolution of the caulimoviridae. *Sci. Rep.* **2018**, *8*, 572. [**Google Scholar**] [**CrossRef**] [**PubMed**] [**Green Version**]
93. Sharma, V.; Lefeuvre, P.; Roumagnac, P.; Filloux, D.; Teycheney, P.-Y.; Martin, D.P.; Maumus, F. Large-scale survey reveals pervasiveness and potential function of endogenous geminiviral sequences in plants. *Virus Evol.* **2020**, *6*, veaa071. [**Google Scholar**] [**CrossRef**] [**PubMed**]
94. Tangherlini, M.; Dell'Anno, A.; Zeigler Allen, L.; Riccioni, G.; Corinaldesi, C. Assessing viral taxonomic composition in benthic marine ecosystems: Reliability and efficiency of

different bioinformatic tools for viral metagenomic analyses. *Sci. Rep.* **2016**, *6*, 28428. [**Google Scholar**] [**CrossRef**] [**Green Version**]

95. Camacho, C.; Coulouris, G.; Avagyan, V.; Ma, N.; Papadopoulos, J.; Bealer, K.; Madden, T.L. BLAST+: Architecture and applications. *BMC Bioinform.* **2009**, *10*, 439. [**Google Scholar**] [**CrossRef**] [**PubMed**] [**Green Version**]
96. Kent, W.J. BLAT—The BLAST-like alignment tool. *Genome Res.* **2002**, *12*, 656–664. [**Google Scholar**] [**CrossRef**] [**Green Version**]
97. Buchfink, B.; Xie, C.; Huson, D.H. Fast and sensitive protein alignment using diamond. *Nat. Methods* **2014**, *12*, 59–60. [**Google Scholar**] [**CrossRef**] [**PubMed**]
98. Langmead, B.; Trapnell, C.; Pop, M.; Salzberg, S.L. Ultrafast and memory-efficient alignment of short DNA sequences to the human genome. *Genome Biol.* **2009**, *10*, R25. [**Google Scholar**] [**CrossRef**] [**Green Version**]
99. Hong, C.; Manimaran, S.; Shen, Y.; Perez-Rogers, J.F.; Byrd, A.L.; Castro-Nallar, E.; Crandall, K.A.; Johnson, W.E. PathoScope 2.0: A complete computational framework for strain identification in environmental or clinical sequencing samples. *Microbiome* **2014**, *2*, 33. [**Google Scholar**] [**CrossRef**]
100. Mistry, J.; Finn, R.D.; Eddy, S.R.; Bateman, A.; Punta, M. Challenges in homology search: HMMER3 and convergent evolution of coiled-coil regions. *Nucleic Acids Res.* **2013**, *41*, e121. [**Google Scholar**] [**CrossRef**] [**PubMed**] [**Green Version**]
101. Skewes-Cox, P.; Sharpton, T.J.; Pollard, K.S.; DeRisi, J.L. Profile hidden Markov models for the detection of viruses within metagenomic sequence data. *PLoS ONE* **2014**, *9*, e105067. [**Google Scholar**] [**CrossRef**] [**PubMed**] [**Green Version**]
102. Bzhalava, Z.; Hultin, E.; Dillner, J. Extension of the viral ecology in humans using viral profile hidden Markov models. *PLoS ONE* **2018**, *13*, e0190938. [**Google Scholar**] [**CrossRef**] [**PubMed**] [**Green Version**]
103. Wood, D.E.; Salzberg, S.L. Kraken: Ultrafast metagenomic sequence classification using exact alignments. *Genome Biol.* **2014**, *15*, R46. [**Google Scholar**] [**CrossRef**] [**PubMed**] [**Green Version**]
104. Wood, D.E.; Lu, J.; Langmead, B. Improved metagenomic analysis with Kraken 2. *Genome Biol.* **2019**, *20*, 257. [**Google Scholar**] [**CrossRef**] [**PubMed**] [**Green Version**]
105. Menzel, P.; Ng, K.L.; Krogh, A. Fast and sensitive taxonomic classification for metagenomics with Kaiju. *Nat. Commun.* **2016**, *7*, 11257. [**Google Scholar**] [**CrossRef**] [**PubMed**] [**Green Version**]
106. Flygare, S.; Simmon, K.; Miller, C.; Qiao, Y.; Kennedy, B.; Di Sera, T.; Graf, E.H.; Tardif, K.D.; Kapusta, A.; Rynearson, S.; et al. Taxonomer: An interactive metagenomics analysis portal for universal pathogen detection and host mRNA expression profiling. *Genome Biol.* **2016**, *17*, 111. [**Google Scholar**] [**CrossRef**] [**PubMed**] [**Green Version**]
107. Baizan-Edge, A.; Cock, P.; MacFarlane, S.; McGavin, W.; Torrance, L.; Jones, S. Kodoja: A workflow for virus detection in plants using k-mer analysis of RNA-sequencing data. *J. Gen. Virol.* **2019**, *100*, 533–542. [**Google Scholar**] [**CrossRef**] [**PubMed**]

108. Tampuu, A.; Bzhalava, Z.; Dillner, J.; Vicente, R. ViraMiner: Deep learning on raw DNA sequences for identifying viral genomes in human samples. *PLoS ONE* **2019**, *14*, e0222271. [**Google Scholar**] [**CrossRef**] [**PubMed**] [**Green Version**]

109. Ren, J.; Song, K.; Deng, C.; Ahlgren, N.A.; Fuhrman, J.A.; Li, Y.; Xie, X.; Poplin, R.; Sun, F. Identifying viruses from metagenomic data using deep learning. *Quant. Biol.* **2020**, *8*, 64–77. [**Google Scholar**] [**CrossRef**] [**Green Version**]

110. Abdelkareem, A.O.; Khalil, M.I.; Elaraby, M.; Abbas, H.; Elbehery, A.H.A. VirNet: Deep attention model for viral reads identification. In Proceedings of the 2018 13th International Conference on Computer Engineering and Systems (ICCES), Cairo, Egypt, 18–19 December 2018; pp. 623–626. [**Google Scholar**]

111. Ren, Y.; Xu, Y.; Lee, W.M.; Di Bisceglie, A.M.; Fan, X. In-depth serum virome analysis in patients with acute liver failure with indeterminate etiology. *Arch. Virol.* **2020**, *165*, 127–135. [**Google Scholar**] [**CrossRef**]

112. Li, H. Minimap2: Pairwise alignment for nucleotide sequences. *Bioinformatics* **2018**, *34*, 3094–3100. [**Google Scholar**] [**CrossRef**] [**PubMed**]

113. Warwick-Dugdale, J.; Solonenko, N.; Moore, K.; Chittick, L.; Gregory, A.C.; Allen, M.J.; Sullivan, M.B.; Temperton, B. Long-read viral metagenomics captures abundant and microdiverse viral populations and their niche-defining genomic islands. *PeerJ* **2019**, *7*, e6800. [**Google Scholar**] [**CrossRef**] [**Green Version**]

114. Lefkowitz, E.J.; Dempsey, D.M.; Hendrickson, R.C.; Orton, R.J.; Siddell, S.G.; Smith, D.B. Virus taxonomy: The database of the international committee on taxonomy of viruses (ICTV). *Nucleic Acids Res.* **2018**, *46*, D708–D717. [**Google Scholar**] [**CrossRef**] [**PubMed**] [**Green Version**]

115. Davison, A.J. Journal of general virology—Introduction to 'ICTV virus taxonomy profiles'. *J. Gen. Virol.* **2017**, *98*, 1. [**Google Scholar**] [**CrossRef**] [**PubMed**] [**Green Version**]

116. Bao, Y.; Chetvernin, V.; Tatusova, T. Improvements to pairwise sequence comparison (PASC): A genome-based web tool for virus classification. *Arch. Virol.* **2014**, *159*, 3293–3304. [**Google Scholar**] [**CrossRef**] [**PubMed**] [**Green Version**]

117. Gibbs, A.J.; Hajizadeh, M.; Ohshima, K.; Jones, R.A.C. The potyviruses: An evolutionary synthesis is emerging. *Viruses* **2020**, *12*, 132. [**Google Scholar**] [**CrossRef**] [**Green Version**]

118. Jones, S.; Baizan-Edge, A.; MacFarlane, S.; Torrance, L. Viral diagnostics in plants using next generation sequencing: Computational analysis in practice. *Front. Plant Sci.* **2017**, *8*, 1770. [**Google Scholar**] [**CrossRef**]

119. Blawid, R.; Silva, J.M.F.; Nagata, T. Discovering and sequencing new plant viral genomes by next-generation sequencing: Description of a practical pipeline. *Ann. Appl. Biol.* **2017**, *170*, 301–314. [**Google Scholar**] [**CrossRef**]

120. Roenhorst, J.W.; de Krom, C.; Fox, A.; Mehle, N.; Ravnikar, M.; Werkman, A.W. Ensuring validation in diagnostic testing is fit for purpose: A view from the plant virology laboratory. *EPPO Bull.* **2018**, *48*, 105–115. [**Google Scholar**] [**CrossRef**]

121. Simmonds, P.; Adams, M.J.; Benk, M.; Breitbart, M.; Brister, J.R.; Carstens, E.B.; Davison, A.J.; Delwart, E.; Gorbalenya, A.E.; Harrach, B.; et al. Consensus statement: Virus taxonomy in the age of metagenomics. *Nat. Rev. Microbiol.* **2017**, *15*, 161–168. [**Google Scholar**] [**CrossRef**] [**PubMed**]

122. Rwahnih, M.A.; Daubert, S.; Úrbez-Torres, J.R.; Cordero, F.; Rowhani, A. Deep sequencing evidence from single grapevine plants reveals a virome dominated by mycoviruses. *Arch. Virol.* **2011**, *156*, 397–403. [**Google Scholar**] [**CrossRef**] [**PubMed**] [**Green Version**]

123. Marzano, S.Y.L.; Domier, L.L. Novel mycoviruses discovered from metatranscriptomics survey of soybean phyllospherephytobiomes. *Virus Res.* **2016**, *213*, 332–342. [**Google Scholar**] [**CrossRef**] [**Green Version**]

124. Kreuze, J. siRNA deep sequencing and assembly: Piecing together viral infections. In *Detection and Diagnostics of Plant Pathogens*; Gullino, M.L., Bonants, P.J.M., Eds.; Springer: Dordrecht, The Netherlands, 2014; pp. 21–38. ISBN 978-94-017-9020-8. [**Google Scholar**]

125. Massart, S.; Candresse, T.; Gil, J.; Lacomme, C.; Predajna, L.; Ravnikar, M.; Reynard, J.S.; Rumbou, A.; Saldarelli, P.; Škoric, D.; et al. A framework for the evaluation of biosecurity, commercial, regulatory, and scientific impacts of plant viruses and viroids identified by NGS technologies. *Front. Microbiol.* **2017**, *8*, 45. [**Google Scholar**] [**CrossRef**] [**PubMed**] [**Green Version**]

126. Kreuze, J.F.; Perez, A.; Gargurevich, M.G.; Cuellar, W.J. Badnaviruses of sweet potato: Symptomless coinhabitants on a global scale. *Front. Plant Sci.* **2020**, *11*, 313. [**Google Scholar**] [**CrossRef**] [**PubMed**]

127. Koloniuk, I.; Thekke-Veetil, T.; Reynard, J.S.; Pleško, I.M.; Přibylová, J.; Brodard, J.; Kellenberger, I.; Sarkisova, T.; Špak, J.; Lamovšek, J.; et al. Molecular characterization of divergent closterovirus isolates infecting Ribes species. *Viruses* **2018**, *10*, 369. [**Google Scholar**] [**CrossRef**] [**PubMed**] [**Green Version**]

128. Sõmera, M.; Kvarnheden, A.; Desbiez, C.; Blystad, D.R.; Sooväli, P.; Kundu, J.K.; Gantsovski, M.; Nygren, J.; Lecoq, H.; Verdin, E.; et al. Sixty years after the first description: Genome sequence and biological characterization of European wheat striate mosaic virus infecting cereal crops. *Phytopathology* **2020**, *110*, 68–79. [**Google Scholar**] [**CrossRef**]

129. Hammond, J.; Adams, I.; Fowkes, A.R.; McGreig, S.; Botermans, M.; van Oorspronk, J.J.A.; Westenberg, M.; Verbeek, M.; Dullemans, A.M.; Stijger, C.C.M.M.; et al. Sequence analysis of 43-year old samples of plantago lanceolata show that plantain virus x is synonymous with actinidia virus X and is widely distributed. *Plant Pathol.* **2020**, 249–258. [**Google Scholar**] [**CrossRef**]

130. Tamisier, L.; Haegeman, A.; Foucart, Y.; Fouillien, N.; Rwahnih, M.A.; Buzkan, N.; Candresse, T.; Chiumenti, M.; De Jonghe, K.; Lefebvre, M.; et al. Semi-artificial datasets as a resource for validation of bioinformatics pipelines for plant virus detection. *Zenodo* **2020**, *4273791*, 1–15. [**Google Scholar**] [**CrossRef**]

131. Martin, D.P.; Murrell, B.; Golden, M.; Khoosal, A.; Muhire, B. RDP4: Detection and analysis of recombination patterns in virus genomes. *Virus Evol.* **2015**, *1*, vev003. [**Google Scholar**] [**CrossRef**] [**Green Version**]

132. Lole, K.S.; Bollinger, R.C.; Paranjape, R.S.; Gadkari, D.; Kulkarni, S.S.; Novak, N.G.; Ingersoll, R.; Sheppard, H.W.; Ray, S.C. Full-length human immunodeficiency virus type 1 genomes from subtype c-infected seroconverters in india, with evidence of intersubtype recombination. *J. Virol.* **1999**, *73*, 152–160. [**Google Scholar**] [**CrossRef**] [**PubMed**] [**Green Version**]

133. Simmonds, P.; Midgley, S. Recombination in the genesis and evolution of hepatitis B virus genotypes. *J. Virol.* **2005**, *79*, 15467–15476. [**Google Scholar**] [**CrossRef**] [**Green Version**]

134. Routh, A.; Johnson, J.E. Discovery of functional genomic motifs in viruses with ViReMa-a virus recombination mapper-for analysis of next-generation sequencing data. *Nucleic Acids Res.* **2014**, *42*, e11. [**Google Scholar**] [**CrossRef**] [**Green Version**]

135. Xu, C.; Sun, X.; Taylor, A.; Jiao, C.; Xu, Y.; Cai, X.; Wang, X.; Ge, C.; Pan, G.; Wang, Q.; et al. Diversity, distribution, and evolution of tomato viruses in china uncovered by small RNA sequencing. *J. Virol.* **2017**, *91*, e00173-17. [**Google Scholar**] [**CrossRef**] [**PubMed**] [**Green Version**]

136. Bertran, A.; Ciuffo, M.; Margaria, P.; Rosa, C.; Resende, R.O.; Turina, M. Host-specific accumulation and temperature effects on the generation of dimeric viral RNA species derived from the S-RNA of members of the Tospovirus genus. *J. Gen. Virol.* **2016**, *97*, 3051–3062. [**Google Scholar**] [**CrossRef**] [**PubMed**]

137. Maliogka, V.I.; Salvador, B.; Carbonell, A.; Sáenz, P.; León, D.S.; Oliveros, J.C.; Delgadillo, M.O.; García, J.A.; Simón-Mateo, C.; Progenika Biopharma, S.A. Virus variants with differences in the p1 protein coexist in a plum pox virus population and display particular host-dependent pathogenicity features. *Mol. Plant Pathol.* **2012**, *13*, 877–886. [**Google Scholar**] [**CrossRef**]

138. da Silva, W.; Kutnjak, D.; Xu, Y.; Xu, Y.; Giovannoni, J.; Elena, S.F.; Gray, S. Transmission modes affect the population structure of potato virus Y in potato. *PLoSPathog.* **2020**, *16*, e1008608. [**Google Scholar**] [**CrossRef**]

139. Saitou, N.; Nei, M. The neighbor-joining method: A new method for reconstructing phylogenetic trees. *Mol. Biol. Evol.* **1987**, *4*, 406–425. [**Google Scholar**] [**CrossRef**]

140. Guindon, S.; Dufayard, J.F.; Lefort, V.; Anisimova, M.; Hordijk, W.; Gascuel, O. New algorithms and methods to estimate maximum-likelihood phylogenies: Assessing the performance of PhyML 3.0. *Syst. Biol.* **2010**, *59*, 307–321. [**Google Scholar**] [**CrossRef**] [**PubMed**] [**Green Version**]

141. Stamatakis, A. RAxML version 8: A tool for phylogenetic analysis and post-analysis of large phylogenies. *Bioinformatics* **2014**, *30*, 1312–1313. [**Google Scholar**] [**CrossRef**] [**PubMed**]

142. Ronquist, F.; Teslenko, M.; Van Der Mark, P.; Ayres, D.L.; Darling, A.; Höhna, S.; Larget, B.; Liu, L.; Suchard, M.A.; Huelsenbeck, J.P. Mrbayes 3.2: Efficient bayesian phylogenetic inference and model choice across a large model space. *Syst. Biol.* **2012**, *61*, 539–542. [**Google Scholar**] [**CrossRef**] [**PubMed**] [**Green Version**]

143. Huson, D.H.; Beier, S.; Flade, I.; Górska, A.; El-Hadidi, M.; Mitra, S.; Ruscheweyh, H.J.; Tappu, R. MEGAN Community edition—Interactive exploration and analysis of large-scale microbiome sequencing data. *PLoSComput. Biol.* **2016**, *12*, e1004957. [**Google Scholar**] [**CrossRef**] [**Green Version**]

144. Suchard, M.A.; Lemey, P.; Baele, G.; Ayres, D.L.; Drummond, A.J.; Rambaut, A. Bayesian phylogenetic and phylodynamic data integration using BEAST 1.10. *Virus Evol.* **2018**, *4*, vey016. [**Google Scholar**] [**CrossRef**] [**Green Version**]

145. Fuentes, S.; Gibbs, A.J.; Adams, I.P.; Wilson, C.; Botermans, M.; Fox, A.; Kreuze, J.; Kehoe, M.A.; Jones, R.A.C. Potato virus A isolates from three continents: Their biological properties, phylogenetics, and prehistory. *Phytopathology* **2021**, *111*, 217–226. [**Google Scholar**] [**CrossRef**] [**PubMed**]

146. Hardy, O.J.; Vekemans, X. SPAGeDI: A versatile computer program to analyse spatial genetic structure at the individual or population levels. *Mol. Ecol. Notes* **2002**, *2*, 618–620. [**Google Scholar**] [**CrossRef**] [**Green Version**]

147. Zheng, Y.; Gao, S.; Padmanabhan, C.; Li, R.; Galvez, M.; Gutierrez, D.; Fuentes, S.; Ling, K.S.; Kreuze, J.; Fei, Z. VirusDetect: An automated pipeline for efficient virus discovery using deep sequencing of small RNAs. *Virology* **2017**, *500*, 130–138. [**Google Scholar**] [**CrossRef**]

148. Lefebvre, M.; Theil, S.; Ma, Y.; Candresse, T. The virannot pipeline: A resource for automated viral diversity estimation and operational taxonomy units assignation for virome sequencing data. *Phytobiomes J.* **2019**, *3*, 256–259. [**Google Scholar**] [**CrossRef**] [**Green Version**]

149. Ho, T.; Tzanetakis, I.E. Development of a virus detection and discovery pipeline using next generation sequencing. *Virology* **2014**, *471–473*, 54–60. [**Google Scholar**] [**CrossRef**] [**Green Version**]

150. Visser, M.; Burger, J.T.; Maree, H.J. Targeted virus detection in next-generation sequencing data using an automated e-probe based approach. *Virology* **2016**, *495*, 122–128. [**Google Scholar**] [**CrossRef**] [**PubMed**]

151. Afgan, E.; Baker, D.; Batut, B.; Van Den Beek, M.; Bouvier, D.; Ech, M.; Chilton, J.; Clements, D.; Coraor, N.; Grüning, B.A.; et al. The galaxy platform for accessible, reproducible and collaborative biomedical analyses: 2018 update. *Nucleic Acids Res.* **2018**, *46*, W537–W544. [**Google Scholar**] [**CrossRef**] [**PubMed**] [**Green Version**]

152. Kalantar, K.L.; Carvalho, T.; de Bourcy, C.F.A.; Dimitrov, B.; Dingle, G.; Egger, R.; Han, J.; Holmes, O.B.; Juan, Y.F.; King, R.; et al. IDseq-An open source cloud-based pipeline and analysis service for metagenomic pathogen detection and monitoring. *Gigascience* **2020**, *9*, giaa111. [**Google Scholar**] [**CrossRef**] [**PubMed**]

Chapter-8

Use of High-Throughput Sequencing and Two RNA Input Methods to Identify Viruses Infecting Tomato Crops

1. Introduction

Tomato (*Solanum Lycopersicum* L.) is one of the most important vegetable crops. The worldwide production of tomato in 2019 was more than 182 million tons (FAOSTAT, 2019; **http://www.fao.org/faostat/**; accessed on September 2020). Spain is one of the world's leading producers of tomato plants for fresh consumption, second in the European Union after Italy (Eurostat 2019; **https://ec.europa.eu/eurostat/371**; accessed on September 2020). Tomato cultivation in Spain is very intensive, with significant acreage devoted to greenhouse production. The major threats to tomato intensive cultivation are viral diseases, which are responsible for significant yield and fruit quality losses, causing important economic damage [**1**]. The main viruses frequently reported to affect tomato crops include tomato yellow leaf curl virus (TYLCV), pepino mosaic virus (PepMV), tomato spotted wilt virus (TSWV) and tomato chlorosis virus (ToCV), with all four of them having been reported in the Iberian Peninsula (EPPO). Nevertheless, emerging viruses, i.e., those recently reported and the incidence or geographic range of which increase rapidly [**2**], often cause the most important problems. A recent example of an emerging virus infecting tomato crops in the Iberian Peninsula is tomato brown rugose fruit virus (ToBRFV) (EPPO). The availability of sensitive and reliable virus discovery and detection techniques is crucial for diagnosing and controlling viral diseases, as well as anticipating problems that are potentially caused by major, emerging, or new viruses.

High-throughput sequencing (HTS) technologies enable, in a relatively short period of time, the characterization of plant viromes, allowing both the detection of known viruses and the discovery of novel ones [**3**]. These technologies have been successfully used with several crop species, including tomato plants [**4,5,6,7,8,9**]. The application of HTS to samples from tomato crops in China revealed the presence of 22 viruses, of which five of them had not been reported previously to infect plants of this species [**8**]. Another study, comparing the diversity of viral populations between tomato plants and neighboring *Solanum nigrum* plants using HTS, showed a large variability in virome richness, but with little overlapping of the viruses found in both species [**9**]. In addition to its detection potential, different works have shown that HTS can increase the resolution of virus population genetics and evolution studies and also allows the determination of the complete or near-complete genomes of novel viruses without any prior knowledge [**3,10**]. To date, different starting materials have been used for metagenomic studies, including double-stranded RNA (dsRNA), total RNA depleted of ribosomal RNA (rRNA-depleted total RNA), virion associated nucleic acids (VANA) and small RNAs (sRNAs). The comparison between methods using different starting materials has shown differences in the spectrum of viruses or viroids that can be detected. Previous works have shown that the outcome of rRNA-depleted total RNA-based methods tends to be virus-dependent; sRNA sequencing is better than rRNA-depleted total RNA for the detection of viroids [**10,11**] and both dsRNA and VANA allowed for the enrichment of virus sequences in the samples [**12**].

Here, we used the rRNA-depleted total RNA and dsRNA approaches to identify the viruses present in tomato samples from plants exhibiting virus-like symptoms. We compared the virus diversity and mapping reads between two replicates from total RNA extractions. In addition, we adopted an in silico approach to determine the minimum sequencing depth needed to detect the less abundant viruses in our sample pools. We identified seven known virus species, three of which are reported for the first time in tomato plants in Spain: *Lettuce ring necrosis virus*, *Olive latent virus 1*, and *Tomato fruit blotch virus*.

2. Materials and Methods

2.1. Plant Material

Twenty samples of tomato leaves exhibiting symptoms suggesting viral infection were collected during the 2015–2020 period from different locations in Spain and Portugal (**Table 1**). A portion of leaf tissue from each sample was placed in 1.5-mL tubes, frozen in liquid nitrogen, and stored at −80 °C until RNA extraction.

2.2. Nucleic Acid Extraction, Library Construction and Sequencing

For the total RNA extraction, 100 mg of leaf tissue from each individual sample was ground until attaining a fine powder with a mortar and pestle in the presence of liquid nitrogen. Total RNA was extracted using TRI Reagent (Sigma-Aldrich, St. Louis, MO, USA) following the manufacturer's protocol. For the dsRNA extraction, 100 mg of leaf tissue from each sample was pooled and ground. dsRNA was purified using the protocol from Valverde et al. [**13**] with Whatman CF-11 cellulose powder (GE Healthcare Life Science Corp., Piscataway, NJ, USA). Both preparations were subjected to RNase-free DNase I treatment (New England Biolabs, Ipswich, MA, USA), following the manufacturer's protocol, to remove traces of DNA, and the dsRNA preparation was also treated with RNase A (Machery-Nagel, Duren, Germany), following the protocol described in [**14**], to remove single-stranded RNA traces. After these treatments, the preparations were cleaned up by phenol/chloroform extraction [**14**] and their integrity was confirmed using gel electrophoresis.

Table 1. Description of samples and confirmation of viral infections by means of qRT-PCR or conventional RT-PCR.

Sample ID	Surveyed	Location	Symptoms	Virus Detected [1]							
				OLV1	TYLCV	ToCV	STV [2]	ToFBV	PepMV-EU	PepMV-CH2	LRNV
R20-01	March 2020	Almería	Vein clearing	–	–	–	+	–	++	+++	+++
R19-12	November 2019	Almería	Necrotic spots on leaves	–	–	+	+	–	+	+++	+
R19-09	October 2019	Almería	Leaf curling, leaf mosaics	–	–	–	+	–	+++	+++	–
R19-08	October 2019	Almería	Leaf curling, leaf mosaics	–	–	–	–	–	+++	++	–
R19-07	September 2019	Almería	Chlorosis, yellow spots on leaves, leaf mosaics	–	–	–	+	++	+++	++	–
R17-01	Febuary 2017	Murcia	Upward curling of leaves, chlorosis on leaves	–	–	–	–	++	+	++	–
H-57	December 2016	Murcia	Leaf mosaics	–	–	–	+	–	++	++	–
H-55	June 2016	Murcia	Leaf distortion	–	–	+	+	–	–	++	–
H-54	May 2016	Murcia	Leaf distortion	–	–	–	+	–	++	++	–
H-53	May 2016	Murcia	Leaf distortion	–	+	+	+	+++	+	+++	–
H-52	May 2016	Murcia	Distortion and mosaic in fruit	–	+	+	+	+++	–	+++	–
H-50	April 2016	Murcia	Leaf distortion	–	–	–	+	–	+	+++	–
H-43	December 2015	Granada	No clear symptoms	–	–	–	+	–	–	+	–
H-42	December 2015	Granada	Leaf curling	–	++	–	–	–	+	+	–
H-31	October 2015	Almería	Yellow mosaic	–	++	+	+	–	–	–	–
H-20	April 2015	Portugal [3]	No clear symptoms	–	–	+	+	++	–	+++	–
H-13	Aprli 2015	Portugal [3]	No clear symptoms	–	+	–	–	–	–	+++	–
H-11	April 2015	Portugal [3]	No clear symptoms	++	–	–	+	–	–	++	–
H-10	April 2015	Almería	Necrosis, yellow mosaic and distortion of leaves	–	+	++	+	–	+	++	+
H-09	April 2015	Almería	Necrosis, yellow mosaic and distortion of leaves	+	+	++	+	-	+	+++	+

The quantity of the total RNA was assessed using a Nanodrop™ One Microvolume UV-Vis Spectrophotometer (Thermo Fisher Scientific, Waltham, MA, USA) and a Qubit™ 3.0 Fluorometer (Thermo Fisher Scientific), and individual samples were normalized to a final concentration of 20 ng/µL. Two identical pools denoted Tom1 and Tom2 were prepared by adding 7.5 µL from each total RNA sample to obtain a final amount of 3 µg of total RNA in 150 µL of sterile MiliQ water. The dsRNA sample (denoted as TomDS) had a final volume of 50 µL in sterile MiliQ water. The samples were sent to Macrogen (Seoul, Korea) for library preparation and sequencing. The quality and quantity of the RNA in the three samples were further analyzed using a 2100 Bioanalyzer (Agilent Technologies, Palo Alto, CA, USA). For the three samples, cDNA libraries were synthesized using a TrueSeq Stranded Total RNA sample preparation kit (Illumina, San Diego, CA, USA) with ribosomal depletion using a Ribo-Zero plant kit (Illumina). Sequencing was performed with the Illumina NovaSeq 6000 platform to obtain 150 bp paired-end reads.

2.3. Bioinformatic Analysis

Raw reads were analyzed using the custom bioinformatic pipeline implemented in the R language, as described in **Figure 1**. Paired-end reads in fastq format served as the input. The quality of the raw reads was screened using the FastQC program (**http://www.bioinformatics.babraham.ac.uk/projects/fastqc**; accessed on May 2020). Adapters and low-quality reads (Phred < 30) were trimmed from each data set using Trimmomatic v0.39 [**15**]. After this step, paired-end reads were repaired using BBMap [**16**]. Host reads were filtered out by aligning reads to the host genome (Tomato genome version SL2.4) using Bowtie2 [**17**]. Unmapped reads were subjected to de novo assembly using Trinity v2.10 [**18**]. For virus detection, contigs were aligned against a custom plant virus nucleotide database using BLASTn [**19**]. To build the virus database, viral sequences fitting the criteria of host: Viridiplantae and sequence length: 800–23,000 nt were downloaded from NCBI Virus (**https://www.ncbi.nlm.nih.gov/labs/virus/vssi/#/**; accessed on May 2020). The database was built using makeblastdb, and low-complexity sequences were filtered out with dustmasker [**20**]. Sequences sharing 98% at both nucleotide (nt) and amino acid (aa) levels were collapsed using cd-hit [**21**]. After BLASTn, viral hits were filtered using the following criteria: contig length between 0.5 to 14 kb, e-value lower than 10^{-4}, and length of alignment between the query and the hit ≥ 300 nt. Pyfasta v0.5.2 (**https://pypi.org/project/pyfasta/**; accessed on May 2020) was used to retrieve the sequences of the reference viruses detected. Only one random accession for each virus was retrieved in the case that more than one accession for the same virus was found by BLASTn. These viral sequences were used as the reference to re-map the filtered reads using BWA with the mem algorithm [**22**]. From these alignments, virus genome coverage and average depth were calculated using SAMtools [**23**] and our own R script. To compare the percentage of identity of a given virus across the three datasets, a consensus sequence was generated using SAMtools and the Seqtk tool [**24**]. In cases where we could not obtain good consensus sequences, the longest contigs were used for these comparisons (**Table S1**). To determine the closest viral isolate, the consensus or the contig sequences of each virus were screened against its taxon using the NCBI database with BLASTn. The presence of divergent viral sequences was investigated by mapping contigs against a custom plant virus protein database using BLASTx [**19**], using the same filters as mentioned above. Although we did not find any novel viruses in this work, our pipeline included a step to analyze potential new viral species: ORF Finder (**https://www.ncbi.nlm.nih.gov/orffinder/**; accessed on May 2020) to predict ORFs of new putative viruses, and BLASTx against non-redundant proteins NCBI database to find the closest virus species. Subsets of reads used to determine the minimum number of reads needed to detect the viruses present in the datasets were obtained with Seqtk [**24**], using different seed values (100 and 120) in case two replicates were generated.

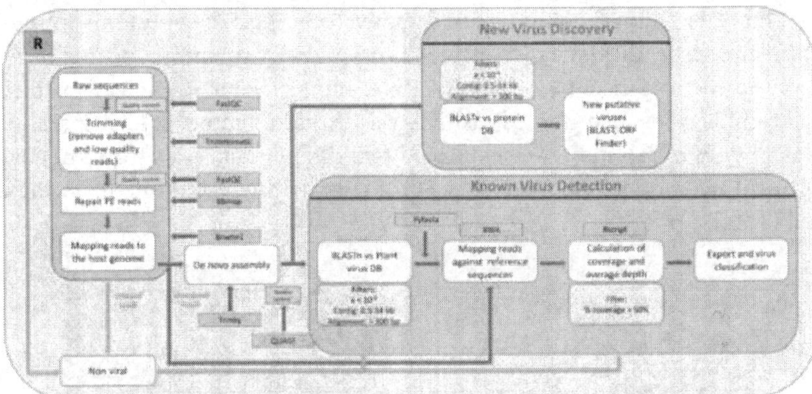

Figure 1. Bioinformatic workflow for the detection of known viruses and for novel virus discovery. Schematic representation of the bioinformatics pipeline followed in this work implemented in the R language. Specific programs (blue rectangles) used for each step (white rectangles) are indicated; applied filters are framed in light blue rectangles.

2.4. Conventional RT-PCR and qRT-PCR

All viruses were detected in individual samples by qRT-PCR, except southern tomato virus (STV), which was detected by conventional RT-PCR using the primer pair described in **Table S2**, Expand Reverse Transcriptase (Roche, Basel, Switzerland) and NXT Taq PCR Kit (EURx, Gdańsk, Poland) according to the manufacturer's instructions. Primers and probes for PepMVwere published elsewhere [**25**]. For the other viruses, forward and reverse primers, together with TaqMan probes, were designed to target conserved regions of the CP gene, except for tomato fruit blotch virus (ToFBV) for which we used the RdRp gene, using the PrimerQuest Tool from IDT (**https://eu.idtdna.com**; acceded on September 2020). The specificity of all the primers and probes was confirmed in silico by a BLASTn search against the NCBI database. Primers' and probes' sequence information and the length of the amplicons are detailed in **Table S2**. For virus detection by qRT-PCR, the KAPA PROBE FAST One-Step qRT-PCR kit (Roche, Basel, SZ) was used. Each 10-μL reaction consisted of 5 μL of 2× Master Mix, 0.2 μL of forward/reverse primers (10 μM) and probe (10 μM), 0.2 μL of 50× RT-Mix, 0.2 μL of 50× ROX high, 2 μL of DEPC-treated water and 2 μL of RNA (20 ng/μL). The performance of the primers/probe pairs was determined by calculating the PCR amplification efficiency of the reaction from a standard curve of five 1:10 serial dilutions of the pooled total RNA sample (200 ng/μL) (**Table S2**).

3. Results

3.1. HTS Using Two Different RNA Inputs: Total RNA and dsRNA

Twenty samples from leaves of tomato plants exhibiting virus-like symptoms were collected from 2015 to 2020 from different locations in the South of Spain and Portugal (**Table 1**). The sampled plants exhibited a wide range of disease symptoms suggestive of viral infection, i.e., vein clearing, leaf distortion, leaf curling, necrotic spots or mosaics on leaves (**Table 1**). **Figure 2** shows representative examples of symptoms found in two different greenhouses, displaying fruit blotching, uneven ripening and necrosis. From these samples, we prepared two different RNA inputs. Total RNA was purified from individual samples and then pooled; dsRNA was extracted in a single preparation from an equivalent pool of samples. We sequenced two different libraries from the total RNA pool, representing two technical replicates (Tom1 and Tom2) and one from the dsRNA extraction (TomDS).

Figure 2. Tomato plants and fruits exhibiting virus-like symptoms. (A,B) correspond to the greenhouse where sample R19-07 was collected in Murcia. Tomato fruits exhibited fruit blotching and discoloration. (C,D) correspond to another greenhouse in Almería where sample R19-12 was collected, and tomato plants exhibited necrosis.

After sequencing the three libraries, we obtained 86,284,538, 84,739,174 and 64,540,826 reads for Tom1, Tom2 and TomDS, respectively (**Table 2**). The raw reads were analyzed following the pipeline described in **Figure 1**. These were trimmed and filtered to remove low-quality bases, and tomato-derived sequences were extracted by mapping against the tomato genome (**Table 2**). As expected, the percentage of reads mapping to the plant genome was much lower using dsRNA (25.7% of the clean reads) than using total RNA as the input (around 48% of the clean reads) (**Table 2**). Hence, a total of 44,121,432, 43,524,150 and 47,320,422 filtered reads corresponding to 51.9%, 52.1% and 74.3% of the clean reads were further used for virus identification in Tom1, Tom2 and TomDS, respectively (**Table 2**). Although the number of raw reads sequenced for the two total RNA samples was higher than for the dsRNA sample (around 80 M compared to around 60 M), the number of reads after the application of different quality filters was similar for the three libraries (**Table 2**).

Table 2. Summary of sequencing and mapping results.

	Tom1		**Tom2**		**TomDS**	
	Reads	**%**	**Reads**	**%**	**Reads**	**%**
Raw reads	86,284,538		84,739,174		64,540,826	
Clean reads	85,026,574	98.54	83,516,356	98.56	63,715,596	98.72
Host mappings	40,905,042	48.11	39,992,206	47.89	16,395,174	25.73
Filter reads	44,121,532	51.89	43,524,150	52.11	47,320,422	74.27
Viral contigs	63		51		55	
Unique viruses	7		7		6	
Viral reads	6,790,296	7.99	7,159,776	8.57	20,491,882	32.16

3.2. Comparison of Viral Species Found in Two Technical Replicates Using Total RNA as the Input

In order to evaluate the reproducibility of the library construction and sequencing, the results from the two independent libraries, Tom1 and Tom2, were compared. The filtered reads were de novo assembled, and long contigs were mapped by BLASTn against our own plant virus database (**Figure 1** and **Table 2**). After filtering the BLASTn results, we obtained a total of 63 and 51 contigs, with an average length of 2788 nt and 3076 nt, that mapped to viral sequences for Tom1 and Tom2, respectively (**Table 2**). To ensure high confidence in the detection of viruses, we set a minimum threshold of the assembled contig of 500 nt in length. In both replicates, we identified the same seven virus species (**Table 3**). No novel viruses were found by BLASTx using our pipeline.

Table 3. Summary of mapping of reads against identified viral genomes.

Virus	Accession	Genome	Segment	Ref. Length	Tom1			Tom2			TomDS		
					Reads	AD	PC	Reads	AD	PC	Reads	AD	PC
OLV1	DQ083996	(+)ssRNA		3702	390	9	97.97	364	9	89.68	678	590	0.62
TYLCV	HF548826	(+)ssDNA		2787	1694	65	100	1596	60	99.64	1344	69	90.17
ToCV	KF018280	(+)ssRNA	RNA1	8596	1106	14	96.92	1076	15	94.16	222,112	3441	98.15
	KJ815045		RNA2	8249	2736	42	99.33	2794	43	99.52	235,672	3759	99.79
STV	KT438549	dsRNA		3463	1782	63	98.84	1812	64	98.64	3,459,440	7319	99.19
ToFBV	MK517477	(+)ssRNA	RNA1	5811	39,452	878	99.78	41,498	930	99.78	255,152	5665	99.88
	MK517478		RNA2	3643	17,810	626	99.75	17,760	628	99.45	79,414	2892	99.56
	MK517479		RNA3	2872	72,830	2096	99.51	81,684	2417	99.65	500,460	6360	99.93
	MK517480		RNA4	1946	47,102	2938	100	51,060	3158	100	317,660	7309	100
PepMV	NC_004067	(+)ssRNA		6450	6,431,722	7687	100	6,809,266	7686	100	15,351,220	7831	100
LRNV	NC_006051	(−)ssRNA	RNA 1	7651	13,116	223	99.76	10,716	183	99.48	14,738	258	99.12
	NC_006052		RNA 2	1830	17,546	1258	99.89	15,668	1124	99.95	10,512	758	99.89
	NC_006053		RNA 3	1527	108,412	6655	98.76	95,402	6507	99.41	38,700	3458	97.12
	NC_006054		RNA 4	1417	34,598	3226	99.86	29,080	2749	98.52	4780	462	96.47

AD: average read depth; PC: percentage of reference sequence covered by reads.

To calculate the average sequencing depth and the genome coverages, the filtered reads were mapped against the reference sequences of the identified viruses. In the cases where multiple accessions were found for the same virus species, the accession of the reference sequence used to map the reads was randomly selected. Viral reads constituted 7.99% and 8.57% of the clean reads for Tom1 and Tom2, respectively (**Table 2**). The number of reads mapping to viral genomes varied from 390 (Tom1) and 364 (Tom2) reads, counted for olive latent virus 1 (OLV1), to 6,431,722 (Tom1) and 6,809,266 (Tom2), for PepMV (**Table 3**). The average sequencing depth varied from nine for OLV1 in either Tom1 and Tom2 to ~7690 for PepMV in either Tom1 and Tom2 (**Table 3**). The lowest percentage of coverage along the corresponding viral genome was found for OLV1 (97.97% in Tom1 and 89.68% in Tom2) and RNA1 from ToCV (96.92% in Tom1 and 94.16% in Tom2) (**Table 3**). For the other virus species, the percentage of coverage was higher than 98% (**Table 3**). The nucleotide sequence identity among the viruses found in the two replicates was higher than 98.9% (**Table S1**). Overall, our results indicate that reproducibility using total RNA after rRNA depletion is very high, as no significant differences were observed among the results obtained here for the two replicates.

3.3. Viral Species Found Using Total RNA or dsRNA as the Input

Procedures for sample preparation and RNA extraction for dsRNA and total RNA were obviously different, but the amount of plant material used for both methods was the same, allowing some comparisons. Since Tom1 and Tom2 are almost identical replicates, only the comparison between TomDS and Tom1 is described. Fifty-five assembled contigs, with an average length of 3698 nt, derived from TomDS, mapped with our plant virus database, representing six virus species previously found in the analysis of the total RNA sample (**Table 2** and **Table 3**). The virus species not detected using dsRNA as the input was OLV1, although we found some mappings when reads were aligned against the OLV1 reference sequence (**Table 3**). For TomDS, the number of reads mapping to the identified reference virus sequences was substantially higher compared to the number of reads in Tom1 (32.16% versus 7.99%) (**Table 2**). This result was expected, as the dsRNA extraction method enriches preparations in virus-specific products, in this case in the replicative form of the ssRNA viruses [**13**]. Accordingly, the number of reads mapping to each viral genome, the average depth of sequencing and the percentage of the viral genome covered by the reads were similar or much higher for TomDS than for Tom1 (**Table 3**). The exceptions were TYLCV, for which the number of reads mapping to its genome and the average depth were slightly lower in TomDS (1344 and 69, respectively) than in Tom1 (1694 and 65, respectively), and RNAs 2, 3 and 4 from lettuce necrosis ringspot virus (LRNV), for which these numbers were significantly lower in TomDS than in Tom1 (**Table 3**). The nucleotide sequence identity between the viruses found in TomDS compared to Tom1 varied from 96.2% for PepMV to 99.9% for RNA4 from ToFBV (**Table S1**). In conclusion, the dsRNA-based method seemed to provide better enrichment in viral reads and the assembly of longer contigs than the total RNA-based method for RNA viruses, with some apparent exceptions.

Table 4. Summary of results obtained after subsetting the raw reads.

	Subset 1 (50%)				Subset 2 (37.5%)				Subset 3 (25%)			
	Tom1		TomDS		Tom1		TomDS		Tom1		TomDS	
	Reads	**%**	**Reads**	**%**	**Reads**	**%**	**Reads**	**%**	**Reads**	**%**	**Reads**	**%**
Subset	40,000,000		30,000,000		30,000,000		22,500,000		20,000,000		15,000,000	
Clean reads	39,416,281	98.54	29,616,416	98.72	29,562,358	98.54	22,212,742	98.72	19,708,041	98.54	14,809,024	98.73
Host mappings	19,544,446	49.58	7,618,386	25.72	14,223,278	48.11	5,714,092	25.72	9,483,289	48.12	3,808,670	25.72
Filter reads	20,455,554	51.90	21,998,030	74.28	15,339,080	51.89	16,498,650	74.28	10,224,752	51.88	11,000,354	74.28
Viral contigs	50		56		43		40		36		40	
Unique viruses	7		6		7		6		7		6	
Viral reads	3,033,136	14.83	9,245,404	42.03	2,275,725	14.84	6,934,690	42.03	1,515,690	14.82	4,622,820	42.02

3.4. In Silico Analysis of the Minimum Sequencing Depth Needed to Detect the Less Abundant Viruses

In an attempt to determine the minimum number of reads needed to detect the viruses infecting our tomato samples, we performed three different in silico simulations by decreasing the number of initial raw reads used in the bioinformatic analysis. For this, three subsets of raw data, consisting of 50% (Subset 1), 37.5% (Subset 2) and 25% (Subset 3) of the original Tom1 and TomDS reads were extracted randomly and analyzed following the same pipeline described previously (**Table 4** and **Table S3**). We next measured three parameters in these datasets: number of mapped reads, percentage of coverage and average read depth along the viral genomes.

After the analysis of the different subsets (**Table 4** and **Table S3**), we found the same viruses as when using the full datasets (seven species for Tom1 and six for TomDS), but there were differences in the estimated parameters. The percentage of unmapped reads against the tomato genome did not vary with the subsetting either for Tom1 or TomDS (**Table 2** and **Table 4**). Although the number of contigs mapping to the plant virus database decreased with the subsets, the average contig length was not always lower, with the Subset 3 in Tom1 being higher (3167 nt) than for the full dataset in Tom1 (2788 nt). The number of reads mapping to the reference viruses decreased across the three subsets (**Table 2** and **Table 4**); however, the percentage of the mapped reads was maintained among the three different subsets as compared to the full datasets (approximately 14.82% for Tom1 and 42% for TomDS) (**Table 4**). In general, the average read depth decreased across the different subsets, with the only exception being PepMV in TomDS, for which the average depth was maintained across the subsets (**Figure 3** and **Table S3**). For the viruses in which the percentage of coverage using the full datasets was higher than 89%, there were only slight differences in the percentage of coverage when decreasing the sequencing depth (**Figure 3** and **Table S3**). Four additional subsets, consisting of 12.50% (Subset 4), 6.25% (Subset 5), 3.13% (Subset 6) and 1.5% (Subset 7), were made (**Table S3**). For OLV1 in Tom1 and using 12.5% (Subset 4) of the total reads, 25 reads were mapped to the viral genome, with an average depth of 1.57 (**Table S3**), but no contigs longer than the minimum threshold could be assembled (**Table S3**). For TYLCV in TomDS, decreasing the raw reads to 3.13% (Subset 6) resulted in 38 reads mapping to its viral genome, with an average depth of 3.58 (**Table S3**), but again no contigs could be assembled. This in silico analysis was repeated using two new random subsets of 37.5% (Subset 2) and 25% (Subset 3) of the full datasets obtaining reproducible results (**Table S3**). In conclusion, to detect viruses infecting our sample pool, we could have decreased the initial sequencing depth to 25% of the full datasets for the total RNA input, from 80 M to 20 M, and to 6.25% for the dsRNA input, from 60 M to 3.75 M, and still identify the same virus species.

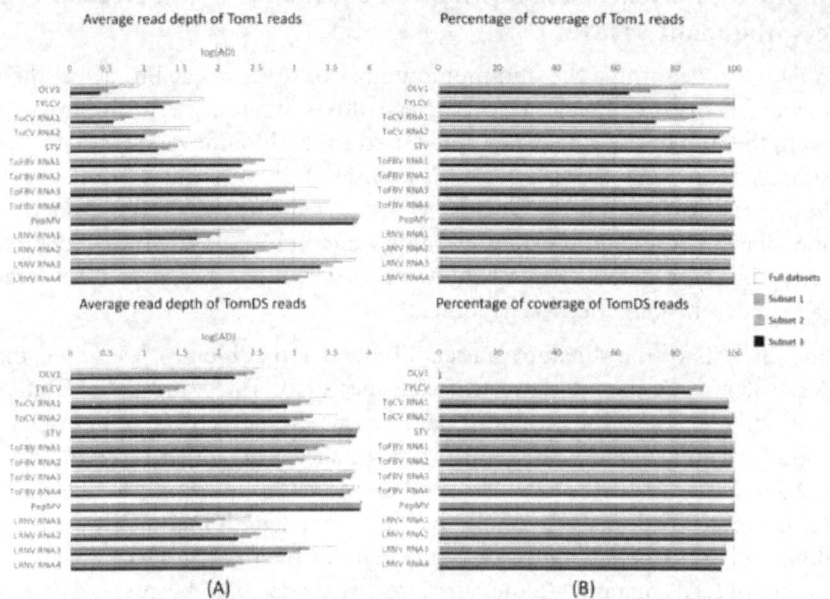

Figure 3. Comparison of average read depth and percentage of viral genomes covered by reads in Tom1 and TomDS. (A) Bar plots showing the logarithm of the average read depth for each viral genome in Tom1 (upper) and TomDS (bottom). (B) Bar plots showing the percentage of viral genomes covered by reads in Tom1 (upper) and TomDS (bottom). Full datasets: white bars; Subset 1: gray bars; Subset 2: dotted bars; Subset 3: black bars.

3.5. Viruses Already Reported to Infect Tomato Plants in Spain

Four viruses that are frequently reported in tomato crops were detected in our samples: PepMV, STV, ToCV and TYLCV. To identify their closest virus isolates in the databases, assembled contigs were further aligned by BLASTn against the specific virus taxon using the NCBI nr/nt database. We performed qRT-PCR or conventional RT-PCR to determine the presence of these viruses in the individual samples used for the pools (**Table 1** and **Table S2**). PepMV (family Alphaflexiviridae, genus *Potexvirus*), a (+)ssRNA virus, was the most abundant virus in our sample pools, with more than 28 M reads across the three datasets: more than 13 M reads in both total RNA samples, and more than 15 M reads in the dsRNA sample (**Table 3**). Thirteen contigs, almost covering the complete genome with the typical genome features described for PepMV, were determined from the three samples, all of them mapping with the highest identity against both CH2 and EU strains. We detected PepMV using qRT-PCR in 19 out of 20 samples, 13 of them in mixed infections with viruses from both strains, and the other six infected only with viruses from the CH2 strain, which seemed to accumulate at higher concentrations in most of the samples (**Table 1**).

STV (family Amalgaviridae, genus *Amalgavirus*), which possesses a dsRNA genome, was the second virus for which a higher number of reads mapped along its genome in the dsRNA sample (more than 3 M), although very few viral reads were obtained from the total RNA samples (1782 in Tom1 and 1812 in Tom2) (**Table 3**). We were able to assemble almost the full viral genome from reads derived from the three datasets, obtaining three different contigs, one per dataset, that shared a percentage of identity higher than 99.9% among them (**Table S1**). These sequences contained the two overlapping ORFs described for STV: a putative CP and the RdRp protein. All the contigs showed a nucleotide identity of more than 99.9% with a sequence from Canada (MK610257.1). STV was detected in 16 out of 20 individual samples using conventional RT-PCR (**Table 1**).

ToCV (family Closteroviridae, genus *Crinivirus*), with a bipartite (+)ssRNA genome, was the fourth most abundant virus among the seven viruses identified, with 457,789 reads in TomDS and around hundred times less in the total RNA samples (3842 in Tom1 and 3870 in Tom2) (**Table 3**). A unique 8585-nt contig, covering almost the complete genome of RNA 1 from ToCV, was determined from the dsRNA sample. It shared 99.9% nucleotide identity with RNA 1 from the ToCV isolate from Spain (KJ200304.1). Multiple contigs were determined from Tom1 and Tom2 with a length shorter than 5777 nt, showing the highest identity with the same ToCV strain. Three contigs, ranging from 8195 nt to 8239 nt in length, that corresponded to the near complete genome of RNA 2 were determined from the three datasets respectively. These contigs showed more than 99.8% nt identity with RNA2 from the ToCV isolate from Spain (KJ200305.1). ToCV was detected in eight out of 20 individual samples by qRT-PCR (**Table 1**).

TYLCV (family Geminiviridae, genus *Begomovirus*), a (+) ssDNA virus, was detected using the two different RNA inputs but, with the exception of OLV1, it was the virus for which we obtained the lowest number of reads mapping to its genome (around 1694 and 1596 in Tom1 and Tom2 and 1344 in TomDS) (**Table 3**). Different contigs ranging from 685 nt to 2212 nt were found to derive from TYLCV across the three datasets. A BLASTn analysis of these sequences revealed a nucleotide similarity above 97% with other TYLCV sequences belonging to the Israel strain (TYLCV-IL). No insertion characterizing the TYLCV-IS76 isolates was detected in the non-coding intergenic region of these sequences [**26**]. Based on qRT-PCR of individual samples, TYLCV was detected in seven out of 20 samples.

3.6. Viruses Not Previously Reported to Infect Tomato Plants in Spain

We also identified three virus species that, to our knowledge, were not previously reported or not frequently reported to infect tomato plants in Spain: LRNV, OLV1 and ToFBV [**27,28**]. To further understand these observations, we determined the consensus sequences from the reads mapping to virus reference genomes. As for the other viruses, we detected the presence of these three viruses in the individual samples using qRT-PCR (**Table 1** and **Table S2**).

LRNV (family Aspiviridae, genus *Ophiovirus*) is a four-segmented (−)ssRNA virus, and was detected in the pooled sample using the two RNA extraction approaches. We determined the consensus sequences for the four LRNV RNA segments (RNA 1 to 4). The RNA 1 consensus sequence (MW594439) was 7604 nt in length, covering 99% of the reference segment (AY535016), sharing a 99.4% identity with it. This RNA comprises two ORFs, which are 582 nt and 6834 nt in length, respectively. The RNA 2 consensus sequence (MW594440) was 1826 nt in length, lacking only 4 nt at the 5′ end compared to the reference RNA2 sequence (AY535017). Our sequence shared 99.2% at the nt level and 99.1% at the amino acid level with its RNA2 reference. The RNA 3 consensus sequence (MW594441) was 1505 nt (lacking 22 nt in total from both ends compared to the reference sequence, AY535018). Its vcRNA has one ORF of 1311 nt that encodes the CP protein. The CP showed a 99.1% amino acid identity with the CP of this reference isolate. The consensus sequence of RNA 4 (MW594442) was 1378 nt in length, covering 97% of the LRNV RNA 4 genome (AY535019) and sharing a 99.1% of nucleotide identity. The presence of LRNV was confirmed in the pooled sample as well as in four out of 20 individual samples by qRT-PCR using specific primers and probes targeting the CP (**Table 1** and **Table S2**). The samples in which the virus was detected were from crops in Almería (Southeastern Spain) collected in 2015, 2019 and 2020, with the highest accumulation of this virus in samples collected in 2020 (**Table 1**), as determined by qRT-PCR.

ToFBV (family Kitaviridae, genus *Blunervirus*) is a four-segmented (+)ssRNA virus, recently discovered to infect tomato plants in Italy and Australia [**27**]. The consensus sequences were determined for the four RNAs, and these were 5779 nt (RNA1, MW594435), 3586 nt (RNA2, MW594436), 2869 nt (RNA3, MW594437), and 1926 nt (RNA4, MW594438) in length, covering the full length of the four ToBFV RNA segments of isolate Fondi2018 from Italy (MK517477, MK517478, MK517479, and MK517480). The percentages of nucleotide identity of the consensus sequences against the reference sequences were 98.8%, 98.8%, 99.6% and 99% for RNAs 1 to 4, respectively. Five out of 20 individual samples were positive according to qRT-PCR using specific primers and probes targeting a conserved region of the RdRp (RNA1) (**Table 1** and **Table S2**). Four samples were collected in Murcia in 2016, 2017 and 2019, and one in Portugal in 2015 (**Table 1**).

OLV1 (family Tombusviridae, genus *Alphanecrovirus*), with a (+)ssRNA genome, was the least abundant virus in our datasets (**Table 3**), and was only detected using the total RNA extraction method. Unlike the other two previous viruses, we compared contigs against the NCBI database to describe this virus, as the determined consensus sequence contained a very high proportion of unknown nucleotides. Four contigs, ranging from 641 nt to 1847 nt, were determined, covering a partial sequence of the RdRp protein and the full sequence of the CP. The contig covering the full CP gene shared 95.2% of nucleotide identity with isolate OLV1 Anhui from China (MK376952.1) and 98.1% amino acids with isolate A4P2 from Portugal (AHE40781.1). Two out of 20 samples were infected with OLV1 by qRT-PCR using specific primers and probes targeting the CP; both samples were collected in 2015 in Almería and Portugal, respectively (**Table 1**).

4. Discussion

HTS was used to identify viruses present in a pool of twenty samples of tomato leaves showing virus-like symptoms collected in areas where tomato cultivation is very important. We first assessed the reproducibility of the sequencing method by generating a replicate of the total RNA with a ribosomal depletion sample; after analysis, we did not find important differences between replicates in terms of virus species detected or the number of reads mapping to each reference virus genome. In contrast, our data suggest that the extraction method seems to have an impact on the viruses that could be identified by HTS, in agreement with previous reports. For instance, Kutnjak et al. [**10**] compared siRNA and VANA approaches and found that both provided highly similar viral mutational landscapes, but VANA allowed for better recovery of complete viral genomes and detection of recombinant genomes [**10**]. In another example, rRNA-depleted total RNA was found to be superior to siRNA for the identification of citrus tristeza virus (CTV) (family Closteroviridae, genus *Closterovirus*) and citrus dwarfing viroid (CDVd) (family Pospiviroidae, genus *Apscaviroid*) infecting grapefruit, rendering better coverage for CTV but not for the viroid [**29**]. Another study showed that the performance of these two approaches tended to be virus-dependent, but in general longer contigs and higher genome coverage were obtained by rRNA-depleted total RNA than by sRNA sequencing [**11**]. Ma et al. [**12**] compared dsRNA and VANA approaches in assessing the virus diversity in wild plant populations. In their experimental system, the dsRNA approach revealed a broader and more comprehensive diversity for RNA viruses than VANA [**12**]. Gallo-García et al. [**30**] used total RNA and dsRNA as inputs to assess the virus populations in cape gooseberry (*Physalis peruviana* L.). They found higher sequence diversity for a specific virus species in total RNA as compared to dsRNA, but the total RNA extraction method failed to detect viruses present at low concentrations [**30**]. Hence, these and other authors have suggested the use of rRNA-depleted total RNA and dsRNA as complementary methods to obtain a comprehensive picture of the viruses present in a field sample [**3**,**30**]. In our case, and contrary to what was described by Gallo-Garcia et al. [**30**], rRNA-depleted total RNA performed better because seven, as opposed to six, viral species were detected using rRNA-depleted total RNA versus dsRNA. However, we cannot rule out the possibility that this difference may be due to the different pooling strategies used here. Although there were reads mapping to OLV1 in the dsRNA sample, the percentage of read coverage along the viral genome was only 0.62% (**Table 3**). This coverage was rendered by a fragment of 22 nt inside a read of around 109 nt that had been sequenced many times, hence the high average depth found (**Table 3**). However, this read had no significant hits against the NCBI database, not even with OLV1. Both methods allowed almost complete coverage of the genomes of the most abundant viruses (PepMV, STV, LRNV, ToFBV), however, the average depth was higher using the dsRNA approach in most of the cases. Interestingly, we noticed that total RNA generated more reads for LRNV, a (−)ssRNA virus, and for TYLCV, a ssDNA virus. This latter result was not surprising, as during TYLCV replication no dsRNA replicative intermediates are formed [**31**].

Some previous works have discussed the possibility of the application of HTS for routine plant virus diagnostics, mentioning different parameters to take into consideration, such as sensitivity, specificity and reproducibility [**32**]. Al Rwahnih et al. [**33**] compared the sensitivity of HTS to biological indexing for plant material certification in grapevines and concluded that it may reach a high sensitivity level, with the advantage of being time effective as compared to conventional methods [**33**]. In addition, Candresse et al. [**34**] used HTS to detect sugarcane white streak virus (family Geminiviridae, genus *Mastervirus*) in two quarantined sugarcane plants, showing the importance of including this method to assess plant health status [**34**]. However, sequencing large numbers of individuals or samples, which is often needed to obtain an overview of the plant viruses present in a population, is still challenging, despite falling HTS costs over the last decade. Furthermore, the need for a high-quality RNA input for library preparation and the complexity of the bioinformatics analyses should be taken into consideration when approaching HTS studies. One of the main methodological decisions for pool sequencing is the balance between the number of samples to be pooled and the sequencing depth required to detect all viruses present in the pool. According to the literature, the pool composition of different metagenomic studies for plant viromes varied from four to 50 individual samples per pool [**9,12,35,36,37,38,39**]. Skums et al. [**40**] provided a mathematical approach for a pooling strategy for the massive sequencing of human viruses, since the complex nature of these samples imposed restrictions on the pool design [**40**]. Here, we tried to address the problem of minimum reads needed to detect the less abundant viruses in a pool by following an in silico approach. The libraries for the two strategies used were sequenced with a different initial sequencing depth: 80 M for the total RNA replicates and 60 M for the dsRNA sample. In theory, this sequencing depth would correspond to 4 M reads per sample for the total RNA and 3 M reads per sample for the dsRNA input. We have demonstrated that we were able to detect the same viruses from the full datasets and from the three different subsets composed of 50%, 37.5% and 25% of the initial reads, and even for two additional subsets composed of 12.5% and 6.25% of the initial reads in the case of dsRNA. The less abundant viruses identified in the full datasets were OLV1 in the case of the total RNA replicates and TYLCV in the case of the dsRNA sample. However, the percentage of coverage decreased in the subsets, so the level of confidence in the detection of these viruses also decreased. TYLCV and OLV1 were detected in seven and two individual samples, respectively, and at very high Ct values (higher than 28 in most of the samples), suggesting the low accumulation of these viral RNAs in the pooled sample. In conclusion, we believe that 1.5 M reads per sample could have been used for assessing the tomato virome when using rRNA-depleted total RNA. In the case of using dsRNA as the input, a minimum of 0.2 M reads per sample could have been used as the initial sequencing depth. Generalizing these results is difficult, as different crops under different environmental conditions infected by different sets of viruses may require other sequencing depths. Using a low number of reads in a de novo assembly, it is possible that bioinformatics analysis fails to build significantly long contigs; hence, some virus derived reads could be disregarded during the bioinformatic analysis. The high output noise generated by this technique and/or possible contaminations demonstrate the necessity of using conventional detection methods as a complementary tool for the confirmation of the presence of a virus.

Three viruses known to infect tomato plants, and to induce important crop losses, were detected in our samples. PepMV was detected in 19 samples by means of qRT-PCR, confirming the HTS results. Two strains, EU and CH2, were detected in mixed infections in 13 out of 19 PepMV-infected samples. This high incidence could be due to the generalization of cross-protection as a means of disease control in the South of Spain [**41**], though generalized single and mixed infections of PepMV isolates of these two strains were already reported in the region before the extended use of cross-protection [**42**]. The detection here of ToCV and TYLCV is not surprising as both are prevalent viruses across the southern and eastern regions of Spain and both are transmitted by whiteflies [**43,44,45,46**]. A survey of STV incidence was conducted on different tomato fields in Spain in 2018, revealing that STV was widespread [**47**]. Moreover, this virus was detected in different tomato varieties and nurseries, but STV-infected tomato plants did not show any disease symptoms [**47**]. Apart from the four viruses mentioned above, there are other viruses known to be widespread and of major concern for tomato plants, including, for instance, TSWV, potato virus Y, cucumber mosaic virus [**1**] and the emergent ToBRFV, but none of these were detected in our samples.

In addition to viruses frequently cited in tomato plants, three viruses that are seldom if ever cited to infect these species were detected in this study. The first report of LRNV was in lettuce (*Lactuca sativa*) crops in 1996, associated with lettuce ring necrosis disease. LRNV is transmitted by the soil-borne fungus *Olpidiumbrassicae* [**48**]. However, no additional information on the distribution and the epidemiological status of the virus could be found in the literature. Although in our study LRNV was detected in samples showing necrosis, vein clearing and yellow mosaic, these symptoms can be hardly associated with it, because it was detected in mixed infections in all the positive samples. OLV1 was isolated for the first time from olive trees in the Apulia region of Southern Italy [**49**]. Infected olive trees had normal flowers and fruits and did not show disease symptoms except for occasional fasciation and bifurcations of leaves and twigs [**49**]. Since then, OLV1 has been reported in different hosts in various countries [**50,51**], and in 2010 it was reported for the first time in tomato plants in Poland, associated with necrotic spots on the leaves [**28**]. The presence of this virus in the plants of different tomato cultivars was restricted to local lesions or to necrotic areas [**28**]. This may suggest that the surrounding crops may constitute the primary natural source of virus inoculum in the greenhouse-grown tomato, and that the daily manipulation of tomato plants by workers could play a key role in its spread. ToFBV is a new blunervirus that was recently reported in Italy and Australia, and it was associated with blotchy ripening and dimpling of the tomato fruits [**27**]. ToFBV could not be transmitted mechanically to either tomato or a set of various herbaceous plants, and thus Koch's postulates have not been fulfilled yet for this virus. Generally, kitaviruses share important epidemiological aspects such as symptomatology, lack of systemic movement and mite-mediated transmission [**52**].

The analysis of the individual samples, carried out to detect the seven viruses, revealed the extent of mixed infections, and the almost universal mixed infections of any of the viruses with CH2 and/or EU isolates of PepMV. Multiple viruses could infect a single plant; for instance, LRNV was detected in plants infected with OLV1, TYLCV and ToCV. ToFBV was detected in mixed infections with TYLCV and ToCV. Mixed infections can affect the virus's replication and movement competence, transmission capacity, virulence, host range and symptom severity [**53**]; therefore, more studies must be conducted to assess the impact of mixed infections in tomato and other crops. Mixed infections also prevented us from associating virus detection with disease symptoms—using the data collected during our sampling, no obvious correlation could be established between observed symptoms and the detection of any of the three viruses discussed above.

HTS is a very powerful technique for virus discovery and detection, and this emerged clearly from our study. We identified viruses that were present in tomato crops several years ago (samplings were conducted in 2015) but which remained unreported. The increased ability to detect new or infrequent viruses using this technique raises several questions relating to how to deal with them from a crop protection point of view, as well as the complexity of their biological characterization, particularly for the newly-identified plant viruses and viroids, and their impact at biosecurity, commercial, regulatory and scientific levels [**54**]. Here, we have reported for the first time the presence of OLV1 in tomato crops in Portugal and the South of Spain, and the presence of LRNV and ToFBV in tomato crops in the South of Spain. However, broader surveys are needed to assess the prevalence and potential impact of these viruses, including surveys of alternative hosts that may serve as virus reservoirs, in order to better understand their epidemiological status.

Supplementary Materials

Table S1. Comparison of percentage of identity of virus sequences among the three samples; Table S2. Primers and probes used for viral detection by qRT-PCR and conventional RT-PCR for STV; Table S3. Summary of results obtained after mapping the reads from different subsets against identified viral genomes.

References

1. Hanssen, I.M.; Lapidot, M. Major tomato viruses in the Mediterranean basin. In *Advances in Virus Research*; Elsevier: Amsterdam, The Netherlands, 2012; Volume 84, pp. 31–66. ISBN 0065-3527. [**Google Scholar**]
2. Woolhouse, M.E.J.; Gowtage-Sequeria, S. Host range and emerging and reemerging pathogens. *Emerg. Infect. Dis.* **2005**, *11*, 1842. [**Google Scholar**] [**CrossRef**] [**PubMed**]
3. Roossinck, M.J.; Martin, D.P.; Roumagnac, P. Plant virus metagenomics: Advances in virus discovery. *Phytopathology* **2015**, *105*, 716–727. [**Google Scholar**] [**CrossRef**] [**PubMed**] [**Green Version**]
4. Li, Y.; Jia, A.; Qiao, Y.; Xiang, J.; Zhang, Y.; Wang, W. Virome analysis of lily plants reveals a new potyvirus. *Arch. Virol.* **2018**, *163*, 1079–1082. [**Google Scholar**] [**CrossRef**]
5. Jo, Y.; Choi, H.; Kim, S.-M.; Kim, S.-L.; Lee, B.C.; Cho, W.K. The pepper virome: Natural co-infection of diverse viruses and their quasispecies. *BMC Genom.* **2017**, *18*, 1–12. [**Google Scholar**] [**CrossRef**] [**Green Version**]
6. Jo, Y.; Cho, W.K. RNA viromes of the oriental hybrid lily cultivar "Sorbonne". *BMC Genom.* **2018**, *19*, 748. [**Google Scholar**] [**CrossRef**] [**PubMed**]
7. Mutuku, J.M.; Wamonje, F.O.; Mukeshimana, G.; Njuguna, J.; Wamalwa, M.; Choi, S.-K.; Tungadi, T.; Djikeng, A.; Kelly, K.; DomelevoEntfellner, J.-B. Metagenomic analysis of plant virus occurrence in common bean (Phaseolus vulgaris) in Central Kenya. *Front. Microbiol.* **2018**, *9*, 2939. [**Google Scholar**] [**CrossRef**] [**PubMed**]
8. Xu, C.; Sun, X.; Taylor, A.; Jiao, C.; Xu, Y.; Cai, X.; Wang, X.; Ge, C.; Pan, G.; Wang, Q. Diversity, distribution, and evolution of tomato viruses in China uncovered by small RNA sequencing. *J. Virol.* **2017**, *91*. [**Google Scholar**] [**CrossRef**] [**Green Version**]
9. Ma, Y.; Marais, A.; Lefebvre, M.; Faure, C.; Candresse, T. Metagenomic analysis of virome cross-talk between cultivated Solanum lycopersicum and wild Solanum nigrum. *Virology* **2020**, *540*, 38–44. [**Google Scholar**] [**CrossRef**]
10. Kutnjak, D.; Rupar, M.; Gutierrez-Aguirre, I.; Curk, T.; Kreuze, J.F.; Ravnikar, M. Deep sequencing of virus-derived small interfering RNAs and RNA from viral particles shows highly similar mutational landscapes of a plant virus population. *J. Virol.* **2015**, *89*, 4760–4769. [**Google Scholar**] [**CrossRef**] [**Green Version**]

11. Pecman, A.; Kutnjak, D.; Gutiérrez-Aguirre, I.; Adams, I.; Fox, A.; Boonham, N.; Ravnikar, M. Next generation sequencing for detection and discovery of plant viruses and viroids: Comparison of two approaches. *Front. Microbiol.* **2017**, *8*, 1998. [**Google Scholar**] [**CrossRef**] [**Green Version**]
12. Ma, Y.; Marais, A.; Lefebvre, M.; Theil, S.; Svanella-Dumas, L.; Faure, C.; Candresse, T. Phytovirome analysis of wild plant populations: Comparison of double-stranded RNA and virion-associated nucleic acid metagenomic approaches. *J. Virol.* **2019**, *94*, e0146219. [**Google Scholar**] [**CrossRef**]
13. Valverde, R.A.; Nameth, S.T.; Jordan, R.L. Analysis of double-stranded RNA for plant virus diagnosis. *Plant Dis.* **1990**, *74*, 255–258. [**Google Scholar**]
14. Marais, A.; Faure, C.; Bergey, B.; Candresse, T. Viral double-stranded RNAs (dsRNAs) from plants: Alternative nucleic acid substrates for high-throughput sequencing. In *Viral Metagenomics*; Springer: Berlin/Heidelberg, Germany, 2018; pp. 45–53. [**Google Scholar**]
15. Bolger, A.M.; Lohse, M.; Usadel, B. Trimmomatic: A flexible trimmer for Illumina sequence data. *Bioinformatics* **2014**, *30*, 2114–2120. [**Google Scholar**] [**CrossRef**] [**PubMed**] [**Green Version**]
16. Bushnell, B. *BBMap: A Fast, Accurate, Splice-Aware Aligner*; Lawrence Berkeley National Lab.(LBNL): Berkeley, CA, USA, 2014. [**Google Scholar**]
17. Langmead, B.; Salzberg, S.L. Fast gapped-read alignment with Bowtie. *Nat. Methods* **2012**, *9*, 357. [**Google Scholar**] [**CrossRef**] [**PubMed**] [**Green Version**]
18. Haas, B.J.; Papanicolaou, A.; Yassour, M.; Grabherr, M.; Blood, P.D.; Bowden, J.; Couger, M.B.; Eccles, D.; Li, B.; Lieber, M. De novo transcript sequence reconstruction from RNA-seq using the Trinity platform for reference generation and analysis. *Nat. Protoc.* **2013**, *8*, 1494–1512. [**Google Scholar**] [**CrossRef**] [**PubMed**]
19. Altschul, S.F.; Gish, W.; Miller, W.; Myers, E.W.; Lipman, D.J. Basic local alignment search tool. *J. Mol. Biol.* **1990**, *215*, 403–410. [**Google Scholar**] [**CrossRef**]
20. Camacho, C.; Madded, T.; Tao, T.; Agarwala, R.; Morgulis, A. *BLAST® Command Line Applications User Manual*; National Center for Biotechnology Information (US): Bethesda, MD, USA, 2018. [**Google Scholar**]
21. Li, W.; Godzik, A. Cd-hit: A fast program for clustering and comparing large sets of protein or nucleotide sequences. *Bioinformatics* **2006**, *22*, 1658–1659. [**Google Scholar**] [**CrossRef**] [**Green Version**]
22. Li, H.; Durbin, R. Fast and accurate short read alignment with Burrows–Wheeler transform. *Bioinformatics* **2009**, *25*, 1754–1760. [**Google Scholar**] [**CrossRef**] [**Green Version**]
23. Li, H.; Handsaker, B.; Wysoker, A.; Fennell, T.; Ruan, J.; Homer, N.; Marth, G.; Abecasis, G.; Durbin, R. The sequence alignment/map format and SAMtools. *Bioinformatics* **2009**, *25*, 2078–2079. [**Google Scholar**] [**CrossRef**] [**Green Version**]
24. Li, H. *Seqtk: A Fast and Lightweight Tool for Processing Sequences*; Broad Inst.: Cambridge, MA, USA, 2016. [**Google Scholar**]

25. Gutiérrez-Aguirre, I.; Mehle, N.; Delić, D.; Gruden, K.; Mumford, R.; Ravnikar, M. Real-time quantitative PCR based sensitive detection and genotype discrimination of Pepino mosaic virus. *J. Virol. Methods* **2009**, *162*, 46–55. [**Google Scholar**] [**CrossRef**]

26. Torre, C.; Donaire, L.; Gómez-Aix, C.; Juárez, M.; Peterschmitt, M.; Urbino, C.; Hernando, Y.; Agüero, J.; Aranda, M.A. Characterization of begomoviruses sampled during severe epidemics in tomato cultivars carrying the Ty-1 Gene. *Int. J. Mol. Sci.* **2018**, *19*, 2614. [**Google Scholar**] [**CrossRef**] [**PubMed**] [**Green Version**]

27. Ciuffo, M.; Kinoti, W.M.; Tiberini, A.; Forgia, M.; Tomassoli, L.; Constable, F.E.; Turina, M. A new blunervirus infects tomato crops in Italy and Australia. *Arch. Virol.* **2020**, *165*, 2379–2384. [**Google Scholar**] [**CrossRef**] [**PubMed**]

28. Borodynko, N.; Hasiów-Jaroszewska, B.; Pospieszny, H. Identification and characterization of an Olive latent virus 1 isolate from a new host: Solanum lycopersicum. *J. Plant Pathol.* **2010**, *92*, 789–792. [**Google Scholar**]

29. Visser, M.; Bester, R.; Burger, J.T.; Maree, H.J. Next-generation sequencing for virus detection: Covering all the bases. *Virol. J.* **2016**, *13*, 1–6. [**Google Scholar**] [**CrossRef**] [**PubMed**] [**Green Version**]

30. García, Y.G.; Montoya, M.M.; Gutiérrez, P.A. Detection of RNA viruses in Cape gooseberry (Physalis peruviana L.) by RNAseq using total RNA and dsRNA inputs. *Arch. Phytopathol. Plant Prot.* **2020**, *53*, 395–413. [**Google Scholar**] [**CrossRef**]

31. Czosnek, H.; Ber, R.; Navot, N.; Antignus, Y.; Cohen, S.; Zamir, D. Tomato yellow leaf curl virus DNA forms in the viral capside, in infected plants and in the insect vector. *J. Phytopathol.* **1989**, *125*, 47–54. [**Google Scholar**] [**CrossRef**]

32. Maree, H.J.; Fox, A.; Al Rwahnih, M.; Boonham, N.; Candresse, T. Application of HTS for routine plant virus diagnostics: State of the art and challenges. *Front. Plant Sci.* **2018**, *9*, 1082. [**Google Scholar**] [**CrossRef**] [**Green Version**]

33. Al Rwahnih, M.; Daubert, S.; Golino, D.; Islas, C.; Rowhani, A. Comparison of next-generation sequencing versus biological indexing for the optimal detection of viral pathogens in grapevine. *Phytopathology* **2015**, *105*, 758–763. [**Google Scholar**] [**CrossRef**] [**Green Version**]

34. Candresse, T.; Filloux, D.; Muhire, B.; Julian, C.; Galzi, S. Appearances Can Be Deceptive: Revealing a Hidden Viral Infection with Deep Sequencing. *PLoS ONE* **2014**, *9*, e102945. [**Google Scholar**] [**CrossRef**] [**PubMed**] [**Green Version**]

35. Akinyemi, I.A.; Wang, F.; Zhou, B.; Qi, S.; Wu, Q. Ecogenomic survey of plant viruses infecting tobacco by next generation sequencing. *Virol. J.* **2016**, *13*, 181. [**Google Scholar**] [**CrossRef**] [**Green Version**]

36. Alcalá-Briseño, R.I.; Casarrubias-Castillo, K.; López-Ley, D.; Garrett, K.A.; Silva-Rosales, L. Network analysis of the papaya oroecological regions of Chiapas, Mexicochardvirome from two agr. *Msystems* **2020**, *5*, e0042319. [**Google Scholar**] [**CrossRef**] [**PubMed**] [**Green Version**]

37. Bernal-Vicente, A.; Donaire, L.; Torre, C.; Gómez-Aix, C.; Sánchez-Pina, M.A.; Juarez, M.; Hernando, Y.; Aranda, M.A. Small RNA-seq to characterize viruses responsible of

Lettuce big vein disease in Spain. *Front. Microbiol.* **2018**, *9*, 3188. [**Google Scholar**] [**CrossRef**] [**PubMed**]

38. Jo, Y.; Bae, J.-Y.; Kim, S.-M.; Choi, H.; Lee, B.C.; Cho, W.K. Barley RNA viromes in six different geographical regions in Korea. *Sci. Rep.* **2018**, *8*, 1–13. [**Google Scholar**] [**CrossRef**] [**PubMed**]
39. Mumo, N.N.; Mamati, G.E.; Ateka, E.M.; Rimberia, F.K.; Asudi, G.O.; Boykin, L.M.; Machuka, E.M.; Njuguna, J.N.; Pelle, R.; Stomeo, F. Metagenomic Analysis of Plant Viruses Associated With Papaya Ringspot Disease in Carica papaya L. in Kenya. *Front. Microbiol.* **2020**, *11*, 205. [**Google Scholar**] [**CrossRef**]
40. Skums, P.; Artyomenko, A.; Glebova, O.; Ramachandran, S.; Campo, D.S.; Dimitrova, Z.; Măndoiu, I.I.; Zelikovsky, A.; Khudyakov, Y. Pooling Strategy for Massive Viral Sequencing. In *Computational Methods for Next Generation Sequencing Data Analysis*; Măndoiu, I., Zelikovsky., A., Eds.; John Wiley & Sons, Inc.: Hoboken, NJ, USA, 2016; pp. 57–83. [**Google Scholar**]
41. Agüero, J.; Gómez-Aix, C.; Sempere, R.N.; García-Villalba, J.; García-Núñez, J.; Hernando, Y.; Aranda, M.A. Stable and broad spectrum cross-protection against Pepino mosaic virus attained by mixed infection. *Front. Plant Sci.* **2018**, *9*, 1810. [**Google Scholar**] [**CrossRef**]
42. Gómez, P.; Sempere, R.N.; Elena, S.F.; Aranda, M.A. Mixed infections of Pepino mosaic virus strains modulate the evolutionary dynamics of this emergent virus. *J. Virol.* **2009**, *83*, 12378–12387. [**Google Scholar**] [**CrossRef**] [**Green Version**]
43. Fiallo-Olivé, E.; Navas-Castillo, J. Tomato chlorosis virus, an emergent plant virus still expanding its geographical and host ranges. *Mol. Plant Pathol.* **2019**, *20*, 1307–1320. [**Google Scholar**] [**CrossRef**] [**Green Version**]
44. Navas-Castillo, J.; López-Moya, J.J.; Aranda, M.A. Whitefly-Transmitted RNA Viruses that Affect Intensive Vegetable Production. *Ann. Appl. Biol.* **2014**, *165*, 155–171. [**Google Scholar**] [**CrossRef**]
45. Orílio, A.F.; Fortes, I.M.; Navas-Castillo, J. Infectious cDNA clones of the crinivirus Tomato chlorosis virus are competent for systemic plant infection and whitefly-transmission. *Virology* **2014**, *464*, 365–374. [**Google Scholar**] [**CrossRef**] [**Green Version**]
46. Navas-Castillo, J.; Fiallo-Olivé, E.; Sánchez-Campos, S. Emerging Virus Diseases Transmitted by Whiteflies. *Annu. Rev. Phytopathol.* **2011**, *49*, 219–248. [**Google Scholar**] [**CrossRef**]
47. Elvira-Gonzalez, L.; Carpino, C.; Alfaro Fernández, A.O.; Font San Ambrosio, M.I.; Peiró Barber, R.M.; Rubio MIgélez, L.; Galipienso-Torregrosa, L. A sensitive real-time RT-PCR reveals a high incidence of Southern tomato virus (STV) in Spanish tomato crops. *Span. J. Agric. Res.* **2018**, *16*. [**Google Scholar**] [**CrossRef**] [**Green Version**]
48. Campbell, R.N.; Lot, H. Lettuce ring necrosis, a viruslike disease of lettuce: Evidence for transmission by Olpidiumbrassicae. *Plant Dis.* **1996**, *80*, 611–615. [**Google Scholar**] [**CrossRef**]

49. Gallitelli, D.; Savino, V. Olive latent virus-1, an isometric virus with a single RNA species isolated from olive in Apulia, Southern Italy. *Ann. Appl. Biol.* **1985**, *106*, 295–303. [**Google Scholar**] [**CrossRef**]

50. Kanematsu, S.; Yumiko, T.; Morikawa, T. Isolation of Olive latent virus 1 from tulip in Toyama Prefecture. *J. Gen. Plant Pathol.* **2001**, *67*, 333–334. [**Google Scholar**] [**CrossRef**]

51. Martelli, G.P.; Yilmaz, M.A.; Savino, V.; Baloglu, S.; Grieco, F.; Güldür, M.E.; Greco, N.; Lafortezza, R. Properties of a citrus isolate of olive latent virus 1, a new necrovirus. *Eur. J. Plant Pathol.* **1996**, *102*, 527–536. [**Google Scholar**] [**CrossRef**]

52. Quito-Avila, D.F.; Freitas-Astúa, J.; Melzer, M.J. Bluner-, Cile-, and higreviruses (kitaviridae). *Encycl. Virol.* **2020**, *3*, 247–251. [**Google Scholar**]

53. Moreno, A.B.; López-Moya, J.J. When viruses play team sports: Mixed infections in plants. *Phytopathology* **2020**, *110*, 29–48. [**Google Scholar**] [**CrossRef**]

54. Massart, S.; Candresse, T.; Gil, J.; Lacomme, C.; Predajna, L.; Ravnikar, M.; Reynard, J.-S.; Rumbou, A.; Saldarelli, P.; Škorić, D. A framework for the evaluation of biosecurity, commercial, regulatory, and scientific impacts of plant viruses and viroids identified by NGS technologies. *Front. Microbiol.* **2017**, *8*, 45. [**Google Scholar**] [**CrossRef**] [**Green Version**]

Chapter-9

Nanopore Sequencing is a Credible Alternative to Recover Complete Genomes of Geminiviruses

1. Introduction

Recent advances in metagenomics applied to viruses has fostered a greater inventory of the viral diversity [**1,2,3,4**]. Hence, the large scale sampling of oceanic water [**5,6**], plants, animals, and humans [**7,8,9,10**], extreme environments [**11**], or the mining of genomic and transcritptomic data [**12,13,14,15**] have completely shifted our understanding of viral diversity and the function of viruses in host populations or even at the global ecosystem scale. However, these inventories remain largely incomplete, and the current knowledge of the virus diversity probably only represents the contour of the extant diversity [**4**]. The collection of hundreds of new genome sequences with, sometimes only remote resemblance to known viruses led to the acceptance of genome from metagenomic studies as genuine and legitimate taxonomic material for the description of new viruses [**16,17,18,19,20,21**], even without knowledge of the phenotype or the host associated to these viruses [**17,22**].

Yet, as access to metagenomics is usually costly and requires sophisticated technical expertise in data management and analysis [**23,24**]; for day-to-day analysis, classical Sanger sequencing remains more common [**25**]. Also, despite the potential of third generation sequencing technique to provide, in real time, hundreds of sequences with read length of more than 15 kb [**26,27**], the resulting assembled genomes remains tainted with doubt. Indeed, assembly of the low base quality reads is required to obtain more accurate contigs but may results in chimeric genomes, or miss less frequent variants [**17,28,29**]. It would thus be beneficial to evaluate the use of third generation sequencing as an alternative to the gold standard of cloning and Sanger sequencing for everyday virus characterization. Indeed, several studies have successfully employed third-generation sequencing for virus detection or full genome recovery. It has been largely used to detect and sequence full genomes of a range of animal and human viruses, including influenza [**30**], Ebola [**31**], Dengue [**32**], Zika [**33**], or SRAS-Cov-2 [**34**]. Recently the method was also successfully applied to plant viruses for some yam infecting viruses [**29**], maize yellow mosaic virus [**35**] and plum pox virus [**36**].

Here, within the context of the study of circular replication-associated protein encoding single stranded (CRESS) DNA viruses (from the *Cressdnaviricota* phylum), we harness the power of third-generation sequencing for routine full genome assembly of viruses. A breakthrough in CRESS DNA virus studies was associated with the development of protocols using isothermal rolling circle amplification (RCA) in frequent association with enzymatic restriction, cloning and Sanger sequencing [**37**]. Metagenomic protocols using RCA were soon developed and greatly improved our knowledge of the CRESS DNA virus diversity [**38,39**]. Yet, the use of Illumina based protocols remains laborious and expensive for day-to-day viral genome characterization and sequences analysis usually requires the use of complex bioinformatics tools. The field would thus benefit from the development of a more convenient protocol on the MinION platform from the Oxford Nanopore Technologies (ONT). Indeed, recent studies have paved the way towards using nanopore sequencing, either from direct sequencing of total DNA extracts or after the application of RCA, to achieve full genome sequencing of CRESS-DNA viruses in general [**40,41,42**] or viruses from the *Geminiviridae* family in particular [**43,44,45**].

Two alternative protocols for use on the MinION bench-top sequencing device were designed in this study and the assemblies of the nanopore-sequenced reads of multiple strains of capulaviruses infecting a *Medicago arborea* plant [**46**] were compared to the Sanger genome sequences. *Capulavirus* is one of the nine genera of the *Geminiviridae* family [**47**]. This family is composed of plant-infecting viruses with genomes comprising one or two circular ssDNA of 2.5–5.2 kb encapsidated in twinned icosahedral (geminate) particles. They are transmitted by a high range of hemipterans (whiteflies, leafhoppers, aphids, and treehoppers) [**48,49**]. Whereas members of the *Geminivirus* family were first described in 1993 (ICTV), the standardization of RCA based protocols lead to an explosion of its known diversity and the family regularly counts new genus-level lineage addition [**47,50,51,52**]. Following the description of the Euphorbia caput-medusae latent virus (EcmLV), the genus capulavirus has been proposed [**53**]. Their genome length ranged between 2.7 and 2.8 kb with two intergenic regions. The replication-associated protein (Rep) is expressed from a spliced complementary-strand transcript. A unique feature of capulavirus genomes is a complex arrangement of possible MP-encoding ORFs located in the 5′ direction from the coat protein gene (*cp*) [**47**]. It is known to be transmitted by only two species of aphids: *Aphis Craccivora* [**49**] and *Dysaphisplantaginea* [**54**].

Our analysis revealed that MinION sequencing followed by read assembly results in genome sequences mostly similar to the consensus of the virus population that was obtained after Sanger sequencing of multiple isolates. The two alternate protocols used, one that does not required knowledge of the viruses present within the sample (without a priori) and the other designed to specifically amplify a given virus (with a priori), were successful for full genome assembly of the two capulavirus strains. MinION assembled consensus sequences present with more differences in the homopolymeric tracks but remain closely related to any sequence of the sample than any Sanger sequence to one another. Overall, both nanopore-based protocols are adapted to the genome size of the CRESS DNA viruses and the cost of the method is on par with the Sanger sequencing approach.

2. Material and Methods

2.1. Sampling and DNA Extraction

Leaf samples of an apparently asymptomatic *Medicago arborea* were collected in November 2019 at Montferrier-sur-Lez (France). The sample was stored at 20− °C before use. Total DNA was extracted using the DNeasy Plant DNA extraction kit (Qiagen, Hilden, Germany), following the manufacturer's instructions. DNA extract was stored at 20− °C before use. From a previous analysis [**46**], using a PCR amplification and Sanger sequencing, two strains of the capulavirus Trifolium virus 1 (TrV1-B and TrV1-C) were identified into the sample.

2.2. Full Genome Cloning and Sanger Sequencing

Pairs of abutting primers were designed to recover the full-length genome of TrV1-B and TrV1-C isolates. A two-step amplification was achieved, including a first amplification step using either the primer pair 3580F-CAPULUZARB-1F: 5′-ACT TGC CTG TCG CTC TAT CTT CTC CCT TGG AGA TGT AAT CTG CCA CGT CAG-3′, and PR2-CAPULUZARB-2R: 5′-TTT CTG TTG GTG CTG ATA TTG CGG AGT TTT TGA GGA ACG AGG AAT ACT TAG AGC TTC A-3′ for amplifying TrV1-B genomes or the primer pair 3580F-CAPUCORO-1F: 5′-ACT TGC CTG TCG CTC TAT CTT CAA CTG TCC TCC CTT TGC AAT GTA GTC AGC C-3′ and PR2-CAPUCORO-2R: 5′-TTT CTG TTG GTG CTG ATA TTG CCG AGG AGC GAG GAC TTC TTA AGG CAA GT-3′for amplifying TrV1-C genomes. Amplification was carried out using the GoTaq® Master Mix Kit (Promega Corporation, Madison, WI, USA) and the following conditions: an initial denaturation at 95 °C for 5 min, 8 cycles at 94 °C for 30 s, 60 °C for 30 s, 72 °C for 3 min, and a final extension step at 72 °C for 10 min. A common second amplification step was then performed using the primer pair (3580F: -′5ACT TGC CTG TCG CTC TAT CTT C′3- and PR-′5 :2TTT CTG TTG GTG CTG ATA TTG C′3) and the GoTaq® Master Mix Kit (Promega Corporation, Madison, WI, USA). The amplification conditions were as followed: an initial denaturation at 95 °C for 5 min, 25 cycles at 94 °C for 30 s, 60 °C for 30 s, 72 °C for 3 min, and a final extension step at 72 °C for 10 min. Amplification products of approximately 2.8–2.7 kb were gel purified, ligated to pGEM-T (Promega, Madison, WI, USA) and sequenced by standard Sanger sequencing using a primer walking approach.

2.3. Minion Sequencing

Two alternative protocols for MinION sequencing were used. In the first protocol, called hereafter the RCA-MinION protocol, a RCA amplification was performed using *Phi*29 DNA polymerase (IllustraTempliPhi Amplification Kit, GE Healthcare, Chicago, IL, USA) by mixing 2 μL of total plant DNA extract with 5 μL of Sample Buffer before incubation at 95 °C during 3 min. After cooling at room temperature, 0.2 μL of enzyme mix and 5 μL of Reaction Buffer were added before incubation at 30 °C for 6 h followed by 20 min of polymerase deactivation at 65 °C. RCA products were cleaned-up using 2× of Sera-Mag Select Size Selection beads (GE Healthcare) and the 10 μL eluate were digested with 1 μL of T7 Endonuclease I (NEB), 2 μL of 5× buffer in a 10 μL reaction volume at 37 °C during 1 h. The fragments were purified with a 1× Sera-Mag Select Size Selection beads and eluate with 10 μL of purified water. Library construction for MinION sequencing was performed using the PCR Barcoding Kit (SQK-PBK004), following the manufacturer's instructions but using SeraMag Select Size Selection beads (GE Healthcare) for DNA purification. Sequencing was performed as described below for the PCR-MinION procedure. Two Flongles (flow cell dongle) FLO-FLG001 were used for sequencing. Whereas in the first Flongle, a single RCA amplicon was sequenced, in the second, three distinct RCA amplicons were multiplexed.

In the second protocol, called hereafter the PCR-MinION protocol, a two-step amplification was carried out. In the first PCR, both sets of abutting primers (3580F-CAPULUZARB-1F/PR2-CAPULUZARB-2R and 3580F-CAPUCORO-1F/PR2-CAPUCORO-2R) described above for respectively amplifying the genomes of strains TrV1-B and TrV1-C and the GoTaq® Master Mix Kit (Promega Corporation, Madison WI, USA) were employed. The amplification conditions were as follows: an initial denaturation at 95 °C for 5 min, 15 cycles at 94 °C for 30 s, 60 °C for 30 s, 72 °C for 3 min, and a final extension step at 72 °C for 10 min. The second amplification step was performed using the cDNA Primer (cPRM) supplied in the PCR-cDNA Sequencing Kit (SQK-PCS109) and the LongAmp Taq 2X Master Mix Kit (New England Biolabs, Evry, France). The amplification conditions were as followed: an initial denaturation at 95 °C for 30 s, 20 cycles at 95 °C for 15 s, 62 °C for 15 s, 65 °C for 3 min, and a final extension step at 65 °C for 6 min. The amplicons were purified using AgencourtAMPure XP beads (Beckman Coulter, Brea, CA, USA) and MinION sequencing library was constructed using the PCR-cDNA Sequencing Kit (SQK-PCS109), following manufacturer's instructions. Sequencing was then performed on Flongle (FLO-FLG001) using MinKNOW software 19.06.8. Three Flongles were used, two for TrV-1B (Flongle 13 and Flongle 14) and another one for TrV-1C (Flongle 15).

2.4. MinION Sequence Assembly

For the reads obtained through the two MinION protocols, accurate basecalling was performed using Guppy (v4.09 or 4.015; [**55**]). Demultiplexing and adapter removal was then performed using Porechop v0.2.4 [**56**]. Quality of the reads was investigated using NanoPlot v1.33.0 [**57**].

The demultiplexed reads obtained from RCA-MinION procedure were assembled with FLYE 2.6 [**58**], using the "meta" and "plasmid" parameters and when possible circularized as monomers using a homemade script. Contigs were then subjected to a BLASTn search against a CRESS DNA reference sequence database obtained from GenBank. CRESS DNA contigs were then polished using Medaka v1.2.2 [**59**]. PCR-MinION sequences higher than 1500 nt for one run (TrV1-C) and 2000 nt for the two other run (TrV1-B), were assembled using Canu v1.8 [**60**]. CRESS DNA contigs were filtered using BLAST as described above. Contigs coverage was estimated after mapping the raw reads back to the assembled CRESS DNA sequences using Minimap2 v2.17 [**61**] (**Figure 1**).

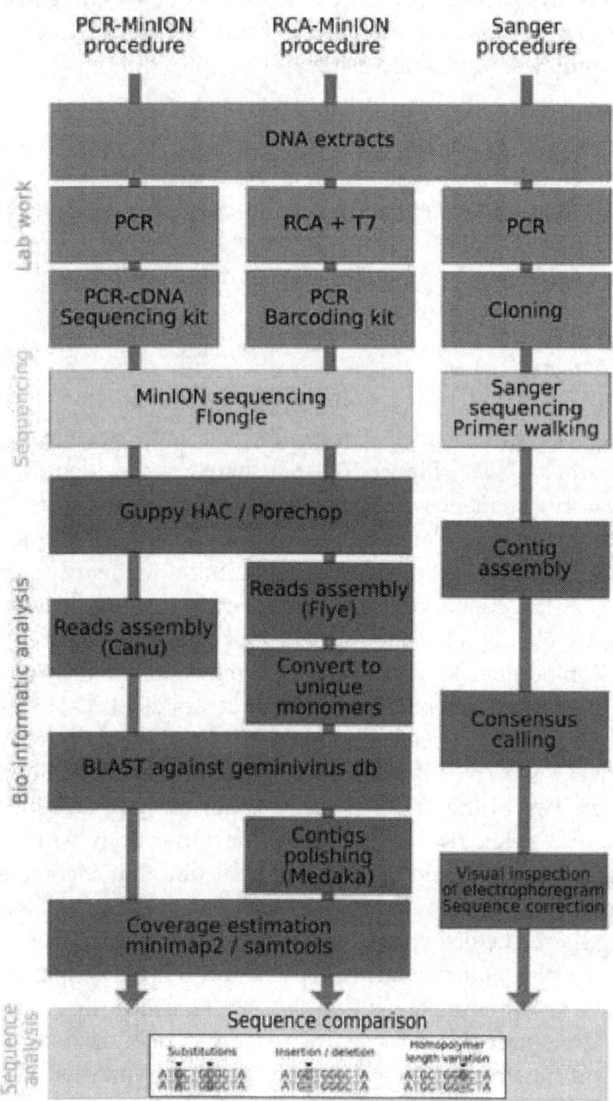

Figure 1. Schematic representation of the three distinct sequencing methodologies used including the molecular biology procedure (purple boxes) the sequencing procedure (light blue boxes), the bio-informatics procedure (red boxes) and the method comparison (orange boxes).

2.5. Sequence Comparison and Phylogenetic Analysis

All the available capulavirus full genome sequences were downloaded from GenBank on 4 March 2021 and aligned with the sequences of strains TrV1-B and TrV1-C obtained after Sanger sequencing using MAFFT v7.475 [**62**]. A maximum likelihood (ML) phylogenetic tree was inferred using FastTree2 [**63**] using the "gtr" and "gamma" parameters. Branch supports were tested using SH-like local supports. Tree edition was performed using the ape R package [**64**]. In order to properly classify the sequences obtained, an analysis that include a subset of representative capulavirus from GenBank and the Sanger sequence obtained in this study was performed using SDT1.2 [**65**].

Sequences obtained using the three distinct protocols (i.e., Sanger, PCR-MinION, and RCA-MinION procedures) were aligned together with MAFFT v7.475 before manual edit of the alignment. A home-made R script was used for sequence comparison and mutation count. Mutations were classified in three categories: substitution, insertion/deletion (INDEL) and homopolymer length variation (HLV) (**Figure 1**). ML trees were inferred from these alignments using FastTree2 as described above.

3. Results and Discussion

3.1. Sanger Sequences

A total of 46 sequences were obtained after cloning and Sanger sequencing. These sequence groups in two distinct clades (**Figure S1**) that share a mean identity of 90.3%. Within the first clade (*n* = 19), sequences present with identity ranging from 99.5 to %100 with each other and were most similar to the isolate BG2_capuz_47 of the B strain of TrV1 (GenBank accession number MW698819) with a minimum identity of %99.7. Within the second clade (*n* = 27), sequences present identity ranging from 99.0 to %100 with each other and were most closely related to the isolate BG2_coro_2–02 of the C strain of TrV1 (GenBank accession number MW698820) with a minimum identity of %99.1. Therefore, our two groups of sequences belong to two distinct strains (TrV-1B and TrV-1C) of the capulavirus *Trifolium virus 1* species (**Figure S1**). It is important to notice here that among the 46 isolates, five of the TrV1-B isolates and seven of the TrV1-C isolates present with defective genomes. Two of the TrV1-B defective isolates have deletions that encompass a fraction of the V3 gene, two other isolates have deletions within the gene of the replication-associated protein and one last has a deletion that encompasses the *rep* gene. Four sequences of the TrV-1C isolates have deletions that encompass a fraction of the CP gene. Among these, the deletions also span a fraction of the V3 gene or the *rep* gene, depending on the isolate. One isolate has a deletion that encompasses a fraction of all the gene encoded in the complement strand. These twelve defective sequences were excluded from downstream analysis. The genomic organization of the remaining sequences confirms the presence of a short intergenic region (SIR) and a long intergenic region (LIR), a characteristic inverted repeat potentially capable of forming a stem–loop structure that included a conserved nonanucleotide sequence TAATATTAC present at almost all geminivirus virion-strand replication origins, the *cp*, a spliced complementary-strand intron-containing transcript which expresses replication-associated protein gene (*rep*), a large complementary-sense ORF (C3) that is completely embedded within the *rep* gene, and a complex arrangement of possible MP-encoding ORFs located in the ′5 direction from the *cp* gene, which is a unique feature of *Capulavirus* genomes [**47**]. Four and three sequences of the TrV1-B and -C strains, present truncated ORFs. The full genome sequences of the isolate without ORFs truncation are available on GenBank under the accession numbers MW698819–MW698821 and MW768713–MW768736.

3.2. Long Read Sequencing and Assembly

The RCA-MinION generated reads that confirmed the presence of both TrV1-B and TrV1-C strains in the *Medicago arborea* sample (**Figure 2**). The raw sequencing statistics are available in **Table 1**. From Flongle 1 and Flongle 2, 188,123 and 273,088 raw reads were obtained from which 110,830 (59%) and 152,076 (56%) barcoded reads passed the quality control (**Figure S2**), respectively. The median read length was 1154 bp with a longest read of 10,491 bp. From Flongle 1, only 27 reads (0.02%) mapped with the capulavirus references. From Flongle 2, barcodes were retrieved from 65,413 reads, 40,242 reads and 46,421 reads for each of the three barcodes, respectively. From 6 to 8.5% of those reads mapped with the capulavirus sequences. Although it was performed on the same DNA extract, RCA amplification yielded more than two order of magnitude less viral sequence for the Flongle 1 amplification than those performed for Flongle 2. It highlights known bias of the RCA amplification [**66,67**] that were already evidenced in the context of CRESS virus amplification [**68**]. All the four distinct sequence sets (one barcode sample from Flongle 1 and three for Flongle 2) were then submitted to the assembly and circularization pipeline. For each of the four barcodes, unique contigs corresponding to the full genome sequences of the two strains were obtained. The four TrV1-B sequence lengths ranged from 2745 to 2769 bp and were at least 99.5% similar to any Sanger sequence. One of the TrV1-B sequences (RCA-Minion_10_2, Flongle 2 barcode 10, **Table 1**) present with a region that seems to be mis-assembled (100 nt in length, see grey tracks in **Figure 3**A). The four TrV1-C sequences length ranged from 2763 to 2771 bp and were at least 99.1% similar to any Sanger sequence. Again, a probable mis-assembly (68 nt in length, grey tracks in **Figure 3**B) was present within one sequence (RCA-Minion_01_1). As none of the raw minion reads supported the presence of this putative recombinant region, it has not been taken into consideration for further analyses.

The PCR-MinION generated reads that also confirmed the presence of TrV1-B and TrV1-C strains after RCA-MinION and Sanger sequencing. From Flongle 13, Flongle 14, and Flongle 15, 492,922, 414,665, and 768,144 raw reads were obtained from which 386,099 (78.3%), 371,700 (89.6%), and 714,730 (93.0%) reads passed the quality control (**Table 1, Figure S2**). The median length of the passed reads was 643 bp with a longest read of 9816 bp. From Flongle 13, 14 and 15, 380,933 reads (98.7%), 367,381 reads (98.8%) and 337,338 reads (47.2%) mapped with the TrV1-B and -C reference sequences. All the three read sets were then submitted to the assembly. For every Flongle, contigs corresponding to the full genome sequence of the TrV1-B and TrV1-C Sanger references were obtained. The two TrV1-B sequences length ranged from 2731 to 2739 bp and were at least 99.7% similar to any Sanger sequence. The TrV1-C sequence is 2754 bp length and were at least 99.3% similar to any Sanger sequence (**Figure 2**).

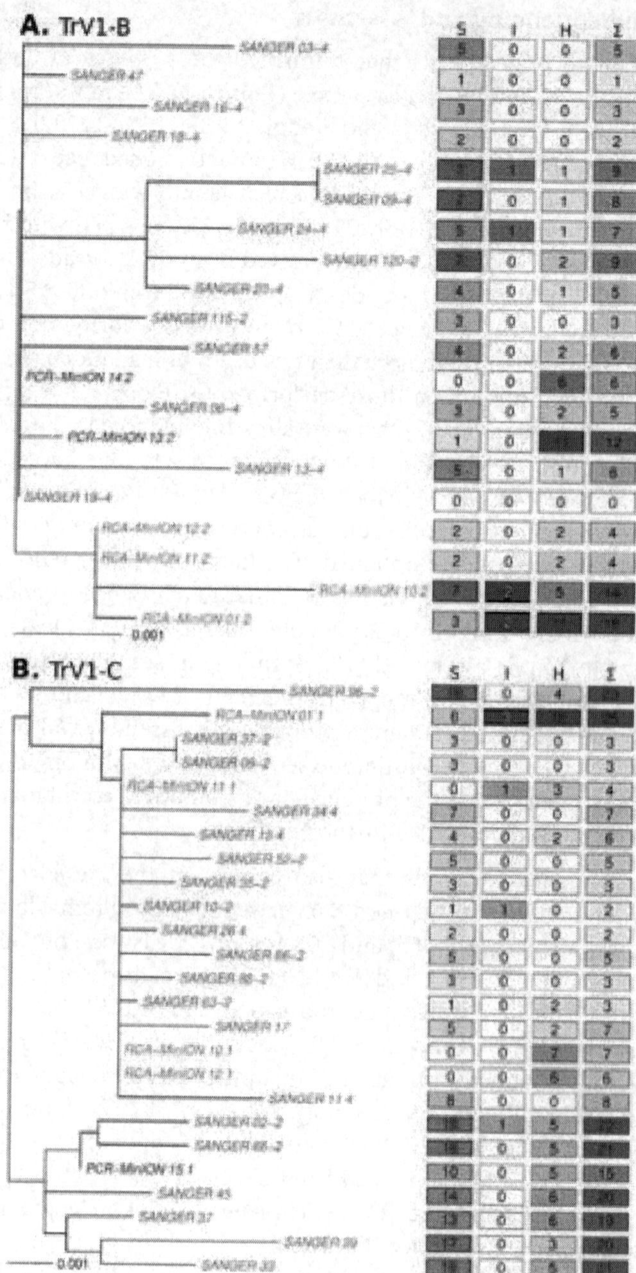

Figure 2. Maximum likelihood phylogenetic tree of all the sequences of the TrV1-B (A) and TrV1-C (B) strains obtained after Sanger sequencing (brown tips) or MinION sequencing followed by read assemblies (blue and purple tips for the RCA-MinION and PCR-MinION procedures respectively) on the left along with a matrix presenting the number of mutation relative to the Sanger consensus on the right. The numbers in the four columns present the substitutions (S), INDELs (I), HLVs (H) and the sum (Σ) of all these variations.

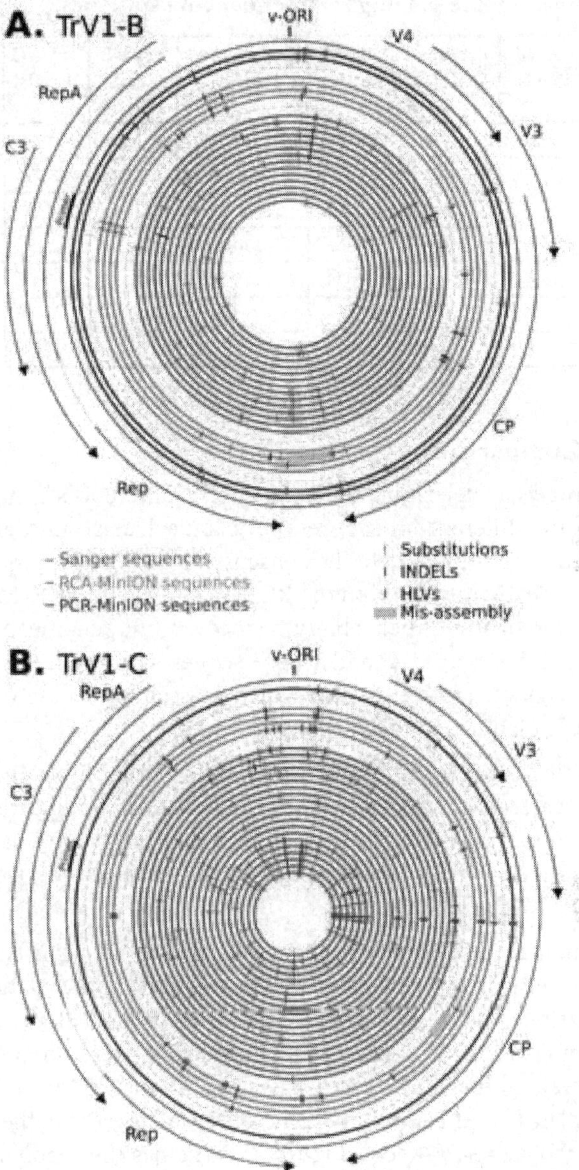

Figure 3. Diagram prese5nting the positions of the mutations along the genome of the TrV1-B (A) and TrV1-C (B) isolates and assemblies. Concentric circles represent each complete genome sequence obtained, with from the center to the outer, Sanger sequences in brown, RCA-MinION in blue and PCR-MinION in purple. Substitutions, INDELs and homopolymer length variations are represented with green, blue and red ticks respectively. Grey areas represent large deletions. Positions are relative to reference sequences MW698819 for TrV1-B and MW698820 for TrV1-C. The origin of replication (v-ORI) is indicated on top and the ORFs are represented on the outside of the figure.

Table 1. Long read sequencing statistics.

Flongle Id	Raw Reads	Passed Reads	Barcode ID	Trimmed Reads	Capulavirus Reads	TrV1-B Reads	TrV1-C Reads	Length TrV1-B Assembly	Length TrV1-C Assembly
1	188,123	162,263	1	110,830	27	14	13	2745	2771
2	273,088	215,143	10	65,413	4809	3230	1579	2769	2764
			11	40,242	3423	2244	1179	2748	2763
			12	46,421	2803	1585	1218	2746	2765
13	492,922	386,099	-	385,029	380,933	380,933	-	2731	-
14	414,665	371,700	-	370,755	367,381	367,381	-	2739	-
15	768,144	714,730	-	602,579	337,338	-	337,338	-	2754

3.3. Sequences Comparison

In order to more precisely determine the nature of the differences between the sequences obtained through the different procedures, per capulavirus strains, all the full genome sequences obtained were compared to the consensus of the Sanger sequences. With more than 99.1% nucleotide identity for both the RCA-MinION and PCR-MinION sequences, the two methods demonstrate their ability to recover full genomes that are accurately assigned to strains TrV1-B and -C and whose sequence are representative of the viral population they originate. However, none of the assemblies obtained from MinION sequences was 100% identical to the Sanger sequence.

Beside mis-assemblies (grey tracks on **Figure 3**), the differences between sequences were classified in three categories with substitution (green ticks on **Figure 3**), INDEL (blue ticks) and HLV (red ticks). It must be noticed here that HLVs are a category of INDELs but were count separately as HLVs are recognized as the main source of errors in nanopore sequencing [**69**,**70**,**71**,**72**].

First, Sanger sequences differ from each other (and to the consensus) mostly with substitutions (accounting for 79.5% of the variations) and more marginally with HLVs (19.4% of the variations) and INDELs (1.5% of the variations) (**Figure 2**). Except for two of the TrV1-C sequences, all the other Sanger sequences were unique and the number of variations between all these sequences was up to 12 and 27 for TrV1-B and TrV1-C, respectively. A total of 46 and 113 polymorphic sites were present in the Sanger sequences for TrV1-B and TrV1-C, respectively (**Figure 3**). Whereas these polymorphic sites tend be more frequently present within non-coding regions (binomial test p-value of 0.090 and 0.016 for TrV1-B and TrV1-C, respectively), these sites had globally more mutations among all the Sanger sequences (binomial test p-value of 3.7×10^{-3} and $^{6-}10 \times 7.7$ for TrV-1B and TrV-1C, respectively). Whereas sequencing errors can explain a fraction of the mutation, with the high accuracy of Sanger sequencing in mind (per base Phred quality score of **73**] 50]), one can expect that most of the variations uncovered during the analysis are genuine and represent the biological variation associated with the diversity of the viral population infecting the plant [**74**,**75**,**76**].

The analysis of the RCA-MinION and PCR-MinION sequences revealed a different pattern of polymorphism. Two sequences, one of TrV1-B and one of TrV1-C, both assembled from a restricted number of reads (Flongle 1, RCA-MinION_01-1 and RCA-MinION_01-2) were, as expected, less accurate. Coverage of the assembly, (e.g., representing the mean number of times every position of an assembly was read) is therefore a good indicator of the reliability of the resulting assembly (**Figure 4**). Among the nine other MinION sequences assembled, seven present with no substitution to the consensus and one with ten substitutions (**Figure 2**). Through the TrV1-B sequences there was few common mutations: only a single HLV was common between Sanger and MinION assemblies (**Figure 3**A). PCR-MinION and RCA-MinION sequences presented four common HLVs (**Figure 3**A). For TrV1-C, Sanger sequences presented one substitution, no INDEL and three HLVs in common with MinION sequences (**Figure 3**B). PCR-MinION and RCA-MinION sequences presented one common substitution, three common HLVs and no common INDELs (**Figure 3**B). As these sequences were obtained after the assembly of reads, we were unable to catch the diversity of the distinct variants forming the viral population but rather to obtain a sequence very similar to the consensus of that population. Conversely to the reduced number of substitutions in comparison to the consensus, the assemblies present with larger numbers of INDELs (from 1 to 3) and HLVs (2 to 16). Some of these variations were also found in the Sanger sequences and it is probable that the assemblies actually represented some of the variability within the population. Nevertheless, for six INDELs and 38 HLVs, no corresponding mutations were found, and most would induce frameshift or protein truncation (**Figure 3**). Despite the use of a dedicated sequence correction program, multiple HLVs remained in the assembled sequences.

3.4. Defective Genome and Sequence Coverage

Defective genomes are frequently detected within geminivirus populations [77]. For instance, twelve of the Sanger sequences displayed large deletions in comparison to full reference genomes. As MinION raw sequences are obtained after direct reads of either PCR or RCA amplicons, they should capture the diversity of defective genomes of the viral populations. Indeed, to obtain full genome assemblies using the PCR-MinION procedure, a selection of the reads approaching full genome size was required. The analysis of coverage of the reads (**Figure 4**) confirms the pervasive nature of defective genomes for both the TrV1-B and -C strains. The highest coverage was obtained for regions encompassing the stem-loop and reads were frequently missing most of the coding regions (see the blue lines for RCA-Minion). Importantly, it must be noticed here that the coverage inferred using the PCR-MinION procedure are not representative of the global population but rather represent the subset of virus that contains the priming site of the abutting primers used in the PCR (indicated with the verticals red dotted lines in **Figure 4**). The results indicate that every position of the TrV1-B strains present with a high coverage, most of the amplicons of TrV1-C were defective for a region encompassing the whole CP gene (from position 147 to 1559).

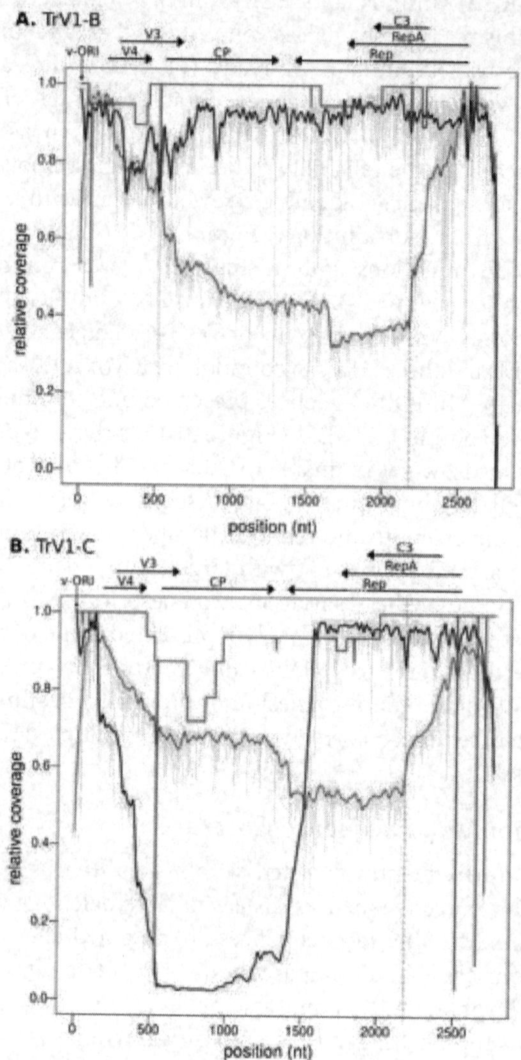

Figure 4. Coverage plot along the reference genome of TrV1-B (A) and TrV1-C (B). The coverage, interpreted here as the fold coverage of a given position divided by the maximum coverage of the whole genome, is represented with a brown line for Sanger sequences, a blue line for RCA-MinION sequences and a purple line for PCR-MinION sequences. MinION coverage are interpolations of the raw coverages presented here in grey on the background. The position of the abutting primers used in the PCR-MinION procedure is indicated by the vertical red dotted line. Positions are relative to the reference genome MW698819 for TrV1-B and MW698820 for TrV1-C. ORFs of these reference genomes are symbolized with horizontal arrows on top of the figure. The Origin of replication is indicated with an arrow. The position 1 of each sequence correspond to the nick-site within the stem loop structure.

4. Conclusions

From an asymptomatic sample of *Medicago arborea*, two distinct strains of the capulavirus TrV1 have been cloned and sequenced by the Sanger methodology. Using both the *a priori* and the agnostic nanopore-based procedures, both TrV-1B and TrV-1C strains were detected, and full genome sequences were assembled. Despite being very similar to the consensus of Sanger sequences, mutations specific to the MinION assemblies were detected mostly within homopolymeric regions of the genomes, which is in agreement with other studies that have also pinpointed higher number of errors associated with homopolymer lengths [**78,79**]. Whereas it could be argued that the HLV errors would be reduced with the increasing accuracy of sequencing, the development of a new base caller technologies and correction algorithm [**80**], current MinION sequences assemblies may be avoided for some specific applications were the exact nucleotide sequence is required. Otherwise, for other applications, such as virus discovery, virus classification, or recombination analysis, MinION assemblies represent a competitive alternative to Sanger sequencing. Although, similarly to other NGS protocols, MinION based studies require the use of sophisticated bioinformatics tools for data management and analysis and despite the specific drawbacks on sequence quality, bench top sequencer such as the MinION would probably become routinely used in the laboratory. It allows a real-time detection and diagnostic of multiple viral strains or virus species in a single run. For niche applications, such as the exploration of geminivirus genomes, which do not exceed 10 kb, nanopore sequencing is poised to push the cost and performance limits of sequencing technologies. The high reactivity offered by the platform would make it more and more democratized as a mobile real-time plant disease diagnostic tool.

Supplementary Materials

Figure S1: Maximum likelihood phylogenetic tree of all the *Capulavirus* genus full genome sequences obtained from NCBI along with the two TrV-1B and TrV-1C genotypes obtained in this study; Supplementary Figure S2: Reads length (x-axis) against quality score (y-axis) for all passed reads (A and C) and capulavirus only passed reads (B and D) obtained using the RCA-MinION and PCR-MinION procedures.

References

1. Dolja, V.V.; Koonin, E.V. Metagenomics reshapes the concepts of RNA virus evolution by revealing extensive horizontal virus transfer. *Virus Res.* **2018**, *244*, 36–52. [**Google Scholar**] [**CrossRef**] [**PubMed**]
2. Greninger, A.L. A decade of RNA virus metagenomics is (not) enough. *Virus Res.* **2018**, *244*, 218–229. [**Google Scholar**] [**CrossRef**]
3. Koonin, E.V.; Dolja, V.V. Metaviromics: A tectonic shift in understanding virus evolution. *Virus Res.* **2018**, *246*, A1–A3. [**Google Scholar**] [**CrossRef**] [**PubMed**]
4. Roux, S.; Matthijnssens, J.; Dutilh, B.E. Metagenomics in Virology. *Encycl. Virol.* **2021**, *1*, 133–140. [**Google Scholar**] [**CrossRef**]
5. Roux, S.; Brum, J.R.; Dutilh, B.E.; Sunagawa, S.; Duhaime, M.B.; Loy, A.; Poulos, B.T.; Solonenko, N.; Lara, E.; Poulain, J.; et al. Ecogenomics and potential biogeochemical

impacts of globally abundant ocean viruses. *Nat. Cell Biol.* **2016**, *537*, 689–693. [**Google Scholar**] [**CrossRef**] [**Green Version**]

6. McMullen, A.; Martinez-Hernandez, F.; Martinez-Garcia, M. Absolute quantification of infecting viral particles by chip-based digital polymerase chain reaction. *Environ. Microbiol. Rep.* **2019**, *11*, 855–860. [**Google Scholar**] [**CrossRef**]
7. Breitbart, M.; Hewson, I.; Felts, B.; Mahaffy, J.M.; Nulton, J.; Salamon, P.; Rohwer, F. Metagenomic Analyses of an Uncultured Viral Community from Human Feces. *J. Bacteriol.* **2003**, *185*, 6220–6223. [**Google Scholar**] [**CrossRef**] [**Green Version**]
8. Wylie, K.M.; Weinstock, G.M.; Storch, G.A. Emerging view of the human virome. *Transl. Res.* **2012**, *160*, 283–290. [**Google Scholar**] [**CrossRef**] [**PubMed**] [**Green Version**]
9. Lecuit, M.; Eloit, M. The diagnosis of infectious diseases by whole genome next generation sequencing: A new era is opening. *Front. Cell. Infect. Microbiol.* **2014**, *4*, 25. [**Google Scholar**] [**CrossRef**] [**PubMed**] [**Green Version**]
10. Maclot, F.; Candresse, T.; Filloux, D.; Malmstrom, C.M.; Roumagnac, P.; Van Der Vlugt, R.; Massart, S. Illuminating an Ecological Blackbox: Using High Throughput Sequencing to Characterize the Plant Virome Across Scales. *Front. Microbiol.* **2020**, *11*, 578064. [**Google Scholar**] [**CrossRef**]
11. Dávila-Ramos, S.; Castelán-Sánchez, H.G.; Martínez-Ávila, L.; Sánchez-Carbente, M.D.R.; Peralta, R.; Hernández-Mendoza, A.; Dobson, A.D.W.; Gonzalez, R.A.; Pastor, N.; Batista-García, R.A. A Review on Viral Metagenomics in Extreme Environments. *Front. Microbiol.* **2019**, *10*, 2403. [**Google Scholar**] [**CrossRef**] [**PubMed**] [**Green Version**]
12. Goodwin, S.; McPherson, J.D.; McCombie, W.R. Coming of age: Ten years of next-generation sequencing technologies. *Nat. Rev. Genet.* **2016**, *17*, 333–351. [**Google Scholar**] [**CrossRef**]
13. Shi, M.; Lin, X.-D.; Tian, J.-H.; Chen, L.-J.; Chen, X.; Li, C.-X.; Qin, X.-C.; Li, J.; Cao, J.-P.; Eden, J.-S.; et al. Redefining the invertebrate RNA virosphere. *Nat. Cell Biol.* **2016**, *540*, 539–543. [**Google Scholar**] [**CrossRef**]
14. Shi, M.; Zhang, Y.-Z.; Holmes, E.C. Meta-transcriptomics and the evolutionary biology of RNA viruses. *Virus Res.* **2018**, *243*, 83–90. [**Google Scholar**] [**CrossRef**]
15. Zhang, Y.-Z.; Shi, M.; Holmes, E.C. Using Metagenomics to Characterize an Expanding Virosphere. *Cell* **2018**, *172*, 1168–1172. [**Google Scholar**] [**CrossRef**] [**PubMed**]
16. Rosario, K.; Duffy, S.; Breitbart, M. A field guide to eukaryotic circular single-stranded DNA viruses: Insights gained from metagenomics. *Arch. Virol.* **2012**, *157*, 1851–1871. [**Google Scholar**] [**CrossRef**] [**PubMed**]
17. Simmonds, P.; Adams, M.J.; Benkő, M.; Breitbart, M.; Brister, J.R.; Carstens, E.B.; Davison, A.J.; Delwart, E.; Gorbalenya, A.E.; Harrach, M.B.B.; et al. Virus taxonomy in the age of metagenomics. *Nat. Rev. Genet.* **2017**, *15*, 161–168. [**Google Scholar**] [**CrossRef**]
18. Batovska, J.; Lynch, S.E.; Rodoni, B.C.; Sawbridge, T.I.; Cogan, N.O. Metagenomic arbovirus detection using MinION nanopore sequencing. *J. Virol. Methods* **2017**, *249*, 79–84. [**Google Scholar**] [**CrossRef**] [**PubMed**]

19. Roux, S.; Adriaenssens, E.M.; Dutilh, B.E.; Koonin, E.V.; Kropinski, A.M.; Krupovic, M.; Kuhn, J.H.; Lavigne, R.; Brister, J.R.; Varsani, A.; et al. Minimum Information about an Uncultivated Virus Genome (MIUViG). *Nat. Biotechnol.* **2019**, *37*, 29–37. [**Google Scholar**] [**CrossRef**] [**PubMed**]

20. Claverie, S.; Ouattara, A.; Hoareau, M.; Filloux, D.; Varsani, A.; Roumagnac, P.; Martin, D.P.; Lett, J.-M.; Lefeuvre, P. Exploring the diversity of Poaceae-infecting mastreviruses on Reunion Island using a viral metagenomics-based approach. *Sci. Rep.* **2019**, *9*, 1–11. [**Google Scholar**] [**CrossRef**] [**Green Version**]

21. International Committee on Taxonomy of Viruses Executive Committee. The new scope of virus taxonomy: Partitioning the virosphere into 15 hierarchical ranks. *Nat. Microbiol.* **2020**, *5*, 668–674. [**Google Scholar**] [**CrossRef**]

22. Roossinck, M.J.; Martin, D.P.; Roumagnac, P. Plant Virus Metagenomics: Advances in Virus Discovery. *Phytopathology* **2015**, *105*, 716–727. [**Google Scholar**] [**CrossRef**] [**Green Version**]

23. Quiñones-Mateu, M.E.; Avila, S.; Reyes-Teran, G.; Martinez, M.A. Deep sequencing: Becoming a critical tool in clinical virology. *J. Clin. Virol.* **2014**, *61*, 9–19. [**Google Scholar**] [**CrossRef**] [**PubMed**] [**Green Version**]

24. Mikheyev, A.S.; Tin, M.M.Y. A first look at the Oxford Nanopore MinION sequencer. *Mol. Ecol. Resour.* **2014**, *14*, 1097–1102. [**Google Scholar**] [**CrossRef**]

25. Sanger, F.; Coulson, A. A rapid method for determining sequences in DNA by primed synthesis with DNA polymerase. *J. Mol. Biol.* **1975**, *94*, 441–448. [**Google Scholar**] [**CrossRef**]

26. Jain, M.; Olsen, H.E.; Paten, B.; Akeson, M. The Oxford Nanopore MinION: Delivery of nanopore sequencing to the genomics community. *Genome Biol.* **2016**, *17*, 1–11. [**Google Scholar**] [**CrossRef**] [**Green Version**]

27. Pomerantz, A.; Peñafiel, N.; Arteaga, A.; Bustamante, L.; Pichardo, F.; Coloma, L.A.; Barrio-Amorós, C.L.; Salazar-Valenzuela, D.; Prost, S. Real-time DNA barcoding in a rainforest using nanopore sequencing: Opportunities for rapid biodiversity assessments and local capacity building. *GigaScience* **2018**, *7*, 1–14. [**Google Scholar**] [**CrossRef**] [**PubMed**] [**Green Version**]

28. Xu, Y.; Lewandowski, K.; Lumley, S.; Pullan, S.; Vipond, R.; Carroll, M.; Foster, D.; Matthews, P.C.; Peto, T.; Crook, D. Detection of Viral Pathogens with Multiplex Nanopore MinION Sequencing: Be Careful With Cross-Talk. *Front. Microbiol.* **2018**, *9*, 2225. [**Google Scholar**] [**CrossRef**]

29. Filloux, D.; Fernandez, E.; Loire, E.; Claude, L.; Galzi, S.; Candresse, T.; Winter, S.; Jeeva, M.L.; Makeshkumar, T.; Martin, D.P.; et al. Nanopore-based detection and characterization of yam viruses. *Sci. Rep.* **2018**, *8*, 1–11. [**Google Scholar**] [**CrossRef**] [**PubMed**] [**Green Version**]

30. Ewang, J.; Moore, N.E.; Edeng, Y.-M.; Eccles, D.A.; Hall, R.J. MinION nanopore sequencing of an influenza genome. *Front. Microbiol.* **2015**, *6*, 766. [**Google Scholar**] [**CrossRef**] [**Green Version**]

31. Quick, J.; Loman, N.J.; Duraffour, S.; Simpson, J.T.; Severi, E.; Cowley, L.; Bore, J.A.; Koundouno, R.; Dudas, G.; Mikhail, A.; et al. Real-time, portable genome sequencing for Ebola surveillance. *Nat. Cell Biol.* **2016**, *530*, 228–232. [**Google Scholar**] [**CrossRef**] [**Green Version**]

32. Yamagishi, J.; Runtuwene, L.R.; Hayashida, K.; Mongan, A.E.; Thi, L.A.N.; Thuy, L.N.; Nhat, C.N.; Limkittikul, K.; Sirivichayakul, C.; Sathirapongsasuti, N.; et al. Serotyping dengue virus with isothermal amplification and a portable sequencer. *Sci. Rep.* **2017**, *7*, 1–10. [**Google Scholar**] [**CrossRef**]

33. Quick, J.; Grubaugh, N.D.; Pullan, S.T.; Claro, I.M.; Smith, A.D.; Gangavarapu, K.; Oliveira, G.; Robles-Sikisaka, R.; Rogers, T.F.; Beutler, N.A.; et al. Multiplex PCR method for MinION and Illumina sequencing of Zika and other virus genomes directly from clinical samples. *Nat. Protoc.* **2017**, *12*, 1261–1276. [**Google Scholar**] [**CrossRef**] [**PubMed**] [**Green Version**]

34. Kim, D.; Lee, J.-Y.; Yang, J.-S.; Kim, J.W.; Kim, V.N.; Chang, H. The Architecture of SARS-CoV-2 Transcriptome. *Cell* **2020**, *181*, 914–921.e10. [**Google Scholar**] [**CrossRef**]

35. Adams, I.P.; Braidwood, L.A.; Stomeo, F.; Phiri, N.; Uwumukiza, B.; Feyissa, B.; Mahuku, G.; Wangi, A.; Smith, J.; Mumford, R.; et al. Characterising Maize Viruses Associated with Maize Lethal Necrosis Symptoms in Sub Saharan Africa. *bioRxiv* **2017**. preprint. [**Google Scholar**] [**CrossRef**] [**Green Version**]

36. Badial, A.B.; Sherman, D.; Stone, A.; Gopakumar, A.; Wilson, V.; Schneider, W.; King, J. Nanopore Sequencing as a Surveillance Tool for Plant Pathogens in Plant and Insect Tissues. *Plant Dis.* **2018**, *102*, 1648–1652. [**Google Scholar**] [**CrossRef**] [**PubMed**] [**Green Version**]

37. Inoue-Nagata, A.K.; Albuquerque, L.C.; Rocha, W.B.; Nagata, T. A simple method for cloning the complete begomovirus genome using the bacteriophage φ29 DNA polymerase. *J. Virol. Methods* **2004**, *116*, 209–211. [**Google Scholar**] [**CrossRef**] [**PubMed**]

38. Rosario, K.; Dayaram, A.; Marinov, M.; Ware, J.; Kraberger, S.; Stainton, D.; Breitbart, M.; Varsani, A. Diverse circular ssDNA viruses discovered in dragonflies (Odonata: Epiprocta). *J. Gen. Virol.* **2012**, *93*, 2668–2681. [**Google Scholar**] [**CrossRef**]

39. Rosario, K.; Schenck, R.O.; Harbeitner, R.C.; Lawler, S.N.; Breitbart, M. Novel circular single-stranded DNA viruses identified in marine invertebrates reveal high sequence diversity and consistent predicted intrinsic disorder patterns within putative structural proteins. *Front. Microbiol.* **2015**, *6*, 696. [**Google Scholar**] [**CrossRef**] [**Green Version**]

40. Theuns, S.; Vanmechelen, B.; Bernaert, Q.; Deboutte, W.; Vandenhole, M.; Beller, L.; Matthijnssens, J.; Maes, P.; Nauwynck, H.J. Nanopore sequencing as a revolutionary diagnostic tool for porcine viral enteric disease complexes identifies porcine kobuvirus as an important enteric virus. *Sci. Rep.* **2018**, *8*, 1–13. [**Google Scholar**] [**CrossRef**] [**Green Version**]

41. Boykin, L.M.; Sseruwagi, P.; Alicai, T.; Ateka, E.; Mohammed, I.U.; Stanton, J.-A.L.; Kayuki, C.; Mark, D.; Fute, T.; Erasto, J.; et al. Tree Lab: Portable genomics for Early

Detection of Plant Viruses and Pests in Sub-Saharan Africa. *Genes* **2019**, *10*, 632. [**Google Scholar**] [**CrossRef**] [**PubMed**] [**Green Version**]

42. Cao, J.; Zhang, Y.; Dai, M.; Xu, J.; Chen, L.; Zhang, F.; Zhao, N.; Wang, J. Profiling of Human Gut Virome with Oxford Nanopore Technology. *Med. Microecol.* **2020**, *4*, 100012. [**Google Scholar**] [**CrossRef**]

43. Chalupowicz, L.; Dombrovsky, A.; Gaba, V.; Luria, N.; Reuven, M.; Beerman, A.; Lachman, O.; Dror, O.; Nissan, G.; Manulis-Sasson, S. Diagnosis of plant diseases using the Nanopore sequencing platform. *Plant Pathol.* **2018**, *68*, 229–238. [**Google Scholar**] [**CrossRef**]

44. Naito, F.Y.B.; Melo, F.L.; Fonseca, M.E.N.; Santos, C.A.F.; Chanes, C.R.; Ribeiro, B.M.; Gilbertson, R.L.; Boiteux, L.S.; Pereira-Carvalho, R.D.C. Nanopore sequencing of a novel bipartite New World begomovirus infecting cowpea. *Arch. Virol.* **2019**, *164*, 1907–1910. [**Google Scholar**] [**CrossRef**]

45. Leiva, A.M.; Siriwan, W.; Alvarez, D.L.; Barrantes, I.; Hemniam, N.; Saokham, K.; Cuellar, W.J. Nanopore-Based Complete Genome Sequence of a Sri Lankan Cassava Mosaic Virus (Geminivirus) Strain from Thailand. *Microbiol. Resour. Announc.* **2020**, *9*, 9. [**Google Scholar**] [**CrossRef**] [**Green Version**]

46. Ma, Y.; Svanella-Dumas, L.; Julian, C.; Galzi, S.; Fernandez, E.; Yvon, M.; Pirolles, E.; Lefebvre, M.; Filloux, D.; Roumagnac, P.; et al. Genome characterization and diversity of trifolium virus 1: Identification of a novel legume-infective capulavirus. *Arch. Virol.* **2021**, in press. [**Google Scholar**]

47. Varsani, A.; Roumagnac, P.; Fuchs, M.; Navas-Castillo, J.; Moriones, E.; Idris, A.; Briddon, R.W.; Rivera-Bustamante, R.; Zerbini, F.M.; Martin, D.P. Capulavirus and Grablovirus: Two new genera in the family Geminiviridae. *Arch. Virol.* **2017**, *162*, 1819–1831. [**Google Scholar**] [**CrossRef**] [**Green Version**]

48. Fauquet, C.M.; Briddon, R.W.; Brown, J.K.; Moriones, E.; Stanley, J.; Zerbini, M.; Zhou, X. Geminivirus strain demarcation and nomenclature. *Arch. Virol.* **2008**, *153*, 783–821. [**Google Scholar**] [**CrossRef**] [**Green Version**]

49. Roumagnac, P.; Granier, M.; Bernardo, P.; Deshoux, M.; Ferdinand, R.; Galzi, S.; Fernandez, E.; Julian, C.; Abt, I.; Filloux, D.; et al. Alfalfa Leaf Curl Virus: An Aphid-Transmitted Geminivirus. *J. Virol.* **2015**, *89*, 9683–9688. [**Google Scholar**] [**CrossRef**] [**Green Version**]

50. Varsani, A.; Navas-Castillo, J.; Moriones, E.; Hernández-Zepeda, C.; Idris, A.; Brown, J.K.; Zerbini, F.M.; Martin, D.P. Establishment of three new genera in the family Geminiviridae: Becurtovirus, Eragrovirus and Turncurtovirus. *Arch. Virol.* **2014**, *159*, 2193–2203. [**Google Scholar**] [**CrossRef**]

51. Zerbini, F.M.; Briddon, R.W.; Idris, A.; Martin, D.P.; Moriones, E.; Navas-Castillo, J.; Rivera-Bustamante, R.; Roumagnac, P.; Varsani, A. ICTV Report Consortium ICTV Virus Taxonomy Profile: Geminiviridae. *J. Gen. Virol.* **2017**, *98*, 131–133. [**Google Scholar**] [**CrossRef**] [**PubMed**]

52. Claverie, S.; Bernardo, P.; Kraberger, S.; Hartnady, P.; Lefeuvre, P.; Lett, J.-M.; Galzi, S.; Filloux, D.; Harkins, G.W.; Varsani, A.; et al. From Spatial Metagenomics to Molecular

Characterization of Plant Viruses: A Geminivirus Case Study. In *Advances in Clinical Chemistry*; Elsevier: Amsterdam, The Netherlands, 2018; Volume 101, pp. 55–83. [**Google Scholar**]

53. Bernardo, P.; Golden, M.; Akram, M.; Naimuddin; Nadarajan, N.; Fernandez, E.; Granier, M.; Rebelo, A.G.; Peterschmitt, M.; Martin, D.P.; et al. Identification and characterisation of a highly divergent geminivirus: Evolutionary and taxonomic implications. *Virus Res.* **2013**, *177*, 35–45. [**Google Scholar**] [**CrossRef**]
54. Susi, H.; Filloux, D.; Frilander, M.J.; Roumagnac, P.; Laine, A.-L. Diverse and variable virus communities in wild plant populations revealed by metagenomic tools. *PeerJ* **2019**, *7*, e6140. [**Google Scholar**] [**CrossRef**] [**PubMed**]
55. Oxford Nanopore Technologies. Available online: **https://nanoporetech.com/** (accessed on 18 March 2021).
56. Porechop. Available online: **https://github.com/rrwick/Porechop** (accessed on 18 March 2021).
57. De Coster, W.; D'Hert, S.; Schultz, D.T.; Cruts, M.; Van Broeckhoven, C. NanoPack: Visualizing and processing long-read sequencing data. *Bioinformatics* **2018**, *34*, 2666–2669. [**Google Scholar**] [**CrossRef**] [**PubMed**]
58. Kolmogorov, M.; Yuan, J.; Lin, Y.; Pevzner, P.A. Assembly of long, error-prone reads using repeat graphs. *Nat. Biotechnol.* **2019**, *37*, 540–546. [**Google Scholar**] [**CrossRef**]
59. Medaka. Available online: **https://github.com/nanoporetech/medaka** (accessed on 18 March 2021).
60. Koren, S.; Walenz, B.P.; Berlin, K.; Miller, J.R.; Bergman, N.H.; Phillippy, A.M. Canu: Scalable and accurate long-read assembly via adaptivek-mer weighting and repeat separation. *Genome Res.* **2017**, *27*, 722–736. [**Google Scholar**] [**CrossRef**] [**PubMed**] [**Green Version**]
61. Li, H. Minimap2: Pairwise alignment for nucleotide sequences. *Bioinformatics* **2018**, *34*, 3094–3100. [**Google Scholar**] [**CrossRef**] [**PubMed**]
62. Rozewicki, J.; Li, S.; Amada, K.M.; Standley, D.M.; Katoh, K. MAFFT-DASH: Integrated protein sequence and structural alignment. *Nucleic Acids Res.* **2019**, *47*, W5–W10. [**Google Scholar**] [**CrossRef**]
63. Price, M.N.; Dehal, P.S.; Arkin, A.P. FastTree 2—Approximately Maximum-Likelihood Trees for Large Alignments. *PLoS ONE* **2010**, *5*, e9490. [**Google Scholar**] [**CrossRef**] [**PubMed**]
64. Paradis, E.; Schliep, K. ape 5.0: An environment for modern phylogenetics and evolutionary analyses in R. *Bioinformatics* **2018**, *35*, 526–528. [**Google Scholar**] [**CrossRef**]
65. Muhire, B.M.; Varsani, A.; Martin, D.P. SDT: A Virus Classification Tool Based on Pairwise Sequence Alignment and Identity Calculation. *PLoS ONE* **2014**, *9*, e108277. [**Google Scholar**] [**CrossRef**]

66. Yilmaz, S.; Allgaier, M.; Hugenholtz, P. Multiple displacement amplification compromises quantitative analysis of metagenomes. *Nat. Methods* **2010**, *7*, 943–944. [**Google Scholar**] [**CrossRef**]

67. Kim, K.-H.; Bae, J.-W. Amplification Methods Bias Metagenomic Libraries of Uncultured Single-Stranded and Double-Stranded DNA Viruses. *Appl. Environ. Microbiol.* **2011**, *77*, 7663–7668. [**Google Scholar**] [**CrossRef**] [**Green Version**]

68. Gallet, R.; Fabre, F.; Michalakis, Y.; Blanc, S. The Number of Target Molecules of the Amplification Step Limits Accuracy and Sensitivity in Ultradeep-Sequencing Viral Population Studies. *J. Virol.* **2017**, *91*, 91. [**Google Scholar**] [**CrossRef**] [**Green Version**]

69. Laehnemann, D.; Borkhardt, A.; McHardy, A.C. Denoising DNA deep sequencing data—high-throughput sequencing errors and their correction. *Brief. Bioinform.* **2016**, *17*, 154–179. [**Google Scholar**] [**CrossRef**] [**PubMed**]

70. Wick, R.R.; Judd, L.M.; Holt, K.E. Performance of neural network basecalling tools for Oxford Nanopore sequencing. *Genome Biol.* **2019**, *20*, 1–10. [**Google Scholar**] [**CrossRef**] [**PubMed**] [**Green Version**]

71. Gargis, A.S.; Cherney, B.; Conley, A.B.; McLaughlin, H.P.; Sue, D. Rapid Detection of Genetic Engineering, Structural Variation, and Antimicrobial Resistance Markers in Bacterial Biothreat Pathogens by Nanopore Sequencing. *Sci. Rep.* **2019**, *9*, 13501–13514. [**Google Scholar**] [**CrossRef**]

72. Seah, A.; Lim, M.C.; McAloose, D.; Prost, S.; Seimon, T.A. MinION-Based DNA Barcoding of Preserved and Non-Invasively Collected Wildlife Samples. *Genes* **2020**, *11*, 445. [**Google Scholar**] [**CrossRef**] [**PubMed**] [**Green Version**]

73. Shendure, J.; Ji, H. Next-generation DNA sequencing. *Nat. Biotechnol.* **2008**, *26*, 1135–1145. [**Google Scholar**] [**CrossRef**]

74. Ge, L.; Zhang, J.; Zhou, X.; Li, H. Genetic Structure and Population Variability of Tomato Yellow Leaf Curl China Virus. *J. Virol.* **2007**, *81*, 5902–5907. [**Google Scholar**] [**CrossRef**] [**Green Version**]

75. Harkins, G.W.; Delport, W.; Duffy, S.; Wood, N.; Monjane, A.L.; Owor, B.E.; Donaldson, L.; Saumtally, S.; Triton, G.; Briddon, R.W.; et al. Experimental evidence indicating that mastreviruses probably did not co-diverge with their hosts. *Virol. J.* **2009**, 6, 104. [**Google Scholar**] [**CrossRef**] [**Green Version**]

76. Domingo, E.; Sheldon, J.; Perales, C. Viral Quasispecies Evolution. *Microbiol. Mol. Biol. Rev.* **2012**, *76*, 159–216. [**Google Scholar**] [**CrossRef**] [**Green Version**]

77. Jeske, H. Barcoding of Plant Viruses with Circular Single-Stranded DNA Based on Rolling Circle Amplification. *Viruses* **2018**, *10*, 469. [**Google Scholar**] [**CrossRef**] [**PubMed**] [**Green Version**]

78. McNaughton, A.L.; Roberts, H.E.; Bonsall, D.; De Cesare, M.; Mokaya, J.; Lumley, S.F.; Golubchik, T.; Piazza, P.; Martin, J.B.; De Lara, C.; et al. Illumina and Nanopore methods for whole genome sequencing of hepatitis B virus (HBV). *Sci. Rep.* **2019**, *9*, 1–14. [**Google Scholar**] [**CrossRef**] [**PubMed**] [**Green Version**]

79. Viehweger, A.; Krautwurst, S.; Lamkiewicz, K.; Madhugiri, R.; Ziebuhr, J.; Hölzer, M.; Marz, M. Direct RNA nanopore sequencing of full-length coronavirus genomes provides novel insights into structural variants and enables modification analysis. *Genome Res.* **2019**, *29*, 1545–1554. [**Google Scholar**] [**CrossRef**] [**PubMed**] [**Green Version**]

80. Chang, J.J.M.; Ip, Y.C.A.; Ng, C.S.L.; Huang, D. Takeaways from Mobile DNA Barcoding with BentoLab and MinION. *Genes* **2020**, *11*, 1121. [**Google Scholar**] [**CrossRef**] [**PubMed**]

Chapter-10

Geminiviral Triggers and Suppressors of Plant Antiviral Immunity

1. Introduction

Geminiviruses are circular single-stranded DNA viruses grouped into one of the largest and most successful families of plant viruses (*Geminiviridae* family) [**1**]. Collectively, the geminiviruses cause devasting diseases in a large variety of economically relevant crops and vegetables, resulting in the most diverse symptoms. The broad-range host spectrum of the viruses from the Geminiviridae family may be associated with the large capacity of geminiviruses to overcome the multilayered antiviral immune system of the plant cell, which is broadly divided into RNA interference (RNAi), pathogen-associated molecular pattern (PAMP)-triggered immunity (PTI), and effector-triggered immunity (ETI) (see Abbreviations) [**2**]. Signaling from the cell surface, PTI is the first line of plant defense mediated by immune pattern recognition receptors (PRRs), which detect and interact with conserved molecular motifs from the pathogens, PAMPs [**3**]. As a second line of defense, ETI is mediated by intracellular immune receptors that specifically recognize, directly or indirectly, viral effectors delivered into the cytosol by the pathogens.

RNAi or RNA-silencing-derived antiviral immunity is a well-characterized plant antiviral immunity mechanism, which has been shown to operate against virtually all plant viruses [**4,5**]. Likewise, ETI, also referred to as resistance (R) gene-mediated immunity, has long been recognized as an efficient plant defense layer against viruses [**6,7**]. Studies with plant viruses pioneered in describing hallmarks in ETI responses, including the hypersensitive response (HR), salicylic acid accumulation, and systemic acquired resistance (SAR) [**8,9,10**]. In addition, several viral effectors (avirulence gene products) and their cognate R proteins have been characterized [**2**]. Emerging evidence has demonstrated that the classical PTI characterized in nonviral pathogen–plant interactions also operates against plant viruses [**11,12**]. These multilayered immune defenses are activated and suppressed by viral components or effectors, functioning as virulence factors in susceptible genotypes and as avirulence (Avr) factors in resistant genotypes.

Like any other plant virus, geminiviruses both activate and suppress RNA-silencing mediated antiviral immunity. Infected hosts accumulate geminivirus-derived small interfering RNA (siRNA) of 21–24 bp, and almost all geminivirus proteins have been shown to suppress critical steps in the RNA-silencing mechanism [**13**]. Evidence that viral PTI functions against geminiviruses includes the finding that PTI upstream receptors are virulence targets of geminivirus proteins, which can also suppress downstream PTI-like responses, fulfilling the concept that PTI must be inhibited for successful infection [**14,15**]. Likewise, some geminiviruses proteins have been shown to activate and suppress ETI-like responses [**16**]. As further evidence that plants deploy ETI to fight geminiviruses, the tomato Ty-1 locus, which confers resistance to tomato yellow leaf curl virus (TYLCV), encodes a nucleotide-binding leucine-rich repeat (NLR) domain-containing protein, a reminiscent structural configuration of typical ETI receptors [**17**]. This review focuses on RNA silencing and the antiviral innate immunity mechanisms that plants deploy to fight viruses and the strategy that geminiviruses evolved to overcome these defense barriers. Virtually all geminiviral proteins, which carry a primary function (movement, replication, encapsidation) required to complete the virus life cycle, evolved to accommodate virulence functions as well. Although not covered in this review, hormone signaling has been shown to be connected with anti-geminiviral immunity. For in-depth information on this topic, the reader is referred to a recent review in the molecular interplay between hormones and geminiviruses [**18**].

2. Structural and Functional Organization of the Geminivirus Genome

The *Geminiviridae* family encompasses circular single-stranded DNA viruses that are packed into icosahedral, geminate virion particles. The family is further divided into nine genera (*Begomovirus, Mastrevirus, Capulavirus, Curtovirus, Becurtovirus, Eragrovirus, Grablovirus, Topocuvirus*, and *Turncurtovirus*), according to the genome organization and phylogenetic relationship of the geminivirus species and types of the transmissible insect vectors [**19**].

The geminiviruses can be either monopartite with a single genomic configuration (DNA-A-like) or bipartite with two genomic components, designated DNA-A and DNA-B, with a coding capacity ranging from 4 to 8 viral proteins (**Figure 1**). Their genome is transcribed into bidirectional transcriptional units from the origin of replication (Ori). The functional structures of Ori include a conserved stem-loop structure and the nonanucleotide sequence TAGTATTAC that constitutes the site of replication initiation by the viral replication initiator protein (Rep), encoded by the complementary-sense strand and, hence, also designated C1 (AC1) [**1**]. The remaining complementary-sense strand-encoded viral proteins are designated C2 (AC2), which functions as a transcriptional activator protein (TrAP); C3 (AC3), or replication enhancer protein (REn); C4 (AC4), a pathogenicity determinant; and C5 (AC5), not identified in all geminivirus genomes. The virion strand encodes V1 (AV1); the coat protein (CP); V2 (AV2), which exhibits movement functions; and V3, not present in all geminivirus genomes.

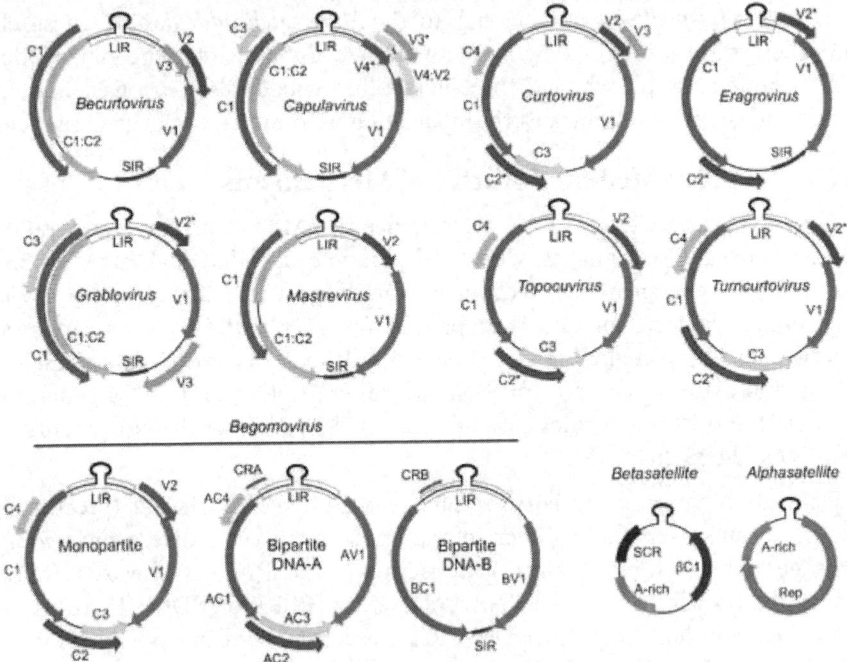

Figure 1. Genomic organization of geminiviruses (Geminiviridae family). The Geminiviridae family includes nine genera represented by monopartite or bipartite species. LIR denotes the long intergenic region; SIR, the short intergenic region; and CR, the common region. The open reading frame (ORF) C1/AC1 encodes the replication initiator protein (Rep) and C3/AC3 encodes the replication enhancer protein (Ren), which are associated with replication. The ORF C2/AC2 encodes a transcriptional activator protein (TrAP) that controls the transcription of viral and host genes; C4/AC4 is a virulence factor. The capsid protein (CP) is indicated in the monopartite and bipartite genomes, as V1 and AV1. In monopartite species, V2 represents the movement protein (MP). V3, present in some genomes, is an inhibitor of gene silencing. ORFs with asterisks (*) have not been functionally assayed. In bipartite begomoviruses, MP (BC1) is encoded by the DNA-B, which also encodes the nuclear shuttle protein, NSP (BV1), which facilitates the nucleocytoplasmic movement of viral DNA. Bipartite begomoviruses are often associated with ssDNA satellites, i.e., the alphasatellites, which encode a replication protein (Rep), and the betasatellites, which encode the virulence-related βC1 protein. A-rich is a conserved adenine rich region of the DNA satellites, and SCR is the satellite conserved region. Adapted from [1]

The B component of bipartite geminiviruses encodes BC1 and BV1. BV1 is a nuclear shuttling protein (NSP) that facilitates the intracellular transport of viral DNA from the nucleus to the cell periphery and assists BV1, the classic movement protein (MP), to move the viral DNA to the adjacent, uninfected cell via plasmodesmata [**20**]. Some geminiviruses are often associated with DNA satellites, designated beta- and alphasatellites, which affect geminiviral pathogenicity and symptom development

[21,22] (**Figure 1**). The genus *Betasatellite* belongs to the family *Tolecusatellitidae*, whereas the *Alphasatellite* genus belongs to the *Alphasatellitidae* family. These circular ssDNA satellites are approximately 1–1.4 kb ssDNA; alphasatellites encode a replication-associated protein (Rep), whereas the betasatellites encode βC1 involved in symptom induction and suppression of transcriptional and post-transcriptional gene silencing.

3. RNA-Silencing-Mediated Antiviral Mechanisms

RNA silencing, also designated RNA interference (RNAi), is a primary antiviral defense mechanism of plant cells that is activated by double-stranded (ds) RNAs [**4**,**23**]. The dsRNA-induced gene-silencing mechanisms are divided into three phases: biogenesis, amplification, and effector phases. The type III RNases dicers (DCLs) carry out the siRNA biogenesis phase by recognizing and processing dsRNA derived from several sources, including viral dsRNA or micro (mi)RNA precursors. DCL2 and DLC4 generate short dsRNAs of 21 and 22 nucleotides (nt), whereas DCL3 processes dsRNA precursors into 24-nt siRNAs [**24**,**25**,**26**,**27**,**28**].

As a plant DNA virus, geminiviruses can generate dsRNA triggers (precursors) via different mechanisms, including overlapping transcripts from a divergent transcription of virus genes, structured (hairpin) transcripts, or dsRNA synthesized from viral mRNA by the host RNA-directed RNA polymerase (RdRP or RDR) (**Figure 2**). These dsRNA precursors are then cleaved by DCLs and converted into virus-derived small interfering RNAs (vsiRNAs) [**29**,**30**]. In the amplification phase, vsiRNAs are amplified by RdRP or RDR and are subsequently stabilized by HUA enhancer 1 (HEN1)-mediated 2′O-methylation, which protects the 3'-end of siRNAs from uridylation activity and subsequent degradation [**31**,**32**,**33**].

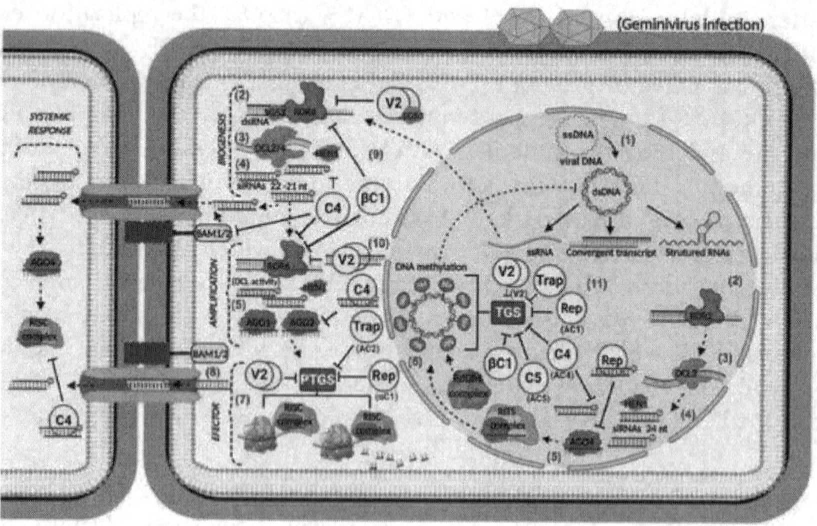

Figure 2. Geminivirus-induced RNA silencing and viral suppressors. Geminivirus particles are delivered into the cytoplasm, where they are uncoated and CP-bound viral ssDNA complexes are imported into the nucleus. (1) In the nucleus, the viral ssDNA is converted into dsDNA, the replicative form. (2) Then, viral ssRNAs are

synthesized and are recognized by the RNA-silencing machinery, initiating the biogenesis phase by converting ssRNA to dsRNA via RDR2 (RNA-dependent RNA polymerase 2 in the nucleus) or RDR6 (in the cytoplasm) with the help of suppressor gene silencing 3 (SGS3), which stabilizes the substrates for (3) dicer (DCL)-mediated processing into 21-, 22-, or 24-nt siRNAs (virus-derived small interfering RNA; vsiRNA). (4) HUA Enhancer 1 (HEN1)-methylated siRNAs are then amplified by RDR1 or RDR6 enhancing siRNA-mediated viral immunity. (5) During the effector phase, Argonaute 4 (AGO4) and AGO1/2 (in the cytoplasm) interact with siRNAs to form RNA-induced transcriptional silencing complex (RITS) and RNA-induced silencing complex (RISC), respectively. (6) In the nucleus, RITS targets the viral transcribed genome and sequesters the RNA-dependent DNA methylation (RdDM) complex, which remodulates the chromatin for transcriptional gene silencing (TGS) of the viral genome. (7) In the cytoplasm, RISC mediates post-transcriptional gene silencing (PTGS) inhibiting the transcription of viral genes via degradation of viral mRNAs. (8) Viral siRNAs generated in the biogenesis and amplification phases are systemically spread via plasmodesma in a barely any meristem 1 (BAM1)- or BAM2-dependent process. For successful infection, geminiviral suppressors of RNA silencing or (VSRs) can act at all levels of TGS and PTGS to suppress siRNA- or RNA interference-mediated antiviral immunity. (9) AC4/C4, Rep, βC1, and V2 inhibit biogenesis of siRNA; (10) βC1, V2, AC2/C2, C4 suppress amplification of siRNA; and (11) βC1, Rep, AC2/C2, and V2 impair the effector phase. Furthermore, the geminivirus suppressors of RNA silencing, Rep, V2, AC4/C4, AC2/C2, and AC5, may interfere with downstream events of TGS. AC4 also interferes with siRNA systemic translocation by targeting BAM1/2. Some geminiviral suppressors interfere with RNA silencing by targeting siRNA (Rep and C4) and long non-coding RNAs (lncRNAs; V2). They also act by activating or inducing the expression of endogenous suppressors of RNA silencing. See Figure 1 for the designations of the viral proteins. The figure was created with BioRender.com.

The effector phase initiates with the assembly of newly synthesized vsiRNAs with one member of the effector AGO (argonaute) family into RNA-induced silencing complexes (RISCs) or RNA-induced transcriptional silencing complexes (RITSs), which target complementary RNA or DNA for silencing (**Figure 2**) [**34,35,36**]. RISC acts at the post-transcriptional level and targets viral RNAs for degradation through cleavage (slicing) or translational arrest, leading to post-transcriptional gene silencing (PTGS). RITS is involved in transcriptional gene silencing (TGS) through DNA or histone methylation and heterochromatin formation.

In PTGS-mediated antiviral defense, DCL2 and DCL4 often process dsRNA precursors into 21- and 22-nt siRNAs that interact with AGO1 and AGO2 [**28,30,37**]. PTGS predominantly involves RNA cleavage via the endonucleolytic activity of AGO1 [**38**]. Recent studies have identified AGO-mediated translational repression as an additional RNA-silencing mechanism against plant viruses [**39,40,41,42**]. Loss-of-function mutations have implicated AGO1 and AGO10 in translational repression, whereas the AGO2-mediated translational suppression has been studied through an in vivo reporter assay [**43,44,45,46**]. The mechanism of AGO-mediated translational repression in plants is still poorly understood, but it seems to depend on the complementary site for siRNA or miRNA on mRNA. Targeting the 5′UTR enables AGO1-RISC to sterically hinder ribosomal recruitment [**45**], whereas siRNA targeting sites within the open reading frame (ORF) may impair ribosome elongation [**45**]. At the 3′ UTR on mRNA, AGO1-RISC may repress translational initiation by interfering with 48S initiation complex formation, resembling the mechanism observed in animal cells [**45,47**].

Plants also deploy siRNA-directed TGS as an antiviral defense against DNA viruses [**48**]. The TGS mechanism is often used by plant cells to control endogenous gene expression. In this case, the RNA polymerase (pol) IV transcript is converted into dsRNA by RNA pol IV-associated RDR2 (RNA-dependent RNA polymerase 2) [**49**]. Then, DCL3 cleaves dsRNA to produce 24-nt siRNA, which is transported to the cytoplasm for AGO4 loading and RITS assembling. The AGO4:siRNA complex is redirected to the nucleus to target the complementary transcript of RNA polymerase V, synthesized in the opposite direction from the RNA pol IV transcript. Therefore, siRNA correctly positions AGO4 that recruits the Novo DNA methyltransferase DRM2-containing RNA-directed DNA methylation (RdDM) complex to methylate DNA on the target locus [**49,50,51,52**]. The RNA-directed DNA methylation is often directed to promoter regions to interfere with gene expression at the transcriptional level.

4. Geminiviral Suppressors of PTGS and TGS

As an adaptive antiviral defense mechanism, RNA silencing is induced during infection, and virus-derived siRNAs accumulate at high levels. Likewise, geminivirus infection induces siRNA accumulation [**24**]. Cabbage leaf curl virus (CabLCV) infection in *Arabidopsis* and African cassava mosaic virus (ACMV) infection in *Nicotiana benthamiana* and cassava have been associated with the biogenesis of -22 ,-21, and -24nt vsiRNAs derived from the coding and the intergenic regions of these geminiviruses [**5,24,53,54**]. In these previous studies, all 24-nt vsiRNAs and a large fraction of 22-nt vsiRNAs were generated by DCL3 and DCL2, respectively, demonstrating the assembly of several distinct silencing pathways in geminivirus–plant interactions [**24**]. In addition to siRNAs, long non-coding RNAs (lncRNAs) accumulate in plants infected by TYLCV [**55**]. These long RNAs can be derived from intergenic sequences and antisense sequences of natural transcripts and may mimic endogenous targets to compete with TYLCV-related siRNAs by a yet unknown mechanism.

Antiviral TGS is considered a primary defense pathway against geminiviruses. Accordingly, reverse genetics studies have demonstrated that loss-of-function mutants of TGS components increase hypersensitivity to geminivirus infection and interfere with host recovery [**48,52**]. DCL3 is essential to combat DNA viruses because antiviral immunity persists in *dcl2* and *dcl4* mutants but not in *dcl3* mutants. Loss of *DCL3* function enhances *Begomovirus* pathogenicity and abolishes -24nt vsiRNAs biogenesis [**24,52**]. In TGS, geminiviral ssDNA is converted into the replicative form dsDNA, which is recognized by RNA polymerase IV, in the nuclei of infected cells [**48**]. The RNA pol IV-transcribed single-stranded RNA is converted to dsRNA by RDR2 and then processed by DCL3 into 24-nt vsiRNAs. AGO4 is loaded with 24-nt vsiRNAs, which target the complementary nascent transcript of the RNA pol IVb complex, recruiting the RdDM complex for viral DNA methylation. The Histone H3 lysine 9 (H3K9) methyltransferase (kyp2/suvh4) is also involved in chromatin modification of the viral genome, resulting in epigenetic modification of the viral genome [**56**].

Many plant viruses have evolved viral suppressors of RNA silencing (VSRs) as a virulence strategy [**57,58,59,60**]. The geminiviral VSRs are multifunctional proteins displaying host defense-suppressing activities and viral life cycle–supporting functions. They can interfere with both PTGS and TGS in all three steps of the processes (**Figure 2**). They also act downstream of TGS, directly or indirectly affecting DNA methylation [**13**]. Mechanistically, geminiviral VSRs either suppress the activity or repress the accumulation (expression) of RNA-silencing machinery components.

Rep, also designated AC1 (bipartite geminiviruses) and C1 (monopartite geminiviruses), has been shown to function as an efficient VSR at both PTGS and TGS (**Figure 2**). Tomato yellow leaf curl Sardinia virus (TYLCSV) Rep directly reduces methyltransferase *1* (*MET1*) and chromomethylase 3 (*CTM3*) expression in infected plants, interfering with the cycle of DNA methylation and TGS (**Figure 2**) [**61**]. Wheat dwarf virus (WDV) Rep suppresses PTGS by binding to 21-nt single-stranded and double-stranded vsiRNAs, thereby preventing their association with the respective AGOs to impair RNA silencing of viral genes [**62**]. Likewise, TrAP (AC2 or C2) suppresses antiviral TGS and PTGS via different mechanisms depending on the cognate TrAP-encoding geminivirus genome. Among the effective mechanisms for suppressing RNA silencing, TrAP either directly interacts with silencing host factors or transactivates expression of host suppressors of RNA silencing; thereby, the TrAP silencing–suppressing function is often coupled to its transcriptional activity that targets and transactivates not only viral gene promoters but also host genes [**63,64,65,66**]. Both AC2 and C2 have been shown to interact with and inhibit common host RNA-silencing factors, including adenosine kinases (ADKs), H3K9me2 histone methyltransferase, SU(VAR)3-9 homolog 4/kryptonite (SUVH4/KYP) involved in TGS [**66,67,68**], AGO1 and RDR6, involved in PTGS [**69**], and calmodulin-like protein (rgs-CaM), an endogenous suppressor of PTGS [**63,65**]. Beet severe curly top virus (BSCTV) C2 also interacts with and stabilizes S-adenosyl-methionine decarboxylase 1 (SAMDC1); thereby, C2 interferes with the host defense mechanism of DNA methylation-mediated gene silencing by attenuating the 26S proteasome-mediated degradation of SAMDC1 [**70**]. In addition to directly interacting with calmodulin-like protein 39 (rgsCaM), AC2 from tomato golden mosaic virus (TGMV) induces the expression of *rgsCaM*, which may target silencing suppressors of RNA viruses for degradation via the autophagy pathway [**63**]. Likewise, AC2 proteins from the mungbean yellow mosaic virus-*Vigna* (MYMV) and ACMV have been shown to target promoters and regulate gene expression of endogenous silencing suppressors, including the Werner exonuclease-like 1 (*WEL1*) gene [**65**].

C4/AC4 is also part of the virus arsenal against antiviral RNA silencing by interacting and sequestering dsRNA precursors from DCL cleavage and siRNA from RISC loading [**71,72**]. C4 interacts with 21-nt vsiRNAs to prevent the spread of vsiRNAs and, hence, suppresses RNA silencing systemic immunity [**72,73**]. TYLCV C4 strongly associates with the intracellular kinase domain of the plasmodesma-localized RLKs, barely any meristem 1 (BAM1) and BAM2, required for systemic immunity [**73**]. C4 from cotton leaf curl Multan virus (CLCuMuV) also interacts with and inhibits *S*-adenosyl methionine (SAM), the universal donor of methyl groups in methylation reactions [**74**]. Therefore, C4 decreases SAM and HEN1 activities indirectly and interferes with the viral genome's methylation and PTGS [**74**]. Additionally, C4 protein accessorizes Rep in the suppression of TGS via downregulation of MET1 and interaction with AGO4 [**61,75**]. The AC5/C5

ORF, located downstream of AC3/C3 in the complementary sense of DNA overlapping a region of the CP sequence, has been recently described as a potent suppressor of RNA silencing [**76**]. Mungbean yellow mosaic India virus (MYMIV) AC5 affects dsRNA production by suppressing sense ssRNA-induced gene silencing and interferes with TGS by inhibiting the expression of a CHH cytosine methyltransferase in *N. benthamiana* [**76**].

AV2 and V2 also suppress antiviral RNA silencing in the amplification phase and interfere with host methylation activities, downstream events of TGS [**77,78,79**]. TYLCV V2 interacts with and inhibits the suppressor of gene silencing 3 (SGS3), the cofactor of RDR6; thereby, affecting vsiRNA amplification [**80**]. V2 may also sequester the RDR6/SGS3 intermediate/substrate dsRNA with 5' overhang ends from SGS3 association, further interfering with vsiRNA amplification [**77,80**]. Likewise, V2 from the curtovirus beet curly top virus (BCTV) has been recently shown to suppress post-transcriptional gene silencing (PTGS) by impairing the RDR6/SGS3 pathway [**77,78**]. CLCuMV V2 interacts with long dsRNA to prevent DCL processing, whereas tomato yellow leaf curl China virus (TYLCCV) V2 interacts with siRNAs to prevent AGO incorporation [**70,81**]. V2 also suppresses TGS by interacting with histone deacetylase 6 (NbHDA6) and interfering with the recruitment of MET1 by HDA6 resulting in decreased methylation of the viral genome and consequent increase in TYLCV pathogenicity [**79**]. Mulberry crinkle leaf virus (MCLV), a newly characterized geminivirus species, encodes a novel viral protein V3, which has been shown to suppress antiviral RNA silencing by a yet unknown mechanism [**82**]. CabLCV BVI (NSP) induces asymmetric leaves 2 (AS2) expression that activates DCP2 decapping activity and, in turn, accelerates mRNA turnover rate and inhibits siRNA accumulation [**83**].

Geminiviruses are often associated with alpha- and betasatellites, which encode efficient suppressors of viral genome methylation in infected plants and PTGS [**84,85**]. The betasatellite genome-encoded βC1 protein functions as an efficient VSR via different mechanisms [**22**]. TYLCCNV βC1 has been shown to inhibit *S*-adenosyl-L-homocysteine hydrolase (SAHH) activity, a methyl cycle enzyme required for TGS, interfering with the host methylation–mediated virus defense pathway [**85**]. Furthermore, βC1 from different betasatellites affects vsiRNA amplification by inducing the calmodulin-like protein (CaM), a negative regulator of RDR6 expression [**86,87**]. This scenario demonstrates that geminiviruses are very efficient in suppressing all steps in RNA-silencing-mediated antiviral mechanisms.

5. Does PAMP-Triggered Immunity Operate against Geminiviruses? What about Effector-Triggered Immunity?

In plant–virus interactions, the host immune system often recognizes viral components or effectors to activate defenses [**2**]. The plant antiviral innate defense consists of a two-level perception system represented by the cell surface receptor, PRR, and the intracellular immune receptors, the resistance (R) proteins [**6**]. PRRs are represented by two classes of transmembrane receptors, the receptor-like kinases (RLKs) and receptor-like proteins (RLPs). These PRRs recognize PAMPs exclusively expressed by pathogens, or endogenous danger signals released by host plants during infection, designated damage-associated molecular patterns (DAMPs) [**88**]. Activation of RLKs and RLPs often requires a coreceptor to form an active immune complex [**89**]. Mechanistically, PAMPs/DAMPs function as ligands that induce dimerization/oligomerization of single-pass transmembrane receptor PRRs with RLK coreceptors to initiate signaling via phosphorylation-induced activation of the immune complexes [**88**]. Downstream events of PTI activation include ROS accumulation, followed by MAP kinase cascade activation, induction of PTI-associated defense genes, ethylene and salicylic acid synthesis, and callose deposition [**90**] (**Figure 3**).

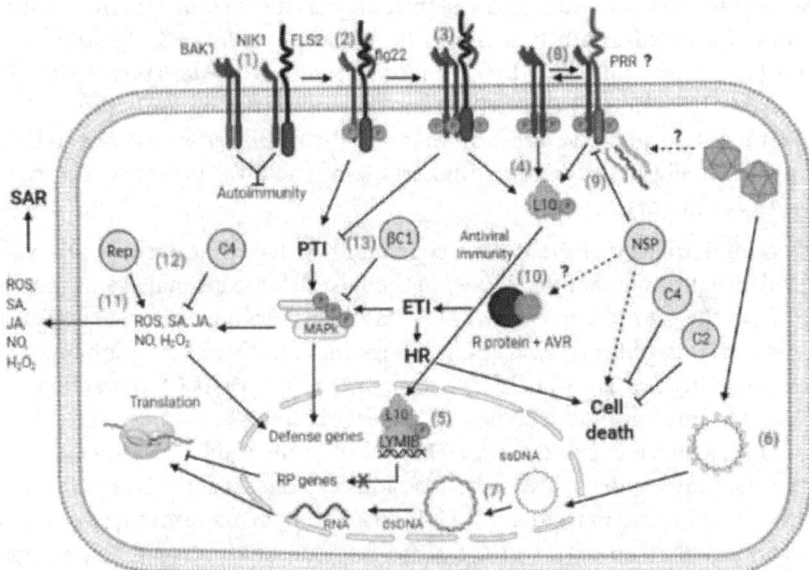

Figure 3. Antiviral innate immunity, interactions with viral suppressors and bacterial PTI (1). Under normal, resting conditions, NSP-interacting kinase 1 (NIK1) is bound to the flagellin receptor FLS2 (Flagellin-sensitive 2) and coreceptor BAK1 (Brassinosteroid insensitive 1-associated receptor kinase 1) to prevent autoimmunity. (2) Upon bacterial infection, the bacterial PAMP (pathogen-associated molecular pattern) flg22 is sensed by FLS2, inducing its oligomerization with BAK1, which results in phosphorylation-mediated activation of the immune complex to initiate PAMP-triggered immunity (PTI), activating the mitogen-activated protein kinase (MAPK) cascade and inducing defense genes. (3) Activated BAK1 also phosphorylates NIK1 at Thr-474, which in turn inhibits further PTI (4) and activates the NIK1-mediated antiviral signaling pathway, mediating ribosomal protein 10 (RPL10) phosphorylation. (5) Phosphorylated RPL10 is translocated to the nucleus, where it interacts with L10-interacting Myb domain-containing protein (LIMYB) to repress protein ribosomal (RB) genes and translational factors, suppressing global translation. (6) In infected cells, the geminivirus particles are disassembled in the cytoplasm and the CP-viral ssDNA complex is directed to the nucleus where (7) the viral ssDNA is converted into dsDNA for replication and transcription of the viral genome. (8) Begomoviruses-derived nucleic acids also function as viral PAMPs inducing NIK1 dimerization with itself or an unidentified viral PAMP recognition receptor (PRR) to transduce the antiviral signal that culminates in translational suppression. Then, viral

mRNA is not efficiently translated impairing infection. (9) The begomovirus nuclear shuttling protein (NSP) counters the activation of this defense pathway by binding to the NIK1 kinase domain and, hence, prevents phosphorylation and transduction of the antiviral signal that otherwise would impair infection. (10) In resistant genotypes, NSP may also function as an avirulence (AVR) factor to activate effector-triggered immunity (ETI) via resistance (R) protein recognition, inducing hypersensitive response (HR) and cell death. (11) The PTI and ETI products (ROS, SA, JA, NO, H_2O_2) can induce defense genes and systemic acquired resistance (SAR). Geminivirus suppressors of innate immunity include (12) C4 that inhibits PTI by interacting and inhibiting FLS2 and along with Rep inhibits SA- and ROS-dependent signaling; C4 and C2 inhibit HR and cell death. (13) Furthermore, the betasatellite βC1 inhibits the MAP kinase cascade. The question marks indicate either events not well clarified or unknown. See Figure 1 for the designations of the viral proteins. The figure was created with BioRender.com.

Although geminiviral PAMPs and their cognate PRRs have not been identified, several lines of evidence indicate that PTI is part of the host defense arsenal against geminivirus infection. First, the C4 protein from TYLCV has been shown to associate with a classical PRR, the bacterial flagellin receptor FLS2 (Flagellin-sensitive 2), and inhibits early PTI responses [**14**] (**Figure 3**). TYLCV C4 also interacts with NSP-interacting kinase 1 (NIK1), an antiviral immune leucine-rich repeat receptor-like kinase (LRR-RLK) that protects plants against begomoviruses [**15**]. NSP from CabLCV associates with the almost universal coreceptor of PRRs, the Brassinosteroid insensitive 1-associated kinase 1 (BAK1) [**91**]. NSP interaction with the BAK1 kinase domain may prevent phosphorylation and activation of the coreceptor in the same fashion as it does with the receptor-like kinase NIK1 [**92**]. Second, geminiviral proteins also activate and suppress downstream immune events of PTI activation (**Figure 3**). Rep from different geminiviruses induces PTI-associated marker genes and SA-dependent defenses [**93,94**]. However, when co-expressed with TYLCV C4 protein, Rep redirects C4 to the chloroplasts, where it acts as a PTI suppressor by reducing SA- and ROS-dependent defense signals [**93**]. The C4 immune-suppressing function co-opts a protein trafficking route directed by protein myristoylation/de-myristoylation processing [**95**]. Upon geminivirus infection, a fraction of the membrane-bound *N*-myristoylated C4 protein is de-myristoylated and translocated to the chloroplast. Inside the chloroplast, non-myristoylated C4 interacts with the thylakoid membrane-bound plant calcium-sensing receptor (CAS) and hampers SA biosynthesis and mediated defenses [**95**]. Finally, the betasatellite βC1 protein from TYLCCV has been shown to interfere with PTI-induced MAPK activation and downstream responses by targeting mitogen-activated protein kinase kinase 2 (MKK2) in *Arabidopsis thaliana* and *N. benthamiana* (**Figure 3**) [**96,97,98**].

The second level of microbe perception by the host plant is mediated by intracellular immune receptors (R proteins), which recognize pathogen avirulent effectors to activate ETI [**3,93,99**]. ETI represents a more specific and robust line of host defense, often associated with programmed cell death, HR that restricts the pathogen to the site of infection [**100**]. Most, but not all, antiviral R proteins harbor a nucleotide-binding leucine-rich repeat (NLR) domain, reminiscent of nonviral intracellular R proteins, which are further classified into coiled-coil (CC)-NLR or Toll/interleukin 1 receptor-like

(TIR)-NLR proteins [**2,6,101**]. The natural Ty-2 resistance locus to TYLCV encodes an NB-LRR protein, named TYNBS1, which might mediate ETI-like resistance. However, the geminiviral effector that would interact specifically with TYNBS1 to activate ETI remains to be determined. Therefore, the underlying mechanism for TYNBS1 activation to mediate resistance is still elusive [**17**].

Compelling evidence that plants deploy ETI against geminiviruses results from infectivity assays demonstrating that geminiviruses induce and suppress HR and ETI-like responses. In *Arabidopsis*, CabLCV infection induces HR- and senescence-related genes without developing a visible cell death phenotype [**102**]. More specifically, TYLCV, cotton leaf curl Kokhran virus (CLCuKoV) and ACMV Rep, bean dwarf mosaic virus (BDMV) and bean golden yellow mosaic virus (BGYMV) NSP, and CLCuKoV V2 have been shown to induce HR (**Figure 3**) [**81,103,104**]. In contrast, the C2 proteins from the papaya leaf curl virus (PaLCuV) and CLCuKoV have been shown to inhibit V2-mediated HR, and C2 from tomato leaf curl New Delhi virus (ToLCNDV) suppresses NSP-mediated HR [**105,106**]. Likewise, TYLCV infection reduces cell death in tomato plants, which is induced by the inactivation of heat shock protein 90 (HSP90) and suppressor of the G2 allele of skp1 (SGT1), via an unknown TYLCV-mediated cell death suppression mechanism [**107**]. In contrast, the underlying mechanism for the cell death-suppressing activity of C4 from tomato leaf curl Yunnan virus (TLCYnV) has been recently unraveled [**16**]. C4 interacts with hypersensitive induced reaction 1 (HIR1), promotes its degradation by impairing HIR1 self-oligomerization, and hence inhibits the HIR1-mediated HR, increasing virus pathogenicity. Although several lines of evidence indicate that both monopartite and bipartite begomoviruses induce and suppress HR, with few exceptions, the mechanisms underlying these viral activities are still far from understood.

6. Transmembrane Receptor-Mediated Antiviral Immunity via Translational Suppression

As obligate intracellular parasites, viruses interact extensively with the host cell functions to complete their life cycle. Independent of the repertoire of viral genome–encoded proteins, all viruses are dependent on the host protein synthesis machinery to translate viral messenger RNAs (mRNAs) [**108,109,110**]. Therefore, many host cell-intrinsic immune defenses target translation factors to inhibit protein synthesis in the infected cells [**108,110**]. Among the translational control-mediated immune defenses, plant cells employ an LRR-RLK to sense viruses at the cell surface and activate a defense pathway that culminates in suppressing host and viral mRNA translation [**12,111,112,113**]. This translational control in antiviral immunity is mediated by the LRR-RLK NIK1, which was first identified as a virulence target of the begomovirus-encoded NSP [**92,114**]. NIK1 belongs to the subfamily II of LRR-RLK, which is further subdivided into phylogenetically related subclades, including a NIK antiviral subclade (NIK1-NIK3) and a somatic embryogenesis receptor kinase (SERK1-5) subclade of coreceptors in innate immunity [**115**]. In *Arabidopsis*, this subfamily of RLKs encompasses14 proteins that harbor four complete LRRs (with 24 residues) and a fifth incomplete LRR (with 16 residues) arranged in a single continuous block within the N-terminal extracellular domain, a single-pass transmembrane segment, and a highly conserved serine/threonine kinase domain at the C-terminal cytosolic side [**116**]. As a common property of the LRR-RLK subfamily II members, they often serve as coreceptors of multiple signaling receptors [**89**]. Accordingly, BAK1/SERK3, the best-characterized member of this subfamily II, acts as a coreceptor for the Brassinosteroid insensitive 1 (BRI-1) receptor in developmental signaling and several different PRRs in innate immunity. In addition to structural conservation, NIK1 also shares other properties with members of the LRR-RLK subfamily II [**117**]. The phosphorylation of the threonine (Thr) residue position 474 constitutes a critical regulatory mechanism for NIK1 activation [**113,118**]. Likewise, the activation site of BAK1/SERK3, SERK4, and SERK1 lies in a conserved position with NIK1 Thr-474 within the activation loop of the kinases [**117**]. Like BAK1/SERK4, which is activated upon PAMP-induced oligomerization of PRRs, NIK1 has been recently shown to be activated by begomoviruses-derived nucleic acids that may act as viral PAMPs. NIK1 may serve as a coreceptor for a yet unknown viral PAMP-sensing PRR.

According to the current model for the NIK1-mediated antiviral signaling, at the onset of infection, *Begomovirus*-derived nucleic acids act as viral a PAMP to activate NIK1 via phosphorylation on Thr-474 (**Figure 3**) [**12,118**]. As a member of the LRR-RLK subfamily II, NIK1 may function as a coreceptor of a yet-to-be-identified immune receptor that may recognize the viral PAMPs for the assembly of the active immune complex. Alternatively, NIK1 may undergo viral PAMP-induced dimerization with itself or its paralog NIK2. This latter hypothesis is based on in vitro phosphorylation assays demonstrating that NIK1 undergoes dimerization and autophosphorylation [**92**]. Phosphorylation-induced activation of NIK1 mediates the phosphorylation of the downstream component ribosomal protein (RP) L10 on the Ser-104 residue, which in turn is redirected to the nucleus [**119,120**]. Activated NIK1 also undergoes sequential autophosphorylation at Thr-469 that antagonistically inhibits NIK1 activation providing a mechanism to fine-tune the extent of RPL10 phosphorylation [**117,118**]. In the nucleus, RPL10 interacts with the transcriptional repressor L10-interacting Myb domain-containing protein (LIMYB) to fully repress the expression of ribosomal protein genes and translational regulatory factors [**113**]. Prolonged repression of translational-machinery-related genes leads to the suppression of global host translation. Begomoviral mRNAs cannot escape this translational regulatory mechanism of plant cells; they are not translated efficiently, impairing infection. NSP from begomoviruses counters this activation mechanism by binding to the NIK1 kinase domain and preventing phosphorylation at Thr-474, thereby enhancing *Begomovirus* pathogenicity [**92,117**]. In infected cells, RPL10 is trapped into the cytosol and is not translocated to the nucleus to mount the defense against begomoviruses. NIK1 overexpression titrates the viral suppressor NSP and restores the RPL10 nuclear localization upon NIK1 activation [**118**].

The NSP binding site on NIK1 was mapped to an 80-amino acid stretch within the activation loop, overlapping the essential Thr-474 for activation [**92**]. Therefore, NSP binding on the NIK1 kinase domain may cause stereochemical constraint on NIK1 phosphorylation at Thr-474, suggesting that NSP inhibition is upstream to NIK1 phosphorylation [**120**]. Consistent with this observation, the replacement of Thr-474 with the phosphomimetic aspartate residue creates a constitutively activated NIK1 mutant, NIK1-T474D, which is barely inhibited by the viral suppressor NSP [**117**]. Furthermore, expression of the gain-of-function mutant NIK1-T474D in *Arabidopsis* enhances resistance to CabLCV, and, in tomato plants, increases resistance to tomato yellow spot virus (ToYSV) and tomato severe rugose virus (ToSRV) [**111,113**]. This enhanced resistance phenotype has been associated with repression of translation-related regulatory genes, reduced loading of coat protein viral mRNA in actively translating polysomes, lower infection rate, and reduced viral DNA accumulation in systemic leaves [**111**]. However, a pitfall in this engineered defense strategy may be the side effects on development from impairing protein synthesis in transgenic crops under field growth conditions. Expression of NIK1-T474D in *Arabidopsis* causes stunted growth, although, in tomato plants, the transgenic lines are phenotypically indistinguishable from the untransformed lines under greenhouse standard conditions [**111,113**]. This difference in developmental performance between these plant species may be due to their differential intrinsic capacity to withstand the deleterious effect of translational inhibition under environmentally controlled conditions, which may not be sustained under field conditions.

Recent studies have shown that NIK1 not only plays an essential role in the translational control of antiviral immunity but regulates bacterial PTI negatively [**121**]. NIK1 interacts with the bacterial PAMP flagellin receptor, FLS2, and its coreceptor BAK1 to prevent autoimmunity under normal, non-infected conditions (**Figure 3**). However, during *Pseudomonas* infection, the bacterial PAMP flagellin induces FLS2 and BAK1 oligomerization and transphosphorylation, forming an active immune complex to initiate PTI signaling. The active FLS2–BAK1 immune complex, in turn, phosphorylates NIK1 on Thr474-, strengthening the NIK1 interaction with FLS2–BAK1 and concurrently activating the NIK1-mediated antiviral signaling. Significantly, bacterial flagellin (PAMP)-induced phosphorylation of NIK1 requires both FLS2 PRR and BAK1 coreceptor, indicating that NIK1 acts downstream of receptor signaling [**121**]. These results may implicate NIK1 as a coreceptor in receptor-mediated antiviral signaling that senses viral PAMPs via yet-unknown viral PRRs [**12**]. Furthermore, they demonstrate that a bacterial PAMP can induce NIK1 activation via the immune complex FLS2–BAK1, thereby allowing bacteria to activate antiviral immunity prior to virus infection. However, the finding that NIK1 suppresses antibacterial immunity further complicates the attempts to target NIK1-mediated antiviral signaling for engineered resistance against begomoviruses.

7. Conclusions

In RNA-silencing-mediated antiviral immunity, two mechanisms are fundamental to protect plants against invading nucleic acids from viruses, PTGS and TGS. PTGS is used as a defense mechanism against RNA and DNA viruses, whereas TGS targets virus DNA. PTGS acts at the post-transcriptional level by directing mRNA targets to degradation or translational suppression. In contrast, TGS represses transcription of target loci by controlling the chromatin methylation status and heterochromatin formation. Geminiviruses have evolved different suppressing strategies of host RNA-silencing-derived antiviral defenses. The geminiviral suppressors of RNA silencing can interfere with PTGS, TGS, and downstream events of TGS. They also act by activating or inducing the expression of endogenous suppressors. The immune-suppressing activities of geminiviruses may account for their broad host range and compromise the success of siRNA-based strategies for engineering host resistance.

The interactions of the plant's innate immune system with geminiviruses are far less understood. Nevertheless, growing evidence has demonstrated that the classical plant PTI limits geminivirus infection similarly to nonviral pathogens. Geminiviral proteins have been shown to target PTI components to both induce and suppress PTI-like responses. Furthermore, begomoviruses-derived nucleic acids act as viral PAMPs and activate the transmembrane receptor NIK1, which shares with PTI coreceptors conserved regulatory mechanisms for activation. However, geminiviral PAMPs and their cognate PRRs have yet to be identified, and hence, the mechanism of geminiviral PTI activation remains unknown.

Likewise, plants may deploy ETI as part of the defense arsenal to combat geminiviruses. Accordingly, geminivirus infection induces ETI-like responses, including cell death, HR, ROS, SA accumulation in resistant genotypes, and at least one isolated resistant gene to TYLCV encodes an NLR protein, although the cognate geminiviral effector (Avr gene) is missing. The current studies designed to uncover geminiviral PTI- and ETI-suppressing functions are limited as they mostly target conserved features between the antiviral and anti-nonviral pathogen immunity. The discovery of geminiviral PAMPs and their cognate PRRs, intracellular R proteins, and the matching geminiviral effectors will potentially shed light on the mechanism of antiviral PTI and ETI activation uncovering specific and conserved features and their specific roles in resistance against viruses.

Abbreviations

ACMV	African cassava mosaic virus
ADKs	Adenosine kinases
AGO	Argonaute
AS2	Asymmetric leaves 2
Avr	Avirulence
BAK1	Brassinosteroid insensitive 1-associated receptor kinase 1
BAM1	Barely any meristem 1
BCTV	Beet curly top virus
BDMV	Bean dwarf mosaic virus
BGYMV	Bean golden yellow mosaic virus
BRI-1	Brassinosteroid insensitive 1
BSCTV	Beet severe curly top virus
CabLCV	Cabbage leaf curl virus
CAS	Calcium-sensing receptor
CC-NLR	Coiled-coil nucleotide-binding leucine-rich repeat
CLCuKoV	Cotton leaf curl Kokhran virus
CLCuMuV	Cotton leaf curl Multan virus
CP	Coat protein
CTM3	Chromomethylase 3
DAMPs	Damage-associated molecular patterns
DCLs	Dicers
dsDNA	Double-stranded DNA

DRM2	DNA (cytosine-5)-methyltransferase 2
dsRNA	Double-stranded RNA
ETI	Effector-triggered immunity
FLS2	Flagellin-sensitive 2
H3K9	Histone H3 lysine 9
HEN1	Hua enhancer 1
HIR1	Hypersensitive induced reaction 1
HR	Hypersensitive response
HSP90	Heat shock protein 90
JA	Jasmonic acid
Kyp2/suvh4	suvh4/kryptonite2
LIR	Long intergenic region
LRR-RLK	Leucine-rich repeat receptor-like kinase
lncRNAs	Long non-coding RNAs
MAPK	Mitogen-activated protein kinase
MET1	Methyltransferase 1
miRNA	MicroRNA
MKK2	Mitogen-activated protein kinase kinase 2
MP	Movement protein
mRNA	Messenger RNA
MYMIV	Mungbean yellow mosaic India
MYMV	Mungbean yellow mosaic virus-*Vigna*
NbHDA6	*Nicotiana benthamiana* histone deacetylase 6
NB-LRR	Nucleotide-binding site-leucine-rich repeat
NIK1	NSP-interacting kinase1
NLR	Nucleotide-binding leucine-rich repeat
NSP	Nuclear shuttling protein
nt	Nucleotides
ORF	Open reading frame
Ori	Origin of replication
PaLCuV	Papaya leaf curl virus
PAMP	Pathogen-associated molecular pattern
pol	Polymerase
PRR	Pattern recognition receptor
PTGS	Post-transcriptional gene silencing
PTI	PAMP-triggered immunity
R	Resistance
RdDM	RNA-directed DNA methylation
RdRP or RDR	RNA-dependent RNA polymerase

REn	Replication enhancer protein
Rep	Replication initiator protein
rgs-CaM	Regulator of G-protein signaling calmodulin-like
RISC	RNA-induced silencing complex
RITS	RNA-induced transcriptional silencing complex
RLK	Receptor-like kinase
RNAi	RNA interference
ROS	Reactive oxygen species
RP	Ribosomal protein
SA	Salicylic acid
SAHH	*S*-adenosyl-L-homocysteine hydrolase
SAM	*S*-adenosyl methionine
SAMDC1	*S*-adenosyl-methionine decarboxylase 1
SAR	Systemic acquired resistance
SERK1-5	Somatic embryogenesis receptor kinase 1–5
SGS3	Suppressor gene silencing 3
SGT1	Suppressor of the G2 allele of skp1
SIR	Short intergenic region
siRNA	Small interfering RNA
ssDNA	Single-stranded DNA
ssRNA	Single-stranded RNA
TGMV	Tomato golden mosaic virus
TGS	Transcriptional gene silencing
Thr	Threonine
TIR-NLR	Toll/interleukin 1 receptor-like nucleotide-binding leucine-rich repeat
TLCYnV	Tomato leaf curl Yunnan virus
ToLCNDV	Tomato leaf curl New Delhi virus
ToSRV	Tomato severe rugose virus
ToYSV	Tomato yellow spot virus
TrAP	Transcriptional activator protein
TYLCCV	Tomato yellow leaf curl China virus
TYLCSV	Tomato yellow leaf curl Sardinia virus
TYLCV	Tomato yellow leaf curl virus
UTR	Untranslated region
vsiRNAs	Virus-derived short interfering RNAs
VSRs	Viral suppressors of RNA silencing
WDV	Wheat dwarf virus
WEL1	Werner exonuclease-like 1

References

1. Kumar, R.V. Plant antiviral immunity against Geminiviruses and viral counter-defense for survival. *Front. Microbiol.* **2019**, *10*, 1460. [**Google Scholar**] [**CrossRef**] [**PubMed**] [**Green Version**]

2. Calil, I.P.; Fontes, E.P.B. Plant immunity against viruses: Antiviral immune receptors in focus. *Ann. Bot.* **2017**, *119*, 711–723. [**Google Scholar**] [**CrossRef**] [**Green Version**]

3. Jones, J.D.; Dangl, J.L. The plant immune system. *Nature* **2006**, *444*, 323–329. [**Google Scholar**] [**CrossRef**] [**Green Version**]

4. Li, F.; Wang, A. RNA-Targeted Antiviral Immunity: More Than Just RNA Silencing. *Trends Microbiol.* **2019**, *27*, 792–805. [**Google Scholar**] [**CrossRef**] [**PubMed**]

5. Pooggin, M.M. Small RNA-Omics for plant virus identification, virome reconstruction, and antiviral defense characterization. *Front. Microbiol.* **2018**, *9*, 2779. [**Google Scholar**] [**CrossRef**] [**PubMed**]

6. Gouveia, B.C.; Calil, I.P.; Machado, J.P.B.; Santos, A.A.; Fontes, E.P.B. Immune receptors and coreceptors in antiviral innate immunity in plants. *Front. Microbiol.* **2017**, *7*, 2139. [**Google Scholar**] [**CrossRef**] [**PubMed**] [**Green Version**]

7. Mandadi, K.K.; Scholthof, K.-B.G. Plant immune responses against viruses: How does a virus cause disease? *Plant Cell* **2013**, *25*, 1489–1505. [**Google Scholar**] [**CrossRef**] [**PubMed**] [**Green Version**]

8. Holmes, F.O. Local lesions in tobacco mosaic. *Bot. Gaz.* **1929**, *87*, 39–55. [**Google Scholar**] [**CrossRef**]

9. Holmes, F.O. Inheritance of resistance to tobacco-mosaic disease in tobacco. *Phytopathoogy* **1938**, *28*, 553–561. [**Google Scholar**]

10. Ross, A.F. Systemic acquired resistance induced by localized virus infections in plants. *Virology* **1961**, *14*, 340–358. [**Google Scholar**] [**CrossRef**]

11. Amari, K.; Niehl, A. Nucleic acid-mediated PAMP-triggered immunity in plants. *Curr. Opin. Virol.* **2020**, *42*, 32–39. [**Google Scholar**] [**CrossRef**]

12. Teixeira, R.M.; Ferreira, M.A.; Raimundo, G.A.; Loriato, V.A.; Reis, P.A.; Fontes, E.P.B. Virus perception at the cell surface: Revisiting the roles of receptor-like kinases as viral pattern recognition receptors. *Mol. Plant Pathol.* **2019**, *20*, 1196–1202. [**Google Scholar**] [**CrossRef**] [**PubMed**]

13. Loriato, V.A.P.; Martins, L.G.C.; Euclydes, N.C.; Reis, P.A.B.; Duarte, C.E.M.; Fontes, E.P.B. Engineering resistance against geminiviruses: A review of suppressed natural defenses and the use of RNAi and the CRISPR/Cas system. *Plant Sci.* **2020**, *292*, 110410. [**Google Scholar**] [**CrossRef**]

14. Gómez, B.G.; Zhang, D.; Díaz, R.T.; Wei, Y.; Macho, A.P.; Lozano-Durán, R. The C4 protein from *Tomato Yellow leaf curl virus* can broadly interact with plant receptor-like kinases. *Viruses* **2019**, *11*, 1009. [**Google Scholar**] [**CrossRef**] [**Green Version**]

15. Macho, A.P.; Lozano-Duran, R. Molecular dialogues between viruses and receptor-like kinases in plants. *Mol. Plant Pathol.* **2019**, *20*, 1191–1195. [**Google Scholar**] [**CrossRef**] [**PubMed**]

16. Mei, Y.; Ma, Z.; Wang, Y.; Zhou, X. Geminivirus C4 antagonizes the HIR1-mediated hypersensitive response by inhibiting the HIR1 self-interaction and promoting degradation of the protein. *New Phytol.* **2020**, *225*, 1311–1326. [**Google Scholar**] [**CrossRef**]

17. Yamaguchi, H.; Ohnishi, J.; Saito, A.; Ohyama, A.; Nunome, T.; Miyatake, K.; Fukuoka, H. An NB-LRR gene, *TYNBS1*, is responsible for resistance mediated by the Ty-2 *Begomovirus* resistance locus of tomato. *Theor. Appl. Genet.* **2018**, *131*, 1345–1362. [**Google Scholar**] [**CrossRef**]

18. Ghosh, D.; Chakraborty, S. Molecular interplay between phytohormones and geminiviruses: A saga of a never-ending arms race. *J. Exp. Bot.* **2021**, in press. [**Google Scholar**] [**CrossRef**]

19. Zerbini, F.M.; Briddon, R.W.; Idris, A.; Martin, D.P.; Moriones, E.; Navas-Castillo, J.; Rivera-Bustamante, R.; Roumagnac, P.; Varsani, A. ICTV Virus Taxonomy Profile: Geminiviridae. *J. Gen. Virol.* **2017**, *98*, 131–133. [**Google Scholar**] [**CrossRef**] [**PubMed**]

20. Martins, L.G.C.; Raimundo, G.A.S.; Ribeiro, N.G.A.; Silva, J.C.F.; Euclydes, N.C.; Loriato, V.A.P.; Duarte, C.E.M.; Fontes, E.P.B. A Begomovirus Nuclear Shuttle Protein-Interacting Immune Hub: Hijacking Host Transport Activities and Suppressing Incompatible Functions. *Front. Plant Sci.* **2020**, *11*, 398. [**Google Scholar**] [**CrossRef**]

21. Briddon, R.W.; Martin, D.P.; Roumagnac, P.; Navas-Castillo, J.; Fiallo-Olivé, E.; Moriones, E.; Lett, J.-M.; Zerbini, F.M.; Varsani, A. Alphasatellitidae: A new family with two subfamilies for the classification of geminivirus- and nanovirus-associated alphasatellites. *Arch. Virol.* **2018**, *163*, 2587–2600. [**Google Scholar**] [**CrossRef**] [**Green Version**]

22. Yang, X.; Guo, W.; Li, F.; Sunter, G.; Zhou, X. Geminivirus-associated betasatellites: Exploiting chinks in the antiviral arsenal of plants. *Trends Plant Sci.* **2019**, *24*, 519–529. [**Google Scholar**] [**CrossRef**] [**PubMed**]

23. Ding, S.W.; Voinnet, O. Antiviral immunity directed by small RNAs. *Cell* **2007**, *130*, 413–426. [**Google Scholar**] [**CrossRef**] [**Green Version**]

24. Akbergenov, R.; Si-Ammour, A.; Blevins, T.; Amin, I.; Kutter, C.; Vanderschuren, H.; Zhang, P.; Gruissem, W.; Meins, F.; Hohn, T.; et al. Molecular characterization of geminivirus-derived small RNAs in different plant species. *Nucleic Acids Res.* **2006**, *34*, 462–471. [**Google Scholar**] [**CrossRef**]

25. Bouché, N.; Lauressergues, D.; Gasciolli, V.; Vaucheret, H. An antagonistic function for Arabidopsis DCL2 in development and a new function for DCL4 in generating viral siRNAs. *EMBO J.* **2006**, *25*, 3347–3356. [**Google Scholar**] [**CrossRef**] [**PubMed**] [**Green Version**]

26. Reinhart, B.J.; Weinstein, E.G.; Rhoades, M.W.; Bartel, B.; Bartel, D.P. MicroRNAs in plants. *Genes Dev.* **2002**, *16*, 1616–1626. [**Google Scholar**] [**CrossRef**] [**Green Version**]

27. Xie, Z.; Johansen, L.K.; Gustafson, A.M.; Kasschau, K.D.; Lellis, A.D.; Zilberman, D.; Jacobsen, S.E.; Carrington, J.C. Genetic and functional diversification of small RNA pathways in plants. *PLoS Biol.* **2004**, *2*, e104. [**Google Scholar**] [**CrossRef**] [**Green Version**]

28. Wang, X.-B.; Jovel, J.; Udomporn, P.; Wang, Y.; Wu, Q.; Li, W.-X.; Gasciolli, V.; Vaucheret, H.; Ding, S.-W. The 21-nucleotide, but not 22-nucleotide, viral secondary small interfering RNAs direct potent antiviral defense by two cooperative Argonautesin *Arabidopsis thaliana*. *Plant Cell* **2011**, *23*, 1625–1638. [**Google Scholar**] [**CrossRef**] [**Green Version**]

29. Baulcombe, D. RNA silencing in plants. *Nature* **2004**, *431*, 356–363. [**Google Scholar**] [**CrossRef**] [**PubMed**]

30. Blevins, T.; Rajeswaran, R.; Aregger, M.; Borah, B.K.; Schepetilnikov, M.; Baerlocher, L.; Farinelli, L.; Meins, F.; Hohn, T.; Pooggin, M.M. Massive production of small RNAs from a non-coding region of Cauliflower mosaic virus in plant defense and viral counter-defense. *Nucleic Acids Res.* **2011**, *39*, 5003–5014. [**Google Scholar**] [**CrossRef**] [**Green Version**]

31. Li, J.; Yang, Z.; Yu, B.; Liu, J.; Chen, X. Methylation protects miRNAs and siRNAs from a 3′-end uridylation activity in Arabidopsis. *Curr. Biol.* **2005**, *15*, 1501–1507. [**Google Scholar**] [**CrossRef**] [**Green Version**]

32. Yang, Z.Y.; Ebright, Y.W.; Yu, B.; Chen, X.M. HEN1 recognizes 21–24 nt small RNA duplexes and deposits a methyl group onto the 2′ OH of the 3′ terminal nucleotide. *Nucleic Acids Res.* **2006**, *34*, 667–675. [**Google Scholar**] [**CrossRef**]

33. Yu, B.; Yang, Z.; Li, J.; Minakhina, S.; Yang, M.; Padgett, R.W.; Steward, R.; Chen, X. Methylation as a crucial step in plant microRNA biogenesis. *Science* **2005**, *307*, 932–935. [**Google Scholar**] [**CrossRef**] [**Green Version**]

34. Bühler, M.; Verdel, A.; Moazed, D. Tethering RITS to a nascent transcript initiates RNAi- and heterochromatin-dependent gene silencing. *Cell* **2006**, *125*, 873–886. [**Google Scholar**] [**CrossRef**] [**PubMed**]

35. Song, J.J.; Smith, S.K.; Hannon, G.J.; Joshua-Tor, L. Crystal structure of Argonaute and its implications for RISC slicer activity. *Science* **2004**, *305*, 1434–1437. [**Google Scholar**] [**CrossRef**] [**PubMed**] [**Green Version**]

36. Verdel, A.; Jia, S.; Gerber, S.; Sugiyama, T.; Gygi, S.; Grewal, S.I.S.; Moazed, D. RNAi-mediated targeting of heterochromatin by the RITS complex. *Science* **2004**, *303*, 672–676. [**Google Scholar**] [**CrossRef**] [**PubMed**] [**Green Version**]

37. Deleris, A.; Gallego-Bartolome, J.; Bao, J.; Kasschau, K.D.; Carrington, J.C.; Voinnet, O. Hierarchical action and inhibition of plant dicer-like proteins in antiviral defense. *Science* **2006**, *313*, 68–71. [**Google Scholar**] [**CrossRef**]

38. Baumberger, N.; Baulcombe, D.C. *Arabidopsis* ARGONAUTE1 is an RNA slicer that selectively recruits microRNAs and short interfering RNAs. *Proc. Natl. Acad. Sci. USA* **2005**, *102*, 11928–11933. [**Google Scholar**] [**CrossRef**] [**PubMed**] [**Green Version**]

39. Bhattacharjee, S.; Zamora, A.; Azhar, M.T.; Sacco, M.A.; Lambert, L.H.; Moffett, P. Virus resistance induced by NB-LRR proteins involves Argonaute4-dependent translational control. *Plant J.* **2009**, *58*, 940–951. [**Google Scholar**] [**CrossRef**] [**PubMed**]

40. Ghoshal, B.; Sanfaçon, H. Symptom recovery in virus-infected plants: Revisiting the role of RNA silencing mechanisms. *Virology* **2015**, *479–480*, 167–179. [**Google Scholar**] [**CrossRef**] [**PubMed**] [**Green Version**]

41. Karran, R.A.; Sanfaçon, H. Tomato ringspot virus coat protein binds to ARGONAUTE 1 and suppresses the translation repression of a reporter gene. *Mol. Plant Microbe Interact.* **2014**, *27*, 933–943. [**Google Scholar**] [**CrossRef**] [**PubMed**] [**Green Version**]

42. Ma, X.; Nicole, M.C.; Meteignier, L.V.; Hong, N.; Wang, G.; Moffett, P. Different roles for RNA silencing and RNA processing components in virus recovery and virus-induced gene silencing in plants. *J. Exp. Bot.* **2015**, *66*, 919–932. [**Google Scholar**] [**CrossRef**] [**Green Version**]

43. Brodersen, P.; Sakvarelidze-Achard, L.; Bruun-Rasmussen, M.; Dunoyer, P.; Yamamoto, Y.Y.; Sieburth, L.; Voinnet, O. Widespread translational inhibition by plant miRNAs and siRNAs. *Science* **2008**, *320*, 1185–1190. [**Google Scholar**] [**CrossRef**]

44. Fátyol, K.; Ludman, M.; Burgyán, J. Functional dissection of a plant Argonaute. *Nucleic Acids Res.* **2016**, *44*, 1384–1397. [**Google Scholar**] [**CrossRef**] [**Green Version**]

45. Iwakawa, H.O.; Tomari, Y. Molecular insights into microRNA-mediated translational repression in plants. *Mol. Cell* **2013**, *52*, 591–601. [**Google Scholar**] [**CrossRef**] [**Green Version**]

46. Lanet, E.; Delannoy, E.; Sormani, R.; Floris, M.; Brodersen, P.; Crété, P.; Voinnet, O.; Robaglia, C. Biochemical evidence for translational repression by *Arabidopsis* microRNAs. *Plant Cell* **2009**, *21*, 1762–1768. [**Google Scholar**] [**CrossRef**] [**Green Version**]

47. Wilczynska, A.; Bushell, M. The complexity of miRNA-mediated repression. *Cell Death Differ.* **2015**, *22*, 22–33. [**Google Scholar**] [**CrossRef**] [**Green Version**]

48. Raja, P.; Sanville, B.C.; Buchmann, R.C.; Bisaro, D.M. Viral genome methylation as an epigenetic defense against geminiviruses. *J. Virol.* **2008**, *82*, 8997–9007. [**Google Scholar**] [**CrossRef**] [**Green Version**]

49. Matzke, M.A.; Moshe, R.A. RNA-directed DNA methylation: An epigenetic pathway of increasing complexity. *Nat. Rev. Genet.* **2014**, *5*, 394–408. [**Google Scholar**] [**CrossRef**]

50. El-Shami, M.; Pontier, D.; Lahmy, S.; Braun, L.; Picart, C.; Vega, D.; Hakimi, M.-A.; Jacobsen, S.E.; Cooke, R.; Lagrange, T. Reiterated WG/GW motifs form functionally and evolutionarily conserved ARGONAUTE-binding platforms in RNAi-related components. *Genes Dev.* **2007**, *21*, 2539–2544. [**Google Scholar**] [**CrossRef**] [**PubMed**] [**Green Version**]

51. Fang, X.; Qi, Y. RNAi in plants: An Argonaute-Centered view. *Plant Cell* **2016**, *28*, 272–285. [**Google Scholar**] [**CrossRef**] [**PubMed**] [**Green Version**]

52. Raja, P.; Jackel, J.N.; Li, S.; Heard, I.M.; Bisaro, D.M. Arabidopsis double-stranded RNA binding protein DRB3 participates in methylation-mediated defense against

geminiviruses. *J. Virol.* **2014**, *88*, 2611–2622. [**Google Scholar**] [**CrossRef**] [**PubMed**] [**Green Version**]

53. Rodríguez-Negrete, E.A.; Carrillo-Tripp, J.; Rivera-Bustamante, R.F. RNA silencing against geminivirus: Complementary action of posttranscriptional gene silencing and transcriptional gene silencing in host recovery. *J. Virol.* **2009**, *83*, 1332–1340. [**Google Scholar**] [**CrossRef**] [**Green Version**]

54. Aregger, M.; Borah, B.K.; Seguin, J.; Rajeswaran, R.; Gubaeva, E.G.; Zvereva, A.S.; Windels, D.; Vazquez, F.; Blevins, T.; Farinelli, L.; et al. Primary and secondary siRNAs in geminivirus-induced gene silencing. *PLoSPathog.* **2012**, *8*, e1002941. [**Google Scholar**] [**CrossRef**]

55. Wang, J.; Yu, W.; Yang, Y.; Li, X.; Chen, T.; Liu, T.; Ma, N.; Yang, X.; Liu, R.; Zhang, B. Genome-wide analysis of tomato long non-coding RNAs and identification as endogenous target mimic for microRNA in response to TYLCV infection. *Sci. Rep.* **2015**, *5*, 16946. [**Google Scholar**] [**CrossRef**] [**PubMed**] [**Green Version**]

56. Sun, Y.W.; Tee, C.S.; Ma, Y.H.; Wang, G.; Yao, X.M.; Ye, J. Attenuation of histone methyltransferase KRYPTONITE-mediated transcriptional gene silencing by Geminivirus. *Sci. Rep.* **2015**, *5*, 16476. [**Google Scholar**] [**CrossRef**] [**PubMed**] [**Green Version**]

57. Csorba, T.; Kontra, L.; Burgyan, J. Viral silencing suppressors: Tools forged to fine-tune host-pathogen coexistence. *Virology* **2015**, *479*, 85–103. [**Google Scholar**] [**CrossRef**] [**Green Version**]

58. Ding, S.W. RNA-based antiviral immunity. *Nat. Rev. Immunol.* **2020**, *10*, 632–644. [**Google Scholar**] [**CrossRef**]

59. Incarbone, M.; Dunoyer, P. RNA silencing and its suppression: Novel insights from in planta analyses. *Trends Plant Sci.* **2013**, *18*, 382–392. [**Google Scholar**] [**CrossRef**]

60. Yang, Z.; Li, Y. Dissection of RNAi-based antiviral immunity in plants. *Curr. Opin. Virol.* **2018**, *32*, 88–99. [**Google Scholar**] [**CrossRef**]

61. Rodríguez-Negrete, E.; Lozano-Durán, R.; Piedra-Aguilera, A.; Cruzado, L.; Bejarano, E.R.; Castillo, A.G. Geminivirus Rep protein interferes with the plant DNA methylation machinery and suppresses transcriptional gene silencing. *New Phytol.* **2013**, *199*, 464–475. [**Google Scholar**] [**CrossRef**] [**Green Version**]

62. Wang, Y.; Dang, M.; Hou, H.; Mei, Y.; Qian, Y.; Zhou, X. Identification of an RNA silencing suppressor encoded by a mastrevirus. *J. Gen. Virol.* **2014**, *95*, 2082–2088. [**Google Scholar**] [**CrossRef**]

63. Chung, H.Y.; Lacatus, G.; Sunter, G. Geminivirus AL2 protein induces expression of, and interacts with, a calmodulin-like gene, an endogenous regulator gene silencing. *Virology* **2014**, *460*, 108–118. [**Google Scholar**] [**CrossRef**] [**PubMed**] [**Green Version**]

64. Jackel, J.N.; Buchmann, R.C.; Singhal, U.; Bisaro, D.M. Analysis of geminivirus AL2 and L2 proteins reveals a novel AL2 silencing suppressor activity. *J. Virol.* **2015**, *89*, 3176–3187. [**Google Scholar**] [**CrossRef**] [**Green Version**]

65. Trinks, D.; Rajeswaran, R.; Shivaprasad, P.V.; Akbergenov, R.; Oakeley, E.J.; Veluthambi, K.; Hohn, T.; Pooggin, M.M. Suppression of RNA silencing by a geminivirus nuclear protein, AC2, correlates with transactivation of host genes. *J. Virol.* **2005**, *79*, 2517–2527. [**Google Scholar**] [**CrossRef**] [**Green Version**]

66. Wang, H.; Buckley, K.J.; Yang, X.; Buchmann, R.C.; Bisaro, D.M. Adenosine kinase inhibition and suppression of RNA silencing by geminivirus AL2 and L2 proteins. *J. Virol.* **2005**, *79*, 7410–7418. [**Google Scholar**] [**CrossRef**] [**Green Version**]

67. Castillo-González, C.; Liu, X.; Huang, C.; Zhao, C.; Ma, Z.; Hu, T.; Sun, F.; Zhou, Y.; Zhou, X.; Wang, X.-J.; et al. Geminivirus-encoded TrAP suppressor inhibits the histone methyltransferase SUVH4/KYP to counter host defense. *Elife* **2015**, *4*, e06671. [**Google Scholar**] [**CrossRef**] [**PubMed**]

68. Wang, H.; Hao, L.; Shung, C.Y.; Sunter, G.; Bisaro, D.M. Adenosine kinase is inactivated by geminivirus AL2 and L2 proteins. *Plant Cell* **2003**, *15*, 3020–3032. [**Google Scholar**] [**CrossRef**] [**PubMed**] [**Green Version**]

69. Kumar, V.; Mishra, S.K.; Rahman, J.; Taneja, J.; Sundaresan, G.; Mishra, N.S.; Mukherjee, S.K. Mungbean yellow mosaic Indian virus-encoded AC2 protein suppresses RNA silencing by inhibiting Arabidopsis RDR6 and AGO1 activities. *Virology* **2015**, *486*, 158–172. [**Google Scholar**] [**CrossRef**] [**PubMed**] [**Green Version**]

70. Zhang, Z.; Chen, H.; Huang, X.; Xia, R.; Zhao, Q.; Lai, J.; Teng, K.; Li, Y.; Liang, L.; Du, Q.; et al. BSCTV C2 attenuates the degradation of SAMDC1 to suppress DNA methylation-mediated gene silencing in *Arabidopsis*. *Plant Cell* **2011**, *23*, 273–288. [**Google Scholar**] [**CrossRef**] [**Green Version**]

71. Amin, I.; Patil, B.L.; Briddon, R.W.; Mansoor, S.; Fauquet, C.M. Comparison of phenotypes produced in response to transient expression of genes encoded by four distinct begomoviruses in *Nicotiana benthamiana* and their correlation with the levels of developmental miRNAs. *Virol. J.* **2011**, *8*, 238. [**Google Scholar**] [**CrossRef**] [**PubMed**] [**Green Version**]

72. Sunitha, S.; Shanmugapriya, G.; Balamani, V.; Veluthambi, K. *Mungbean yellow mosaic virus* (MYMV) *AC4* suppresses post-transcriptional gene silencing and an *AC4* hairpin RNA gene reduces MYMV DNA accumulation in transgenic tobacco. *Virus Genes* **2013**, *46*, 496–504. [**Google Scholar**] [**CrossRef**] [**PubMed**]

73. Rosas-Diaz, T.; Zhang, D.; Fan, P.; Wang, L.; Ding, X.; Jiang, Y.; Jimenez-Gongora, T.; Medina-Puche, L.; Zhao, X.; Feng, Z.; et al. A virus-targeted plant receptor-like kinase promotes cell-to-cell spread of RNAi. *Proc. Natl. Acad. Sci. USA* **2018**, *115*, 1388–1393. [**Google Scholar**] [**CrossRef**] [**Green Version**]

74. Ismayil, A.; Haxim, Y.; Wang, Y.; Li, H.; Qian, L.; Han, T.; Chen, T.; Jia, Q.; Liu, A.Y.; Zhu, S.; et al. *Cotton Leaf Curl Multan virus* C4 protein suppresses both transcriptional and post-transcriptional gene silencing by interacting with SAM synthetase. *PLoSPathog.* **2018**, *14*, e1007282. [**Google Scholar**] [**CrossRef**] [**PubMed**]

75. Vinutha, T.; Kumar, G.; Garg, V.; Canto, T.; Palukaitis, P.; Ramesh, S.; Praveen, S. Tomato geminivirus encoded RNAi suppressor protein, AC4 interacts with host AGO4 and precludes viral DNA methylation. *Gene* **2018**, *678*, 184–195. [**Google Scholar**] [**CrossRef**]

76. Li, F.; Xu, X.; Huang, C.; Gu, Z.; Cao, L.; Hu, T.; Ding, M.; Li, Z.; Zhou, X. The AC5 protein encoded by *Mungbean yellow mosaic India virus* is a pathogenicity determinant that suppresses RNA silencing-based antiviral defenses. *New Phytol.* **2015**, *208*, 555–569. [**Google Scholar**] [**CrossRef**] [**Green Version**]

77. Luna, A.P.; Rodríguez-Negrete, E.A.; Morilla, G.; Wang, L.; Lozano-Durán, R.; Castillo, A.G.; Bejarano, E.R. V2 from a curtovirus is a suppressor of post-transcriptional gene silencing. *J. Gen. Virol.* **2017**, *98*, 2607–2614. [**Google Scholar**] [**CrossRef**] [**PubMed**]

78. Luna, A.P.; Romero-Rodríguez, B.; Rosas-Díaz, T.; Cerero, L.; Rodríguez-Negrete, E.A.; Castillo, A.G.; Bejarano, E.R. Characterization of Curtovirus V2 protein, a functional homolog of begomovirus V2. *Front. Plant Sci.* **2020**, *11*, 835. [**Google Scholar**] [**CrossRef**] [**PubMed**]

79. Wang, B.; Yang, X.; Wang, Y.; Xie, Y.; Zhou, X. Tomato yellow leaf curl virus V2 interacts with host histone deacetylase 6 to suppress methylation-mediated transcriptional gene silencing in plants. *J. Virol.* **2018**, *92*, e0036-1. [**Google Scholar**] [**CrossRef**] [**Green Version**]

80. Glick, E.; Zrachya, A.; Levy, Y.; Mett, A.; Gidoni, D.; Belausov, E.; Citovsky, V.; Gafni, Y. Interaction with host SGS3 is required for suppression of RNA silencing by tomato yellow leaf curl virus V2 protein. *Proc. Natl. Acad. Sci. USA* **2008**, *105*, 157–161. [**Google Scholar**] [**CrossRef**] [**Green Version**]

81. Amin, I.; Hussain, K.; Akbergenov, R.; Yadav, J.S.; Qazi, J.; Mansoor, S.; Hohn, T.; Fauquet, C.M.; Briddon, R.W. Suppressors of RNA silencing encoded by the components of the cotton leaf curl begomovirus-beta satellite complex. *Mol. Plant Microbe Interact.* **2011**, *24*, 973–983. [**Google Scholar**] [**CrossRef**] [**PubMed**] [**Green Version**]

82. Lu, Q.Y.; Yang, L.; Huang, J.; Zheng, L.; Sun, X. Identification and subcellular location of an RNA silencing suppressor encoded by mulberry crinkle leaf virus. *Virology* **2019**, *526*, 45–51. [**Google Scholar**] [**CrossRef**] [**PubMed**]

83. Yang, X.; Xie, Y.; Raja, P.; Li, S.; Wolf, J.N.; Shen, Q.; Bisaro, D.M.; Zhou, X. Geminivirus activates *ASYMMETRIC LEAVES 2* to accelerate cytoplasmic DCP2-mediated mRNA turnover and weakens RNA silencing in *Arabidopsis*. *PLoSPathog.* **2015**, *11*, e1005196. [**Google Scholar**]

84. Saeed, M.; Briddon, R.; Dalakouras, A.; Krczal, G.; Wassenegger, M. Functional analysis of *Cotton Leaf Curl Kokhran Virus/Cotton Leaf Curl Multan* beta satellite RNA silencing suppressors. *Biology* **2015**, *4*, 697–714. [**Google Scholar**] [**CrossRef**] [**PubMed**] [**Green Version**]

85. Yang, X.; Xie, Y.; Raja, P.; Li, S.; Wolf, J.N.; Shen, Q.; Bisaro, D.M.; Zhou, X. Suppression of methylation-mediated transcriptional gene silencing by βC1-SAHH protein interaction during geminivirus-beta satellite infection. *PLoSPathog.* **2011**, *7*, e1002329. [**Google Scholar**] [**CrossRef**] [**PubMed**] [**Green Version**]

86. Kamal, H.; Minhas, F.-U.-A.A.; Tripathi, D.; Abbasi, W.A.; Hamza, M.; Mustafa, R.; Khan, M.Z.; Mansoor, S.; Pappu, H.R.; Amin, I. βC1, pathogenicity determinant encoded by Cotton leaf curl Multan beta satellite, interacts with calmodulin-like protein

11 (Gh-CML11) in *Gossypium hirsutum*. *PLoS ONE* **2019**, *14*, e0225876. [**Google Scholar**] [**CrossRef**]

87. Li, F.; Huang, C.; Li, Z.; Zhou, X. Suppression of RNA Silencing by a plant DNA virus satellite requires a host calmodulin-like protein to repress RDR6 expression. *PLoSPathog.* **2014**, *10*, e1003921. [**Google Scholar**] [**CrossRef**] [**PubMed**]
88. Macho, A.P.; Zipfel, C. Targeting of plant pattern recognition receptor-triggered immunity by bacterial type-III secretion system effectors. *Curr. Opin. Microbiol.* **2015**, *23*, 14–22. [**Google Scholar**] [**CrossRef**]
89. Ma, X.; Xu, G.; He, P.; Chan, L. SERKing coreceptors for Receptors. *Trends Plant Sci.* **2016**, *21*, 1017–1033. [**Google Scholar**] [**CrossRef**] [**PubMed**]
90. Bigeard, J.; Colcombet, J.; Hirt, H. Signaling mechanisms in pattern-triggered immunity (PTI). *Mol. Plant* **2015**, *8*, 521–539. [**Google Scholar**] [**CrossRef**] [**Green Version**]
91. Sakamoto, T.; Deguchi, M.; Brustolini, O.J.; Santos, A.A.; Silva, F.F.; Fontes, E.P.B. The tomato RLK superfamily: Phylogeny and functional predictions about the role of the LRRII-RLK subfamily in antiviral defense. *BMC Plant Biol.* **2012**, *12*, 229. [**Google Scholar**] [**CrossRef**] [**Green Version**]
92. Fontes, E.P.B.; Santos, A.A.; Luz, D.F.; Waclawovsky, A.J.; Chory, J. The geminivirus NSP acts as a virulence factor to suppress an innate transmembrane receptor kinase-mediated defensesignaling. *Genes Dev.* **2004**, *18*, 2545–2556. [**Google Scholar**] [**CrossRef**] [**Green Version**]
93. Aguilar, E.; Gomez, B.G.; Lozano-Duran, R. Recent advances on the plant manipulation by geminiviruses. *Curr. Opin. Plant Biol.* **2020**, *56*, 56–64. [**Google Scholar**] [**CrossRef**] [**PubMed**]
94. Wang, D.; Zhang, X.; Yao, X.; Zhang, P.; Fang, R.; Ye, J. A 7-amino-acid motif of Rep protein essential for virulence is critical for triggering host defense against Sri *Lankan cassava mosaic virus*. *Mol. Plant Microbe Interact.* **2020**, *33*, 78–86. [**Google Scholar**] [**CrossRef**]
95. Medina-Puche, L.; Tan, H.; Dogra, V.; Wu, M.; Rosas-Diaz, T.; Wang, L.; Ding, X.; Zhang, D.; Fu, X.; Kim, C.; et al. A Defense Pathway Linking Plasma Membrane and Chloroplasts and Co-opted by Pathogens. *Cell* **2020**, *182*, 1109–1124.e25. [**Google Scholar**] [**CrossRef**] [**PubMed**]
96. Li, F.; Yang, X.; Bisaro, D.M.; Zhou, X. The βC1 Protein of Geminivirus-Betasatellite Complexes: A Target and Repressor of Host Defenses. *Mol. Plant* **2018**, *11*, 1424–1426. [**Google Scholar**] [**CrossRef**] [**Green Version**]
97. Cui, X.; Tao, X.; Xie, Y.; Fauquet, C.M.; Zhou, X. A DNAβ associated with *Tomato yellow leaf curl* China virus is required for symptom induction. *J. Virol.* **2004**, *78*, 13966–13974. [**Google Scholar**] [**CrossRef**] [**Green Version**]
98. Hu, T.; Huang, C.; He, Y.; Castillo-González, C.; Gui, X.; Wang, Y.; Zhang, X.; Zhou, X. βC1 protein encoded in geminivirus satellite concertedly targets MKK2 and MPK4 to counter host defense. *PLoSPathog.* **2019**, *15*, e1007728. [**Google Scholar**] [**CrossRef**] [**PubMed**] [**Green Version**]

99. Decroocq, V.; Salvador, B.; Sicard, O.; Glasa, M.; Cosson, P.; Svanella-Dumas, L.; Revers, F.; García, J.A.; Candresse, T. The determinant of potyvirus ability to overcome the RTM resistance of *Arabidopsis thaliana* maps to the N-terminal region of the coat protein. *Mol. Plant Microbe Interact.* **2009**, *22*, 1302–1311. [**Google Scholar**] [**CrossRef**] [**PubMed**] [**Green Version**]

100. Dalio, R.J.D.; Paschoal, D.; Arena, G.D.; Magalhães, D.M.; Oliveira, T.S.; Merfa, M.V.; Maximo, H.J.; Machado, M.A. Hypersensitive response: From NLR pathogen recognition to cell death response. *Ann. Appl. Biol.* 2020; in press. [**Google Scholar**] [**CrossRef**]

101. Ando, S.; Miyashita, S.; Takahashi, H. Plant defense systems against cucumber mosaic virus: Lessons learned from CMV—*Arabidopsis* interactions. *J. Gen. Plant Pathol.* **2019**, *85*, 174–181. [**Google Scholar**] [**CrossRef**]

102. Ascencio-Ibáñez, J.T.; Sozzani, R.; Lee, T.J.; Chu, T.M.; Wolfinger, R.D.; Cella, R.; Hanley-Bowdoin, L. Global analysis of *Arabidopsis* gene expression uncovers a complex array of changes impacting pathogen response and cell cycle during geminivirus infection. *Plant Physiol.* **2008**, *148*, 436–454. [**Google Scholar**] [**CrossRef**] [**PubMed**] [**Green Version**]

103. Garrido-Ramirez, E.R.; Sudarshana, M.R.; Lucas, W.J.; Gilbertson, R.L. Bean dwarf mosaic virus BV1 protein is a determinant of the hypersensitive response and avirulence in *Phaseolus vulgaris*. *Mol. Plant Microbe Interact.* **2000**, *13*, 1184–1194. [**Google Scholar**] [**CrossRef**] [**PubMed**] [**Green Version**]

104. Selth, L.A.; Randles, J.W.; Rezaian, M.A. Host responses to transient expression of individual genes encoded by Tomato leaf curl virus. *Mol. Plant Microbe Interact.* **2004**, *17*, 27–33. [**Google Scholar**] [**CrossRef**] [**PubMed**] [**Green Version**]

105. Hussain, M.; Mansoor, S.; Iram, S.; Zafar, Y.; Briddon, W. The hypersensitive response to Tomato leaf curl New Delhi virus nuclear shuttle protein is inhibited by transcriptional activator protein. *Mol. Plant Microbe Interact.* **2007**, *20*, 1581–1588. [**Google Scholar**] [**CrossRef**] [**Green Version**]

106. Mubin, M.; Amin, I.; Amrao, L.; Briddon, R.W.; Mansoor, S. The hypersensitive response induced by the V2 protein of a monopartite begomovirus is countered by the C2 protein. *Mol. Plant Pathol.* **2010**, *11*, 245–254. [**Google Scholar**] [**CrossRef**]

107. Moshe, A.; Gorovits, R.; Liu, Y.; Czosnek, H. Tomato plant cell death induced by inhibition of HSP90 is alleviated by tomato yellow leaf curl virus infection. *Mol. Plant Pathol.* **2016**, *17*, 247–260. [**Google Scholar**] [**CrossRef**]

108. Machado, J.P.B.; Calil, I.P.; Santos, A.A.; Fontes, E.P.B. Translational control in plant antiviral immunity. *Genet. Mol. Biol.* **2017**, *40*, 292–304. [**Google Scholar**] [**CrossRef**] [**PubMed**] [**Green Version**]

109. Mohr, I.; Sonenber, N. Host Translation at the Nexus of Infection and Immunity. *Cell Host Microbe* **2012**, *12*, 470–483. [**Google Scholar**] [**CrossRef**] [**Green Version**]

110. Stern-Ginossar, N.; Thompson, S.R.; Mathews, M.B.; Mohr, I. Translational Control in Virus-Infected Cells. *Cold Spring Harb. Perspect Biol.* **2019**, *11*, a03300. [**Google Scholar**] [**CrossRef**] [**Green Version**]

111. Brustolini, O.J.; Machado, J.P.B.; Condori-Apfata, J.A.; Coco, D.; Deguchi, M.; Loriato, V.A.; Pereira, W.A.; Alfenas-Zerbini, P.; Zerbini, F.M.; Inoue-Nagata, A.K.; et al. Sustained NIK-mediated antiviral signalling confers broad-spectrum tolerance to begomoviruses in cultivated plants. *Plant Biotechnol. J.* **2015**, *13*, 1300–1311. [**Google Scholar**] [**CrossRef**] [**PubMed**] [**Green Version**]

112. Machado, J.P.B.; Brustolini, O.J.B.; Mendes, G.C.; Santos, A.A.; Fontes, E.P.B. NIK1, a host factor specialized in antiviral defense or a novel general regulator of plant immunity? *Bioessays* **2015**, *37*, 1236–1242. [**Google Scholar**] [**CrossRef**]

113. Zorzatto, C.; Machado, J.P.B.; Lopes, K.V.G.; Nascimento, K.J.T.; Pereira, W.A.; Brustolini, O.J.B.; Reis, P.A.B.; Calil, I.P.; Deguchi, M.; Sachetto-Martins, G.; et al. NIK1-mediated translation suppression functions as a plant antiviral immunity mechanism. *Nature* **2015**, *520*, 679–682. [**Google Scholar**] [**CrossRef**] [**Green Version**]

114. Mariano, A.C.; Andrade, M.O.; Santos, A.A.; Carolino, S.M.; Oliveira, M.L.; Baracat-Pereira, M.C.; Brommonshenkel, S.H.; Fontes, E.P. Identification of a novel receptor-like protein kinase that interacts with a geminivirus nuclear shuttle protein. *Virology* **2004**, *318*, 24–31. [**Google Scholar**] [**CrossRef**] [**PubMed**] [**Green Version**]

115. Hosseini, S.; Schmidt, E.D.L.; Bakker, F.T. Leucine-rich repeat receptor-like kinase II phylogenetics reveals five main clades throughout the plant kingdom. *Plant J.* **2020**, *103*, 547–560. [**Google Scholar**] [**CrossRef**] [**Green Version**]

116. Zhang, X.S.; Choi, J.H.; Heinz, J.; Chetty, C.S. Domain-specific positive selection contributes to the evolution of Arabidopsis leucine rich repeat receptor-like kinase (LRR RLK) genes. *J. Mol. Evol.* **2006**, *63*, 612–621. [**Google Scholar**] [**CrossRef**]

117. Santos, A.A.; Lopes, K.V.G.; Apfata, J.A.C.; Fontes, E.P.B. NSP-interacting kinase, NIK: A transducer of plant defence signalling. *J. Exp. Bot.* **2010**, *61*, 3839–3845. [**Google Scholar**] [**CrossRef**] [**Green Version**]

118. Santos, A.A.; Carvalho, C.M.; Florentino, L.H.; Ramos, H.J.O.; Fontes, E.P.B. Conserved threonine residues within the A-loop of the receptor NIK differentially regulate the kinase function required for antiviral signaling. *PLoS ONE* **2009**, *4*, e5781. [**Google Scholar**] [**CrossRef**] [**PubMed**] [**Green Version**]

119. Carvalho, C.M.; Santos, A.A.; Pires, S.R.; Rocha, C.S.; Saraiva, D.I.; Machado, J.P.B.; Mattos, E.C.; Fietto, L.G.; Fontes, E.P.B. Regulated nuclear trafficking of rpL10A mediated by NIK1 represents a defense strategy of plant cells against viruses. *PLoSPathog.* **2008**, *4*, e1000247. [**Google Scholar**] [**CrossRef**] [**Green Version**]

120. Rocha, C.S.; Santos, A.A.; Machado, J.P.B.; Fontes, E.P.B. The ribosomal protein L10/QM-like protein is a component of the NIK-mediated antiviral signaling. *Virology* **2008**, *380*, 165–169. [**Google Scholar**] [**CrossRef**] [**Green Version**]

121. Li, B.; Ferreira, M.A.; Huang, M.; Camargos, L.F.; Yu, X.; Teixeira, R.M.; Carpinetti, P.A.; Mendes, G.C.; Gouveia-Mageste, B.C.; Liu, C.; et al. The receptor-like kinase NIK1 targets FLS2/BAK1 immune complex and inversely modulates antiviral and antibacterial immunity. *Nat. Commun.* **2019**, *10*, 4996. [**Google Scholar**] [**CrossRef**]

Chapter-11

Variability, Functions and Interactions of Plant Virus Movement Proteins: What do we know so far?

1. Introduction

The process of infection of plants with viruses is broadly divided into two types; local and systemic. Local infection is often subliminal and is characterized by intracellular confinement of virus within and nearby the site of infection, while the systemic infection is progressive throughout the host and involves first, the short-distance movement of the virus from the infected cell to surrounding cells followed by long-distance movement using the host vasculature preferably the phloem tissue [**1**]. However, some phloem-limited viruses often skip the first route and are directly injected into the phloem by their vector interested in the phloem sap. In general, the plant viruses show a "symplastic" life cycle, i.e., from entry into the host cell to it accumulates in multiple copies; all occur in the cell symplast. The local intercellular virus spread in the host largely occurs through the symplastically connected cells via the plasmodesmata (PD, **Figure 1**) until they encounter the host vasculature for long-distance systemic infection [**1,2**].

The intercellular transport through PD depends upon its size exclusion limit (SEL), defined by the size of the largest molecule, which can pass through. Although solely dependent upon the hydrodynamic stokes radius [**3**], the PD SEL is highly dynamic and varies with cell-type, ambient light, temperature and developmental stages, being higher in some newly formed mesophyll cells compared to the fully differentiated mature cells [**4**]. Not surprisingly, the size of most viruses, (e.g., tobacco mosaic virus, TMV = 300 × 18 nm) or their nucleic acids, whether in the free or folded state, exceeds the SEL of most PDs (~60 nm) [**5,6**]. This indicates that the SEL of PD is modified during virus infection leading to systemic movement of viral progenies. Most of the plant viruses, if not all, encode a class of structurally diverse protein(s) known as movement protein (MP) that facilitates such intercellular adjustments for virus movements. These MPs have been reported to interact with various host factors to ensure successful virus movement across cells [**5**]. In general, the MPs perform the function of interacting with the viral genome, targeting them to PDs and modifying the SEL of PD, a process often termed as "gating". The idea of synchronous coupling of virus replication and intercellular movement of viral genomes through MP-assisted PD-gating has further divided the movement into substages involving post replication interactions, movement of an infection unit from the replication site to the gated PDs and the actual intercellular transport. This has further expanded the list of MP interacting partners [**1**,**2**,**5**,**6**].

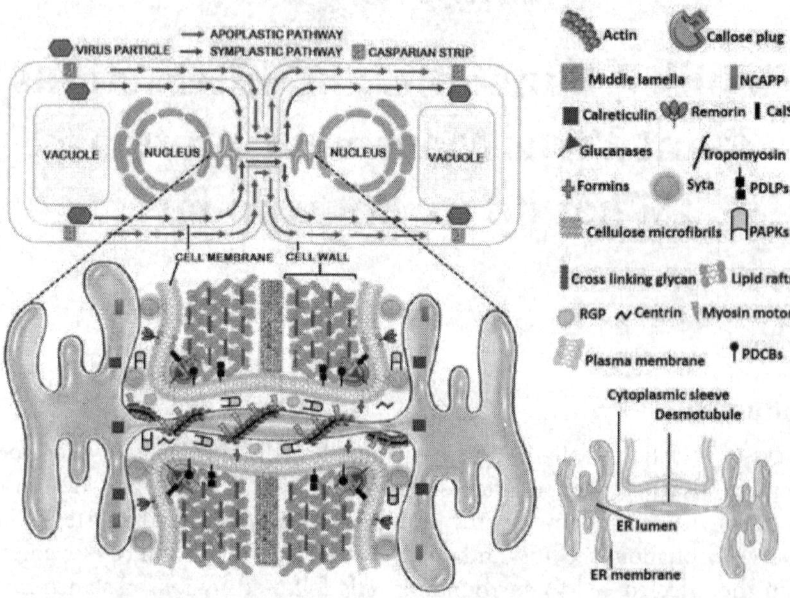

Figure 1. The symplastic and apoplastic pathways between adjacent plant cells and the model illustrating plasmodesmata (PD) structure with-associated cytoskeletal components and proteins governing PD permeability. Abbreviations: CalS—callose synthases; NCAPP—non-cell-autonomous pathway proteins; PAPK—plasmodesmal-associated protein kinase; PDCB—plasmodesmata-associated callose-binding proteins; PDLP—plasmodesmata-located proteins; RGP—reversibly glycosylated polypeptide; Syta—synaptotagmin A.

Despite sharing a common function, there exists a large variation among MPs and their mechanisms of action. In recent years, a number of functions-associated with MPs, namely the mechanisms of modifying the PD SEL, formation of a viral transport complex, interaction with host components and suppression of RNA silencing, have been revealed. In addition, there are other viral proteins that mimic the functionality of MPs. In this review, we have chosen to bring together recent findings on MPs, interactions that they share and the division of labor that they show with other ancillary proteins required for the intercellular movement of plant viruses. Several other previous in-depth reviews present an exhaustive discussion on various aspects governing the intra-, and intercellular movement of viruses and the reader is guided to these for a more holistic understanding of this topic [**7**,**8**,**9**,**10**,**11**,**12**,**13**,**14**]. Additionally, we request readers to go through the reviews [**5**,**15**,**16**,**17**,**18**] to get an overview of the historical perspective and earlier findings, specifically on MPs and their interactions.

2. PD and PD Associated Proteins (PDAPs): The Facilitators of Virus Movement

PDs (**Figure 1**) are membrane-lined interconnected channels formed primarily upon entrapment of components of the endoplasmic reticulum (ER) in the course of phragmoplast formation during cytokinesis. PDs are centrally occupied by the appressed ER (desmotubule; DT), which brings about cytoplasmic continuity at the interface of adjacent cells. Surrounding the DT is a continuous channel of variable size called cytoplasmic sleeve, which is delimited by the plasma membrane (PM) [**19**] and is available for intercellular trafficking. Various PD-associated proteins (PDAPs; **Figure 1**) regulate PD permeability [**20**] and thus can affect the intercellular movement of viruses. The cytoskeleton proteins actin and myosin connect DT to PM through helically arranged globular particles and spoke-like tubular structures, thus, regulating the size of the cytoplasmic sleeve [**21**]. The SEL is negatively regulated by the deposition of callose (β-1,3-glucan). Callose is synthesized by callose synthase (CalS) and degraded by glucanases [**22**]. Calreticulin, a highly conserved Ca^{2+} sequestering chaperone protein, restricts the exit of misfolded proteins from the ER and accumulates in the PD [**23**]. The *Arabidopsis* calreticulin1- (AtCRT1) shows a 22 amino acid (aa) PD-localizing signal sequence and accumulates around DT in PD, thereby blocking the movement of molecules across PD cytoplasmic sleeves [**23,24**]. Centrin is a Ca^{2+}-binding contractile nano-filament protein localized in the PD-neck region, which negatively regulates the PD permeability upon dephosphorylation. Formin is an evolutionarily conserved integral membrane protein that regulates permeability by stabilizing and tethering the actin filaments to the PD membrane [**25**]. Lipid rafts are sterol- and sphingolipid-rich PM microdomains that regulate the callose homeostasis across PD [**26**]. Other PD-associated proteins affecting PD permeability include non-cell-autonomous pathway proteins (NCAPP) located near the orifice [**27**], PD-associated protein kinase (PAPK), carries out phosphorylation of NCAPP and other-associated proteins [**9,27**], PD-associated callose-binding proteins (PDCBs) abundant at PD neck region near the callose deposition site and affecting the callose metabolism [**28**], plasmodesmata-located proteins (PDLPs) promote callose deposition [**29**], lipid raft anchored protein (remorin) and reversibly glycosylated polypeptide (RGP) [**30,31**]. The *Arabidopsis* synaptotagmin A (SYTA) tethers the ER-PM contact sites across PD where it helps in endocytic recycling [**32**], tropomyosin binds to actin and regulates the myosin–actin binding, and myosin motor proteins act for the establishment of actin cables across PD and driving the cytoskeletal movements [**33**]. The locations of the above are depicted diagrammatically in **Figure 1**.

3. Types of MPs

With the pioneering study on temperature-sensitive mutants of TMV and the discovery of a virus-encoded nonstructural 30 kDa MP that assists the virus for its local spread [**5,34**], a large number of MPs have been discovered in several virus genera (**Table 1**). The criteria for a protein to qualify as MP is based upon: (a) loss of virus spread upon removal or mutation of the putative MP encoding gene fragment (b) use of transgenic plants expressing viral MPs or use of MPs of other related viruses for complementing viral movement where MP is removed or made non-functional (c) localization of MP-reporter gene fusions to the PD and (d) comparison of its sequence with a previously known viral MP.

Table 1. General details of movement proteins of some of the most worked out viruses across genera.

Family/Genus	Species	MP	MP—Properties	MP—Mode of Virus Intercellular Movement	Assisting Viral Proteins and Their Putative Function	Interacting Host Proteins and Their Putative Function	References
Virgaviridae/ Tobamovirus	TMV	30 kDa (P30)	α-Helical domain rich Binds ssRNA PD-localized Membrane-bound	PD-gating (increases SEL) Helps in the formation of replication complex and, along with p126 (silencing suppressor), participates in intracellular transport	Replicase—binds to vRNP and along with facilitates its movement across PD	PME—cell wall receptor, PD delivery PAPK, CK2, RIO kinase—MP phosphorylation KELP, MBF1—transcriptional co-activators MPB2C—subcellular localization by the microtubular association of MP Actin—movement of vRNA along ER increase PD SEL EB1a—microtubular association of MP for vRNP movement ANK—cytoplasmic receptor for MP Tubulin and γ-tubulin—movement of vRNA NtMPIP1—DnaJ-like chaperone assisting movement Calreticulin—movement of the viral ribonucleoprotein SYTA—recognizes the MP PLS, remodeling of the PD permeability	[1,11,15,22,35,36, 37,38,39,40,41,42, 43,44,45,46,47,48, 49,50,51,52,53,54, 55,56,57,58,59]
Tombusviridae/ Dianthovirus	RCNMV	35 kDa (P35)	Binds ssRNA Have localization domains for both cell wall and ER Silencing suppressor Host range determinant	PD-gating (increases SEL)	Viral replicase complexes formed with RNA1—recruits MP to punctate cortical structures of ER, which is essential for intercellular movement MP—interacts with CP for long-distance systemic movement	NbGAPDH—A intercalates between VRC and MP and facilitates intercellular movement	[60,61,62,63,64]

Tombusviridae/ Carmovirus MPs-(often termed double gene block proteins)	TCV	P8 and P9	P8—binds ssRNA P8—nuclear-localized P9—cytosolic and ER membrane-localized	PD-gating (increase SEL)	CP—only for long-distance systemic movement through an assemblage of the virus particle and supportive silencing suppression activity	P8—interacts with Atp8 with two "RGD" sequences-cytoskeleton trafficking interactions for virus movement	[65,66]
	CarMV	P7 and P9	P7—binds ssRNA P9—no RNA-binding activity P7—cytosolic initially, later localized near the cell wall P9—probably localized to ER membranes	PD-gating (increase SEL)	NK	NK	[67]
	PFBV	P7 and P12	P12—binds RNA P12—localized to ER membranes	PD-gating (increase SEL)	NK	NK	[68]
	MNSV	P7A and P7B	P7A—binds RNA P7B—no RNA-binding activity P7B—probably localized to ER membranes, silencing suppression activity	PD-gating (increase SEL) P7B accumulates on Golgi and modulates actin filaments to PD	CP—R2-subdomain, also a VSR	Movement is energy-dependent on unknown host protein (s)	[69,70]
Kitaviridae/Higrivirus	HGSV	BMB1 and BMB2	BMB1-binds RNA BMB2—no RNA-binding activity BMB2—an integral ER protein, induces ER constriction, acquire W-like topology like reticulons and also forms a PD-associated replication compartment BMB2—can mediate the transport of BMB1 to and through plasmodesmata	PD-gating (increase SEL)	NK	NK	[71,72]

Tombusviridae/ Tombusvirus	TBSV	P22	Binds RNA Have regulatory sequences for RNA accumulation Induces HR-like necrotic local lesions on *Nicotiana edwardsonii*	PD-gating (increases SEL)	P19—assist systemic movement through silencing suppression	HFI22—leucine zipper homeodomain protein interacts with P22 for delivery of P22/ RNA complexes through PD for intercellular movement	[73]
Closteroviridae/ Closterovirus	BYV CTV	P6, P64, CP, CPm and HSP70h	P6—RER-associated PD-localized (HSP70h) MT-binding (HSP70h) Virion assembly (CP, CPm, and HSP70h)	PD-gating (increase SEL)	P20—interacts with HSP70h for long-distance systemic movement	Class VII myosins—motility and targeting of HSP70h to PD	[74]
Closteroviridae/ Crinivirus	LIYV	P6, P64, CP, CPm and HSP70h P26 (essential for virus systemic infection)	PD-localized Forms conical PM deposits (PLDs) at PM over PD pit fields	PD-gating (increases SEL)	P9—unknown function	NK	[75]
Potyviridae/Potyvirus	TEV, TuMV, SMV, BCNMV, LMV	P3N-PIPO, CP (TEV/TuMV/ SMV) CP and HC-Pro (BCNMV/ LMV)	CP—N terminal and central core domain participates in the movement HC—Pro-silencing suppressor P3N-PIPO—localized to PM and PD	PD-gating (increase SEL)	CI—directed to PD by P3N-PIPO and forms conical structure aiding intercellular movement CP— (*cis* expressed) associates the RNP complexes (RNPs) P3—recruits a small portion of P3N-PIPO to the 6 K2 aggregates	pCAP1—binds to P3N-PIPO aids its localization to the PD and intercellular movement	[76,77,78,79,80]
Luteoviridae/ Polerovirus	PLRV, TuYV, PeVYV	17 kDa (P17) 175 kDa) P4 (PeVYV)	Binds RNA PD-localized Phosphorylated (host-dependent)	PD-gating (increase SEL)	P3a—localization to the outer membrane of mitochondria and plastid	PKC—related membrane-associated protein kinase phosphorylation Actin—intracellular trafficking and PD localization	[81,82,83]

Luteoviridae/ Luteovirus	BYDV	17 kDa (P4)	Binds RNA PD-localized Silencing suppressor Cause PCD Nuclear membrane targeting Self -interactive	PD-gating (increases SEL)	P3a—localization to the outer membrane of mitochondria and plastid	NK	[84]
Geminiviridae/ Begomovirus	BDMV AbMV, ToLCNDV	BC1 (BL1)	Binds ss/ds DNA PM, cell wall and nucleus localized Influence the symptom severity Phosphorylated ToLCNV (BC1 MP) is a determinant of mechanical transmissibility	PD-gating (increases SEL)	BV1-NSP—replicated viral genome delivery from the nucleus to the cytoplasm Formation of a nucleoprotein complex CP—for long-distance movement	PAPK—MP phosphorylation Histone H3—formation of DNA-H3-NSP-MP complex in BDMV Hsp-70—assist AbMV movement through stromules and plastids SCD-2—AbMV movement	[85,86]
Rhabdoviridae/ Nucleorhadbdovirus	RYSV, SYNV, MMV, MFSV, PYDV, EMDV, CaLCuV	P3 (RYSV and MMV) P3 or sc4 (SYNV) P4 (MFSV) Y (PYDV and EMDV)	Secondary structure similarity with 30 K superfamily PD-targeted P3 and sc4 binds RNA nonspecifically sc4 is membrane-associated	PD-gating (increase SEL)	M (matrix) and G (glyco) protein—formation of movement complex (PYDV) G (glyco) protein—formation of movement complex (SYNV)	P3 or sc4 (SYNV) —interacts with MT localized sc4i21 and sc4i17 (homologs of the *Arabidopsis* phloem-associated transcription activator AtVOZ1) —for intracellular trafficking and aiding in the intercellular movement	[87,88]
Rhabdoviridae/ Cytorhabdovirus	LNYV, ADV, TYMaV	4b (LNVY) P3 (ADV and TYMaV)	P3—PD-localized 4b shows nuclear localization Membrane-associated (4b and P3)	PD-gating (increase SEL)	P (phospho) protein—formation of movement complex (ADV)	P3 (LNVY) —interacts with MT-associated VOZ1-like transcriptional activator—for anchoring the virus movement complexes to the MT network for intracellular trafficking and aiding in the intercellular movement	[87,88]

Alphaflexiviridae/ Potexvirus and Betaflexiviridae/ Foveavirus[a]	PVX GRSPaV	TGB1 TGB2 TGB3 TGB1 TGB2 TGB3 Some members have TMV 30 K-like MP	TGB1—25 kDa binds to ssRNA, ATPase/helicase activity, RNA silencing suppressor, translation activator, organization of VRC (X-bodies) through ER/actin remodeling TGB2—12 kDa, ER transmembrane protein TGB3—8 kDa, ER transmembrane protein, induces PCD in *N. benthamiana*	PD-gating (increase SEL) TGB2—provides the environment for robust virus replication	TGB1—CP TGB2—viral RdRp	TGB1- Actin—organization of PVX-X bodies Remorin—interaction impairs the virus movement Fibrillarin and coilin—assist in vRNPs formation CK2-like kinase—for MP phosphorylation TGB2- TIP—interacts with BG1 for regulating callose accumulation NbCPIP2a and NbCPIP2b—interacts with PVX RNA and CP for replication and movement	[12, 18, 37, 89, 90, 91,92,93,94]
Virgaviridae/ Hordeivirus	BSMV	TGB1 TGB2 TGB3	TGB1—42–63 kDa, have nucleolar and nuclear localization signals, binds to RNA, NTPase/ helicase activity causing unwinding of virus RNA duplex TGB2—13–14 kDa ER transmembrane protein TGB3—17 kDa ER transmembrane protein, PD targeting	PD-gating (Increase SEL) TGB3 interacts with TGB1 and TGB2 and provides a basic framework for RNP formation	TGB1—interacts with CP TGB2—replicase γb (also a VSR)-enhances the ATPase-mediated assembly of BSMV-RNP complex	Fibrillarin—formation of RNP for intercellular BSMV movement CK2—protein kinase, phosphorylates TGB1 for intercellular BSMV movement Actin—intercellular BSMV movement and TGB3 localization to the cell wall	[18,89,90,91,92,93]
Bunyaviridae/ Benyviruses	BNYVV	TGB1 TGB2 TGB3	TGB1—has nucleic acid-binding activity at its N-terminal TGB1—contains ATP/ GTP-dependent SF1 helicase type consensus sequence motifs TGB1, TGB2 and TGB3 are localized on ER-derived peripheral membrane bodies	PD-gating (increase SEL) TGB 2 and 3 facilitate the targeting of TGB1 to PD-associated peripheral punctate bodies	NK	NK	[12]

Tombusviridae/ Umbravirus	PEMV-2	P26 and P27	Bind RNA, interact with PEMV-1 *Enamovirus* during natural coinfections and assist its movement Protect viral and host transcript from nonsense-mediated decay	Tubule formation	NK	NK	[95]
Caulimoviridae / Caulimovirus	CaMV	38 kDa MP (P1)	Binds RNA	Tubule formation	P6—forms virus inclusion bodies that serve as translation sites for other proteins, including MP P3—for MP-CP interaction at PD	PRA1—vesicle trafficking regulation AtSRC2.2 and PDLPs—recruitment of MP to PD CHUP1—virion delivery to PD MP17—a rab acceptor-like vesicle-associated protein and PME-putative role in intercellular movement. PDLPs—recruitment of MP to PD	[96,97,98,99]
Secoviridae/Nepovirus	GFLV	38 kDa MP	Binds RNA	Tubule formation	CP	KNOLLE—vesicle trafficking of MP Myosin—MP delivery to PD Calreticulin and PDLPs—recruitment of MP to PD	[100,101,102]
Comoviridae/ Comovirus	CPMV	48 kDa MP	Has large CP, GTP, ssRNA, ssDNA-binding regions	Tubule formation	Large CP (CPL)-37 kDa	PDLPs-recruitment of MP to PD	[103, 104, 105, 106,107]
Bunyaviridae/ Tospovirus	TSWV	NSm protein	Binds ssRNA Avirulence determinant for Sw-5 and RTSW resistance gene Associated with the ER membrane	Tubule formation	TSWV N—protein-recognition of nucleocapsid structures	DNA J-like chaperone proteins, e.g., AtA39—may act as a putative molecular motor for intercellular transport NbSGT1—a molecular co-chaperone, interacts with NSm for intercellular and systemic infection	[96, 97,98, 99, 108, 109, 110]

Tombusviridae/ Umbravirus	GRV	28 kDa, MP (P4)	PD-localized Binds to both ssDNA and ssRNA	Tubule formation	ORF3 protein assist for vRNP formation for cell-to-cell movement	Fibrillarin—vRNP complex formation for transport, interacts with virus ORF3 protein	[111,112]
Bromoviridae/ Alfamovirus	AMV	32 kDa MP (P3)	Bind RNA PD-localized ER-associated	PD-gating (increases SEL) Tubule formation	CP	Patellin 3 (atPATL3) and Patellin 6 (atPATL6) —inhibitory effect on intercellular movement Host kinases—for MP phosphorylation	[113]
Bromoviridae/ Bromovirus	BMV	32 kDa MP (3a)	Binds RNA PD-localized	PD-gating (increases SEL)/ Tubule formation (BMV-MI?)	CP (BMV-MI strain)	NbNACa1—PD localization	[114,115,116]
Bromoviridae/ Cucumovirus	CMV	32 to 36 kDa MP (3a) 2b-for long-distance movement	Binds RNA PD located suppresses the PAMP-triggered immune responses of the host 2b is a VSR	PD-gating (increases SEL)/ Tubule formation	CP (for tubule formation?)	Ascorbate oxidase—movement of vRNP-MP to PD Actin—PD-gating RIO kinase—MP phosphorylation	[117,118,119,120]
Bromoviridae/Ilarvirus	PNRSV	32 to 36 kDa MP (3a)	Binds RNA PD located	PD-gating (increased SEL)/ Tubule formation	CP	NK	[121]

[a] The TGB proteins of GRSPaV (Foveavirus) is highly similar to their counterparts in potato virus X (PVX) [**94**]. Abbreviations: ADV—alfalfa dwarf virus; AMV—alfalfa mosaic virus; AtSRC2.2—Arabidopsis thaliana soybean response to cold; BCNMV—bean common mosaic necrosis virus; BMV—brome mosaic virus; BNYVV—beet necrotic yellow vein virus; BSMV—barley stripe mosaic virus; BYDV—barley yellow dwarf virus; BYV—beet yellows virus; CarMV—carnation mottle virus; CHUP1—chloroplast unusual positioning 1; CK2—casein kinase 2; CMV—cucumber mosaic virus; Eb1a—end-binding protein 1a; EMDV—eggplant mottled dwarf virus; GRSPaV—Grapevine rupestris stem pitting-associated virus; HGSV—hibiscus green spot virus; HSP70h—heat shock protein 70 homolog; LNYV—lettuce necrotic yellows virus; LMV—lettuce mosaic virus; LIYV—lettuce infectious yellows virus; MBF-1—multiprotein bridging factor 1; MPB2C—movement protein-binding 2C; MNSV—melon necrotic spot virus; MMV—maize mosaic virus MFSV—maize fine streak virus; NK—not known; NtMPIP1—Nicotiana tabacum MP interacting protein 1; PAMP—pathogen-associated molecular patterns; PAPK—plasmodesmal-associated protein kinase; PCD—programmed cell death; PEMV-2—pea entaion mosaic virus-2; PeVYV—pepper vein yellows virus; PFBV—pelargonium flower break Carmovirus; PLRV—potato leafroll virus; PME—pectin methylesterase; PNRSV—prunus necrotic ringspot virus; PRA-1—prenylated rab acceptor1; PVX—potato virus X; PYDV—potato yellow dwarf virus; RCNMV—red clover necrotic mosaic virus; RIO kinase—Serine/threonine-protein kinase; RTSW—TSWV resistance locus; RYSV—rice yellow stunt virus; SCD-2—stomatal cytokinesis defective 2; SMV—soybean mosaic virus; SYNV—Sonchus yellow net virus; TBSV—tomato bushy stunt virus; TCV—turnip crinkle virus; TEV—tobacco etch virus; TMV—tobacco mosaic virus; ToLCNDV—tomato leaf curl New Delhi virus; TuYV—turnip yellows virus; TYMaV—tomato yellow mottle-associated virus; vRNA—viral RNA; vRNP—viral ribonucleoprotein; VSR—viral suppressor of RNA silencing.

Functionally, one or more than one viral proteins act as MP and facilitate virus intercellular movement by primarily influencing the host cellular systems for the regulation of PD permeability. On this basis, MPs can be divided into three major types: (a) PD-gating MPs that modify the PD SEL without any structural modifications in the PD (b) Tubule forming MPs that bring about structural modifications in PD by aggregating and forming "multisubunit tubular structures" that serve as conduits for the passage of virus particles and (c) MPs that facilitate virus movement across the cell by either/both gating the PDs and forming tubules. In addition, some of the viral proteins act as suppressors of RNA silencing (VSRs).

3.1. Viral Proteins (MPs) That Increase PD SEL without Any Structural Modifications in PD

The phenomenon of increasing the SEL of PD to allow macromolecular exchange, including the movement of virus particles, is termed as "gating". Although the exact mechanism of gating still needs to be deciphered, to date, many models have been proposed. The PD-gating was initially observed in TMV MP-expressing transgenic plants, which showed a movement of much larger dextrans (9400 kDa) across PD compared to the SEL in control plants (700–800 kDa) [**35**]. In addition, the fluorescent dextrans were observed even far away from the primary microinjected site indicating that either the TMV MP itself moves across PD or initiates a series of diffusible signals that dilate PD at the distant site [**122**]. PD-gating may occur by a pore size increase caused by callose degradation upon action by β-1,3-glucanases (BGs). This causes an increase in the radius of the cytoplasmic sleeve, allowing virus movement [**123**]. There is also the possibility that the decrease in callose is attributed to downregulation or complete suppression of the callose synthase gene [**124,125**]. It is also proposed that the gating through MP may be due to its interaction with some PD-associated host proteins. An increased rate of TMV systemic spread and cell-to-cell movement was observed in tobacco plants with increased ankyrin ANK1 protein levels. ANK1 is normally a cytoplasmic protein, which upon virus infection is recruited to PD by the TMV MP. ANK, when coexpressed with TMV MP, caused reduction of callose deposits, thereby relaxing the callose sphincters at PD [**36**]. Among the PDLPs, which are known to be recruited by the tubule forming viruses (explained later), the PDLP5 initiates PD closure through callose deposition by stimulating callose synthase. The *pdlp5-1* mutants in tobacco reported an increase in cell-to-cell movement of TMV, although the actual interaction still needs further investigation [**126**]. In addition, the stress and damage to the cell wall upon virus entry causes methanol release from existing and newly synthesized pectin methylesterase (PME), which digests pectin and releases methanol vapors that activate the methanol-inducible genes (MIGs), including BG and NCAPP. BG promotes callose degradation, while NCAPP is a cellular factor that stimulates intracellular trafficking important for MP functioning [**20,27**]. New reports of PD lacking cytoplasmic sleeves [**127**] and lack of unambiguous electron microscopic evidence of PD dilation upon virus infection have made the existence of alternate options for intercellular movement of viruses a distinct possibility. Among the various virus genera, the intercellular movement through PD-gating is observed in viruses belonging to the genera *Dianthovirus*, *Carmovirus*, *Closterovirus*, *Luteovirus*, *Potyvirus*, *Tobamovirus*, *Tombusvirus* and some geminiviruses (e.g., Begomovirus). In addition, the PD-gating is also observed in genera

Benyvirus and *Rhabdovirus* and in some members of the family *Alphaflexiviridae, Betaflexiviridae,* and *Virgaviridae* (**Table 1**).

3.1.1. Characteristics of the TMV MP-The "30 K" Superfamily MPs

Tobamoviral MPs

With around 37 species, the Tobamoviruses constitute the largest genus of the family *Virgaviridae,* having TMV as their type genus [**128**]. TMV requires a 30 kDa movement protein for their intercellular movement (**Table 1**; **Figure 2**).

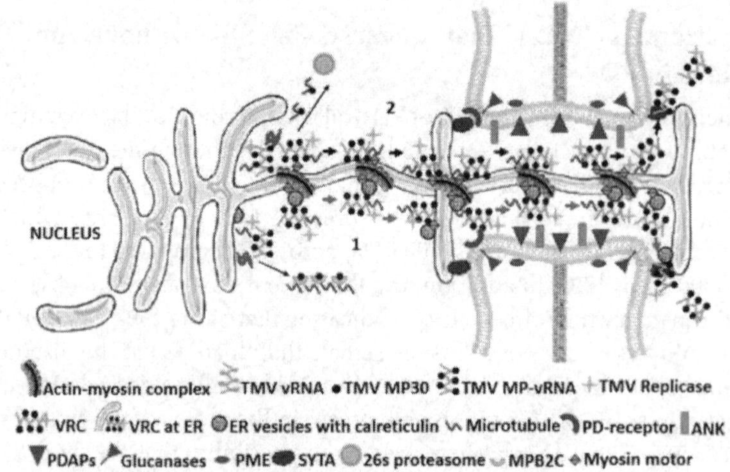

Figure 2. Model for intercellular movement of TMV: The TMV movement protein (MP) at the peripheral endoplasmic reticulum (ER) binds to viral RNA to form a vRNP complex (TMV MP-vRNA), which is joined by replicase to form a VRC. The VRC is delivered to PD either through calreticulin-containing ER-derived vesicle gliding through cell cytoskeleton constituted by the microtubule and ER-actin network (1; red arrows) [8,11] or under the influence of an MP-PLS [32,47,48], the VRC moves along the ER to PD, rafting over cytoskeleton driven by the myosin motor proteins (2; black arrows). Once the VRC reaches the PD, several PDAPs and other viral and host factors cumulatively work for PD "gating" [129]. Gating may occur by MP-mediated severing of actin microfilaments or by recruitment of specific β-1,3-glucanases for callose degradation. Additionally, the MP also interacts with the ANK host factor for downregulating callose. The cell wall-associated PME cause PD targeting of MP and assist gating [36,130]. The MP-PLS is recognized by SYTA, a tethering protein across ER-PM contact sites. These sites are recruited by MP for gating [32,49]. The microtubule near VRC may cause MP degradation via 26 s proteasome [6]. The MPB2C, a microtubule-associated plant factor, causes microtubular accumulation and binds to TMV-MP at the late infection stage to hinder its intercellular movement function [131]. Abbreviations: ANK—ankyrin repeat-containing; MPB2C—movement protein-binding 2C; PDAPs—PD-associated proteins; PLS—plasmodesmata localization signal; PME—pectin methyltransferase; SYTA—synaptotagmin A; VRC—viral replication complex; vRNP—viral ribonucleoprotein.

Many other movement proteins have properties similar to that of TMV P30, and on the basis of this and their predicted secondary structure, they are grouped as the "30 K" superfamily [**37**]. Thus, a detailed study of the TMV MP generally serves as a blueprint for characterizing other similar MPs. Transmission electron microscopic studies revealed TMV MP to be localized in PD [**38**]. TMV MP binds to the vRNA in a sequence non-specific manner and facilitates its intercellular movement through PD in the form of ribonucleoprotein complexes (RNPs). In addition to PD, it also localizes to the peripheral ER membrane, where it forms viral replication centers (VRCs) [**39**,**40**]. The viral replicase acts synergistically with TMV MP, binds to RNP and facilitates its spread in PD desmotubule by lateral diffusion [**41**]. The presence of MP at ER cytosolic face, its association with the MT network for facilitating the VRC movement and the presence of the whole VRC-MP complex at the PD later was also demonstrated by fluorescence immunolabeling and electron microscopy [**42**]. The subcellular distribution and activation of TMV MP gating function is host-dependent and is regulated by C- terminal phosphorylation at Thr-256, Ser-257, Ser-261, and Ser-263 with either of the two host kinases, casein kinase 2 (CK2) and plasmodesmal-associated protein kinase (PAPK) sharing identical molecular weights. The regulation is, however, sequential, with initial activation of TMV MP gating function through phosphorylation at a single site, leading to successful virus intercellular movement followed by its inactivation by further phosphorylation events. This also provides a rationale behind the TMV infecting the plants (*N. tabacum*) even in the presence of the phosphorylation-based inactivation mechanism [**43**,**44**,**45**,**46**]. The plasmodesmal localization signal (PLS) in TMV MP resides in the 50-aa at the N-terminal end and is the first example of PLS in plant virus MP [**47**]. Other less efficient regions (from aa positions 61 to 80 and from 147 to 170) that functionally mimic PLS are also reported [**47**,**48**]. The MP PLS binds to the plant SYTA, which is localized to the plasma membrane, but more concentrated across PD where it tethers the ER-PM contact sites. These sites are then recruited by MP for remodeling the PD permeability [**32**,**48**,**49**,**50**]. This concept of MP-PLS to reach PD via SYTA interaction, however, does not support the ER-actin cytoskeletal involvement in the delivery of MP to PD since the PLS with its hydrophilic stretch of 50 aa lacks the features displayed by proteins interacting with the ER [**40**]. TMV RNP movement through ER membrane occurs by simple diffusion assisted by the host cytoskeletal elements. TMV spread depends upon myosin XI-2 [**51**]. TMV MP localization at PD was inhibited when the actin and ER membrane networks were disrupted. However, disruption of MT had no effect upon TMV intercellular movement, although, at permissive temperatures, TMV MP showed interaction with MT. In addition, the mutant plants with reduced MT dynamics were less susceptible to TMV [**11**]. These contradictory findings of PD localization either through MP-PLS or through ER-actin network indeed need further investigation. Moreover, there is still no direct evidence to show the interaction of MP with PD [**11**]. However, reports of indirect interaction of MP with callose through callose metabolism enzymes [**52**] and PDAPs for regulating the PD aperture [**20**] appear convincing. The TMV MP interacts with *A. thaliana* calreticulin1-(AtCRT1) both in vivo and in vitro and co-resides with it at PD. Under stress and under overexpressed conditions, the increased calreticulin levels interfere with the TMV MP PD localization and instead directs it to the microtubular network away from PD [**23**]. As explained earlier, the PME initiates MIGs, including BG and NCAPP, that aid PD permeability and the intercellular movement of viruses. TMV MP interacts with PME in vitro, and deletion of domains responsible for this interaction hindered the ability of MP to facilitate the spread of viral infection [**53**]. Interestingly, plants expressing anti-sense PME showed delayed systemic spread of TMV [**54**].

In the case of TMV infection, it is observed that once the virus from the initially infected cell starts moving to the adjacent cells, the spread is even faster, and it seems that the MP initiates favorable conditioning of the surrounding cells before the cell is actually approached by the vRNA [**55**]. This prior conditioning event can be attributed to the early synthesis of TMV MP from the genomic vRNA and movement of TMV MP to the surrounding cells through PD-gating without even associating with the VRC. The former is possible due to the presence of an internal ribosome entry site that initiates prior translation of MP straightaway from the genomic RNA before even the synthesis of subgenomic RNA [**56,57,58**], while the latter can be explained by the interactome repertoire of MP, which includes the various PDAPs and other factors regulating the PD SEL [**20**]. The TMV-MP-mediated enhancement of systemic RNA silencing by facilitating the movement of siRNA across PD [**59**] also supports the function of TMV MP as an independent cellular conditioner for virus movement.

Dianthoviral MPs

The RNA2 component of red clover necrotic mosaic virus (RCNMV), a Dianthovirus encodes for a 35 kDa MP (P35, **Table 1**) that belongs to the 30 K superfamily, and like TMV, it does not require capsid protein for intercellular movement [**60**]. Localization studies of the GFP-fused MP mutants established a correlation between MP targeting to the cell wall and intercellular movement [**61**]. In addition, complementation studies of mutant MP with wild-type, alanine scanning mutagenesis and turnip crinkle virus-based GFP assay revealed definite domains for movement, complementarity and cell wall localization, PD-gating, cooperativity and RNA silencing suppression in RCNMV MP [**62**]. Using GFP-fused MP, Kaido et al. [**63,64**] showed colocalization of MP and replicase to the ER, for which C-terminal 70 aa of MP was crucial.

Geminiviral MPs

Among geminiviruses (**Table 1**), the DNA from the nuclei is shuttled to the cytoplasm via coat protein (CP; V1 ORF) in monopartite geminiviruses (e.g., Begomoviruses and Mastreviruses), while in bipartite, it is through the DNA B-encoded nuclear shuttle protein (NSP/BV1) [**85,86**]. Subsequent cytoplasmic localization to PD and further intercellular movement in monopartite geminiviruses is through an MP encoded by the V2 ORF or V3 ORF (e.g., beet curly top virus, a Curtovirus) or by the DNA-B encoded MP (BC1) in bipartite ones. Functional variation occurs in MP among monopartite viruses, for e.g., MPs in monopartite Begomoviruses has DNA-binding property, while MPs encoded by monopartite Mastreviruses do not bind to DNA [**85,86**]. Hence, it needs the viral CP to form a complex with the DNA, and the CP-DNA complex is then carried across PD by the MP. Such differences in the delivery of DNA from the nucleus to PD and across it are also found among bipartite Begomoviruses, for, e.g., abutilon mosaic virus (AbMV) shows "couple skating model" where NSP remains intact with the DNA from the nuclei and the NSP-DNA complex is then carried to PD and across it by the MP [**132**], while BDMV shows "relay race model" where the NSP hands over the viral DNA to MP from the nucleus into the cytoplasm, which is then carried to and across PD [**133**].

Similar to TMV, the MPs encoded by geminiviruses, squash leaf curl virus (SLCV) and cabbage leaf curl virus (CaLCuV) interact with *Arabidopsis* SYTA protein. Mutation in the *SYTA* gene hampered the intercellular movement of these viruses. It is proposed that the interactions of both MPs, encoded by CaLCuV and TMV with SYTA directs them to be loaded on early endosomes, which are then carried away by a recapture pathway to dock at the PD [**134**].

3.1.2. The Triple Gene Block (TGB) Proteins

The triple gene block (TGB1, TGB2 and TGB3) is a specialized and evolutionary conserved group of nonstructural viral movement proteins found in 9 genera of plant viruses belonging to the families *Alphaflexiviridae*, *Benyviridae* and *Betaflexiviridae* (**Table 1**). Their structural features, interactions and role in the intercellular movement of the virus have been extensively reviewed ([**12,18**], and references therein). In addition to PD-gating and assisting the intercellular movement of plant viruses (**Figure 3**), the TGB proteins display a myriad of other functions in host cells [**89**]. The Potexvirus TGBp1 localizes partially in the nucleus and nucleolus, where it interacts with nucleolar proteins fibrillarin and coilin. These proteins may take part in the formation of viral cytoplasmic RNPs and thus take part in the long-distance movement of the viruses. The TGBp1 protein is also known to remodel the host actin and microtubular arrangement and putatively assist the virus movement-independent host protein trafficking to plasmodesmata [**89**]. The hordeiviral TGBp1 binds to host BSr1 R-protein and elicits hypersensitive response [**90**]. The potexviral TGBp1s have strong RNA interference suppression activities [**91**], a property described in detail later. The pepino mosaic virus (PepMV) TGBp1 interacts with host ROS scavenging enzyme catalase 1, resulting in the weakening of the ROS-mediated plant defense mechanism [**92**]. The potato virus X (PVX) TGBp2 plays a role as the molecular adaptor in viral replication by interacting with the C-terminal domain of RdRp and forming chain-mail-like aggregates around RdRp that further localizes TGBp3 aggregates to enhance viral replication [**93**]. The PVX TGBp3 is responsible for virus-induced unfolded protein response under ER stress and upregulates the ER-resident and ubiquitin ligase chaperones [**89**].

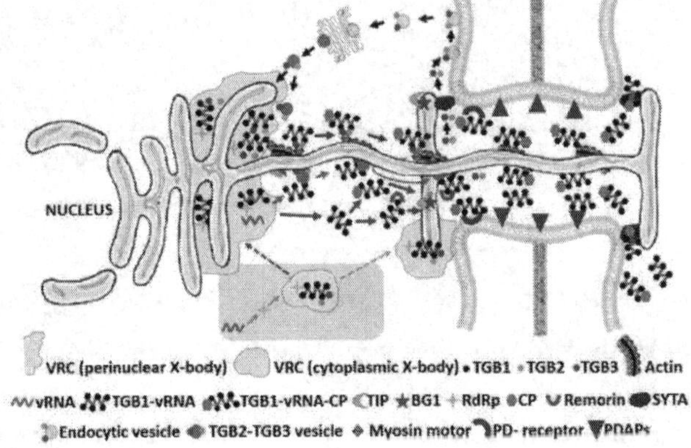

Figure 3. Model for TGB-mediated movement of PVX: The TGB1 organizes the 'X-body' (protective center of virus replication and assembly) and recruits TGB2 and

TGB3 to it. TGB2 bridges the RdRp/dsRNA-TGBp3 interaction in the X-body. The vRNA in the X body replicates and forms VRC with ribosomes and viral RdRp [9]. The TGB1 alone or with CP binds to the vRNA at the VRC to form a vRNP complex (TGB1-vRNA/TGB1-vRNA-CP) that either directly reaches PD (orange arrows) or binds to the TGB2 and TGB3-associated vesicles and reaches PD along the ER through actin and myosin motor proteins guided by TGB3 (blue arrows) [8,17,129]. Alternatively, the VRC can be delivered to PD by TGB2 and TGB3 without vesicles (brown arrows). TGB2 facilitates the VRC fusion to PD. TGB1 and TGB2 perform PD-gating by interacting with remorin and β-1,3-glucanase-associated host factor TIP1, respectively, causing callose reduction. Subsequently, the vRNP complex is delivered to the adjacent cell leaving back TGB2 and TGB3 for recycling via endocytic pathway (black arrows) [18]. According to a recent alternative model (blue background) [93], the vRNA binds to RdRp in the cytoplasm forming the core replication unit later joined by TGB1, TGB2, TGB3 and CP to form a "cytoplasmic X-body", which either joins the ER-associated perinuclear X-body (red dashed arrow) or directly reach the PD through TGB3 guided movement forming cap like complexes at PD (green dashed arrow). Abbreviations: BG1—β-1,3-glucanases; CP—coat protein; PDAPs—PD-associated proteins; RdRp—RNA-dependent RNA polymerase; TGB—triple gene block; TIP—TGB12K-interacting proteins; VRC—virus replication complex; vRNA—viral RNA; vRNP—viral ribonucleoprotein.

3.1.3. Potyviral MPs

PD-gating for intercellular movement is also observed in members of genus *Potyvirus*, the largest group of flexible filamentous viruses, where virus-encoded proteins destined for different functions helps in virus cell-to-cell movement. The PD located protein P3N-PIPO is a dedicated MP in the Potyvirus, turnip mosaic virus (TuMV) [**76**]; it directs the viral cylindrical inclusion protein (CI) to form PD-associated conical structures that assist in intercellular movement (**Figure 4**).

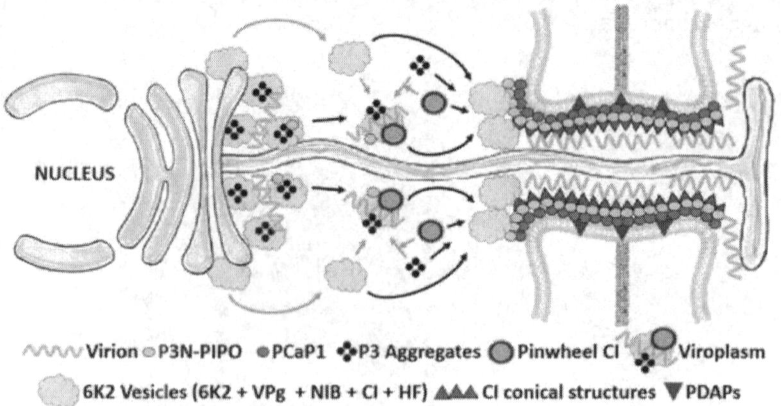

Figure 4. Model for movement of TuMV: The TuMV virion movement involves a multifaceted interaction between virus-encoded proteins 6 K2, VpG, NIb, CI and P3N-PIPO and host-encoded PM localized Ca^{2+}-binding protein PCaP1. The TuMV

MP P3N-PIPO interacts with PCaP1 for its localization to PD. After translation, the virus-encoded proteins for replication along with the assisting host proteins get assembled into clusters of 6 K2 vesicles. The ER-derived P3 aggregates interact with 6 K2 via the P3C domain and accumulate in the 6 K2 vesicle cluster. Along with PD, the P3N-PIPO also localizes to the 6 K2 clusters via P3. The CI also gets recruited to two sites- the PD, by interaction with P3N-PIPO-associated with PCaP1 at PD and-6 K2 clusters by interaction with P3-colocalized P3N-PIPO where probably they participate in replication. The PD recruited CI forms a CI-P3N-PIPO complex to which more CI molecules join via self-interaction resulting in the formation of conical structures. The CI also self-integrates to form a cytosolic pinwheel-like structure. The CK2 vesicle cluster either combines with CI, P3 and P3N-PIPO to form a virus-induced cytosolic viroplasm or is directly delivered to PD and docked at the P3N-PIPO-associated conical CI structures. Upon reaching the PD, the viral RNA is encapsidated by CP to form an intact virion or RNP complex, which is delivered to the neighboring cell by the CI. Adapted and modified from [80]. Abbreviations: CI—cylindrical inclusion; HF—host factor; NIb—nuclear inclusion protein; PDAPs—PD-associated proteins; PM—plasma membrane; RdRp—RNA-dependent RNA polymerase; SYTA—synaptotagmin A; TuMV—turnip mosaic virus; VpG—viral protein genome-linked.

It has been recently shown that the CP with its vRNA interacting conserved core and C-terminal domain also participates in intercellular trafficking of the virion [**77**]. The well-known potyviral RNA silencing suppressor HC-Pro helps in Potyvirus movement by increasing the PD SEL in coordination with CP [**78**]. TuMV P3N–PIPO recruits the PM-associated Ca^{2+}-binding protein 1 (PCaP1) to PD. PCaP1 can sever actin filaments, which is required for the intercellular movement of the virus [**79**]. Notably, both P3N and PIPO domains of TuMV P3N–PIPO are essential for intercellular movement. The PIPO domain is important for its interaction with CI, while the P3N domain for its interaction with P3. The Potyvirus 6 kDa (6 K2) membrane protein interacts with the host ER for the biogenesis of cytoplasmic membranous vesicles, the site for virus replication. The shared N terminus of P3N-PIPO interacts with P3, which recruits the 6 K2-induced vesicles to the PD, where they are docked at the CI-induced conical structures that assist in intercellular movement [**80**].

3.2. Viral Proteins (MPs) That Increase PD SEL by Structural Modifications Caused by Tubular Aggregates

The intercellular movement in some viruses involves intact virions to be moved across the cell [**9**]. These viruses modify the normal PD architecture for their intercellular movement by forming specialized structures (tubules) across cells by the oligomerization of their MP and, sometimes, CP [**135**]. In the transiently transfected protoplasts and/or insect cells, the MPs of these viruses undergo oligomerization and form tubular structures that protrude out from the plasma membrane indicating that in general, the MP in itself is capable of tubule formation. Neither CP nor any other host cellular structure (e.g., PDs) is needed for the formation of tubules [**96,108,109,136**]. The tubule formation for intercellular movement is found in viruses belonging to the genera *Caulimovirus*, *Tospovirus*, *Umbravirus* and members of subfamily *Comovirinae* of the family *Secoviridae* (**Table 1**).

3.2.1. Caulimoviral MP

In Caulimoviruses, the most studied movement protein is that of its type species cauliflower mosaic virus (CaMV). Its tubule forming RNA-binding MP is a 38 kDa protein encoded by the ORF1 (P1; **Table 1**; **Figure 5**) [**96**,**97**]. The tubule-forming capacity of CaMV MP was also demonstrated in infected protoplasts and insect cells [**96**]. CaMV MP interacts with host PDLPs, the MP receptors at PD [**98**], but the actual mechanism of tubule formation involving PD desmotubule replacement, increase in SEL and oligomerization of MP to form tubule is still unclear. Intact virions traversing the PD through a tubular structure have been clearly revealed by electron microscopy [**99**].

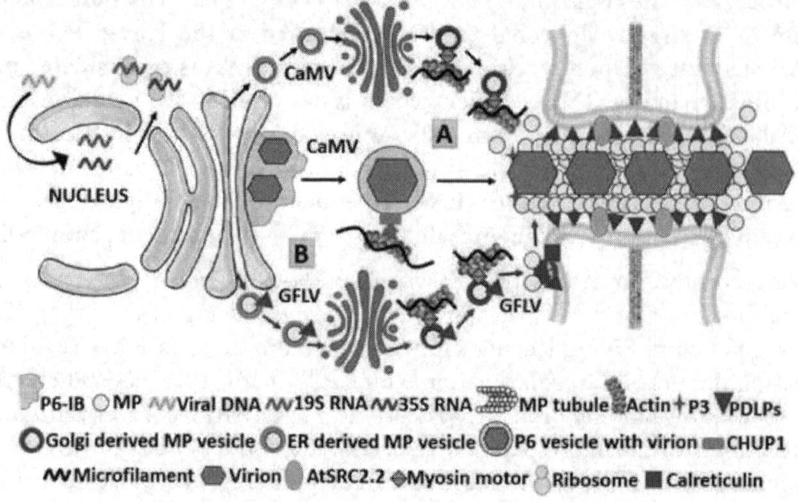

Figure 5. Model for tubule formation by MP in CaMV and GFLV (A) The CaMVvDNA transcribes in the nucleus to 19S and 35S RNA, which translate in the cytoplasm. The 19S RNA encodes a 58 kDa P6 protein that forms ribosomes rich IB (virus factory), the center for other virus proteins translation (e.g., P1-MP, P3, P4-CP, etc.) by 35S RNA. The MP reaches PD through vesicular transport via secretory pathway (black arrows), resulting in multiple MP copies that form tubule across PD. Once the virus attains a threshold copy number in IB, the P6 protein detaches as a vesicle and is assisted by CHUP1 to move over the actin filament network. After the virion reaches PD, the MP interacts with the virion (vDNA+56 kDa CP) through P3. PDLPs (also an MP receptor) and AtSRC2.2 at the PD help in the virion delivery. The virion delivery to the next cell putatively occurs through a tread-milling mechanism where there is unidirectional addition of vDNA-P3-CP-MP subunits assembly at one end and subsequent disassembly at the other, causing the virion delivery [7,137]. (B) The virion delivery is similar in GFLV, but here the MP first reaches the calreticulin-rich sites on the PM and thereafter, it reaches the PD by diffusion (red arrows). The PDLP is also delivered to PD by diffusion through PM, where it reaches via a secretory pathway in association with class XI, XI-K and XI-2 myosin [11,129]. Abbreviations: AtSRC2.2—*Arabidopsis thaliana* soybean response to cold; CaMV—cauliflower mosaic virus; CHUP1—chloroplast unusual positioning protein; GFLV—grapevine

fanleaf Nepovirus IB—inclusion bodies; PDLP—plasmodesmata localized protein; vDNA—viral DNA.

3.2.2. Tospoviral MP

In the tomato spotted wilt virus (TSWV), a Tospovirus, the NS_M protein is the MP as displayed by its properties of oligomerization and formation of tubular structures in infected protoplasts and insect cells. It forms and extends the tubule across PDs, assisting the intercellular movement of the non-enveloped nucleocapsids [**108**]. In the thrips vector *Frankliniella occidentalis*, TSWV NS_M protein is functionally linked with autophagic pathways, not involved in assisting the intercellular movements [**110**].

3.2.3. Umbraviral MP

The ORF-4 of the groundnut rosette virus (GRV), an Umbravirus, encodes for 28 kDa protein that is PD-localized, binds to both ssDNA and ssRNA cooperatively in a sequence nonspecific manner and induces tubule formation on the protoplasts surfaces in *N. benthamiana* [**111,112**]. Umbraviruses do not encode for their own CP and require a helper Luteoviridae member for virion formation and vector transmission. This reduces MP dependency on CP, and as indicated by the protoplast experiments, among Umbravirus members, MP in itself is self-sufficient for tubule formation and intercellular movement of virion [**138**].

3.2.4. MPs of Comoviruses, Fabaviruses, and Nepoviruses

As indicated by the mutagenesis studies, the tubule forming MPs in members of sub-family *Comovirinae,* Comoviruses, Fabaviruses, and Nepoviruses have molecular weights of 58/48 kDa, 37 kDa and 43–38 kDa, respectively (**Table 1**) and are products of polyproteins encoded through their RNA-2 [**96**]. Tubule formation by the MP of cowpea mosaic virus (CPMV), a Comovirus, was demonstrated in both infected protoplasts/ insect cells and infected plant material where clear tubular structures containing virus particles superseding the PD desmotubule were observed under an electron microscope. Studies using antisera and mutagenesis confirmed the 48 kDa moiety of the 58 K/48 K protein to be crucial for tubule formation [**136,139**]. Tubule formation was also observed in grapevine fanleaf Nepovirus (GFLV; **Figure 5**) and broad bean wilt virus 2 (BBWV-2) a Fabavirus [**100,101,109**]. The MPs of these viruses have different binding affinities for CP. Both CPMV and BBWV-2 encode a large CP of 35–40 kDa and a small CP of 20–25 kDa, but the MP of the former binds to the large CP, whereas in the latter, it binds to the small CP [**140,141**]. The C-terminal domain of MP was found to be crucial for this interaction; a CPMV MP with C-terminal deletion showed an empty tubule with no virions [**103**], a similar deletion in GFLV MP abolished the interaction with CP and restricted the systemic spread [**102**]. However, whether the C-terminal deletion abolished the interaction of CPMV MP with the CP is not clear. Combining both, it was postulated that in both CPMV and GFLV, the C terminal domain of MP is crucial for association with CP and virion formation so that an intact virion moves intercellularly via tubules. The tubule formation was later-associated with the N-terminal and central region of the CPMV MP, while the virion incorporation in the tubules was defined as a function of its C-terminal domain [**104,105**]. Deletion and point mutation studies in CPMV MP showed that initially, the MP is targeted to plasma membrane involving its oligomerization (involving aa 228–251); later, it accumulates in spots after possible interaction with some host protein (involving aa 252–276). Finally, tubules assemble (involving aa 293–298 of MP) from the spots, which traverse the PD, replacing the desmotubules. The process culminates with the delivery of the virion in an adjacent uninfected cell and disassembly of the tubule [**106,107**]. The host PDLPs have no role in the transfer of viral particles, but they may act as PD receptors for MP, assisting them in oligomerization to form tubules [**98**].

3.3. MPs That Increase PD SEL with or without Any Structural Modifications across PD

Apart from the exclusive nature of either gating or tubule forming, members of the family Bromoviridaeharbor certain MPs that apparently perform both gating as well as tubule forming functions (**Table 1**). The MPs of viruses belonging to genera *Alfamovirus, Bromovirus, Cucumovirus,* and *Ilarvirus* of family *Bromoviridae* increase the PD-SEL, are PD-localized, bind to RNA and show molecular weights between 32 and 36 kDa. They are encoded by the 3a gene fragment of RNA**117**] 3,**142**].

Among these members, a significant variation is observed between species and strains regarding the requirement of CP for the movement-related functions of MPs [**113,114,143**].

The intercellular movement of the alfalfa mosaic virus (AMV) requires both MP and CP. The MP induces tubule formation and exhibits RNA-binding capacity in a sequence nonspecific manner. An intact virion formation is not necessary for the intercellular movement of the virus; however, the association of MP with CP is crucial, as observed by restricted intercellular movement following mutation of CP [**113**].

While most of the brome mosaic virus (BMV) strains do not require CP for intercellular movement [**115**], the tubule forming M1 strain [**114**] requires CP in addition to its MP when its intercellular movement was tracked in *Chenopodium* sp. The MP interacts with *N. benthamiana* protein (NbNACa1), which apparently regulates its localization on PD [**116**].

In another Bromovirus, the cowpea chlorotic mottle virus (CCMV), MP is sufficient (no CP required) for intercellular movement. Interestingly, the rate of intercellular movement is host-dependent, being faster in an experimental host (*N. benthamiana*) than its natural host (cowpea), as demonstrated when CP was replaced by an enhanced GFP [**144**]. When the MP gene of BMV and CCMV were exchanged, and the CP expression was impaired, the recombinant CCMV harboring BMV MP showed restricted movement, while no effect on intercellular movement was observed in the case of recombinant BMV harboring CCMV MP. This qualifies MP as the chief determinant of the virus-specific CP functions for intercellular movement in Bromovirus [**145**]. However, considering the above facts, the CP and MP interaction and dependency need to be further examined in light of the differential response observed between virus strains and species as well as their host.

The cucumber mosaic virus (CMV) tubule forming MP is PD-localized and binds to ssRNA [**118**]. CMV MP requires CP assistance for intercellular movement. However, the CP-dependency was abolished when a truncated MP was used, having 33 aa removed from C-terminal [**119**]. The truncated MP, as observed through atomic force microscopy, showed a strong binding affinity for viral RNA and formed a more condensed RNP complex compared to a native MP requiring CP [**120**]. The CMV MP showed tubule formation on the protoplast surface; however, no such tubules were observed in CMV-infected *N. clevelandii* as observed by quantitative immunogold labeling of the MP of CMV. However, the MP was observed at both the entry and the central cavity of the PD pore, as well as the distant connecting sieve elements. In addition, CMV having a mutation in the 3a gene encoding the MP was still able to infect tobacco both locally and systemically, although protoplasts containing such mutants showed no tubule formation [**117**].

The MP of prunus necrotic ringspot virus (PNRSV), an Ilarvirus, also binds to ss RNA in a sequence-nonspecific manner. A 33 aa domain at the N-terminal of MP is crucial for ss RNA binding as found out by deletion mutagenesis followed by Northwestern analysis. The N-terminal location of the RNA-binding domain in MPs encoded by Ilarviruses and Alfamoviruses are different from MPs encoded by other members of the family *Bromoviridae* where the similar domain lies at the C-terminal. This may have evolutionary implications showing phylogenetic divergence among viruses belonging to the family *Bromoviridae* [**121**].

Members of the family *Bromoviridae* also show high functional variability in the dependence of their MPs on CP as well as tubule formation. With the current findings, a correlation between CP interaction and tubule formation by the MP can be established. MPs requiring CP for virus movement may form tubule, and the ones, which do not require CP may move by gating the PD. However, it requires further conclusive studies involving both the variability in virus species and strains as well as the host range to establish this functional relationship.

3.4. MPs as RNA Silencing Suppressors

RNA interference (RNAi) is a fundamental mechanism of regulation of gene expression in eukaryotes both at the transcriptional and post-transcriptional level by specific mRNA degradation through complementary small RNA [**146**]. RNAi is used by plants as an important anti-viral defense through degradation of the viral RNA [**147**]. Some viruses mount a counter-defense at the sites of VRC from RNAi by compartmentalization using subcellular structures, e.g., ER spherules or by speeding up the replication and systemic infection process outpacing the mobile RNAi signals [**148**]. Most viruses, however, mount counter-defense by expressing proteins called viral suppressors of RNA silencing (VSRs) [**147**]. The VSRs are generally multifunctional in nature, and some of them, in addition to the inhibition of specific steps of the RNAi pathway, is also involved in virus replication and movement. Interestingly, a large number of MPs have also been shown to have VSR activities (**Table 1**). Although there is a report that the TMV MP controls its own propagation via promoting the spread of host RNA silencing [**149**], most of the other viral MPs act as VSR and suppress the host defense system (**Table 1**; **Figure 6**).

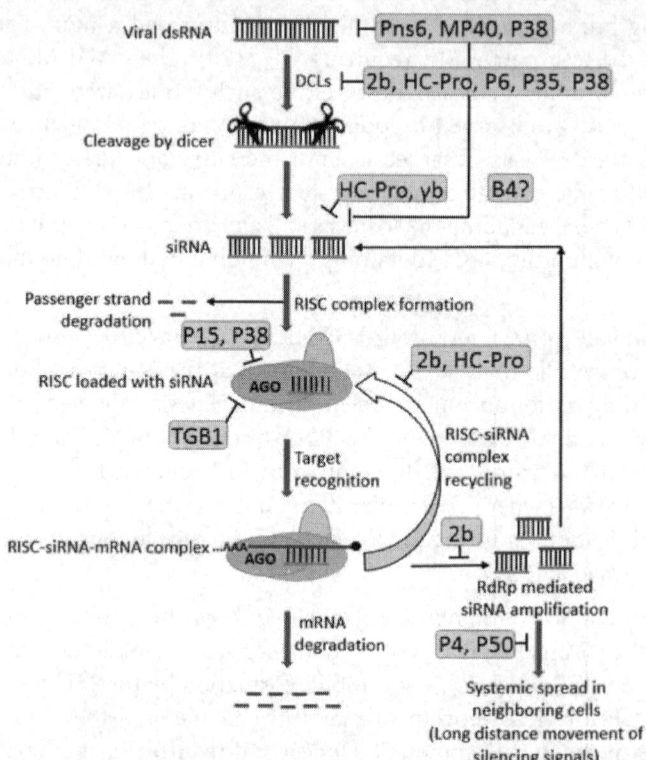

Figure 6. Schematic illustration of the action points of the MPs or movement assisting proteins acting as VSRs over basic RNAi mechanism layout: The PVX TGB1 targets AGO1 and causes its degradation through the proteasome pathway [150]. The RRSV Pns6 targets the upstream step of viral dsRNA formation [151]. The ACLSV P50 is a suppressor of systemic silencing by inhibiting the systemic movement of silencing signals [152]. The CLBV MP40 acts as a local silencing suppressor by putatively affecting the dSRNA and siRNA generation [153]. The P4 movement protein in Luteovirus has been recently identified as a systemic RNA silencing suppressor [84]. The P6 protein of CaMV acts as a silencing suppressor by indirectly blocking the DCL4 [154]. The 2b protein of CMV act as a silencing suppressor by interacting with DCL1, AGO1 and 4, siRNA biogenesis and RdRp downregulation [155]. The Potyvirus HC-Pro selectively binds to siRNA of different sizes, blocks HEN1 methyltransferase, binds and inhibits HEN1, prevents AGO1 loading, downregulates AGO1 by upregulating its corresponding micro RNA and may be involved in AGO3 cleavage, interact with RAV2 factor, thus blocking the siRNA biogenesis [155]. The BSMV γb binds to siRNA [156]. The PCV P15 interacts with AGO1 and prevents the siRNA binding [156]. The RCNMV P35 suppresses RNAi probably by sequestering DCL1 and using its helicase activity for its own replication [62]. The P38 CP, which assists TCV movement, is an RNAi suppressor, which binds to AGO 1 and 2, upregulating the DCL1 for antagonizing the functions of DCL3 and 4, binding to dsRNA and preventing primary siRNA biogenesis and upregulating AGO1 specific miRNAs [155]. The B4 protein of

BBTV is a silencing suppressor; however, the exact target for suppression is not found yet [157]. Abbreviations: VSRs—viral suppressors of RNA silencing; PVX—potato virus X—RRSV—rice ragged stunt virus; ACLSV—apple chlorotic leafspot virus; CLBV—citrus leaf blotch virus; CaMV—cauliflower mosaic virus; CMV—cucumber mosaic virus; DCL—dicer-like; AGO—Argonaute; siRNA—small interfering RNA; RdRp—RNA-dependent RNA polymerase; HC-Pro—helper component proteinase; HEN1—HUA enhancer 1 (small RNA methyltransferase); BSMV—barley stripe mosaic virus; PCV—peanut clump virus; RCNMV—red clover necrotic mosaic virus; TCV—turnip crinkle virus; CP—coat protein; BBTV—banana bunchy top virus.

In this regard, it is important to note that either the MP in itself or another movement-associated protein may act as VSR (e.g., Luteoviral P4 or Potyviral HC-Pro) [**84,158**] or the replicase-type proteins involved in the formation of transport complexes, such as VRCs (e.g., 126 kDa TMV replicase) [**159**]. The synchronicity of transfer of VSRs with the viral genome also plays a critical role in preventing the viral genome from degradation through RNAi and thus promotes pathogenicity. If the MPs having VSR activity get transferred to the new cell first and then the viral genome enters it through plasmodesmata, the former can condition the new cell for the proper proliferation of the latter. If the MPs are a part of VRCs (e.g., 126 kDa TMV replicase), the VSR activity should go hand-in-hand with the viral infection [**130,160,161**].

4. Conclusions

It is abundantly clear that MPs have evolved diverse mechanisms to accomplish their goal of ensuring the transport of viruses across the host imposed natural barriers (e.g., rigid cell walls) and provoke a successful, productive systemic infection. Although highly variable, MPs interact with a plethora of viral as well as host factors, facilitating the movement of the viral genomes to and through PD. Studies on viral MPs indicate that the virus movement is spatially and temporally coupled with replication, encapsidation and suppression of host RNAi-mediated defense, and all this involves a close association of viral MPs with the host PD, cell cytoskeleton, endomembrane system and the secretory pathway.

Further progress in our understanding of their actions will probably be achieved by revealing finer details of their structures and interactions with components of PD and other viral proteins. Whether MPs also play a role in the tissue tropism of certain viruses may be revealed by better understanding the functions of various PD types in tissues and their interactions with MP. A new aspect of the function of MPs is emerging when one considers their role in non-cell-autonomous functions in plants, a subject which may have implications in viral pathogenicity. Several outstanding questions that can be put forward are:

What is the role of the tertiary structure of MPs on their interactions with key proteins of PD? How does the specific function of MP change upon post-translational modifications? Are there variants of PD-localization signal in MPs structurally different from each other? If yes, then how do the various membrane contact sites/receptors differentiate between these signals? Is it possible that the limited non-cell-autonomous functions known in plants are intimately modified by MPs? Is cellular conditioning at the infection front before the virus spread a general phenomenon acquired by viruses of the "30 K" superfamily clan, or the mechanism extends to other viruses with non-related MPs too? How does the role of MPs known in RNA silencing suppression impact their effect on the intercellular spread of viruses? In tubule-forming viruses, what is the mechanism by which MP modifies the PD components in forming the tubules?

Finally, one can look forward to further developments in super-resolution- and fluorescent probe-based 3-D microscopic techniques for a clearer picture of the interactome of MPs. In addition, the detailed study on MP interaction and function can yield a significant amount of information about the response of the host cellular machinery, not only towards viral infection but also may provide a novel handle for the better understanding of cellular machinery for intercellular movement of macromolecules and other substances (both cellular and foreign) across PD.

References

1. Navarro, J.A.; Sanchez-Navarro, J.A.; Pallas, V. Key checkpoints in the movement of plant viruses through the host. In *Advances in Virus Research*; Elsevier: Amsterdam, The Netherlands, 2019; Volume 104, pp. 1–64. [**Google Scholar**]
2. Kappagantu, M.; Collum, T.D.; Dardick, C.; Culver, J.N. Viral Hacks of the Plant Vasculature: The Role of Phloem Alterations in Systemic Virus Infection. *Annu. Rev. Virol.* **2020**, *7*, 351–370. [**Google Scholar**] [**CrossRef**] [**PubMed**]
3. Terry, B.R.; Robards, A.W. Hydrodynamic radius alone governs the mobility of molecules through plasmodesmata. *Planta* **1987**, *171*, 145–157. [**Google Scholar**] [**CrossRef**] [**PubMed**]
4. Liarzi, O.; Epel, B.L. Development of a quantitative tool for measuring changes in the coefficient of conductivity of plasmodesmata induced by developmental, biotic, and abiotic signals. *Protoplasma* **2005**, *225*, 67–76. [**Google Scholar**] [**CrossRef**] [**PubMed**]
5. Lucas, W.J. Plant viral movement proteins: Agents for cell-to-cell trafficking of viral genomes. *Virology* **2006**, *344*, 169–184. [**Google Scholar**] [**CrossRef**] [**Green Version**]
6. Hong, J.-S.; Ju, H.-J. The plant cellular systems for plant virus movement. *Plant Pathol. J.* **2017**, *33*, 213. [**Google Scholar**] [**CrossRef**]
7. Schoelz, J.E.; Angel, C.A.; Nelson, R.S.; Leisner, S.M. A model for intracellular movement of Cauliflower mosaic virus: The concept of the mobile virion factory. *J. Exp. Bot.* **2016**, *67*, 2039–2048. [**Google Scholar**] [**CrossRef**] [**Green Version**]
8. Kumar, D.; Kumar, R.; Hyun, T.K.; Kim, J.-Y. Cell-to-cell movement of viruses via plasmodesmata. *J. Plant Res.* **2015**, *128*, 37–47. [**Google Scholar**] [**CrossRef**]
9. Heinlein, M. Plant virus replication and movement. *Virology* **2015**, *479*, 657–671. [**Google Scholar**] [**CrossRef**] [**Green Version**]
10. Pitzalis, N.; Heinlein, M. The roles of membranes and associated cytoskeleton in plant virus replication and cell-to-cell movement. *J. Exp. Bot.* **2018**, *69*, 117–132. [**Google Scholar**] [**CrossRef**]
11. Reagan, B.C.; Burch-Smith, T.M. Viruses reveal the secrets of plasmodesmal cell biology. *Mol. Plant-Microbe Interact.* **2020**, *33*, 26–39. [**Google Scholar**] [**CrossRef**]

12. Morozov, S.Y.; Solovyev, A.G. Small hydrophobic viral proteins involved in intercellular movement of diverse plant virus genomes. *Aims Microbiol.* **2020**, *6*, 305. [**Google Scholar**] [**CrossRef**]

13. Levy, A.; Tilsner, J. Creating contacts between replication and movement at plasmodesmata–a role for membrane contact sites in plant virus infections? *Front. Plant Sci.* **2020**, *11*, 862. [**Google Scholar**] [**CrossRef**] [**PubMed**]

14. Wu, X.; Cheng, X. Intercellular movement of plant RNA viruses: Targeting replication complexes to the plasmodesma for both accuracy and efficiency. *Traffic* **2020**, *21*, 725–736. [**Google Scholar**] [**CrossRef**] [**PubMed**]

15. Deom, C.M.; Lapidot, M.; Beachy, R.N. Plant virus movement proteins. *Cell* **1992**, *69*, 221–224. [**Google Scholar**] [**CrossRef**]

16. Kellmann, J.-W. Identification of plant virus movement-host protein interactions. *Zeitschrift Für Naturforschung C* **2001**, *56*, 669–679. [**Google Scholar**] [**CrossRef**] [**PubMed**]

17. Folimonova, S.Y.; Tilsner, J. Hitchhikers, highway tolls and roadworks: The interactions of plant viruses with the phloem. *Curr. Opin. Plant Biol.* **2018**, *43*, 82–88. [**Google Scholar**] [**CrossRef**] [**Green Version**]

18. Verchot-Lubicz, J.; Torrance, L.; Solovyev, A.G.; Morozov, S.Y.; Jackson, A.O.; Gilmer, D. Varied movement strategies employed by triple gene block–encoding viruses. *Mol. Plant-Microbe Interact.* **2010**, *23*, 1231–1247. [**Google Scholar**] [**CrossRef**] [**Green Version**]

19. Nicolas, W.J.; Grison, M.S.; Bayer, E.M. Shaping intercellular channels of plasmodesmata: The structure-to-function missing link. *J. Exp. Bot.* **2018**, *69*, 91–103. [**Google Scholar**] [**CrossRef**] [**Green Version**]

20. Dorokhov, Y.L.; Ershova, N.M.; Sheshukova, E.V.; Komarova, T.V. Plasmodesmata Conductivity Regulation: A Mechanistic Model. *Plants* **2019**, *8*, 595. [**Google Scholar**] [**CrossRef**] [**PubMed**] [**Green Version**]

21. Radford, J.E.; White, R.G. Localization of a myosin-like protein to plasmodesmata. *Plant J.* **1998**, *14*, 743–750. [**Google Scholar**] [**CrossRef**] [**Green Version**]

22. Zavaliev, R.; Levy, A.; Gera, A.; Epel, B.L. Subcellular dynamics and role of Arabidopsis β-1, 3-glucanases in cell-to-cell movement of tobamoviruses. *Mol. Plant-Microbe Interact.* **2013**, *26*, 1016–1030. [**Google Scholar**] [**CrossRef**] [**PubMed**] [**Green Version**]

23. Chen, M.-H.; Tian, G.-W.; Gafni, Y.; Citovsky, V. Effects of calreticulin on viral cell-to-cell movement. *Plant Physiol.* **2005**, *138*, 1866–1876. [**Google Scholar**] [**CrossRef**] [**PubMed**] [**Green Version**]

24. Baluška, F.; Šamaj, J.; Napier, R.; Volkmann, D. Maize calreticulin localizes preferentially to plasmodesmata in root apex. *Plant J.* **1999**, *19*, 481–488. [**Google Scholar**] [**CrossRef**] [**Green Version**]

25. Diao, M.; Ren, S.; Wang, Q.; Qian, L.; Shen, J.; Liu, Y.; Huang, S. Arabidopsis formin 2 regulates cell-to-cell trafficking by capping and stabilizing actin filaments at plasmodesmata. *Elife* **2018**, *7*, e36316. [**Google Scholar**] [**CrossRef**]

26. Iswanto, A.B.B.; Kim, J.-Y. Lipid raft, regulator of plasmodesmal callose homeostasis. *Plants* **2017**, *6*, 15. [**Google Scholar**] [**CrossRef**] [**Green Version**]

27. Lee, J.-Y.; Yoo, B.-C.; Rojas, M.R.; Gomez-Ospina, N.; Staehelin, L.A.; Lucas, W.J. Selective trafficking of non-cell-autonomous proteins mediated by NtNCAPP1. *Science* **2003**, *299*, 392–396. [**Google Scholar**] [**CrossRef**]

28. Simpson, C.; Thomas, C.; Findlay, K.; Bayer, E.; Maule, A.J. An Arabidopsis GPI-anchor plasmodesmal neck protein with callose binding activity and potential to regulate cell-to-cell trafficking. *Plant Cell* **2009**, *21*, 581–594. [**Google Scholar**] [**CrossRef**] [**PubMed**] [**Green Version**]

29. Lim, G.-H.; Shine, M.B.; de Lorenzo, L.; Yu, K.; Cui, W.; Navarre, D.; Hunt, A.G.; Lee, J.-Y.; Kachroo, A.; Kachroo, P. Plasmodesmata localizing proteins regulate transport and signaling during systemic acquired immunity in plants. *Cell Host Microbe* **2016**, *19*, 541–549. [**Google Scholar**] [**CrossRef**] [**Green Version**]

30. Raffaele, S.; Bayer, E.; Lafarge, D.; Cluzet, S.; Retana, S.G.; Boubekeur, T.; Leborgne-Castel, N.; Carde, J.-P.; Lherminier, J.; Noirot, E. Remorin, a solanaceae protein resident in membrane rafts and plasmodesmata, impairs potato virus X movement. *Plant Cell* **2009**, *21*, 1541–1555. [**Google Scholar**] [**CrossRef**] [**Green Version**]

31. Burch-Smith, T.M.; Cui, Y.; Zambryski, P.C. Reduced levels of class 1 reversibly glycosylated polypeptide increase intercellular transport via plasmodesmata. *Plant Signal. Behav.* **2012**, *7*, 62–67. [**Google Scholar**] [**CrossRef**] [**Green Version**]

32. Yuan, C.; Lazarowitz, S.G.; Citovsky, V. The plasmodesmal localization signal of TMV MP is recognized by plant synaptotagmin SYTA. *MBio* **2018**, *9*, e01314-18. [**Google Scholar**] [**CrossRef**] [**PubMed**] [**Green Version**]

33. Diao, M.; Huang, S. An Update on the Role of the Actin Cytoskeleton in Plasmodesmata: A Focus on Formins. *Front. Plant Sci.* **2021**, *12*, 191. [**Google Scholar**] [**CrossRef**] [**PubMed**]

34. Nishiguchi, M.; Motoyoshi, F.; Oshima, N. Behaviour of a temperature sensitive strain of tobacco mosaic virus in tomato leaves and protoplasts. *J. Gen. Virol.* **1978**, *39*, 53–61. [**Google Scholar**] [**CrossRef**]

35. Wolf, S.; Lucas, W.J.; Deom, C.M.; Beachy, R.N. Movement protein of tobacco mosaic virus modifies plasmodesmatal size exclusion limit. *Science* **1989**, *246*, 377–379. [**Google Scholar**] [**CrossRef**]

36. Ueki, S.; Spektor, R.; Natale, D.M.; Citovsky, V. ANK, a host cytoplasmic receptor for the Tobacco mosaic virus cell-to-cell movement protein, facilitates intercellular transport through plasmodesmata. *PLoSPathog.* **2010**, *6*, e1001201. [**Google Scholar**] [**CrossRef**] [**PubMed**]

37. Melcher, U. The '30K'superfamily of viral movement proteins. *J. Gen. Virol.* **2000**, *81*, 257–266. [**Google Scholar**] [**PubMed**]

38. Moore, P.J.; Fenczik, C.A.; Deom, C.M.; Beachy, R.N. Developmental changes in plasmodesmata in transgenic tobacco expressing the movement protein of tobacco mosaic virus. *Protoplasma* **1992**, *170*, 115–127. [**Google Scholar**] [**CrossRef**]

39. Beachy, R.N.; Heinlein, M. Role of P30 in replication and spread of TMV. *Traffic* **2000**, *1*, 540–544. [**Google Scholar**] [**CrossRef**]
40. Peiró, A.; Martínez-Gil, L.; Tamborero, S.; Pallás, V.; Sánchez-Navarro, J.A.; Mingarro, I. The Tobacco mosaic virus movement protein associates with but does not integrate into biological membranes. *J. Virol.* **2014**, *88*, 3016–3026. [**Google Scholar**] [**CrossRef**] [**Green Version**]
41. Guenoune-Gelbart, D.; Elbaum, M.; Sagi, G.; Levy, A.; Epel, B.L. Tobacco mosaic virus (TMV) replicase and movement protein function synergistically in facilitating TMV spread by lateral diffusion in the plasmodesmal desmotubule of Nicotiana benthamiana. *Mol. Plant-Microbe Interact.* **2008**, *21*, 335–345. [**Google Scholar**] [**CrossRef**] [**PubMed**] [**Green Version**]
42. Kawakami, S.; Watanabe, Y.; Beachy, R.N. Tobacco mosaic virus infection spreads cell to cell as intact replication complexes. *Proc. Natl. Acad. Sci. USA* **2004**, *101*, 6291–6296. [**Google Scholar**] [**CrossRef**] [**Green Version**]
43. Matsushita, Y.; Hanazawa, K.; Yoshioka, K.; Oguchi, T.; Kawakami, S.; Watanabe, Y.; Nishiguchi, M.; Nyunoya, H. In vitro phosphorylation of the movement protein of tomato mosaic tobamovirus by a cellular kinase. *J. Gen. Virol.* **2000**, *81*, 2095–2102. [**Google Scholar**] [**CrossRef**] [**PubMed**]
44. Matsushita, Y.; Ohshima, M.; Yoshioka, K.; Nishiguchi, M.; Nyunoya, H. The catalytic subunit of protein kinase CK2 phosphorylates in vitro the movement protein of Tomato mosaic virus. *J. Gen. Virol.* **2003**, *84*, 497–505. [**Google Scholar**] [**CrossRef**]
45. Lee, J.-Y.; Taoka, K.; Yoo, B.-C.; Ben-Nissan, G.; Kim, D.-J.; Lucas, W.J. Plasmodesmal-associated protein kinase in tobacco and Arabidopsis recognizes a subset of non-cell-autonomous proteins. *Plant Cell* **2005**, *17*, 2817–2831. [**Google Scholar**] [**CrossRef**] [**PubMed**] [**Green Version**]
46. Trutnyeva, K.; Bachmaier, R.; Waigmann, E. Mimicking carboxyterminal phosphorylation differentially effects subcellular distribution and cell-to-cell movement of Tobacco mosaic virus movement protein. *Virology* **2005**, *332*, 563–577. [**Google Scholar**] [**CrossRef**] [**Green Version**]
47. Yuan, C.; Lazarowitz, S.G.; Citovsky, V. Identification of a functional plasmodesmal localization signal in a plant viral cell-to-cell-movement protein. *MBio* **2016**, *7*, e02052-15. [**Google Scholar**] [**CrossRef**] [**PubMed**] [**Green Version**]
48. Liu, Y.; Huang, C.; Zeng, J.; Yu, H.; Li, Y.; Yuan, C. Identification of two additional plasmodesmata localization domains in the tobacco mosaic virus cell-to-cell-movement protein. *Biochem. Biophys. Res. Commun.* **2020**, *521*, 145–151. [**Google Scholar**] [**CrossRef**] [**PubMed**]
49. Levy, A.; Zheng, J.Y.; Lazarowitz, S.G. Synaptotagmin SYTA forms ER-plasma membrane junctions that are recruited to plasmodesmata for plant virus movement. *Curr. Biol.* **2015**, *25*, 2018–2025. [**Google Scholar**] [**CrossRef**] [**PubMed**] [**Green Version**]
50. Ishikawa, K.; Tamura, K.; Fukao, Y.; Shimada, T. Structural and functional relationships between plasmodesmata and plant endoplasmic reticulum–plasma membrane contact sites consisting of three synaptotagmins. *New Phytol.* **2020**, *226*, 798–808. [**Google Scholar**] [**CrossRef**]

51. Harries, P.A.; Park, J.-W.; Sasaki, N.; Ballard, K.D.; Maule, A.J.; Nelson, R.S. Differing requirements for actin and myosin by plant viruses for sustained intercellular movement. *Proc. Natl. Acad. Sci. USA* **2009**, *106*, 17594–17599. [**Google Scholar**] [**CrossRef**] [**Green Version**]
52. Amsbury, S.; Kirk, P.; Benitez-Alfonso, Y. Emerging models on the regulation of intercellular transport by plasmodesmata-associated callose. *J. Exp. Bot.* **2018**, *69*, 105–115. [**Google Scholar**] [**CrossRef**] [**Green Version**]
53. Chen, M.; Sheng, J.; Hind, G.; Handa, A.K.; Citovsky, V. Interaction between the tobacco mosaic virus movement protein and host cell pectin methylesterases is required for viral cell-to-cell movement. *Embo J.* **2000**, *19*, 913–920. [**Google Scholar**] [**CrossRef**]
54. Chen, M.; Citovsky, V. Systemic movement of a tobamovirus requires host cell pectin methylesterase. *Plant J.* **2003**, *35*, 386–392. [**Google Scholar**] [**CrossRef**] [**Green Version**]
55. Oparka, K.J.; Prior, D.A.M.; Cruz, S.S.; Padgett, H.S.; Beachy, R.N. Gating of epidermal plasmodesmata is restricted to the leading edge of expanding infection sites of tobacco mosaic virus (TMV). *Plant J.* **1997**, *12*, 781–789. [**Google Scholar**] [**CrossRef**] [**PubMed**] [**Green Version**]
56. Komarova, T.V.; Skulachev, M.V.; Ivanov, P.A.; Klyushin, A.G.; Dorokhov, Y.L.; Atabekov, J.G. Internal ribosome entry site from crucifer tobamovirus promotes initiation of translation in Escherichia coli. *Dokl. Biochem. Biophys.* **2003**, *389*, 118–121. [**Google Scholar**] [**CrossRef**]
57. Zvereva, S.D.; Ivanov, P.A.; Skulachev, M.V.; Klyushin, A.G.; Dorokhov, Y.L.; Atabekov, J.G. Evidence for contribution of an internal ribosome entry site to intercellular transport of a tobamovirus. *J. Gen. Virol.* **2004**, *85*, 1739–1744. [**Google Scholar**] [**CrossRef**]
58. Dorokhov, Y.L.; Ivanov, P.A.; Komarova, T.V.; Skulachev, M.V.; Atabekov, J.G. An internal ribosome entry site located upstream of the crucifer-infecting tobamovirus coat protein (CP) gene can be used for CP synthesis in vivo. *J. Gen. Virol.* **2006**, *87*, 2693–2697. [**Google Scholar**] [**CrossRef**] [**PubMed**]
59. Vogler, H.; Kwon, M.-O.; Dang, V.; Sambade, A.; Fasler, M.; Ashby, J.; Heinlein, M. Tobacco mosaic virus movement protein enhances the spread of RNA silencing. *PLoSPathog.* **2008**, *4*, e1000038. [**Google Scholar**] [**CrossRef**] [**PubMed**] [**Green Version**]
60. Xiong, Z.; Kim, K.H.; Giesman-Cookmeyer, D.; Lommel, S.A. The roles of the red clover necrotic mosaic virus capsid and cell-to-cell movement proteins in systemic infection. *Virology* **1993**, *192*, 27–32. [**Google Scholar**] [**CrossRef**]
61. Tremblay, D.; Vaewhongs, A.A.; Turner, K.A.; Sit, T.L.; Lommel, S.A. Cell wall localization of Red clover necrotic mosaic virus movement protein is required for cell-to-cell movement. *Virology* **2005**, *333*, 10–21. [**Google Scholar**] [**CrossRef**] [**PubMed**] [**Green Version**]

62. Powers, J.G.; Sit, T.L.; Heinsohn, C.; George, C.G.; Kim, K.-H.; Lommel, S.A. The Red clover necrotic mosaic virus RNA-2 encoded movement protein is a second suppressor of RNA silencing. *Virology* **2008**, *381*, 277–286. [**Google Scholar**] [**CrossRef**] [**Green Version**]

63. Kaido, M.; Tsuno, Y.; Mise, K.; Okuno, T. Endoplasmic reticulum targeting of the Red clover necrotic mosaic virus movement protein is associated with the replication of viral RNA1 but not that of RNA2. *Virology* **2009**, *395*, 232–242. [**Google Scholar**] [**CrossRef**] [**PubMed**] [**Green Version**]

64. Kaido, M.; Funatsu, N.; Tsuno, Y.; Mise, K.; Okuno, T. Viral cell-to-cell movement requires formation of cortical punctate structures containing Red clover necrotic mosaic virus movement protein. *Virology* **2011**, *413*, 205–215. [**Google Scholar**] [**CrossRef**] [**PubMed**] [**Green Version**]

65. Cao, M.; Ye, X.; Willie, K.; Lin, J.; Zhang, X.; Redinbaugh, M.G.; Simon, A.E.; Morris, T.J.; Qu, F. The capsid protein of Turnip crinkle virus overcomes two separate defense barriers to facilitate systemic movement of the virus in Arabidopsis. *J. Virol.* **2010**, *84*, 7793–7802. [**Google Scholar**] [**CrossRef**] [**PubMed**] [**Green Version**]

66. Martínez-Gil, L.; Johnson, A.E.; Mingarro, I. Membrane insertion and biogenesis of the Turnip crinkle virus p9 movement protein. *J. Virol.* **2010**, *84*, 5520–5527. [**Google Scholar**] [**CrossRef**] [**PubMed**] [**Green Version**]

67. Vilar, M.; Saurí, A.; Marcos, J.F.; Mingarro, I.; Pérez-Payá, E. Transient structural ordering of the RNA-binding domain of carnation mottle virus p7 movement protein modulates nucleic acid binding. *Chembiochem* **2005**, *6*, 1391–1396. [**Google Scholar**] [**CrossRef**] [**PubMed**]

68. Martínez-Turiño, S.; Hernández, C. A membrane-associated movement protein of Pelargonium flower break virus shows RNA-binding activity and contains a biologically relevant leucine zipper-like motif. *Virology* **2011**, *413*, 310–319. [**Google Scholar**] [**CrossRef**] [**PubMed**] [**Green Version**]

69. Genovés, A.; Navarro, J.A.; Pallás, V. The intra-and intercellular movement of Melon necrotic spot virus (MNSV) depends on an active secretory pathway. *Mol. Plant-Microbe Interact.* **2010**, *23*, 263–272. [**Google Scholar**] [**CrossRef**] [**Green Version**]

70. Serra-Soriano, M.; Antonio Navarro, J.; Pallás, V. Dissecting the multifunctional role of the N-terminal domain of the Melon necrotic spot virus coat protein in RNA packaging, viral movement and interference with antiviral plant defence. *Mol. Plant Pathol.* **2017**, *18*, 837–849. [**Google Scholar**] [**CrossRef**]

71. Lazareva, E.A.; Lezzhov, A.A.; Komarova, T.V.; Morozov, S.Y.; Heinlein, M.; Solovyev, A.G. A novel block of plant virus movement genes. *Mol. Plant Pathol.* **2017**, *18*, 611–624. [**Google Scholar**] [**CrossRef**] [**PubMed**]

72. Lazareva, E.A.; Lezzhov, A.A.; Chergintsev, D.A.; Golyshev, S.A.; Dolja, V.V.; Morozov, S.Y.; Heinlein, M.; Solovyev, A.G. Reticulon-like properties of a plant virus-encoded movement protein. *New Phytol.* **2021**, *229*, 1052–1066. [**Google Scholar**] [**CrossRef**] [**PubMed**]

73. Chu, M.; Park, J.-W.; Scholthof, H.B. Separate regions on the tomato bushy stunt virus p22 protein mediate cell-to-cell movement versus elicitation of effective resistance responses. *Mol. Plant-Microbe Interact.* **1999**, *12*, 285–292. [**Google Scholar**] [**CrossRef**]

74. Agranovsky, A.A. Closteroviruses: Molecular biology, evolution and interactions with cells. In *Plant Viruses: Evolution and Management*; Springer: Singapore, 2016; pp. 231–252. [**Google Scholar**]

75. Qiao, W.; Medina, V.; Kuo, Y.-W.; Falk, B.W. A distinct, non-virion plant virus movement protein encoded by a crinivirus essential for systemic infection. *MBio* **2018**, 9. [**Google Scholar**] [**CrossRef**] [**Green Version**]

76. Cui, X.; Yaghmaiean, H.; Wu, G.; Wu, X.; Chen, X.; Thorn, G.; Wang, A. The C-terminal region of the Turnip mosaic virus P3 protein is essential for viral infection via targeting P3 to the viral replication complex. *Virology* **2017**, *510*, 147–155. [**Google Scholar**] [**CrossRef**] [**PubMed**]

77. Dai, Z.; He, R.; Bernards, M.A.; Wang, A. The cis-expression of the coat protein of turnip mosaic virus is essential for viral intercellular movement in plants. *Mol. Plant Pathol.* **2020**, *21*, 1194–1211. [**Google Scholar**] [**CrossRef**]

78. Rojas, M.R.; Zerbini, F.M.; Allison, R.F.; Gilbertson, R.L.; Lucas, W.J. Capsid protein and helper component-proteinase function as potyvirus cell-to-cell movement proteins. *Virology* **1997**, *237*, 283–295. [**Google Scholar**] [**CrossRef**] [**Green Version**]

79. Cheng, G.; Yang, Z.; Zhang, H.; Zhang, J.; Xu, J. Remorin interacting with PCaP1 impairs Turnip mosaic virus intercellular movement but is antagonised by VPg. *New Phytol.* **2020**, *225*, 2122–2139. [**Google Scholar**] [**CrossRef**] [**PubMed**]

80. Chai, M.; Wu, X.; Liu, J.; Fang, Y.; Luan, Y.; Cui, X.; Zhou, X.; Wang, A.; Cheng, X. P3N-PIPO interacts with P3 via the shared N-terminal domain to recruit viral replication vesicles for cell-to-cell movement. *J. Virol.* **2020**, *94*, e01898-19. [**Google Scholar**] [**CrossRef**]

81. Lee, L.; Palukaitis, P.; Gray, S.M. Host-dependent requirement for the Potato leafroll virus 17-kda protein in virus movement. *Mol. Plant-Microbe Interact.* **2002**, *15*, 1086–1094. [**Google Scholar**] [**CrossRef**]

82. DeBlasio, S.L.; Xu, Y.; Johnson, R.S.; Rebelo, A.R.; MacCoss, M.J.; Gray, S.M.; Heck, M. The interaction dynamics of two potato leafroll virus movement proteins affects their localization to the outer membranes of mitochondria and plastids. *Viruses* **2018**, *10*, 585. [**Google Scholar**] [**CrossRef**] [**Green Version**]

83. Li, S.; Su, X.; Luo, X.; Zhang, Y.; Zhang, D.; Du, J.; Zhang, Z.; OuYang, X.; Zhang, S.; Liu, Y. First evidence showing that Pepper vein yellows virus P4 protein is a movement protein. *BMC Microbiol.* **2020**, *20*, 1–6. [**Google Scholar**] [**CrossRef**]

84. Fusaro, A.F.; Barton, D.A.; Nakasugi, K.; Jackson, C.; Kalischuk, M.L.; Kawchuk, L.M.; Vaslin, M.F.S.; Correa, R.L.; Waterhouse, P.M. The luteovirus P4 movement protein is a suppressor of systemic RNA silencing. *Viruses* **2017**, *9*, 294. [**Google Scholar**] [**CrossRef**] [**Green Version**]

85. Kleinow, T.; Happle, A.; Kober, S.; Linzmeier, L.; Rehm, T.M.; Fritze, J.; Buchholz, P.C.F.; Kepp, G.; Jeske, H.; Wege, C. Phosphorylations of the Abutilon Mosaic Virus Movement Protein Affect Its Self-Interaction, Symptom Development, Viral DNA Accumulation, and Host Range. *Front. Plant Sci.* **2020**, *11*, 1155. [**Google Scholar**] [**CrossRef**]

86. Lee, C.; Zheng, Y.; Chan, C.; Ku, H.; Chang, C.; Jan, F. A single amino acid substitution in the movement protein enables the mechanical transmission of a geminivirus. *Mol. Plant Pathol.* **2020**, *21*, 571–588. [**Google Scholar**] [**CrossRef**] [**Green Version**]

87. Mann, K.S.; Bejerman, N.; Johnson, K.N.; Dietzgen, R.G. Cytorhabdovirus P3 genes encode 30K-like cell-to-cell movement proteins. *Virology* **2016**, *489*, 20–33. [**Google Scholar**] [**CrossRef**] [**PubMed**] [**Green Version**]

88. Zhou, X.; Lin, W.; Sun, K.; Wang, S.; Zhou, X.; Jackson, A.O.; Li, Z. Specificity of plant rhabdovirus cell-to-cell movement. *J. Virol.* **2019**, *93*, e00296-19. [**Google Scholar**] [**CrossRef**] [**Green Version**]

89. Solovyev, A.; Kalinina, N.; Morozov, S. Recent advances in research of plant virus movement mediated by triple gene block. *Front. Plant Sci.* **2012**, *3*, 276. [**Google Scholar**] [**CrossRef**] [**PubMed**] [**Green Version**]

90. Lee, M.Y.; Yan, L.; Gorter, F.A.; Kim, B.Y.T.; Cui, Y.; Hu, Y.; Yuan, C.; Grindheim, J.; Ganesan, U.; Liu, Z. Brachypodiumdistachyon line Bd3-1 resistance is elicited by the barley stripe mosaic virus triple gene block 1 movement protein. *J. Gen. Virol.* **2012**, *93*, 2729–2739. [**Google Scholar**] [**CrossRef**] [**PubMed**]

91. Perraki, A.; Binaghi, M.; Mecchia, M.A.; Gronnier, J.; German-Retana, S.; Mongrand, S.; Bayer, E.; Zelada, A.M.; Germain, V. StRemorin1. 3 hampers Potato virus X TGBp1 ability to increase plasmodesmata permeability, but does not interfere with its silencing suppressor activity. *FEBS Lett.* **2014**, *588*, 1699–1705. [**Google Scholar**] [**CrossRef**] [**PubMed**] [**Green Version**]

92. Mathioudakis, M.M.; Veiga, R.S.L.; Canto, T.; Medina, V.; Mossialos, D.; Makris, A.M.; Livieratos, I. P epino mosaic virus triple gene block protein 1 (TGBp1) interacts with and increases tomato catalase 1 activity to enhance virus accumulation. *Mol. Plant Pathol.* **2013**, *14*, 589–601. [**Google Scholar**] [**CrossRef**]

93. Wu, X.; Liu, J.; Chai, M.; Wang, J.; Li, D.; Wang, A.; Cheng, X. The Potato virus X TGBp2 protein plays dual functional roles in viral replication and movement. *J. Virol.* **2019**, *93*, e01635-18. [**Google Scholar**] [**CrossRef**] [**Green Version**]

94. Mann, K.; Meng, B. The triple gene block movement proteins of a grape virus in the genus Foveavirus confer limited cell-to-cell spread of a mutant Potato virus X. *Virus Genes* **2013**, *47*, 93–104. [**Google Scholar**] [**CrossRef**] [**PubMed**]

95. May, J.P.; Johnson, P.Z.; Ilyas, M.; Gao, F.; Simon, A.E. The Multifunctional Long-Distance Movement Protein of Pea Enation Mosaic Virus 2 Protects Viral and Host Transcripts from Nonsense-Mediated Decay. *Mbio* **2020**, 11. [**Google Scholar**] [**CrossRef**] [**Green Version**]

96. Kasteel, D.T.J.; Perbal, M.-C.; Boyer, J.-C.; Wellink, J.; Goldbach, R.W.; Maule, A.J.; Van Lent, J.W.M. The movement proteins of cowpea mosaic virus and cauliflower mosaic virus induce tubular structures in plant and insect cells. *J. Gen. Virol.* **1996**, *77*, 2857–2864. [**Google Scholar**] [**CrossRef**] [**PubMed**]

97. Thomas, C.L.; Perbal, C.; Maule, A.J. A mutation of cauliflower mosaic virus gene I interferes with virus movement but not virus replication. *Virology* **1993**, *192*, 415–421. [**Google Scholar**] [**CrossRef**]

98. Amari, K.; Boutant, E.; Hofmann, C.; Schmitt-Keichinger, C.; Fernandez-Calvino, L.; Didier, P.; Lerich, A.; Mutterer, J.; Thomas, C.L.; Heinlein, M. A family of plasmodesmal proteins with receptor-like properties for plant viral movement proteins. *PLoSPathog.* **2010**, *6*, e1001119. [**Google Scholar**] [**CrossRef**] [**PubMed**] [**Green Version**]

99. Kitajima, E.W.; Lauritis, J.A. Plant virions in plasmodesmata. *Virology* **1969**, *37*, 681–685. [**Google Scholar**] [**CrossRef**]

100. Ritzenthaler, C.; Schmit, A.C.; Michler, P.; Stussi-Garaud, C.; Pinck, L. Grapevine fanleaf nepovirus P38 putative movement protein is located on tubules in vivo. *Mol. Plant-Microbe Interact.* **1995**, *8*, 379–387. [**Google Scholar**] [**CrossRef**]

101. Laporte, C.; Vetter, G.; Loudes, A.-M.; Robinson, D.G.; Hillmer, S.; Stussi-Garaud, C.; Ritzenthaler, C. Involvement of the secretory pathway and the cytoskeleton in intracellular targeting and tubule assembly of Grapevine fanleaf virus movement protein in tobacco BY-2 cells. *Plant Cell* **2003**, *15*, 2058–2075. [**Google Scholar**] [**CrossRef**] [**PubMed**] [**Green Version**]

102. Belin, C.; Schmitt, C.; Gaire, F.; Walter, B.; Demangeat, G.; Pinck, L. The nine C-terminal residues of the grapevine fanleaf nepovirus movement protein are critical for systemic virus spread. *J. Gen. Virol.* **1999**, *80*, 1347–1356. [**Google Scholar**] [**CrossRef**] [**Green Version**]

103. Lekkerkerker, A.; Wellink, J.; Yuan, P.; Van Lent, J.; Goldbach, R.; Van Kammen, A.B. Distinct functional domains in the cowpea mosaic virus movement protein. *J. Virol.* **1996**, *70*, 5658–5661. [**Google Scholar**] [**CrossRef**] [**Green Version**]

104. Bertens, P.; Wellink, J.; Goldbach, R.; van Kammen, A. Mutational analysis of the cowpea mosaic virus movement protein. *Virology* **2000**, *267*, 199–208. [**Google Scholar**] [**CrossRef**] [**Green Version**]

105. Bertens, P.; Heijne, W.; Van der Wel, N.; Wellink, J.; Van Kammen, A. Studies on the C-terminus of the Cowpea mosaic virus movement protein. *Arch. Virol.* **2003**, *148*, 265–279. [**Google Scholar**] [**CrossRef**]

106. Pouwels, J.; Kornet, N.; van Bers, N.; Guighelaar, T.; van Lent, J.; Bisseling, T.; Wellink, J. Identification of distinct steps during tubule formation by the movement protein of Cowpea mosaic virus. *J. Gen. Virol.* **2003**, *84*, 3485–3494. [**Google Scholar**] [**CrossRef**]

107. Pouwels, J.; van der Velden, T.; Willemse, J.; Borst, J.W.; van Lent, J.; Bisseling, T.; Wellink, J. Studies on the origin and structure of tubules made by the movement protein of Cowpea mosaic virus. *J. Gen. Virol.* **2004**, *85*, 3787–3796. [**Google Scholar**] [**CrossRef**]

108. Storms, M.M.H.; Kromelink, R.; Peters, D.; Van Lent, J.W.M.; Goldbach, R.O.B.W. The nonstructuralNSm protein of tomato spotted wilt virus induces tubular structures in plant and insect cells. *Virology* **1995**, *214*, 485–493. [**Google Scholar**] [**CrossRef**] [**Green Version**]

109. Liu, C.; Ye, L.; Lang, G.; Zhang, C.; Hong, J.; Zhou, X. The VP37 protein of Broad bean wilt virus 2 induces tubule-like structures in both plant and insect cells. *Virus Res.* **2011**, *155*, 42–47. [**Google Scholar**] [**CrossRef**]

110. Wijkamp, I.; van Lent, J.; Kormelink, R.; Goldbach, R.; Peters, D. Multiplication of tomato spotted wilt virus in its insect vector, Frankliniella occidentalis. *J. Gen. Virol.* **1993**, *74*, 341–349. [**Google Scholar**] [**CrossRef**]

111. Ryabov, E.V.; Oparka, K.J.; Santa Cruz, S.; Robinson, D.J.; Taliansky, M.E. Intracellular location of two groundnut rosette umbravirus proteins delivered by PVX and TMV vectors. *Virology* **1998**, *242*, 303–313. [**Google Scholar**] [**CrossRef**] [**Green Version**]

112. Nurkiyanova, K.M.; Ryabov, E.V.; Kalinina, N.O.; Fan, Y.; Andreev, I.; Fitzgerald, A.G.; Palukaitis, P.; Taliansky, M. Umbravirus-encoded movement protein induces tubule formation on the surface of protoplasts and binds RNA incompletely and non-cooperatively. *J. Gen. Virol.* **2001**, *82*, 2579–2588. [**Google Scholar**] [**CrossRef**]

113. Tenllado, F.; Bol, J.F. Genetic dissection of the multiple functions of alfalfa mosaic virus coat protein in viral RNA replication, encapsidation, and movement. *Virology* **2000**, *268*, 29–40. [**Google Scholar**] [**CrossRef**] [**Green Version**]

114. Kasteel, D.T.; Van der Wel, N.N.; Jansen, K.A.; Goldbach, R.W.; Van Lent, J.W. Tubule-forming capacity of the movement proteins of alfalfa mosaic virus and brome mosaic virus. *J. Gen. Virol.* **1997**, *78*, 2089–2093. [**Google Scholar**] [**CrossRef**] [**Green Version**]

115. Takeda, A.; Kaido, M.; Okuno, T.; Mise, K. The C terminus of the movement protein of Brome mosaic virus controls the requirement for coat protein in cell-to-cell movement and plays a role in long-distance movement. *J. Gen. Virol.* **2004**, *85*, 1751–1761. [**Google Scholar**] [**CrossRef**] [**PubMed**]

116. Kaido, M.; Inoue, Y.; Takeda, Y.; Sugiyama, K.; Takeda, A.; Mori, M.; Tamai, A.; Meshi, T.; Okuno, T.; Mise, K. Downregulation of the NbNACa1 gene encoding a movement-protein-interacting protein reduces cell-to-cell movement of Brome mosaic virus in Nicotiana benthamiana. *Mol. Plant-Microbe Interact.* **2007**, *20*, 671–681. [**Google Scholar**] [**CrossRef**] [**PubMed**]

117. Canto, T.; Palukaitis, P. Are tubules generated by the 3a protein necessary for cucumber mosaic virus movement? *Mol. Plant-Microbe Interact.* **1999**, *12*, 985–993. [**Google Scholar**] [**CrossRef**] [**Green Version**]

118. Palukaitis, P.; García-Arenal, F. Cucumoviruses. *Adv. Virus Res.* **2003**, *62*, 241–323. [**Google Scholar**] [**PubMed**]

119. Nagano, H.; Mise, K.; Furusawa, I.; Okuno, T. Conversion in the requirement of coat protein in cell-to-cell movement mediated by the cucumber mosaic virus movement protein. *J. Virol.* **2001**, *75*, 8045–8053. [**Google Scholar**] [**CrossRef**] [**Green Version**]

120. Andreev, I.A.; Kim, S.H.; Kalinina, N.O.; Rakitina, D.V.; Fitzgerald, A.G.; Palukaitis, P.; Taliansky, M.E. Molecular interactions between a plant virus movement protein and RNA: Force spectroscopy investigation. *J. Mol. Biol.* **2004**, *339*, 1041–1047. [**Google Scholar**] [**CrossRef**]

121. Herranz, M.C.; Pallas, V. RNA-binding properties and mapping of the RNA-binding domain from the movement protein of Prunus necrotic ringspot virus. *J. Gen. Virol.* **2004**, *85*, 761–768. [**Google Scholar**] [**CrossRef**]

122. Waigmann, E.; Lucas, W.J.; Citovsky, V.; Zambryski, P. Direct functional assay for tobacco mosaic virus cell-to-cell movement protein and identification of a domain involved in increasing plasmodesmal permeability. *Proc. Natl. Acad. Sci. USA* **1994**, *91*, 1433–1437. [**Google Scholar**] [**CrossRef**] [**Green Version**]

123. Epel, B.L. Plant viruses spread by diffusion on ER-associated movement-protein-rafts through plasmodesmata gated by viral induced host beta-1,3-glucanases. *Semin Cell Dev. Biol.* **2009**, 20. [**Google Scholar**] [**CrossRef**]

124. Zavaliev, R.; Sagi, G.; Gera, A.; Epel, B.L. The constitutive expression of Arabidopsis plasmodesmal-associated class 1 reversibly glycosylated polypeptide impairs plant development and virus spread. *J. Exp. Bot.* **2010**, *61*, 131–142. [**Google Scholar**] [**CrossRef**] [**PubMed**] [**Green Version**]

125. Adkar-Purushothama, C.R.; Brosseau, C.; Giguère, T.; Sano, T.; Moffett, P.; Perreault, J.-P. Small RNA derived from the virulence modulating region of the potato spindle tuber viroid silences callose synthase genes of tomato plants. *Plant Cell* **2015**, *27*, 2178–2194. [**Google Scholar**] [**CrossRef**] [**Green Version**]

126. Cui, W.; Lee, J.-Y. Arabidopsis callose synthases CalS1/8 regulate plasmodesmal permeability during stress. *Nat. Plants* **2016**, *2*, 1–9. [**Google Scholar**] [**CrossRef**]

127. Yan, D.; Yadav, S.R.; Paterlini, A.; Nicolas, W.J.; Petit, J.D.; Brocard, L.; Belevich, I.; Grison, M.S.; Vaten, A.; Karami, L. Sphingolipid biosynthesis modulates plasmodesmal ultrastructure and phloem unloading. *Nat. Plants* **2019**, *5*, 604–615. [**Google Scholar**] [**CrossRef**]

128. Adams, M.J.; Adkins, S.; Bragard, C.; Gilmer, D.; Li, D.; MacFarlane, S.A.; Wong, S.-M.; Melcher, U.; Ratti, C.; Ryu, K.H. ICTV virus taxonomy profile: Virgaviridae. *J. Gen. Virol.* **2017**, *98*, 1999. [**Google Scholar**] [**CrossRef**]

129. Rojas, M.R.; Maliano, M.R.; de Souza, J.O.; Vasquez-Mayorga, M.; de Macedo, M.A.; Ham, B.-K.; Gilbertson, R.L. Cell-to-cell movement of plant viruses: A diversity of mechanisms and strategies. In *Current Research Topics in Plant Virology*; Springer: Cham, Switzerland, 2016; pp. 113–152. [**Google Scholar**]

130. Sheshukova, E.V.; Ershova, N.M.; Kamarova, K.A.; Dorokhov, Y.L.; Komarova, T.V. The Tobamoviral Movement Protein: A "Conditioner" to Create a Favorable Environment for Intercellular Spread of Infection. *Front. Plant Sci.* **2020**, *11*, 959. [**Google Scholar**] [**CrossRef**]

131. Curin, M.; Ojangu, E.-L.; Trutnyeva, K.; Ilau, B.; Truve, E.; Waigmann, E. MPB2C, a microtubule-associated plant factor, is required for microtubular accumulation of tobacco mosaic virus movement protein in plants. *Plant Physiol.* **2007**, *143*, 801–811. [**Google Scholar**] [**CrossRef**] [**Green Version**]

132. Kleinow, T.; Tanwir, F.; Kocher, C.; Krenz, B.; Wege, C.; Jeske, H. Expression dynamics and ultrastructural localization of epitope-tagged Abutilon mosaic virus nuclear shuttle and movement proteins in Nicotiana benthamiana cells. *Virology* **2009**, *391*, 212–220. [**Google Scholar**] [**CrossRef**] [**Green Version**]

133. Levy, A.; Tzfira, T. Bean dwarf mosaic virus: A model system for the study of viral movement. *Mol. Plant Pathol.* **2010,** *11,* 451–461. [**Google Scholar**] [**CrossRef**] [**Green Version**]

134. Kumar, R.V. Plant antiviral immunity against geminiviruses and viral counter-defense for survival. *Front. Microbiol.* **2019**, *10*, 1460. [**Google Scholar**] [**CrossRef**] [**Green Version**]

135. Lazarowitz, S.G.; Beachy, R.N. Viral movement proteins as probes for intracellular and intercellular trafficking in plants. *Plant Cell* **1999**, *11*, 535–548. [**Google Scholar**] [**CrossRef**] [**Green Version**]

136. van Lent, J.; Storms, M.; van der Meer, F.; Wellink, J.; Goldbach, R. Tubular structures involved in movement of cowpea mosaic virus are also formed in infected cowpea protoplasts. *J. Gen. Virol.* **1991,** *72,* 2615–2623. [**Google Scholar**] [**CrossRef**] [**PubMed**]

137. Carluccio, A.V.; Zicca, S.; Stavolone, L. Hitching a ride on vesicles: Cauliflower mosaic virus movement protein trafficking in the endomembrane system. *Plant Physiol.* **2014**, *164*, 1261–1270. [**Google Scholar**] [**CrossRef**] [**Green Version**]

138. Taliansky, M.E.; Robinson, D.J. Molecular biology of umbraviruses: Phantom warriors. *J. Gen. Virol.* **2003**, *84*, 1951–1960. [**Google Scholar**] [**CrossRef**] [**PubMed**]

139. Ritzenthaler, C.; Hofmann, C. Tubule-Guided Movement of Plant Viruses. In *Viral Transport in Plants*; Waigmann, E., Heinlein, M., Eds.; Springer: Berlin/Heidelberg, Germany, 2007; Volume 7. [**Google Scholar**] [**CrossRef**]

140. Carvalho, C.M.; Wellink, J.; Ribeiro, S.G.; Goldbach, R.W.; Van Lent, J.W.M. The C-terminal region of the movement protein of Cowpea mosaic virus is involved in binding to the large but not to the small coat protein. *J. Gen. Virol.* **2003,** *84,* 2271–2277. [**Google Scholar**] [**CrossRef**] [**PubMed**]

141. Liu, C.; Meng, C.; Xie, L.; Hong, J.; Zhou, X. Cell-to-cell trafficking, subcellular distribution, and binding to coat protein of Broad bean wilt virus 2 VP37 protein. *Virus Res.* **2009**, *143*, 86–93. [**Google Scholar**] [**CrossRef**] [**PubMed**]

142. Zheng, H.; Wang, G.; Zhang, L. Alfalfa mosaic virus movement protein induces tubules in plant protoplasts. *Mol. Plant-Microbe Interact.* **1997**, *10*, 1010–1014. [**Google Scholar**] [**CrossRef**]

143. Callaway, A.; Giesman-Cookmeyer, D.; Gillock, E.T.; Sit, T.L.; Lommel, S.A. The multifunctional capsid proteins of plant RNA viruses. *Annu. Rev. Phytopathol.* **2001**, *39*, 419–460. [**Google Scholar**] [**CrossRef**]

144. Rao, A.L.N.; Cooper, B. Capsid protein gene and the type of host plant differentially modulate cell-to-cell movement of cowpea chlorotic mottle virus. *Virus Genes* **2006**, *32*, 219–227. [**Google Scholar**] [**CrossRef**]

145. Sasaki, N.; Arimoto, M.; Nagano, H.; Mori, M.; Kaido, M.; Mise, K.; Okuno, T. The movement protein gene is involved in the virus-specific requirement of the coat protein in cell-to-cell movement of bromoviruses. *Arch. Virol.* **2003**, *148*, 803–812. [**Google Scholar**] [**CrossRef**] [**PubMed**]

146. Muhammad, T.; Zhang, F.; Zhang, Y.; Liang, Y. RNA interference: A natural immune system of plants to counteract biotic stressors. *Cells* **2019**, *8*, 38. [**Google Scholar**] [**CrossRef**] [**Green Version**]

147. Pumplin, N.; Voinnet, O. RNA silencing suppression by plant pathogens: Defence, counter-defence and counter-counter-defence. *Nat. Rev. Microbiol.* **2013**, 11. [**Google Scholar**] [**CrossRef**] [**PubMed**]

148. Schwartz, M.; Chen, J.; Janda, M.; Sullivan, M.; den Boon, J.; Ahlquist, P. A positive-strand RNA virus replication complex parallels form and function of retrovirus capsids. *Mol. Cell* **2002**, *9*, 505–514. [**Google Scholar**] [**CrossRef**]

149. Díaz-Pendón, J.A.; Ding, S.W. Direct and indirect roles of viral suppressors of RNA silencing in pathogenesis. *Annu. Rev. Phytopathol* **2008**, *46*, 303–326. [**Google Scholar**] [**CrossRef**]

150. Chiu, M.; Chen, I.; Baulcombe, D.C.; Tsai, C. The silencing suppressor P25 of Potato virus X interacts with Argonaute1 and mediates its degradation through the proteasome pathway. *Mol. Plant Pathol.* **2010**, *11*, 641–649. [**Google Scholar**] [**CrossRef**] [**PubMed**]

151. Wu, J.; Du, Z.; Wang, C.; Cai, L.; Hu, M.; Lin, Q.; Wu, Z.; Li, Y.; Xie, L. Identification of Pns6, a putative movement protein of RRSV, as a silencing suppressor. *Virol. J.* **2010**, *7*, 1–6. [**Google Scholar**] [**CrossRef**] [**PubMed**] [**Green Version**]

152. Yaegashi, H.; Takahashi, T.; Isogai, M.; Kobori, T.; Ohki, S.; Yoshikawa, N. Apple chlorotic leaf spot virus 50 kDa movement protein acts as a suppressor of systemic silencing without interfering with local silencing in Nicotiana benthamiana. *J. Gen. Virol.* **2007**, *88*, 316–324. [**Google Scholar**] [**CrossRef**]

153. Renovell, A.; Vives, M.C.; Ruiz-Ruiz, S.; Navarro, L.; Moreno, P.; Guerri, J. The Citrus leaf blotch virus movement protein acts as silencing suppressor. *Virus Genes* **2012**, *44*, 131–140. [**Google Scholar**] [**CrossRef**]

154. Zvereva, A.S.; Golyaev, V.; Turco, S.; Gubaeva, E.G.; Rajeswaran, R.; Schepetilnikov, M.V.; Srour, O.; Ryabova, L.A.; Boller, T.; Pooggin, M.M. Viral protein suppresses oxidative burst and salicylic acid-dependent autophagy and facilitates bacterial growth on virus-infected plants. *New Phytol.* **2016**, *211*, 1020–1034. [**Google Scholar**] [**CrossRef**] [**PubMed**] [**Green Version**]

155. Csorba, T.; Kontra, L.; Burgyán, J. Viral silencing suppressors: Tools forged to fine-tune host-pathogen coexistence. *Virology* **2015**, *479–480*, 85–103. [**Google Scholar**] [**CrossRef**] [**PubMed**] [**Green Version**]

156. Mérai, Z.; Kerényi, Z.; Kertész, S.; Magna, M.; Lakatos, L.; Silhavy, D. Double-stranded RNA binding may be a general plant RNA viral strategy to suppress RNA silencing. *J. Virol.* **2006**, *80*, 5747–5756. [**Google Scholar**] [**CrossRef**] [**PubMed**] [**Green Version**]

157. Niu, S.; Wang, B.; Guo, X.; Yu, J.; Wang, X.; Xu, K.; Zhai, Y.; Wang, J.; Liu, Z. Identification of two RNA silencing suppressors from banana bunchy top virus. *Arch. Virol.* **2009**, *154*, 1775. [**Google Scholar**] [**CrossRef**] [**PubMed**]

158. Kasschau, K.D.; Carrington, J.C. Long-distance movement and replication maintenance functions correlate with silencing suppression activity of potyviral

HC-Pro. *Virology* **2001**, *285*, 71–81. [**Google Scholar**] [**CrossRef**] [**PubMed**] [**Green Version**]

159. Ding, X.S.; Liu, J.; Cheng, N.-H.; Folimonov, A.; Hou, Y.-M.; Bao, Y.; Katagi, C.; Carter, S.A.; Nelson, R.S. The Tobacco mosaic virus 126-kDa protein associated with virus replication and movement suppresses RNA silencing. *Mol. Plant-Microbe Interact.* **2004**, *17*, 583–592. [**Google Scholar**] [**CrossRef**] [**Green Version**]

160. Bayne, E.H.; Rakitina, D.V.; Morozov, S.Y.; Baulcombe, D.C. Cell-to-cell movement of potato potexvirus X is dependent on suppression of RNA silencing. *Plant J.* **2005**, *44*, 471–482. [**Google Scholar**] [**CrossRef**]

161. Liu, J.Z.; Blancaflor, E.B.; Nelson, R.S. The tobacco mosaic virus 126-kilodalton protein, a constituent of the virus replication complex, alone or within the complex aligns with and traffics along microfilaments. *Plant Physiol.* **2005**, *138*, 1853–1865. [**Google Scholar**] [**CrossRef**] [**Green Version**]

Chapter-12

Membrane Association and Topology of Citrus Leprosis Virus C_2 Movement and Capsid Proteins

1. Introduction

Citrus leprosis disease, caused by viruses belonging to the genera *Cilevirus* and *Dichorhavirus*, has economic importance in the Americas, especially in citrus groves in Brazil. The infection is characterized by the formation of chlorotic and necrotic circular localized lesions in citrus leaves, fruits, and stems [**1**].

Citrus leprosis virus C (CiLV-C) is a member of genus *Cilevirus,* family *Kitaviridae,* and a prevalent citrus leprosis-associated virus in South and Central America [**1**], with the exception of Colombia. Although CiLV-C was reported in that country more than 10 years ago, this virus is currently rarely found in Colombian citrus orchards, where it was replaced by citrus leprosis virus C2 (CiLV-C2) [**2,3,4,5**]. CiLV-C and CiLV-C2 are the only accepted members of genus *Cilevirus.*

Cileviruses have a bisegmented positive ssRNA genome carrying a 5′-cap structure and a 3′-poly (A) tail. For CiLV-C, its RNA1 codes for the protein precursor of the RNA dependent RNA polymerase (RdRp) and the capsid protein (p29) [**6,7,8**], while the RNA 2 codes for an RNA silencing suppressor (RSS) protein (p15) [**9**], the viral movement protein (p32) [**10**], and the putative glyco (p61) and matrix (p24) proteins [**7,11**]. In addition to p15 RSS activity, the p29 and p61 are proteins that also show the ability to suppress RNA silencing [**9**]. CiLV-C2 RNA1 and RNA 2 present a nucleotide identities of 58 and 50%, respectively, compared to the corresponding CiLV-C RNA sequences. Additionally, CiLV-C2 RNAs have a longer 3′ UTR and an extra ORF (p7) in RNA2 [**2**].

Membrane-associated viral proteins can induce substantial cellular remodeling in the processes of viral replication and virion assembly. This intracellular disorder is usually associated with changes in organelles to form viral replication sites [**12,13**]. Viral proteins can also interact with cell membranes to facilitate the intracellular viral spread [**14,15,16**]. Association between viral movement components and host membranes seems to bc an essential factor for virus transport, being a feature constantly identified in this class of viral proteins [**7,17,18,19,20,21**]. Despite the fact that recent studies have clarified some functional aspects for the CiLV-C proteins, no similar information is available for other accepted or tentative cileviruses, as passion fruit green spot virus (PfGSV). This group of viruses has been rather poorly investigated and thus, generating new data about other species of agronomic importance, such as CiLV-C2, can represent an important advance in understanding the citrus leprosis pathosystem.

Therefore, in the present study, we examined biochemical properties of the capsid (p29) and movement (MP) proteins encoded by CiLV-C2. Here, we reported the association of CiLV-C2 MP and p29 with biological membranes. Our findings obtained by computer predictions, chemical treatments after membrane fractionation, and biomolecular fluorescence complementation assay revealed that the p29 is peripherally associated, while the MP is integrally bound to the cell membranes. Topological analyses revealed that p29 and MP expose their N- and C-termini to the cell cytoplasmic compartment. These results allowed us to propose a topological model of the association of CiLV-C2 p29 and MP with cell membranes.

2. Materials and Methods

2.1. DNA Manipulation

For the membrane association assay, the *p29* and *MP* genes [**10**] were amplified with specific primers carrying the *Nco*I/*Nhe*I sites using the TaKaRa Taq DNA polymerase (TaKaRa Bio Inc) following the manufacturer's specifications. All amplified genes were fused at their C-terminus with the human influenza hemagglutinin (HA) epitope. To do this, the genes were cloned into the vector pSK35-TSWVNSm:HA-PoPit [**18**], replacing the *NSm* gene. The correct in-frame insertions were confirmed by plasmid DNA sequencing. The expression cassettes corresponded to each CiLV-C2 isolated gene flanked by the 35S constitutive promoter from the cauliflower mosaic virus (CaMV) and the terminator from the potato proteinase inhibitor (PoPit) [**18**]. Next, the expression cassettes were cloned into the pMOG800 binary vector by using the *Hind*III restriction site.

For bimolecular fluorescence complementation (BiFC) assay, the *p29* and *MP* genes were amplified with specific primers, containing the *Nco*I/*Nhe*I restriction sites, and cloned into the pSKBiFC plasmids [**18,22**], which permitted the fusion of the N- or C-yellow fluorescent protein (YFP) fragments at their N- or C-termini. The resultant expression cassettes were subcloned into the pMOG800 binary vector as previously reported [**7**].

The pMOG vectors carrying the chrysanthemum stem necrosis orthotospovirus (CSNV) NSm, tobacco mosaic virus (TMV) 30K MP, and leader peptidase (Lep) proteins with the HA epitope fused at their C-terminus and the unfused green fluorescent protein (GFP) were described in previous works [**7,18**].

2.2. Computer Analysis

Computer analysis from deduced amino acid sequences of the CiLV-C2 p29 (YP_009508071.1) and MP (YP_009508075.1), CiLV-C MP (ABC75825.1), CSNV NSm (AII20574.1), TMV 30K MP (AAD19281.1), and also for the Lep (WP_112485673.1) and GFP, were performed using computer tools MPEx [**23**], TMHMM server version 2.0 (**http://www.cbs.dtu.dk/services/TMHMM/**), ΔG prediction server version 1.0 (**http://dgpred.cbr.su.se/index.php?p=home**), HMMTOP version 2.0 (**http://www.enzim.hu/**

hmmtop/index.php), OCTOPUS (**https://octopus.cbr.su.se**), SOUSI (**http://harrier.nagahama-i-bio.ac.jp/sosui/**), and TMpred (**https://embnet.vital-it.ch/software/TMPRED_form.html**).

2.3. Subcellular Fractionation, Chemical Treatment, and Western Blot Analyses

Nicotiana benthamiana leaves were agroinfiltrated (OD_{600} = 0.5) with the pMOG800 construct carrying the CiLV-C2 *p29* or *MP* ORFs fused to the HA epitope. As controls, we used *N. benthamiana* leaves agroinfiltrated with constructs expressing the HA-tagged Lep protein, unfused GFP, and HA-tagged NSm of CSNV. At 3 days post-infiltration (dpi), approximately 1.5 g of *N. benthamiana* leaves expressing the tested proteins were ground in lysis buffer (20 mM de HEPES, pH 150 ;6.8 mM potassium acetate; 250 mM mannitol; 1 mM $MgCl_2$, and 50 μL of protease inhibitor cocktail for plant cell, Sigma–Aldrich, St. Louis, MO, USA). The homogenate was clarified by low centrifugation at 3000× *g* for 10 min at 4 °C, then the obtained supernatant was ultracentrifuged at 40,000× *g* for 40 min at 4 °C to yield the soluble (S30) and the crude (P30) microsomal fraction as reported previously [**7**,**24**]. Next, the P30 membrane-rich fraction was resuspended in lysis buffer, divided in six equal parts, and subjected to the chemical treatments with lysis buffer, 100 mM Na_2CO_3 (pH 4 ,2 ,(11, or 8 M urea, and held for 30 min on ice. All fractions were analyzed by western blot in %12 SDS-PAGE gels as referred to previously [**18**] using an anti-HA antibody or an anti-GFP antibody (Thermo Fischer Scientific, Waltham, MA, USA).

For Triton X-114 analysis, the P30 fraction was resuspended in the lysis buffer containing 1% Triton X-114 and incubated on ice for 30 min. The mixture was clarified by centrifugation at 10,000× *g* for 20 min at 4 °C. The resultant supernatant was incubated at 37 °C for 10 min and centrifuged at 10,000× *g* for 10 min at room temperature to form the aqueous (AP) and organic phases (OP). Finally, the OP was resuspended in lysis buffer with the same volume obtained in the aqueous phase, and the fractions were analyzed by western blot in %12 SDS-PAGE gels, as above mentioned. Percentage values of relative protein accumulation was measured using Fiji ImageJ program Ver 2.0 with ISAC plugin. A reference value equivalent to %100 was given based on the pixel quantification of the controls treated with lysis buffer.

2.4. Bimolecular Fluorescence Complementation Assays

Agrobacterium tumefaciens cultures (C58) transformed with the constructs carrying the p29 or MP with the N-YFP or C-YFP fragments were co-infiltrated (OD_{600} = 0.4) with agrobacterium cultures carrying BiFC constructs containing the counterparts of the YFP addressed to cytosol/nucleus (N-YFPcyt and C-YFPcyt) or to the lumen of the endoplasmic reticulum (ER) (C-YFPer and N-YFPer) in *N. benthamiana* leaves. The plants were maintained in a FITOTRON growth chamber under conditions of 24 °C day/18 °C night at 16 h light/8 h dark and %70 humidity regime. The preparation for agroinfiltration was conducted as described previously [7]. At 4 dpi, fluorescence reconstitution was observed. To increase the expression, in order to allow a better visualization of the fluorescence signal, all protein pair combinations were co-expressed with the silencing suppressor HCPro from the tobacco etch virus.

2.5. Confocal Laser Scanning Microscopy

Fluorescence images of the leaf discs from *N. benthamiana* were captured with the aid of a confocal laser scanning microscope Zeiss LSM 780 model. YFP was excited at 514 nm and emission was captured at 560–520 nm. The images were prepared using Fiji ImageJ program (version 2.0r).

3. Results and Discussion

3.1. The CiLV-C2 p29 and MP Are Membrane-Associated Proteins

Computer analysis from deduced amino acid sequences of the CiLV-C2 p29 and MP was performed using several computer tools. In the analysis, we also included the deduced amino acid sequences of the CSNV NSm and TMV 30K MP, which have been shown to be peripheral membrane proteins [**18,19**], as well as for the Lep, CiLV-C MP and GFP, two integral membrane-, and one non-membrane-associated proteins, respectively [**7,19,24**]. Highlighted, they revealed the presence of three hydrophobic regions (HR) for CiLV-C2 MP that encompass the residues 79–97 (HR1), 113–131 (HR2), and 167–185 (HR3), and two for p29: 3–21 (HR1) and 178–196 (HR2) (**Figure 1** and **Table S1**). Two transmembrane-spanning domains were predicted for the CiLV-C2 MP: 76–98 (TM1) and 160–170 (TM2), which overlap in part with the predicted HR1 and HR3, respectively (**Figure S1** and **Table S1**). Complete predictions for all proteins are presented in **Table S1**. Taken together, these analyses suggested that CiLV-C2 p29 and MP are potentially membrane-associated proteins.

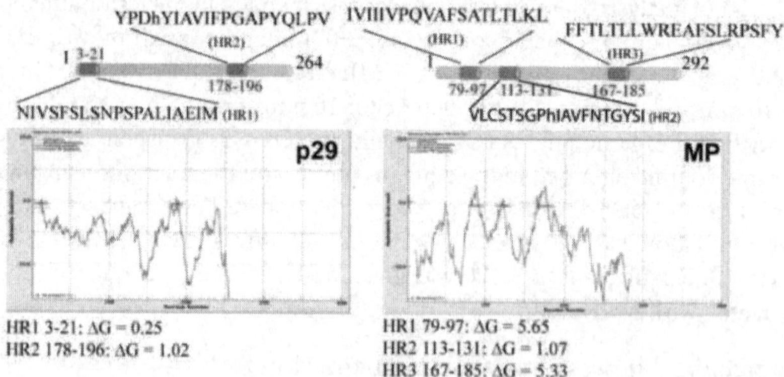

Figure 1. Hydrophobic prediction analyses of the citrus leprosis virus C2 (CiLV-C2) movement protein (MP) and capsid protein (p29). Hydrophobic regions (HR) were predicted for p29 and MP. A schematic representation of the proteins highlighting the HRs can be found at the top of each picture (in orange). Hydrophobic profile of the proteins is shown in the graphics generated with MPEx tool. The red lines show the mean values using a window of 19 residues and the yellow line indicates the predicted HRs. Values of ΔG of the HR are indicated.

In order to confirm these predictions and examine the membrane association of CiLV-C2 p29 and MP, we prepared subcellular microsomal fraction from *N. benthamiana* leaves transiently expressing the CiLV-C2 p29 or MP ORFs fused to the HA epitope. As controls, we used *N. benthamiana* leaves agroinfiltrated with constructs expressing the HA-tagged

Lep protein, unfused GFP, and HA-tagged NSm of CSNV, which corresponded to the transmembrane-, cytosolic-, and membrane-associated proteins, respectively. High-speed ultracentrifugation was performed to separate the plant leaf lysed extract, containing the abovementioned proteins, into pellet (P30) and supernatant (S30) fractions. Treatment with Na_2CO_3 is known to render microsomes into membrane sheets, releasing soluble luminal proteins [**20**], while urea treatment should release all polypeptides bound to the membrane, except for the integral membrane proteins [**21,24**]. We observed that the HA-tagged p29 and MP remained mostly associated with the P30 membranous fraction after Na_2CO_3 treatment (**Figure 2**A, P30 78% for p29, and P30 99% for MP), suggesting that these proteins are tightly associated with the membrane. The Lep and NSm controls remained in the membranous fraction (**Figure 2**A, P30 100% Na_2CO_3) and the majority of the GFP protein remained in the soluble fraction (**Figure 2**A, P30 79% Na_2CO_3), as expected. After the urea treatments and specially after the more aggressive treatment with 8 M urea, the HA-tagged MP and Lep proteins remained in the pellet fraction (100% for both proteins), indicating that the CiLV-C2 MP is fully integrated to cell membranes. On the other hand, a considerable proportion of the HA-tagged p29 protein was observed to be associated with the soluble fraction (43%) after 8 M urea treatment. Similar behavior was observed for the CSNV NSm control (79%, 8 M) suggesting that p29 is peripherally associated with the membrane.

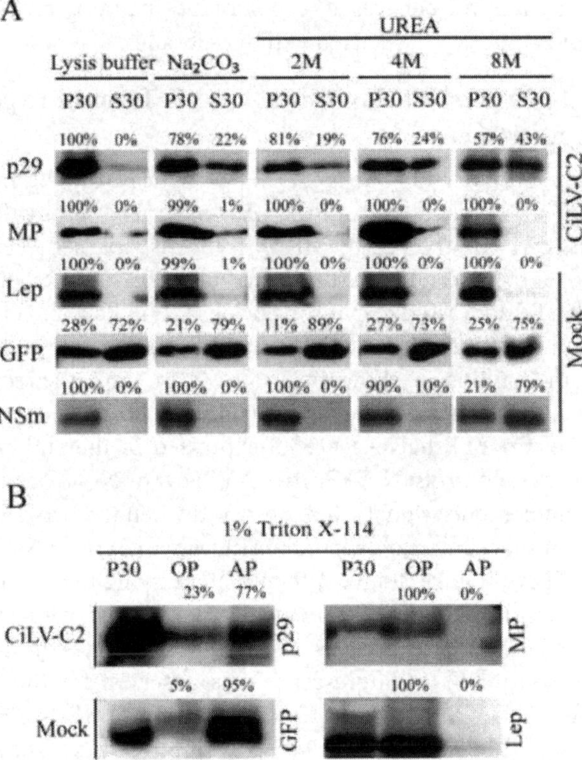

Figure 2. Membrane association analysis of CiLV-C2 p29 and MP proteins. (A) Segregation into membranous and soluble fraction of p29 and MP proteins expressed

in planta. All analyzed proteins were expressed in *Nicotiana benthamiana* leaves by agroinfiltration. As controls, we used leaf protein extracts containing free enhanced green fluorescent protein (eGFP) (non-membrane), the hemagglutinin (HA)-tagged Lep (leader peptidase) (integral membrane), and HA-tagged NSm of chrysanthemum stem necrosis orthotospovirus (CSNV) (peripheral membrane) proteins. The supernatant from ultracentrifugation after membrane fractioning (S30) and comparable pellet (P30), untreated or submitted to alkaline or urea (4 ,2, or 8 M) treatments, were analyzed by western blot using anti-NtGFP antibody or anti-HA antibody (Thermo Fisher Scientific). Relative quantification values are presented. (B) Triton X114- partitioning assay of CiLV-C2 p29 and MP. The P30 fractions subjected to treatment with Triton X114- were separated in aqueous (AP) and organic (OP) phases. Equivalent amounts of fractions were analyzed by western blot. The GFP and Lep proteins were used as controls.

To confirm the cell membrane association of the p29 and MP proteins, the P30 membrane-rich fraction was treated with Triton X-114. This treatment generated two phases, the aqueous (AP) and organic phases (OP), in which the integral membrane proteins should be portioned into the OP, meanwhile the AP should contain non-integral membrane proteins and soluble proteins [**25**]. As expected for an integral membrane protein, the Lep and MP were recovered from the OP (**Figure 2**B); meanwhile, p29 and GFP were mostly recovered from the AP (77% for p29 and 95% for GFP). Taken together, these findings indicate that the MP behaves as a membrane-spanning protein and the p29 is physically—but not integrally—associated with membranes.

3.2. The CiLV-C2 p29 and MP Have the N- and C-Termini Exposed to the Cell Cytoplasmic Compartment

In the next step, the subcellular compartments in which the N- or C-termini of the p29 and MP proteins are exposed were analyzed by BiFC. *A. tumefaciens* cultures (C58) transformed with the constructs carrying the p29 or MP with the N-YFP or C-YFP fragments were co-infiltrated with agrobacterium cultures carrying BiFC constructs containing the counterparts of the YFP addressed to cytosol/nucleus (N-YFPcyt and C-YFPcyt) or to the lumen of the endoplasmic reticulum (C-YFPer and N-YFPer) in *N. benthamiana* leaves. All protein pair combinations are shown in **Table S2**. At 4 dpi, the reconstitution of the fluorescent-competent YFP structure was visualized. Fluorescence reconstitution was visualized when the two YFP halves were co-expressed in the endoplasmic reticulum (ER) or cytosol (positive controls: N-YFPcyt + C-YFPcyt or N-YFPer + C-YFPer; **Figure 3**i,ii); however, no fluorescence signals were observed when the two YFP halves were co-expressed in different subcellular compartments (negative controls: N-YFPcyt + C-YFPer or N-YFPer + C-YFPcyt; **Figure 3**iii,iv). When p29, carrying the N-YFP fused at its N- (N-YFP-p29) or C-terminus (p29-N-YFP), was co-expressed with the C-YFPcyt, the reconstitution of the YFP fluorescence was observed in aggregates through the cytoplasm (**Figure 3**v,vi). Cytoplasmic YFP fluorescence was observed for the MP, carrying the N-YFP fragment fused at its N- (N-YFP-MP) or C-terminus (MP-N-YFP-MP), when co-expressed with C-YFPcyt in *N. benthamiana* leaves (**Figure 3**ix,x). The fluorescent signal was also visualized into the nucleus for the MP-N-YFP + C-YFPcyt combination (**Figure 3**ix, red arrows). No fluorescence signal was visualized when the p29 and MP constructs were co-expressed with the N-YFP or C-YFP fragments targeted to the ER (**Figure** 3vii,xi). Taken together, these findings indicate that both the N- and C-termini of p29 and MP are exposed to the cytoplasmic face. Furthermore, when MP is present within the nucleus, its C-terminus is exposed to the inner nuclear membrane, corroborating the previous findings of the cileviruses MP, to access the cell nucleus [**10**].

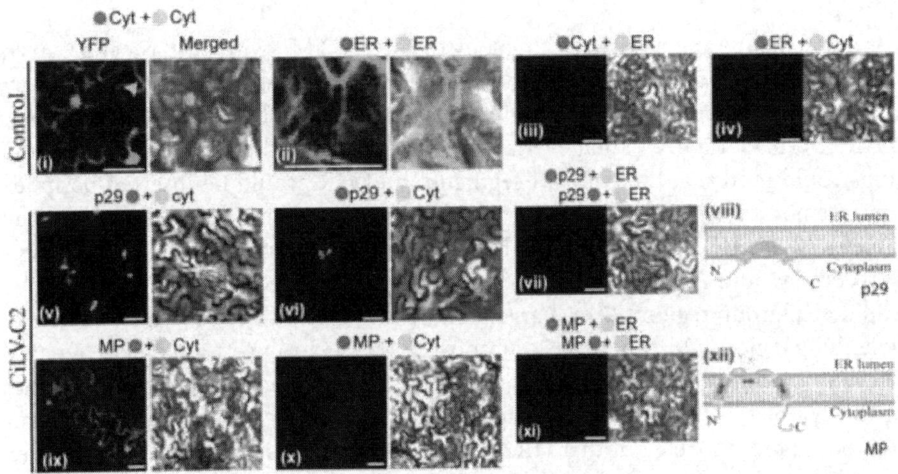

Figure 3. Membrane association and topology of the CiLV-C2 p29 and MP proteins. Subcellular localization (cytosolic face orER lumen) of the N- or C-termini of the CiLV-C2 p29 and MP proteins. The proteins carrying the N-terminal (•) or C-terminal (•) YFP fragments fused at their N-(•/•-ORFs) or C-termini (ORFs-•/•) were transiently co-expressed in *N. benthamiana* leaves with the corresponding complementary yellow fluorescent protein (YFP) fragment addressed to the cytosol face (•-cyt or •-cyt) or the lumen of the ER (•-ER or •-ER). Positive and negative controls are presented in the pictures (i–iv). Images reveal the topology of the C-termini (v,ix) or N-termini (vi,x) of the respective CiLV-C2 proteins. Negative YFP signal was observed when the CiLV-C2 p29 and MP carrying the N-YFP fragment at their N- or C-termini were co-expressed with counterpart YFP fragment (C-YFP) addressed to the ER lumen (vii,xi). Blue and red arrows indicate the cell cytoplasm and nucleus, respectively. Hypothetical topologic models for the p29 and MP proteins are represented (viii,xii). HR, hydrophobic region. All images contain two pictures corresponding to the YFP signal or merged with bright field. The fluorescence was monitored at four days post-infiltration using a confocal Zeiss LSM 780 model. Bars correspond to 50 μm.

These results, together with the subcellular fractionation assays and prediction analysis, allowed us to propose a topological model of association of the CiLV-C2 p29 and MP with cell membranes. Our model posited that the p29 is peripherally associated with the membrane, while the MP is an integral protein, whereby the full-length molecules for both proteins are oriented towards the cytoplasmic face of the biological membrane (**Figure** 3viii,xii).

The membrane coupling capacity shown for CiLV-C2 p29, also observed previously for the CiLV-C p29, could justify the capacity of the cilevirus p29 to move along the ER system [7], suggesting their putative involvement in the intracellular viral spread and/or the initiation of viral replication in the newly infected cells [**10**].

3.3. The MPs of CiLV-C2 and CiLV-C Could Have a Similar Membrane Topology

The integral membrane-associated topology of CiLV-C2 MP with both N- and C-termini exposed to cytosolic face suggests the presence of at least two membrane-spanning domains. It is possible that the HR1 and HR3 strongly hydrophobic regions (HR1 ΔG 5.65 and HR3 ΔG 5.33, **Figure 1**) that overlap in part with the predicted transmembrane (TM) regions could represent the two TM segments, and HR2 could be inserted peripherally to cell membranes. In this sense, it is interesting to mention the presence of charged residues in TM2/HR3, which are rare but can be found in membrane-spanning regions [**26**] where Lys and Arg, in particular, can be accommodated by their "snorkeling" into the lipid headgroup region [**27**]. Charged amino acids are also consistently located at the TM flanking regions [**26**] where basic amino acids act as stronger topological signals than acidic amino acids [**28,29**], but both types of charged residues are predominantly located near the cytoplasmic end of the TM segments [**26**], as observed in the proposed model. However, we did not rule out that HR2 could also be a TM segment; thus, we presented all possible models of association of the MP with the membrane (**Figure S2**). In contrast to CiLV-C2 MP membrane orientation, Leastro et al. [7] hypothesized that CiLV-C MP could be a multi-pass membrane protein with three TM segments, exposing the N- and C-termini to the ER-lumen and the cytosol, respectively. This hypothesis was suggested due to the absence of YFP reconstitution signal from the CiLV-C MP N-terminus to any of the evaluated subcellular compartments, probably due to an incompatible right orientation of the two YFP fragments, and by the prediction of three hydrophobic domains in the protein [7]. Given the cytoplasmic exposure of the N-terminus of CiLV-C2 MP (suggesting the presence of two TM regions) and the partial similarity between both cileviruses MPs (see alignment, **Figure S1**), it is more likely that CiLV-C MP topology behaves just like the models proposed herein for CiLV-C2 MP.

3.4. Cilevirus MPs Are the First Members of the 30K Superfamily Showing a Transmembrane Association Pattern

The CiLV-C2 MP belongs to the 30K superfamily [**10**]. This family includes MPs from DNA and RNA viruses that show conserved motives, corresponding to seven predicted β-strands connected by putative loops with different sequence patterns and a conserved aspartic acid residue, called the "D motif", at the end of strand 3 [**30**]. Although there is similarity between the CiLV-C2 MP topology with other 30K MPs [**18,19,21,31,32**], the transmembrane association-pattern identified here is rarely noticed for members of this family. Almost all 30K MPs have been presented as proteins peripherally associated with the membrane [**18,19,21**]; the exception is the genetically related CiLV-C MP [7]. Interestingly, the transmembrane association pattern observed for the MP of both cileviruses has been noticed for viral movement factors not belonging to the 30K superfamily [**14,16,20**]. The open question is to understand why this model diverges from the rest of the MPs assigned to the 30K superfamily and if this feature results or not in a biological fitness or benefit. It is worth mentioning that, although both CiLV-C and CiLV-C2 MPs have been proven to be functional to rescue the defective alfalfa mosaic virus (AMV), turnip crinkle virus (TCV), and TMV cell-to-cell movement-mutants, including the systemic transport of AMV [**10**,**24**], the cileviruses do not move systemically in their natural or experimental plant hosts [**4**]. It has been hypothesized that members of this genus evolved from an ancestor arthropod virus that became capable of infecting plants after acquiring the movement protein from one or more plant virus(es) [**4**]. Although we have strong evidence that the cileviruses systemic movement limitation is not due the functional restrictions in their MPs [**10**], why cileviruses do not have the ability to move systemically within their plant hosts remains unanswered.

4. Conclusions

The capsid protein (p29) of CiLV-C2 is peripherally associated with cell membranes with the N- and C-termini exposed to the cytosol. The movement protein (MP) of CiLV-C2 is a transmembrane protein with the N- and C-termini exposed to the cytosol. The amino acid sequence analysis suggested that both CiLV-C2 and CiLV-C MPs could have a similar transmembrane topology. The cilevirus MPs are the first members of the 30K superfamily showing a transmembrane association pattern.

Supplementary Materials

The following are available online at **https://www.mdpi.com/2076-2607/9/2/418/s1**.

References

1. Bastianel, M.; Novelli, V.; Kitajima, E.; Kubo, K.; Bassanezi, R.B. Citrus leprosis: Centennial of an unusual mite virus pathosystem. *Plant Dis.* **2010**, *94*, 284–292. [**Google Scholar**] [**CrossRef**] [**PubMed**] [**Green Version**]

2. Roy, A.; Choudhary, N.; Guillermo, L.M.; Shao, J.; Govindarajulu, A.; Achor, D.; Wei, G.; Picton, D.D.; Levy, L.; Nakhla, M.K.; et al. A novel virus of the genus Cilevirus causing symptoms similar to citrus leprosis. *Phytopathology* **2013**, *103*, 488–500. [**Google Scholar**] [**CrossRef**] [**Green Version**]

3. Roy, A.; Hartung, J.S.; Schneider, W.L.; Shao, J.; Leon, G.; Melzer, M.J.; Beard, J.J.; Otero-Colina, G.; Bauchan, G.R.; Ochoa, R.; et al. Role Bending: Complex Relationships between Viruses, Hosts, and Vectors Related to Citrus Leprosis, an Emerging Disease. *Phytopathology* **2015**, *105*, 1013–1025. [**Google Scholar**] [**CrossRef**] [**PubMed**] [**Green Version**]

4. Freitas-Astua, J.; Ramos-Gonzalez, P.L.; Arena, G.D.; Tassi, A.D.; Kitajima, E.W. Brevipalpus-transmitted viruses: Parallelism beyond a common vector or convergent evolution of distantly related pathogens? *Curr. Opin. Virol.* **2018**, *33*, 66–73. [**Google Scholar**] [**CrossRef**] [**PubMed**]

5. Leon, M.G.; Becerra, C.H.; Freitas-Astua, J.; Salaroli, R.B.; Kitajima, E.W. Natural Infection of Swingleaglutinosa by the Citrus leprosis virus Cytoplasmic Type (CiLV-C) in Colombia. *Plant Dis.* **2008**, *92*, 1364. [**Google Scholar**] [**CrossRef**] [**PubMed**]

6. Pascon, R.C.; Kitajima, J.P.; Breton, M.C.; Assumpcao, L.; Greggio, C.; Zanca, A.S.; Okura, V.K.; Alegria, M.C.; Camargo, M.E.; Silva, G.G.; et al. The complete nucleotide sequence and genomic organization of Citrus Leprosis associated Virus, Cytoplasmatic type (CiLV-C). *Virus Genes* **2006**, *32*, 289–298. [**Google Scholar**] [**CrossRef**] [**PubMed**]

7. Leastro, M.O.; Kitajima, E.W.; Silva, M.S.; Resende, R.O.; Freitas-Astua, J. Dissecting the Subcellular Localization, Intracellular Trafficking, Interactions, Membrane Association, and Topology of Citrus Leprosis Virus C Proteins. *Front. Plant Sci.* **2018**, *9*, 1299. [**Google Scholar**] [**CrossRef**] [**PubMed**] [**Green Version**]

8. Locali-Fabris, E.C.; Freitas-Astua, J.; Souza, A.A.; Takita, M.A.; Astua-Monge, G.; Antonioli-Luizon, R.; Rodrigues, V.; Targon, M.L.; Machado, M.A. Complete nucleotide sequence, genomic organization and phylogenetic analysis of Citrus leprosis virus cytoplasmic type. *J. Gen. Virol.* **2006**, *87*, 2721–2729. [**Google Scholar**] [**CrossRef**]

9. Leastro, M.O.; Castro, D.Y.O.; Freitas-Astua, J.; Kitajima, E.W.; Pallas, V.; Sanchez-Navarro, J.A. Citrus Leprosis Virus C Encodes Three Proteins with Gene Silencing Suppression Activity. *Front. Microbiol.* **2020**, *11*, 1231. [**Google Scholar**] [**CrossRef**]

10. Leastro, M.O.; Freitas-Astua, J.; Kitajima, E.W.; Pallas, V.; Sanchez-Navarro, J. Unravelling the involvement of cilevirus p32 protein in the viral transport. *Sci. Rep.* **2021**, *11*, 2943. [**Google Scholar**] [**CrossRef**]

11. Kuchibhatla, D.B.; Sherman, W.A.; Chung, B.Y.; Cook, S.; Schneider, G.; Eisenhaber, B.; Karlin, D.G. Powerful sequence similarity search methods and in-depth manual analyses can identify remote homologs in many apparently "orphan" viral proteins. *J. Virol.* **2014**, *88*, 10–20. [**Google Scholar**] [**CrossRef**] [**Green Version**]

12. Laliberte, J.F.; Sanfacon, H. Cellular remodeling during plant virus infection. *Annu. Rev. Phytopathol.* **2010**, *48*, 69–91. [**Google Scholar**] [**CrossRef**]

13. Schwartz, M.; Chen, J.; Lee, W.M.; Janda, M.; Ahlquist, P. Alternate, virus-induced membrane rearrangements support positive-strand RNA virus genome replication. *Proc. Natl. Acad. Sci. USA* **2004**, *101*, 11263–11268. [**Google Scholar**] [**CrossRef**] [**PubMed**] [**Green Version**]

14. Genovés, A.; Pallás, V.; Navarro, J.A. Contribution of topology determinants of a viral movement protein to its membrane association, intracellular traffic, and viral cell-to-cell movement. *J. Virol.* **2011**, *85*, 7797–7809. [**Google Scholar**] [**CrossRef**] [**PubMed**] [**Green Version**]

15. Schepetilnikov, M.V.; Solovyev, A.G.; Gorshkova, E.N.; Schiemann, J.; Prokhnevsky, A.I.; Dolja, V.V.; Morozov, S.Y. Intracellular targeting of a hordeiviral membrane-spanning movement protein: Sequence requirements and involvement of an unconventional mechanism. *J. Virol.* **2008**, *82*, 1284–1293. [**Google Scholar**] [**CrossRef**] [**Green Version**]

16. Martinez-Gil, L.; Johnson, A.E.; Mingarro, I. Membrane insertion and biogenesis of the Turnip crinkle virus p9 movement protein. *J. Virol.* **2010**, *84*, 5520–5527. [**Google Scholar**] [**CrossRef**] [**PubMed**] [**Green Version**]

17. Pitzalis, N.; Heinlein, M. The roles of membranes and associated cytoskeleton in plant virus replication and cell-to-cell movement. *J. Exp. Bot.* **2017**, *69*, 117–132. [**Google Scholar**] [**CrossRef**] [**PubMed**]

18. Leastro, M.O.; Pallas, V.; Resende, R.O.; Sanchez-Navarro, J.A. The movement proteins (NSm) of distinct tospoviruses peripherally associate with cellular membranes and interact with homologous and heterologous NSm and nucleocapsid proteins. *Virology* **2015**, *478*, 39–49. [**Google Scholar**] [**CrossRef**] [**Green Version**]

19. Peiro, A.; Martinez-Gil, L.; Tamborero, S.; Pallas, V.; Sanchez-Navarro, J.A.; Mingarro, I. The Tobacco mosaic virus movement protein associates with but does not integrate into biological membranes. *J. Virol.* **2014**, *88*, 3016–3026. [**Google Scholar**] [**CrossRef**] [**PubMed**] [**Green Version**]

20. Peremyslov, V.V.; Pan, Y.W.; Dolja, V.V. Movement protein of a closterovirus is a type III integral transmembrane protein localized to the endoplasmic reticulum. *J. Virol.* **2004**, *78*, 3704–3709. [**Google Scholar**] [**CrossRef**] [**Green Version**]

21. Martínez-Gil, L.; Sánchez-Navarro, J.A.; Cruz, A.; Pallás, V.; Pérez-Gil, J.; Mingarro, I. Plant virus cell-to-cell movement is not dependent on the transmembrane disposition of its movement protein. *J. Virol.* **2009**, *83*, 5535–5543. [**Google Scholar**] [**CrossRef**] [**Green Version**]

22. Leastro, M.O.; Pallas, V.; Resende, R.O.; Sanchez-Navarro, J.A. The functional analysis of distinct tospovirus movement proteins (NSM) reveals different capabilities in tubule formation, cell-to-cell and systemic virus movement among the tospovirus species. *Virus. Res.* **2017**, *227*, 57–68. [**Google Scholar**] [**CrossRef**]

23. Snider, C.; Jayasinghe, S.; Hristova, K.; White, S.H. MPEx: A tool for exploring membrane proteins. *Protein Sci. Publ. Protein Soc.* **2009**, *18*, 2624–2628. [**Google Scholar**] [**CrossRef**] [**Green Version**]

24. Leastro, M.O.; Freitas-Astua, J.; Kitajima, E.W.; Pallas, V.; Sanchez-Navarro, J.A. Dichorhaviruses Movement Protein and Nucleoprotein form a Protein Complex That May be Required for Virus Spread and Interacts in vivo with Viral Movement-Related Cilevirus Proteins. *Front. Microbiol.* **2020**, *11*, 571807. [**Google Scholar**] [**CrossRef**] [**PubMed**]

25. Bordier, C. Phase separation of integral membrane proteins in Triton X-114 solution. *J. Biol. Chem.* **1981**, *256*, 1604–1607. [**Google Scholar**] [**CrossRef**]

26. Baeza-Delgado, C.; Marti-Renom, M.A.; Mingarro, I. Structure-based statistical analysis of transmembrane helices. *Eur. Biophys. J. EBJ* **2013**, *42*, 199–207. [**Google Scholar**] [**CrossRef**] [**PubMed**] [**Green Version**]

27. Ojemalm, K.; Higuchi, T.; Lara, P.; Lindahl, E.; Suga, H.; von Heijne, G. Energetics of side-chain snorkeling in transmembrane helices probed by nonproteinogenic amino acids. *Proc. Natl. Acad. Sci. USA* **2016**, *113*, 10559–10564. [**Google Scholar**] [**CrossRef**] [**Green Version**]

28. Sauri, A.; Tamborero, S.; Martinez-Gil, L.; Johnson, A.E.; Mingarro, I. Viral membrane protein topology is dictated by multiple determinants in its sequence. *J. Mol. Biol.* **2009**, *387*, 113–128. [**Google Scholar**] [**CrossRef**]

29. Nilsson, I.; von Heijne, G. Fine-tuning the topology of a polytopic membrane protein: Role of positively and negatively charged amino acids. *Cell* **1990**, *62*, 1135–1141. [**Google Scholar**] [**CrossRef**]

30. Mushegian, A.R.; Elena, S.F. Evolution of plant virus movement proteins from the 30K superfamily and of their homologs integrated in plant genomes. *Virology* **2015**, *476*, 304–315. [**Google Scholar**] [**CrossRef**] [**PubMed**] [**Green Version**]

31. Melcher, U. The '30K' superfamily of viral movement proteins. *J. Gen. Virol.* **2000**, *81*, 257–266. [**Google Scholar**] [**CrossRef**] [**PubMed**]

32. Zhou, X.; Lin, W.; Sun, K.; Wang, S.; Zhou, X.; Jackson, A.O.; Li, Z. Specificity of Plant Rhabdovirus Cell-to-Cell Movement. *J. Virol.* **2019**, *93*. [**Google Scholar**] [**CrossRef**] [**Green Version**]

Chapter-13

Implication of the Whitefly Protein VPS Twenty Associated 1 (Vta1) in the Transmission of Cotton Leaf Curl Multan Virus

1. Introduction

Plant viruses pose considerable threats to the production of many crops in modern agriculture [**1**]. The majority of plant viruses are transmitted by insect vectors (vector borne) [**2**]. In the past decades, geminiviruses, a subgroup of vector-borne plant viruses, have caused extensive epidemics in many crops, most notably in developing countries [**3**]. Among the nine genera in the family *Geminiviridae*, *Begomovirus* is the largest genus, containing over 400 species [**4,5**]. Begomoviruses are transmitted by whiteflies of the *Bemisiatabaci* complex, which comprises over 40 cryptic species, in a persistent circulative manner [**2,3,6**].

So far, studies concerning whitefly transmission of begomoviruses have been mostly conducted with tomato yellow leaf curl viruses (TYLCV). As learned from these studies, once TYLCV is acquired by the whitefly, the virus goes through the food canal to reach the filter chamber, from where it crosses the midgut wall and reaches the whitefly hemolymph; the virus then infects whitefly primary salivary glands and is secreted with saliva during feeding [**7**]. During this process, TYLCV hijacks clathrin-mediated endocytosis and the endosomal network to cross the midgut barrier of the whitefly vector [**8,9**]. Moreover, the coat protein (CP) of TYLCV may interact with many whitefly proteins, thereby facilitating virus transmission [**3,10,11,12**]. Two recent reviews on whitefly transmission of begomoviruses indicate that the transmission efficiency of a given virus may vary with different whitefly species, and different viruses may be transmitted with disparate efficiencies by a given whitefly species [**3,10**]. These variations indicate that the transmission mechanisms among different whitefly–begomovirus combinations may vary, highlighting the need for unravelling transmission mechanisms with previously unexplored whitefly species or begomoviruses.

Cotton leaf curl Multan virus (CLCuMuV) is one of the major casual agents of cotton leaf curl disease, one of the most significant constraints in cotton production in South

Asia [**13**]. CLCuMuV was the major virus causing cotton leaf curl disease in South Asia in the 1990s and seemed to have been displaced by the Burewala strain of the cotton leaf curl Kokhran virus (CLCuKoV-Bur) at the beginning of this century [**13**,**14**]. In recent years, however, field surveys in India have revealed the rebound of CLCuMuV and the association of the recombinant variants of this virus with the breakdown of resistance in cotton [**15**,**16**]. The field surveys also indicate that in some regions in northwest India, CLCuMuV became the dominant virus in the cotton field, a sign of displacement of CLCuKoV-Bur by CLCuMuV [**17**].

Laboratory studies on the transmission of CLCuMuV by different whitefly species show that the virus can be efficiently transmitted by Asia II 1 and Asia II 7, two indigenous species of whiteflies from Asia, but can be hardly transmitted by other species of whiteflies, including MEAM1, Mediterranean (MED) and Asia 1 [**18**,**19**]. In this study, first we used split-ubiquitin yeast two-hybrid assay to identify Asia II 1 whitefly proteins that putatively interact with the CP of CLCuMuV. Next, we used yeast two-hybrid and pull-down assay to detect the interaction between CLCuMuV CP and several putative whitefly proteins, including vacuolar protein sorting-associated protein (Vps) twenty associated 1 (Vta1). We then used RNA interference to investigate the role of Vta1 in the acquisition and transmission of CLCuMuV by Asia II 1. In addition, we examined the function of Vta1 in MEAM1, a poor vector of CLCuMuV, in its transmission of the virus. Our findings provide new insights into the transmission of CLCuMuV by whiteflies.

2. Material and Methods

2.1. Plants, Insects, and Viruses

For plants, cotton (*Gossypium hirsutum* L. cv. Zhemian 1793 and Xinhai 21) and tobacco (*Nicotiana tabacum* L. cv. NC89) were used. Plants were grown in insect-proof greenhouses under natural lighting at controlled temperatures of 3 ± 25 °C and 14 h light/10 h darkness. Infectious clones of CLCuMuV isolate GD37 (GenBank accession number: JN968573) with its conjugated beta-satellite (GenBank accession number: JN968574) were introduced into 4–3 true-leaf-stage tobacco plants. Next, Asia II 1 transmission was used to obtained CLCuMuV-infected tobacco plants. Infection of plants was verified by symptom inspection and PCR detection of the virus using primers CLCuMuV-PCR-F and CLCuMuV-PCR-R (**Table S1**).

For insects, two whitefly species, namely Asia II 1 (mt*COI* GenBank accession number: DQ309077) and MEAM1 (mt*COI* GenBank accession number: KM821540), were used. Cultures of the two species were originally established from whiteflies collected from the field and have been maintained on cotton plants (cv. Zhemian 1793). Maintenance of whitefly cultures and all experiments were conducted in climate chambers at 2 ± 26 °C, in 14 h light/10 h darkness, and at %80–60 relative humidity. The purity of each of the whitefly cultures was monitored every 2 months using the mt*COI* PCR-RFLP technique and sequencing, as previously reported [**20**]. All female adult whiteflies were within 7 days post-emergence when used in experiments.

2.2. Yeast Two-Hybrid System

The split-ubiquitin yeast two-hybrid system (Dualsystems Biotech, Zurich, Switzerland) was used to identify Asia II 1 whitefly proteins that interact with the CLCuMuV CP [**11**].

A cDNA library of Asia II 1 whitefly was constructed in the prey plasmid, SfiI-digested pPR3-N, with the EasyClone cDNA library construction kit (Dualsystems Biotech, Zurich, Switzerland). The quality of the cDNA library was determined as per the kit manual. The titer of the cDNA library of Asia II 1 was over 2×10^6 cfu/19.5 μL, with an average insert size of over 1.0 kb, meeting the requirements of a standard cDNA library. The CLCuMuV CP gene was ligated into the bait plasmid pDHB1 using primers CLCuMuV-CP-pDHB-1 infusion-F and CLCuMuV-CP-pDHB-1infusion-R (**Table S1**). The recombinant plasmid pDHB1-CLCuMuV CP was introduced into yeast strain NMY51, and the expression of the CLCuMuV CP in yeast was verified by Western blotting using anti-TYLCV CP mouse monoclonal antibodies (mAb) (provided by Professor Xue-Ping Zhou, Institute of Biotechnology, Zhejiang University). Next, the cDNA library was introduced into yeast cells containing the pDHB1-CLCuMuV CP. Yeast clones were selected on triple dropout (TDO) medium (S.D./-His/-Leu/-Trp) containing 2.5 mM of 3-aminotriazole (3-AT). The yeast cells were then resuspended in 0.9% NaCl solution (to OD600 = 1.0) and later restreaked on quadruple dropout (QDO) medium (S.D./-Ade/-His/-Leu/-Trp) containing 2.5 mM of 3-AT to verify interactions. In addition, a yeast beta-Gal assay kit (Thermo Scientific, Waltham, MA, USA) was used to examine the interactions by detecting beta-galactosidase activity in yeast clones. Finally, plasmids were recovered from yeast and transformed into *Escherichia. coli* strain DH5α and then sequenced.

For the verification of interaction, plasmids recovered from yeast clones were transformed into yeast cells containing the pDHB1-CLCuMuV CP using the method described above. The full length of whitefly genes was cloned into the plasmid pPR3-N with primers Vta1-pPR3-N-infusion-F, Vta1-pPR3-N-infusion-R, pPR3-N-infusion-F, and pPR3-N-infusion-R (**Table S1**) and analyzed using procedures as described above.

2.3. Bioinformatics Analysis

The whitefly genes whose coding proteins were verified to interact with the CLCuMuV CP were annotated using BLAST (**http://blast.st-va.ncbi.nlm.nih.gov/Blast.cgi**). Gene Ontology (GO) enrichment analysis was then performed using the OmicShare tools (**http://www.omicshare.com/tools**).

2.4. Cloning of Vta1 in Asia II 1 and MEAM1

The predicted full length of MEAM1 *Vta1* was found on the NCBI database (Genbank accession code: LOC109029924). Therefore, we cloned MEAM1 *Vta1* and submitted it to the NCBI database under accession number NW380743. The full length of *Vta1* in the Asia II 1 whitefly was amplified by the SMARTer RACE '3/'5 kit (Clontech, Kyoto Japan), as per the manufacturer's protocol, with primers '5 race-Vta'5 ,1 race-Vta1, and '3 race-Vta-1CS1 (**Table S1**). Total RNAs extracted from the Asia II 1 whitefly using TRIzol reagent (Ambion, Waltham, MA, USA) were used in RACE. Next, Asia II 1 *Vta1* was cloned and submitted to the NCBI database under accession number MW346674.

2.5. Pull-Down Assay

The full length of the CLCuMuV CP gene was cloned into pGEX-6p-1 for fusion with glutathione S-transferase (GST) using primers CLCuMuV-CP-pGEX-6p-1-F and CLCuMuV-CP-pGEX-6p-1-R (**Table S1**). The full length of *Vta1* was cloned into pMAL-c5x for fusion with maltose-binding protein (MBP) using primers Vta-1pMAL-c5x-F and Vta-1pMAL-c5x-R (**Table S1**). Recombinant proteins were expressed in *E. coli* strain BL21. After purification, the GST-CLCuMuV CP and GST (control) were allowed to bind to glutathione agarose beads (GE Healthcare, Boston, MA, USA) for 2 h at 4 °C. The beads were then washed and incubated with MBP-Vta1 or MBP (control) at 4 °C for 4 h. Next, the beads were washed and boiled, and the bead-bound proteins were separated using SDS-PAGE and detected using Western blotting with anti-MBP rabbit polyclonal antibodies (pAb) (Abcam, Cambridge, UK).

2.6. Double Strand RNA (DsRNA) Synthesis and Membrane Feeding

DNA templates for dsRNA synthesis were amplified by PCR using primers with the T7 promoter at both ends, namely *Vta1* (Asia II 1)-T-7F, *Vta1* (Asia II 1)-T-7 R, *Vta1* (MEAM1)-T-7F, and *Vta1* (MEAM1)-T-7R (**Table S1**). DsRNA synthesis was conducted with a T7 high-yield RNA transcription kit (Vazyme, Nanjing, China). Next, dsRNA was purified, and the quality and concentration were determined using agarose gel electrophoresis and Nanodrop (Thermo Fisher, Waltham, MA, USA). For membrane feeding, dsRNA-targeting *Vta1* or *GFP* (control) was added to %15 sucrose solution to make the final concentration 200 ng/μL. Whiteflies were collected and released into artificial diet feeding chambers, as described before [**8**]. The duration of membrane feeding was 48 h.

2.7. Analysis of Gene Expression Level

Total RNAs of the whitefly were extracted using TRIzol reagent as per the manufacturer's instructions. cDNA was synthesized from 1 μg of RNA using the PrimeScript RT reagent kit (Takara, Kyoto, Japan). qPCR was performed on the CFX96™ Real-Time PCR Detection System (Bio-Rad, Hercules, CA, USA) with SYBR Premix Ex Taq II (TaKaRa, Kyoto, Japan). Primers β-actin-qPCR-F and β-actin-qPCR-R were used as internal controls. Vta1 (Asia II 1)-qPCR-F and Vta1 (Asia II 1)-qPCR-R were used for Asia II 1 *Vta1*, and Vta1 (MEAM1)-qPCR-F and Vta1 (MEAM1)-qPCR-R were used for MEAM1 *Vta1*. All primes are listed in **Table S1**.

2.8. Virus Acquisition and Quantification of CLCuMuV in the Whitefly

For virus acquisition, whiteflies were collected and allowed to feed on CLCuMuV-infected tobacco plants. Forty-eight hours later, virus quantification was performed. For the whitefly whole body, whitefly adults were collected as groups of 15 and lysed in lysis buffer (50 mM KCl, 10 mM Tris, 0.45% Tween 20, 0.2% gelatin, 0.45% NP40, 60 mg/mL proteinase K, with pH 8.4). As for organs, 4 midguts or primary salivary glands were collected as one sample, and hemolymph from 4 whiteflies was treated as one sample. The collection and preparation of whitefly organs were conducted, as described before [**18**]. qPCR was conducted, as described above, with primers β-actin-qPCR-F, β-actin-qPCR-R, CLCuMuV-qPCR-F, and CLCuMuV-qPCR-R (**Table S1**).

2.9. Virus Transmission

Plants of tobacco (cv. NC89) and cotton (cv. Xinhai 21) were used. When tobacco plants were used, female adult whiteflies that had fed on virus-infected plants for 48 h were collected as groups of 10 and then transferred to be placed on a 3–4 true-leaf-stage tobacco seedling using clip cages [**21**] to feed for 48 h. Three replicates were conducted, with each containing 6–10 plants. The whiteflies were then removed, and the plants were sprayed with imidacloprid (20 mg/L) to kill all the eggs. The infection status of test plants was examined by symptom inspection and PCR detection of viral DNAs, as described above, 30 days post-inoculation.

When cotton plants were used, female adult whiteflies that had fed on virus-infected plants for 72 h were collected as groups of 10 and then transferred to feed on a 1–2 true-leaf-stage cotton seedling (enclosed in leaf-clip cages) for 72 h. Three replicates were conducted, with each containing 6–10 plants of both tobacco and cotton. After that, observations were conducted using the same procedure as that for tobacco except that the infection status of the plants was conducted 70 days post-inoculation.

2.10. Statistical Analysis

All qPCR data were calculated using $2^{-\Delta Ct}$ as normalized to whitefly *actin*. For the comparison of gene expression level and quantity of viruses, an independent *t*-test was used. For the comparison of transmission efficiency, percentage data were arcsine-square-root-transformed for statistical analysis using an independent *t*-test and back-transformed for presentation. All statistical analyses were conducted using SPSS 20.0 Statistics (IBM, Armonk, NY, USA) and Microsoft Excel (version 2016, Microsoft, Redmond, WA, USA).

3. Results

3.1. Identification of Asia II 1 Whitefly Proteins That Interact with the CLCuMuV CP

We used the split-ubiquitin yeast two-hybrid system to identify proteins in the Asia II 1 whitefly that potentially interact with the CLCuMuV CP. The expression of the bait plasmid pDHB1-CLCuMuV CP in yeast was verified using Western blotting (**Figure S1**), showing the functionality of the bait plasmid pDHB1-CLCuMuV CP. After screening of the Asia II 1 whitefly cDNA library using this yeast two-hybrid system, more than 300 positive clones were isolated and the sequencing results indicated that plasmids in these positive clones encode 200 unique proteins. To examine the interactions between the CLCuMuV CP and whitefly proteins encoded by the plasmids recovered from positive clones, we chose 70 proteins to examine their interaction with the CLCuMuV CP using the yeast two-hybrid system. Of these proteins, 54 were found to interact with the CLCuMuV CP. A BLAST search of the NCBI database was then conducted for the 54 proteins, and the names of the protein sequences that share the highest homology were obtained (**Table 1**).

Table 1. List of Asia II 1 proteins that putatively interact with the CLCuMuV CP as identified by the yeast two-hybrid system.

No.	Accession	Protein Name
1	XP_018912286.1	gelsolin-related protein of 125 kDa-like
2	XP_018897713.1	complement component 1 Q subcomponent-binding protein, mitochondrial isoform X2
3	XP_018906592.1	aquaporin AQPcic-like
4	XP_018902985.1	dnaJ homolog subfamily C member 7

No.	Accession	Protein Name
5	XP_018896724.1	myosin regulatory light chain 2
6	XP_018913048.1	transmembrane protein 189
7	XP_018912886.1	biogenesis of lysosome-related organelles complex 1 subunit 5 isoform X4
8	XP_018902682.1	tubulin beta-1 chain
9	XP_018908832.1	vesicle-associated membrane protein 7 isoform X1
10	XP_018909667.1	ugar transporter SWEET1-like
11	XP_018905417.1	ornithine decarboxylase antizyme 1
12	XP_018903031.1	calmodulin isoform X1
13	XP_018896738.1	protein lifeguard 4-like
14	XP_018906525.1	hsp70-Hsp90 organizing protein 3-like
15	XP_018917594.1	glycosylphosphatidylinositol anchor attachment 1 protein
16	XP_018899531.1	vesicle-trafficking protein SEC22b
17	XP_018898178.1	chloride intracellular channel exc-4
18	XP_018901458.1	thioredoxin-2-like
19	XP_018911715.1	matrix metalloproteinase-14 isoform X1
20	XP_018905472.1	microtubule-associated protein futsch-like
21	XP_018899425.1	CD9 antigen
22	XP_018902269.1	protein YIPF6
23	XP_018917580.1	transport and Golgi organization protein 11 isoform X1
24	XP_018911339.1	translocation protein SEC62
25	XP_018900653.1	heat shock 70 kDa protein cognate 3
26	ADG03467.1	heat shock protein 20
27	XP_018903943.1	FAS-associated factor 1
28	XP_018914247.1	calcium-transporting ATPase sarcoplasmic/endoplasmic reticulum type isoform X1
29	XP_018910594.1	ABC transporter G family member 20-like
30	XP_018911729.1	dnaJ homolog subfamily C member 25-like
31	XP_018908958.1	heat shock 70 kDa protein cognate 4
32	XP_018903618.1	phosphate carrier protein, mitochondrial-like
33	XP_018907161.1	transmembrane protein 104 homolog
34	XP_018907534.1	elongation factor Tu-like
35	XP_018902665.1	vesicle transport through interaction with t-SNAREs homolog 1A
36	XP_018904067.1	vesicle-associated membrane protein 2-like isoform X1
37	XP_018908256.1	probable RNA polymerase II nuclear localization protein SLC7A6OS
38	XP_018896173.1	vacuolar protein sorting-associated protein vacuolar protein sorting-associated protein (Vps) twenty associated 1 homolog
39	XP_018898014.1	calreticulin
40	XP_018918053.1	microtubule-associated protein RP/EB family member 1-like

No.	Accession	Protein Name
41	XP_018896547.1	protein jagunal
42	XP_018901184.1	nucleolar GTP-binding protein 2
43	XP_018901406.1	splicing factor 45
44	XP_018913448.1	ras-related protein Rab6
45	XP_018897712.1	complement component 1 Q subcomponent-binding protein, mitochondrial isoform X1
46	XP_018908611.1	uncharacterized protein LOC109038112
47	XP_018911543.1	uncharacterized protein LOC109040175
48	XP_018910467.1	uncharacterized protein LOC109039442 isoform X2
49	XP_018916418.1	uncharacterized protein LOC109043611
50	XP_018906808.1	uncharacterized protein LOC109036858
51	XP_018897050.1	uncharacterized protein LOC109030509
52	XP_018910780.1	uncharacterized protein LOC109039648
53	XP_018905235.1	uncharacterized protein LOC109035881
54	XP_018908284.1	uncharacterized protein LOC109037882

3.2. Bioinformatics Analysis of Whitefly Proteins That Interact with the CLCuMuV CP

Using GO analysis, the whitefly proteins that putatively interact with the CLCuMuV CP were assigned to 13 biological processes (BPs) or seven molecular functions (MFs) or two cellular components (CCs) (**Figure 1**). Specifically, of the 13 BPs, cellular process contained the majority of proteins (19). Further analysis of these proteins indicated that some proteins can be further classified into vesicle-mediated transport (GO: 0016192), including calmodulin (CALM), vesicle transport protein SEC22 (SEC22), vacuolar protein sorting-associated protein (Vps) twenty associated 1 (Vta1), Ras-related protein Rab-6A (RAB6A), and vesicle-associated membrane protein 7 (VAMP7).

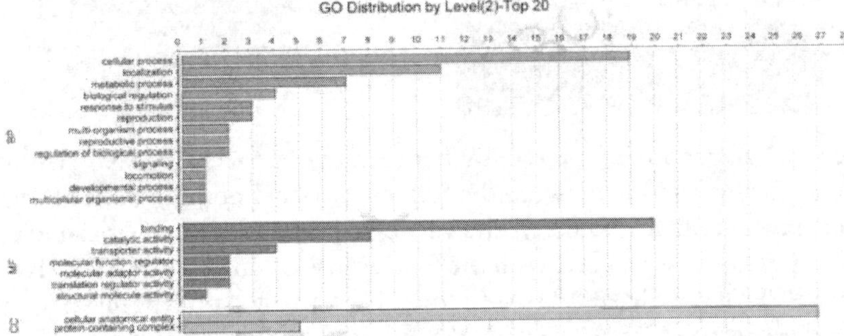

Figure 1. Gene Ontology (GO) analysis of the 54 Asia II 1 whitefly proteins that putatively interact with the CLCuMuV CP, as indicated by the yeast two-hybrid system. Different colors represent different GO categories (BP: biological process; MF: molecular function; CC: cellular components). GO annotation was conducted using Blast2GO software, and the figure was generated using OmicShare tools (http://www.omicshare.com/tools).

3.3. Verification of Interaction Using Yeast Two-Hybrid and Pull-Down Systems

For verification of the interaction between the CLCuMuV CP and prey proteins, we selected the five proteins in the GO category vesicle-mediated transport (GO:0016192) for further analysis using the yeast two-hybrid system. The full-length open reading frames of the five genes were amplified and ligated into pPR3-N, and then these prey plasmids and the bait plasmid pDHB1-CLCuMuV CP were co-transformed into yeast cells. Among the five proteins, the full lengths of calmodulin (CALM), vesicle transport protein SEC22 (SEC22), and vacuolar protein sorting-associated protein (Vps) twenty associated 1 (Vta1) were found to interact with the CLCuMuV CP (**Figure 2**A). Since Vta1 has been reported to regulate virus–host interactions, we subjected it to further analysis [**22,23**]. Analysis of beta-gal activity confirmed the interaction between Vta1 and the CLCuMuV CP (**Figure 2**B). And in the pull-down assay, when the fusion protein GST-CLCuMuV CP was used as a bait protein and the fusion protein MBP-Vta1 used as the prey protein, the prey protein could co-elute with the GST-fused CLCuMuV CP (**Figure 2**C). These results suggest that Vta1 from Asia II 1 whitefly can interact with the CLCuMuV CP both in vivo and in vitro.

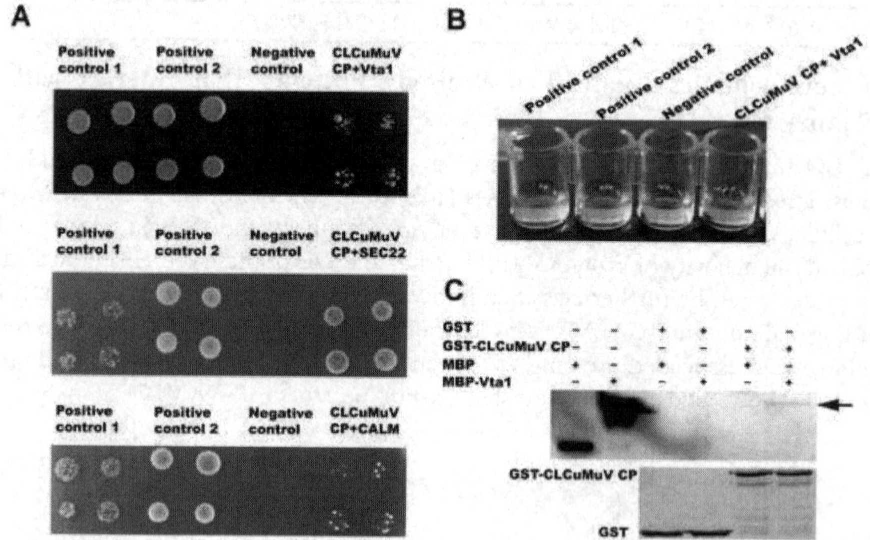

Figure 2. Verification of interaction between the CLCuMuV CP and whitefly proteins. The interaction between the full length of calmodulin (CALM), vesicle transport protein SEC22 (SEC22), and vacuolar protein sorting-associated protein (Vps) twenty associated 1 (Vta1) with the CLCuMuV CP in yeast (A). Positive control 1: pDHB1-CLCuMV-CP+pOst1-NubI; positive control 2: pDHB1-large T+pDSL-p53; negative control: pDHB1-CLCuMV-CP+pPR3-N. Beta-gal activity in yeast cells (B). Pull-down assay between CLCuMuV CP and Vta1 (C).

3.4. Functional Characterization of Vta1 in Asia II 1 Transmission of CLCuMuV

To examine the function of Vta1, Asia II 1 whiteflies were fed with *Vta1* dsRNA. Following dsRNA feeding, the expression of *Vta1* in whiteflies was down-regulated by %27.4 as compared to controls (**Figure 3**A). Next, the whiteflies were transferred to feed on CLCuMuV-infected tobacco plants for 48 h for virus acquisition. Knockdown of *Vta1* resulted in a significant decrease in the relative virus quantity in the whiteflies' whole body, midgut, and hemolymph but did not cause a significant change in the relative virus quantity in primary salivary glands (**Figure** 3B). Further, transmission trails were performed. When tobacco plants were used as test plants, knockdown of *Vta1* resulted in a significant decrease in CLCuMuV transmission, as shown by percentages of plants with viral symptoms but not by viral detection using PCR (**Figure** 4A,B). Similar results were found when cotton plants were used as test plants (**Figure** 4C,D). These results suggest that *Vta1* plays an important role in CLCuMuV acquisition and transmission by Asia II 1.

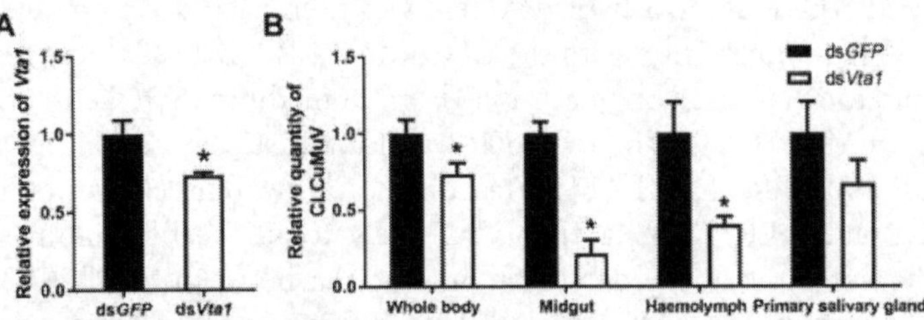

Figure 3. Effects of *Vta1* knockdown on CLCuMuV acquisition by Asia II 1. Gene expression level of *Vta1* following dsRNA feeding (*n* = 4 for ds*Vta1* or ds*GFP*) (A). Following feeding, virus quantity in whitefly whole body and organs (*n* = 8–3 for whole body and midgut, 22–19 for hemolymph, and 12 for primary salivary glands) (B). * stands for significant difference (independent *t*-test, $p < 0.05$).

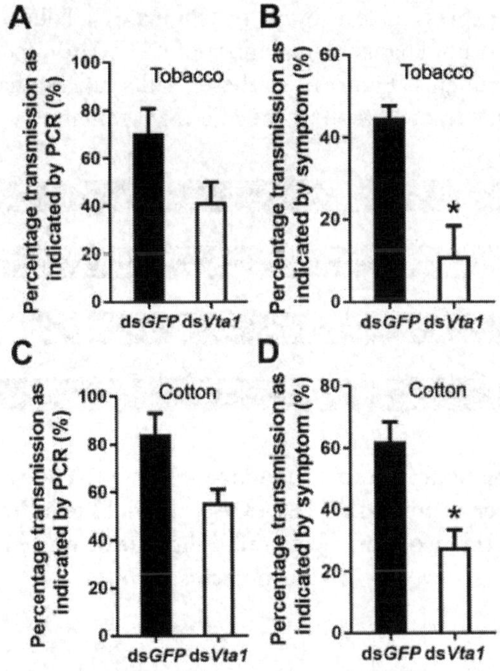

Figure 4. Effects of *Vta1* knockdown on CLCuMuV transmission by Asia II 1. Whiteflies that had acquired CLCuMuV were collected and transferred to feed on

tobacco (A,B) or cotton (C,D). On each of the two test plants, 3 replicates were conducted for both ds*GFP* and ds*Vta1*, with each replicate containing 6–10 plants. * stands for significant difference (independent *t*-test, $p < 0.05$).

3.5. The Role of Vta1 in CLCuMuV Transmission by MEAM1

Asia II 1 was able to readily transmit CLCuMuV, while MEAM1 can only transmit this virus with very low efficiency [**18**]. To explore whether *Vta1* plays a role in CLCuMuV transmission by MEAM1, we first compared the amino acid sequence of MEAM1 Vta1 with that of Asia II 1 Vta1 and found that six amino acids are different between the two Vta1s (**Figure 5**). The yeast two-hybrid system was then used to compare Asia II 1 Vta1 and MEAM1 Vta1 in interaction with the CLCuMuV CP. Yeast cells were resuspended to certain ODs and then cultured on quadruple dropout medium (SD/-Leu/-Trp/-His/-Ade) containing 2.5 mM of 3-AT. When OD600 was 1.0, yeast cells containing the pDHB1-CLCuMuV CP and pPR3-N-MEAM1 Vta1 did not grow, but cells containing pDHB1-CLCuMuV CP and pPR3-N-Asia II 1 Vta1 grew to noticeable colonies. To further determine whether there is any detectable interaction between the CLCuMuV CP and MEAM1 Vta1, we adjusted OD600 to 2.0. Under this condition, yeast cells containing the pDHB1-CLCuMuV CP and pPR3-N-MEAM1 Vta1 grew to colonies, but they were much smaller than those containing the pDHB1-CLCuMuV CP and pPR3-N-Asia II 1 Vta1 (**Figure 6**A). Analysis of beta-gal activity in yeast cells containing different combination of plasmids revealed that the interaction between MEAM1 Vta1 and the CLCuMuV CP was very weak, if any, as judged by the unappreciable yellow color in the solution (**Figure 6**B). Knockdown of MEAM1 *Vta1* resulted in significant down-regulation of the *Vta1* expression level by %44.4 (**Figure 6**C). Following virus acquisition, knockdown of *Vta1* did not change the quantity of CLCuMuV acquired by MEAM1 in two independent experiments (**Figure 6**D). These results suggest that *Vta1* plays a minor, if any, role in CLCuMuV transmission by the MEAM1 whitefly.

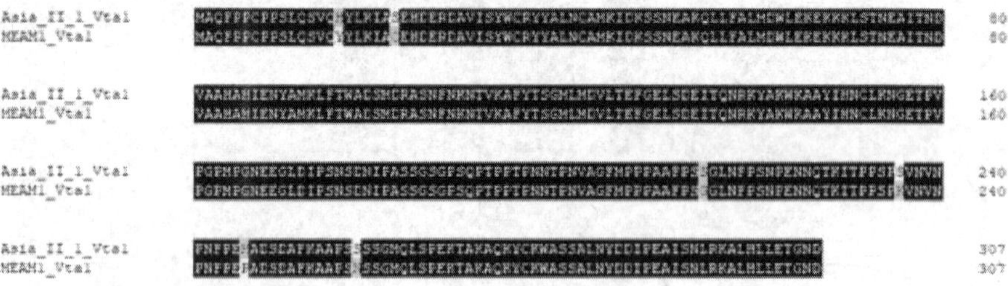

Figure 5. Comparison of amino acid sequences of Asia II 1 Vta1 and MEAM1 Vta1. Alignments were performed using DNAMAN. Dark blue indicates consensus in amino acids between the two sequences, and light blue or white indicates divergence in amino acids.

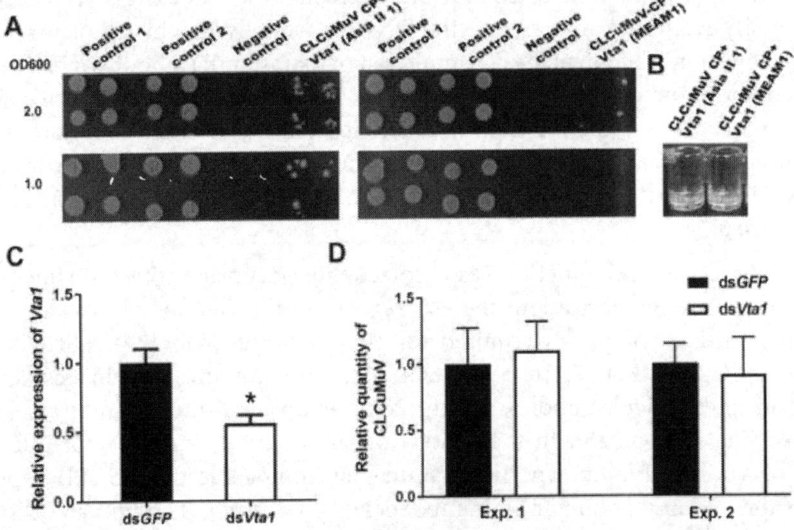

Figure 6. The role of *Vta1* in CLCuMuV transmission by MEAM1. The yeast two-hybrid assay was performed to compare the affinity of the CLCuMuV CP to Asia II 1 Vta1 and MEAM1 Vta1. Yeast cells containing different combination of plasmids were resuspended to OD600 being 1.0 and 2.0 and then cultured on quadruple dropout medium containing 2.5 mM of -3AT (A). Positive control 1: pDHB-1 CLCuMV-CP+pOst-1NubI; positive control 2: pDHB-1large T+pDSL-p53; negative control: pDHB-1CLCuMV-CP+pPR-3N. Yeast beta-gal assay kit was used to examine the interactions (B). Further, knockdown of MEAM1 *Vta1* was performed, and the whiteflies were then set to acquire CLCuMuV for 48 h. MEAM1 *Vta1* expression level ($n = 4$ for ds*GFP* or ds*Vta1*) (C), and quantity of CLCuMuV in whitefly whole body in two experiments ($n = 5$–4 for ds*GFP* or ds*Vta1* in each experiment) (D). * stands for significant difference (independent *t*-test, $p < 0.05$ for C,D).

4. Discussion

In this study, using the yeast two-hybrid system, we identified 54 candidate Asia II 1 proteins that putatively interact with the CLCuMuV CP (**Table 1**). GO enrichment analysis showed that these proteins may be responsible for 13 different biological processes (**Figure 1**). Based on the molecular function of the identified proteins, we selected five proteins to verify their interactions with the CLCuMuV CP using the yeast two-hybrid system and confirmed the interaction of Vta1 with the CLCuMuV CP using the GST pull-down assay (**Figure 2**). Next, we found that RNA interference of *Vta1* in Asia II 1 reduced the virus quantity in the whitefly and the efficiency of virus transmission (**Figure 3** and **Figure 4**). It should be noted that decreases in virus quantity were significant in the whitefly midgut and hemolymph but not in primary salivary glands upon *Vta1* silencing (**Figure 3**). This might be due to the fact that the silencing efficiency of *Vta1* is lower in primary salivary glands than that in the midgut and hemolymph. Moreover, knockdown of *Vta1* in Asia II 1 results in significant decreases in the virus transmission efficiency, as indicated by symptom inspection, but the decrease is not significant when examined by

PCR (**Figure 4**). The possible reason for the discrepancy is that PCR is very sensitive, so it may amplify even trace amounts of viral DNAs in whitefly-inoculated plants. However, symptom appearance requires the accumulation of a substantial amount of viruses, which is a better indicator of successful inoculation. In addition, we showed that MEAM1 Vta1, the sequence of which is slightly different from that of Asia II 1 Vta1, exhibited much lower affinity to the CLCuMuV CP than Asia II 1 Vta1, and RNA interference of MEAM1 *Vta1* did not affect the quantity of virus acquired by MEAM1 (**Figure 5** and **Figure 6**).

In eukaryotic cells, Vta1 functions as a cofactor of vacuolar protein sorting 4 (Vps4) by impacting its oligomerization, thereby regulating the activity of Vps4 to modulate endosomal sorting complexes required for transport (ESCRT) [**24**]. ESCRTs include ESCRT-0, ESCRT-I, ESCRT-II, and ESCRT-III and are involved in regulating the function of multivesicular bodies, which are an endosomal-membrane-trafficking and protein-sorting station [**25**]. In the context of virus–host or virus–vector interactions, the role of Vta1 is little known. The mammalian homologue of Vta1, LIP5, positively regulates the budding of human immunodeficiency virus type 1 in human cells [**23**]. In addition, deletion of MIT domains of *Spodoptera frugiperda* Vta1 reduces the replication of *Autographa californica* multiple nucleopolyhedrovirus [**22**]. In this study, we found that Vta1 positively regulates CLCuMuV transport across the midgut of Asia II 1 whiteflies following viral acquisition, as well as the efficiency of virus transmission. This may be due to the fact that Vta1 affects the ESCRT machinery via its action on Vps4, and the ESCRT machinery regulates vesicle trafficking, which has been shown to regulate whitefly transmission of begomoviruses [**8,9**]. Further, our recent findings show that the early endosome plays an important role in begomoviruses intracellular transport and an endocytic receptor can regulate vesicle transport of begomoviruses in whitefly midgut cells [**9,12**]. These results suggest that endosomal trafficking is important for the transport of both CLCuMuV and TYLCV across epithelial cells in the whitefly midgut. However, the unambiguous dissection of the role of endosomes and Vta1 in Asia II 1 transmission of CLCuMuV warrants further investigations.

For a given begomovirus, different whitefly species may transmit with disparate efficiencies [**3,10**]. Case studies have shown that this can be attributed to the differential capacity of the virus to cross barriers within the body of different whitefly species [**18,26,27**]. For example, the relatively high and low efficiencies in transmitting CLCuMuV by Asia II 1 and MEAM1 were found to be associated with the relatively high and low efficiencies of the virus to cross the midgut wall of the two species of whiteflies [**18**]. Here, we found that Vta1 from Asia II 1 positively regulates CLCuMuV transport across the midgut and transmission, and MEAM1 Vta1 does not seem to play a role in CLCuMuV transmission. Additionally, Asia II 1 Vta1 displays a stronger affinity to the CLCuMuV CP than MEAM1 Vta1. Hence, we propose that Vta1 may be a significant factor in determining the disparate efficiencies of CLCuMuV transmission by Asia II 1 and MEAM1, possibly through its different functions in facilitating the virus to cross physiological barriers such as the midgut in the vector body. Interestingly, a comparison of the amino acid sequences of MEAM1 Vta1 and Asia II 1 Vta1 showed that they differ in only six amino acids. The divergence at the six amino acids may directly contribute to the differential transmission of CLCuMuV by MEAM1 and Asia II 1. Of course, more empirical studies are needed to verify this hypothesis.

In summary, we identified 54 proteins from the Asia II 1 whitefly that putatively interact with the CLCuMuV CP. We showed that Asia II 1 Vta1 positively regulates the acquisition and transmission of CLCuMuV by the Asia II 1 whitefly. We further found that Vta1 from MEAM1, a poor vector of CLCuMuV, interacts with the CLCuMuV CP weakly and does not seem to play a role in CLCuMuV transmission by this whitefly. Taken together, our findings indicate that the protein Vta1 may be an important factor in determining the efficiency of CLCuMuV transmission by a given species of whitefly, and provide new insight into the interaction between single strand DNA (ssDNA) viruses and their insect vectors.

Supplementary Materials

Figure S1: The expression of pDHB1-CLCuMuV CP in yeast. Expression of CLCuMuV CP bait fusion protein was detected by western blot using anti-TYLCV CP antibodies; Table S1: Primers used in this study.

References

1. Lefeuvre, P.; Martin, D.P.; Elena, S.F.; Shepherd, D.N.; Roumagnac, P.; Varsani, A. Evolution and ecology of plant viruses. *Nat. Rev. Microbiol.* **2019**, *17*, 632–644. [**Google Scholar**] [**CrossRef**] [**PubMed**]
2. Hogenhout, S.A.; Ammar, E.; Whitfield, A.E.; Redinbaugh, M.G. Insect vector interactions with persistently transmitted viruses. *Annu. Rev. Phytopathol.* **2008**, *46*, 327–359. [**Google Scholar**] [**PubMed**] [**Green Version**]
3. Wang, X.; Blanc, S. Insect transmission of plant single-stranded DNA viruses. *Annu. Rev. Phytopathol.* **2021**, *66*, 389–405. [**Google Scholar**] [**CrossRef**] [**PubMed**]
4. Zerbini, F.M.; Briddon, R.W.; Idris, A.; Martin, D.P.; Moriones, E.; Navas-Castillo, J.; Rivera-Bustamante, R.; Roumagnac, P.; Varsani, A. ICTV Virus taxonomy profile: Geminiviridae. *J. Gen. Virol.* **2017**, *98*, 131–133. [**Google Scholar**] [**CrossRef**] [**PubMed**]
5. International Committee on Taxonomy of Viruses ICTV. Available online: **https://talk.ictvonline.org/ictv-reports/ictv_online_report/ssdna-viruses/w/geminiviridae** (accessed on 25 November 2020).
6. Kanakala, S.; Ghanim, M. Implication of the whitefly *Bemisiatabaci* cyclophilin B protein in the transmission of Tomato yellow leaf curl virus. *Front. Plant. Sci.* **2016**, *7*, 1702. [**Google Scholar**] [**CrossRef**] [**PubMed**] [**Green Version**]
7. Ghanim, M.; Morin, S.; Czosnek, H. Rate of Tomato yellow leaf curl virus translocation in the circulative transmission pathway of its vector, the whitefly Bemisiatabaci. *Phytopathology* **2001**, *91*, 188–196. [**Google Scholar**] [**CrossRef**] [**PubMed**] [**Green Version**]
8. Pan, L.L.; Chen, Q.F.; Zhao, J.J.; Guo, T.; Wang, X.W.; Hariton-Shalev, A.; Czosnek, H.; Liu, S.S. Clathrin-mediated endocytosis is involved in Tomato yellow leaf curl virus transport across the midgut barrier of its whitefly vector. *Virology* **2017**, *502*, 152–159. [**Google Scholar**] [**CrossRef**]

9. Xia, W.Q.; Liang, Y.; Chi, Y.; Pan, L.L.; Zhao, J.; Liu, S.S.; Wang, X.W. Intracellular trafficking of begomoviruses in the midgut cells of their insect vector. *PLoSPathog.* **2018**, *14*, e1006866. [**Google Scholar**] [**CrossRef**] [**Green Version**]

10. Fiallo-Olive, E.; Pan, L.L.; Liu, S.S.; Navas-Castillo, J. Transmission of begomoviruses and other whitefly-borne viruses: Dependence on the vector species. *Phytopathology* **2020**, *110*, 10–17. [**Google Scholar**] [**CrossRef**]

11. Zhao, J.; Guo, T.; Lei, T.; Zhu, J.; Wang, F.; Wang, X.W.; Liu, S.S. Proteomic analyses of whitefly-begomovirus interactions reveal the inhibitory role of tumorous imaginal discs in viral retention. *Front. Immunol.* **2020**, *11*, 1596. [**Google Scholar**] [**CrossRef**]

12. Zhao, J.; Lei, T.; Zhang, X.J.; Yin, T.Y.; Wang, X.W.; Liu, S.S. A vector whitefly endocytic receptor facilitates the entry of begomoviruses into its midgut cells via binding to virion capsid proteins. *PLoSPathog.* **2020**, *16*, e1009053. [**Google Scholar**] [**CrossRef**] [**PubMed**]

13. Sattar, M.N.; Kvarnheden, A.; Saeed, M.; Briddon, R.W. Cotton leaf curl disease—An emerging threat to cotton production worldwide. *J. Gen. Virol.* **2013**, *94*, 695–710. [**Google Scholar**] [**CrossRef**] [**PubMed**]

14. Briddon, R.W. Cotton leaf curl disease, a multicomponent begomovirus complex. *Mol. Plant. Pathol.* **2003**, *4*, 427–434. [**Google Scholar**] [**CrossRef**]

15. Datta, S.; Budhauliya, R.; Das, B.; Gopalakrishnan, R.; Sharma, S.; Chatterjee, S.; Vanlalhmuaka; Raju, P.S.; Veer, V. Rebound of Cotton leaf curl Multan virus and its exclusive detection in cotton leaf curl disease outbreak, Punjab (India), 2015. *Sci. Rep.* **2017**, *7*, 17361. [**Google Scholar**] [**CrossRef**] [**PubMed**] [**Green Version**]

16. Chakrabarty, P.K.; Kumar, P.; Kalbande, B.B.; Chavhan, R.L.; Koundal, V.; Monga, D.; Pappu, H.R.; Roy, A.; Mandal, B. Recombinant variants of cotton leaf curl Multan virus is associated with the breakdown of leaf curl resistance in cotton in northwestern India. *Virus Dis.* **2020**, *31*, 45–55. [**Google Scholar**] [**CrossRef**]

17. Biswas, K.K.; Bhattacharyya, U.K.; Palchoudhury, S.; Balram, N.; Kumar, A.; Arora, R.; Sain, S.K.; Kumar, P.; Khetarpal, R.K.; Sanyal, A.; et al. Dominance of recombinant cotton leaf curl Multan-Rajasthan virus associated with cotton leaf curl disease outbreak in northwest India. *PLoS ONE* **2020**, *15*, e231886. [**Google Scholar**] [**CrossRef**]

18. Pan, L.L.; Cui, X.Y.; Chen, Q.F.; Wang, X.W.; Liu, S.S. Cotton leaf curl disease: Which whitefly is the vector? *Phytopathology* **2018**, *108*, 1172–1183. [**Google Scholar**] [**CrossRef**] [**Green Version**]

19. Chen, T.; Saeed, Q.; He, Z.; Lu, L. Transmission efficiency of *Cotton leaf curl Multan virus* by three cryptic species of *Bemisiatabaci* complex in cotton cultivars. *Peer J.* **2019**, *7*, e7788. [**Google Scholar**] [**CrossRef**] [**Green Version**]

20. Qin, L.; Wang, J.; Bing, X.L.; Liu, S.S. Identification of nine cryptic species of *Bemisiatabaci* (Hemiptera: Aleyrodidae) from China by using the mt*COI* PCR-RFLP technique. *Acta Entomol. Sin.* **2013**, *56*, 186–194, (In Chinese with English abstract). [**Google Scholar**]

21. Ruan, Y.M.; Luan, J.B.; Zang, L.S.; Liu, S.S. Observing and recording copulation events of whiteflies on plants using a video camera. *Entomol. Exp. Appl.* **2007**, *124*, 229–233. [**Google Scholar**] [**CrossRef**]

22. Sun, Y.; Li, Y.; Wang, S.; Yu, Q.; Li, Z. Effects of deletion of MIT domains of host Vta1 on replication of Autographa californica multiple nucleopolyhedrovirus. *Acta Microbiol. Sin.* **2019**, *59*, 247–257, (In Chinese with English abstract). [**Google Scholar**]

23. Ward, D.M.; Vaughn, M.B.; Shiflett, S.L.; White, P.L.; Pollock, A.L.; Hill, J.; Schnegelberger, R.; Sundquist, W.I.; Kaplan, J. The role of LIP5 and CHMP5 in multivesicular body formation and HIV-1 budding in mammalian cells. *J. Biol. Chem.* **2005**, *280*, 10548–10555. [**Google Scholar**] [**CrossRef**] [**PubMed**] [**Green Version**]

24. Hurley, J.H. The ESCRT complexes. *Crit. Rev. Biochem. Mol.* **2010**, *45*, 463–487. [**Google Scholar**] [**CrossRef**] [**PubMed**]

25. Hanson, P.I.; Cashikar, A. Multivesicular body morphogenesis. *Annu. Rev. Cell Dev. Bi.* **2012**, *28*, 337–362. [**Google Scholar**] [**CrossRef**] [**PubMed**]

26. Wei, J.; Zhao, J.J.; Zhang, T.; Li, F.F.; Ghanim, M.; Zhou, X.P.; Ye, G.Y.; Liu, S.S.; Wang, X.W. Specific cells in the primary salivary glands of the whitefly *Bemisiatabaci* control retention and transmission of begomoviruses. *J. Virol.* **2014**, *88*, 13460–13468. [**Google Scholar**] [**CrossRef**] [**Green Version**]

27. Guo, T.; Zhao, J.; Pan, L.L.; Geng, L.; Lei, T.; Wang, X.W.; Liu, S.S. The level of midgut penetration of two begomoviruses affects their acquisition and transmission by two species of *Bemisiatabaci*. *Virology* **2018**, *515*, 66–73. [**Google Scholar**] [**CrossRef**]

Chapter-14

Persistent Southern Tomato Virus (STV) Interacts with Cucumber Mosaic and/or Pepino Mosaic Virus in Mixed- Infections Modifying Plant Symptoms, Viral Titer and Small RNA Accumulation

1. Introduction

Southern tomato, Pepino mosaic, and Cucumber mosaic viruses (STV, PepMV, and CMV, respectively) infect tomato *(Solanum lycopersicum)* crops worldwide. CMV and PepMV are two pathogenic or acute viruses that are responsible for important economic losses [**1**,**2**]. STV is a persistent double-stranded RNA (dsRNA) virus belonging to the genus *Amalgavirus* (family *Amalgaviridae*), whose genome is 3.5 kb in length, which contains two overlapping open reading frames (ORFs): ORF 1 encodes for the 42 kDa putative coat protein (CP or p42) and ORF 2 encodes for the RNA-dependent RNA-polymerase (RdRp) by +1 ribosomal frameshifting [**3**,**4**]. STV is only transmitted by seed, with rates up to 80% and no viral particles have been detected until now [**3**,**5**,**6**]. High virus incidence has recently been reported in two important Spanish tomato producer areas, such as the Gran Canarias and Valencian Community [**5**,**7**]. Despite that STV was first associated with some fruit symptoms, such as lack of maturation and color alterations, it was recently shown that the virus is not responsible of any apparent plant symptom in tomato plants infect by only STV [**5**,**6**]. Hence, the reported symptoms could be induced by other pathogenic or acute viruses, such as PepMV, CMV, or Tomato mosaic virus (ToMV), which frequently appear in mixed infection with STV in tomato crops or by interaction of STV with other viruses [**8**,**9**].

PepMV is a (+) polarity single stranded RNA (ssRNA+) virus belonging to the genus *Potexvirus* (family *Flexiviridae*) whose genome is 6.4 kb in length and contains five ORFs: ORF 1 encodes for the RdRp, ORF 2, 3, and 4 for the triple gene block proteins (TGB), involved in virus movement, and ORF 5 for the CP [**10**,**11**,**12**]. PepMV induces symptoms of leaf mosaic and alteration of fruit color and maturation, but the symptom severity depends on several factors, such as the virus strain and crop conditions. PepMV is transmitted by contact and by seed with very low rates up to 0.06% [**13,14**]. No commercial tomato varieties with natural resistances against PepMV are available, so disease control has only been achieved by cross protection with mild PepMV strains [**15,16**].

CMV is a tripartite ssRNA+ virus that belongs to the genus *Cucumovirus* (family *Bromoviridae*): RNA 1 is 3.4 kb in length that contains the ORF 1a encoding a RdRp subunit; RNA 2 is 3.1 kb in length and contains the overlapping ORFs 2a y 2b, encoding the other RdRp subunit and the RNA silencing suppressor (VSR) 2b protein; RNA 3 is 2.2 kb in length and it contains the two separated ORFs 3a and 3b encoding for the cell-to-cell movement protein and CP, respectively [**17**]. CMV infects a broad spectrum of plants species (more than 1200 plant species in 100 families), including tomato and pepper *(Capsicum annuum)* and the main way of virus transmission is by aphids in a semi-persistent manner. Symptoms that are induced by CMV depend on the host species and the presence of RNA satellite molecules: in tomato, the most common symptoms induced by CMV are plant stunting, mosaic, and leaf deformation, but the presence of the CARNA5- satellite enhances the disease severity, inducing leaf and fruit necrosis and plant death [**18,19**]. There are no commercial tomato varieties with natural resistances against CMV, and the only manner to minimize the CMV impact is by controlling the aphid populations into the crops.

When plants are infected by RNA viruses, viral dsRNAs (generated during virus replication) activate the post transcriptional gene silencing (PTGS), a plant defense mechanism that produces the degradation of invasive RNAs in small molecules of 21–24 nt (Virus small interfering RNAs, vsiRNAs). PTGS is also involved in the degradation of highly structured plant mRNA rendering micro RNAs (miRNAs), which are small RNA molecules that are equivalent in length to vsiRNAs. miRNAs are involved in the regulation of gene expression in many crucial plant processes, such as development, reproduction, and stress. The modification of the miRNA expression level could lead to disease development [**20,21**]. In addition, vsiRNAs that are derived from viruses could mimic plant miRNAs by sequence homology targeting and regulating post-transcriptionally some host genes [**22,23**]. In the case of persistent viruses, the information about the effect of viral infection regarding on both vsiRNA and miRNAs populations is scarce. The low production of vsiRNAs in plants infected with STV has recently been reported, but the virus can modify the populations of some miRNAs in tomato plants [**6**].

Mixed-infections with two or more plant viruses are frequent in fields and they can interact in multiple and complicate ways [**24**]. The interaction can be synergistic, increasing the replication of at least one of the viruses and/or enhancing symptoms. Synergistic interactions are known to be predominantly produced by unrelated viruses that infect the same host cells. The mechanism underlying the synergistic relationships are not well determined, but numerous viral and/or host products might be involved. The best characterized are those involving potyviruses (genus *Potyvirus*, family *Potyviridae*) as one of the viral partners. In this case, potyviral VSRs are involved in the increase of multiplication and plant symptom enhancing of other viral partner [**25,26**]. In the opposite site, antagonistic interactions between closely viruses (cross protection or mutual exclusion) may occur. In the cross protection, a previous infection with one (protecting) virus prevents or interferes with the subsequent infection by other homologous (challenging) virus [**16,27**] whereas, in the mutual exclusion, two or more viruses infect simultaneously a plant. Several mechanisms have been proposed for the cross protection phenomenon, such as the CP of the protecting virus can prevent the CP disassembly of the challenging virus, which is necessary for infection or the sequence-specific degradation of the challenging virus RNA as consequence of PTGS activation by the protecting virus [**28,29**]. The mechanism for mutual exclusion is still obscure, but it has been proposed that a plant might be considered to be an environment structured spatially for plant virus infections, and cells could only become infected by only one virus [**30**].

The number of studies on viruses in mixed infection has increased lately, providing valuable knowledge that may be useful in controlling complex diseases. However, information regarding interactions between persistent and acute viruses in is very scarce. In this work, we studied the interactions of the persistent STV and the acute PepMV and CMV in tomato. Plant symptoms, virus RNA accumulation, and miRNA and vsiRNA accumulation were assessed in single, double, and triple infections.

2. Materials and Methods

2.1. Plant Material, Virus Infection Assay and Sample Preparation

Tomato seedlings var. Roque were analyzed by RT-qPCR to determinate the presence and/or absence of STV [**5**]. The absence of ToMV and PepMV, the main tomato seed-borne viruses, was assessed by conventional RT-PCR and RT-qPCR, respectively [**31,32**]. PepMV and CMV isolates (kindly provided by Drs. A. Alfaro and M.I. Font) were collected in tomato fields from Southern Spain in 2015 and 2016, respectively, and maintained in *Nicotiana benthamiana* plants. To exclude possible mixed infections with other viruses, the plants that were infected with these CMV and PepMV isolates were tested by ELISA for the most common viruses infecting tomato in the collection region, such as CMV, ToMV, Tomato spotted wilt virus (TSWV), and Parietaria mottle virus (PMoV) [**33,34,35,36,37**].

Mechanical inoculation was performed by the homogenization of 1 g of CMV or PepMV infected *N. benthamiana* plants in inoculation buffer (0.01M Na_2HPO_2 and 0.01M Na_2HPO_4, pH 7.2) and rub-inoculation by using carborundum in the two first tomato (var. Roque) true leaf [**38**]. For double infection with CMV and PepMV, the tomato plants were mechanically inoculated with an equivalent mix (*w*/*w*) of *N. benthamiana* plants infected with each virus. The assay consisted of a total of 71 tomato plants with the following virus combinations: five, eight, and 10 plants were single-infected with PepMV, CMV, and STV, respectively; 10 plants were co-infected with STV and PepMV, 10 were co-infected with STV and CMV, and eight were co-infected with CMV and PepMV, and, finally, 10 plants were triple-infected with STV, CMV, and PepMV. As control, 10 plants were mock-inoculated by using only the inoculation buffer. Because STV is not a mechanically transmitted virus, tomato plants that tested positive for STV by RT-qPCR were used as STV-single infected plants or were inoculated with CMV and/or PepMV to obtain the corresponding co- or triple-infections. Tomato plants were kept in a greenhouse with ventilation and the presence and accumulation of STV, CMV, and PepMV was evaluated by RT-qPCR at five, 15 ,10, and 20 days post inoculation (dpi). Plant symptoms consisting in leaf deformation and mosaic were recorded in this period. A scale of symptom severity was established scoring from 0 to 3, where 0 corresponded to no symptomless, and 2 ,1, and 3 to mild, moderate, and severe symptoms, respectively (**Figure 1**). Plant height and weight were measured at the end of the experiment (20 dpi).

Figure 1. Mosaic and deformation leaf symptoms showed by tomato plants infected with Southern tomato, Pepino mosaic, and Cucumber mosaic viruses (STV, PepMV, and CMV, respectively) in single and mixed infections. Symptoms were considered as mild, moderate or severe (Panels B, C, and D, respectively). The right part of (Panel D) shows a strong leaf deformation in the plant shoots in tomato plants with severe symptoms. Panel A shows a symptomless leaf corresponding to a mock-inoculated tomato plant. (Panel E) shows three tomato leaves showing mild, moderate, or severe symptoms (from left to the right).

For sample preparation, 0.1 g of apical leaves were ground in a power homogenizer TissueLyser (Qiagen, Germany) with liquid nitrogen. The total RNA was extracted by using a phenol:chloroform:isoamyl alcohol standard protocol followed by ethanol precipitation [**39**].

2.2. Conventional RT-PCR and RT-qPCR Assays

CMV and PepMV conventional RT-PCR was performed from the total RNA extracts. The RNA extracts were denatured in the presence of 0.8 μM of the corresponding reverse primer and cDNA was obtained with the SuperScript IV kit (ThermoFisher, Waltham, MA, USA) at 55 °C for 20 min. and 80 °C for 10 min. PCR was done with 0.5 μM of the corresponding forward and reverse for each virus and the Taq polymerase kit (ThermoFisher, Waltham, MA, USA) following the manufacturer's instructions. The PCR conditions were cDNA denaturation at 94 °C for 5 min., 35 cycles of DNA amplification at 94 °C for 30 s, 55 °C for 30 s, and 72 °C for 40 s, and a final DNA chain extension of 72 °C for 5 min. The amplification products were separated by electrophoresis in 2% agarose gels and visualized by UV after staining with GelRed (Sigma-Aldrich, San Luis, MS, USA). Specific PCR products were purified with Qiagen minElute PCR purification kit (Qiagen, Hilden, Germany) and sequenced by Sanger with an ABI 3130 XL capillary sequencer (Applied Biosystems, Foster City, CA, USA). CMV and PepMV nucleotide sequences were deposited in GenBank under the accession numbers MT785769 and MT785770, respectively.

STV quantification was performed by RT-qPCR with primers and TaqMan probe set previously designed in the CP (1189–1257 nts) region [**5**]. PepMV quantification was done using a primers and TaqMan probe set that was designed in a TGB2 conserved region (5126–5213 nts) that allowed for amplifying all virus isolates [**32**]. For CMV quantification, primers and TaqMan probe were designed by using the software Primer Express (ThermoFisher, USA) on basis of the CP nucleotide sequence (1533–1610 nts) that was obtained from conventional RT-PCR. RT-qPCR was performed with the One step PrimeScript RT-PCR Kit (TaKaRa, Shiga, Japan) in LightCyler 480 (Roche, Basilea, Switzerland) following the manufacturer instructions with some modifications. The total RNAs extracts (50 ng) were denaturalized in presence of 0.2 μM of both forward and reverse primers 95 °C for 5 min. Subsequently, a mix containing the 10 μL one-step RT-PCR buffer III, 2 U Ex Taq HS, 0.4 μL PrimeScript RT Enzyme Mix II, and 0.2 μM specific TaqMan probe was added to a final volume of 20 μL. The thermal cycling conditions were: reverse transcription at 42 °C for 15 min., incubation at 94 °C for 10 s, and 40 cycles of DNA amplification at 94 °C for 5 s and 60 °C for 20 s. The total RNA extracts of mock-inoculated tomato plants were used as negative RT-qPCR control. The specificity of all virus primer and probe sets were assessed to avoid unspecific cross-amplifications.

Table S1 shows all the primers and probe sequences and their respective applications.

2.3. Preparation of RNA Transcripts and Standard Curve

The templates for in vitro transcription were obtained by conventional RT-PCR from total nucleic acid extracts of STV-infected tomato and CMV- or PepMV-infected *N. benthamiana* plants, as described in **Section 2.2**. A modified version of the reverse primers (the T7 promoter sequence was added at the 5'-terminus) used for RT-qPCR were used for in vitro transcription (**Table S1**). The transcription reaction was done with the Megascript T7 Kit (ThermoFisher, Waltham, MA, USA) following the manufacturer's instructions. To eliminate the contaminant cDNA, the RNA transcript reaction was treated twice with RNasa free DNasa set (ThermoFisher, Waltham, MA, USA) and then purified by the phenol:chloroform:isoamyl method [**39**]. The final transcript concentration was estimated with a nanodrop 1000 spectrophotometer (ThermoFisher, Waltham, MA, USA), and molarity was assessed with the formula: pmol of ssRNA = μg of ssRNA × (106 pg/1 μg) × (1 pmol/340 pg) × (1/Nb), in which 340 is the average molecular weight of a ribonucleotide and Nb the number of bases of the transcript. The Avogadro's constant (6.023×10^{23} molecules/mol) was used to calculate the number of RNA transcript copies. In order to generate external standard curves, -10fold serial dilutions containing 10^{11}–10^{1} RNA copies of each transcript in total RNA extracts from mock-inoculated tomato plants were analyzed by RT-qPCR. For each dilution, three repeats (technical replicates) were done, and the Ct mean value was calculated. Quantitative optimal range were obtained from 10^{11} to 10^{4} virus RNA copies/ng of total RNA for STV, from 10^{10} to 10^{4} virus RNA copies/ng of total RNA for CMV and from 10^{11} to 10^{3} copies/ng of total RNA for PepMV. For all of the viruses, standard curves showed a strong linear relationship with very high correlation coefficients of $R^2 = 0.99$, low variation coefficient (<%0.5), and high amplification efficiencies (>%99).

2.4. High-Throughput Small RNA Sequencing

For the elaboration of the small RNA libraries, three independent biological replicates were used from tomato plants that were infected with STV, CMV, or PepMV, or the different virus combinations. Each biological replicate consisted of a mix of total RNA extracts that were obtained from two or three tomato plants at 15 dpi. As control, small RNA libraries from mock-inoculated plants were synthetized. RNA concentration and purity were determined using the Qubit® RNA assay Kit in a Qubit® 3.0 Fluorometer (ThermoFisher, Waltham, MA, USA) and the NanoPhotometer® spectrophotometer (IMPLEN, Los Angeles, CA, USA), respectively. The RNA integrity was determined in the Agilent Bioanalyzer 2100 system with the RNA Nano 6000 assay Kit (Agilent Technologies, Santa Clara, CA, USA). cDNA was obtained from 1 µg of total RNA of each biological replicate by using the NEBNext® Multiplex Small RNA library Prep Set for Illumina® (Sigma Aldrich, San Luis, MS, USA) and then sequenced by using the Illumina NextSeq550 platform (Illumina, San Diego, CA, USA). cDNA libraries were uploaded to the NCBI platform and published under the Bioproject PRJNA625104 and PRJNA574043. The reads were cleaned by trimming the sequencing adapters and low-quality reads were filtered using SeqTrimNext software applying the standard parameters for Illumina short reads [**40**]. The biological replicate distribution was analyzed by Principal Component Analysis (PCA) to reduce the dimensionality of the dataset. The length of the reads was restricted from 21 to 24 nts. The identification and quantitation of miRNAs were performed through Oasis 2.0 pipeline analysis (**https://tools4mirs.org/software/precursor_prediction/oasis/**, Access date, 7 March 2021): reads were aligned with the STAR program in the database RNAbase 2.1 (**ftp://mirbase.org/pub/mirbase/** Access date, 7 March 2021), the known miRNAs were quantified with the FeatureCounts program (**https://www.biostars.org/p/259542/** Access date, 7 March 2021), whereas the prediction and quantification of novel miRNA were done with the miRDeep2 program (**http://www.bioconductor.org/packages/release/bioc/vignettes/DESeq2/inst/doc/DESeq2.html** Access date, 7 March **41**] (2021]. For vsiRNA, the total clean reads were aligned with the different virus sequences of STV, CMV, and PepMV (GenBank accession numbers KJ174690.1, AB188234 and KJ018164).

2.5. Statistical Analysis

For plant symptoms, weight and height, and virus titer, the data were statistically analyzed using a mixed model PROC MIXED in the SAS software. Plant effect was included as a random effect, whereas time or inoculation was included as a fixed effect. Least Square Difference (LSD) was used for mean comparisons. The assumption of normal distribution of data was assessed using the normal probability plot of the residuals and the assumption of homoscedasticity using the Levene's test. A 95% of confidence interval was considered in all cases. For miRNA differential expression analysis, a FDR adjusted p-value < 0.05 corresponding to a log Fold-change > 0.56 was considered to be statistically significant.

3. Results

3.1. Characterization of Field CMV and PepMV Isolates Used in this Work

The CMV and PepMV isolates showed nucleotide identities of 100% with the Japanese CM95 isolate (GenBank accession no. AB188236.1) in the 325 nt CP amplified region and with the European EU_CAHN8 isolate (GenBank accession no. JQ314457.1) in the 545 nt TGB3 amplified region, respectively. ELISA results analysis showed that PepMV and CMV isolates were no infected with other viruses, such as ToMV, TSWV, and PMoV: three replicates were used of each virus isolate and negative absorbance values were observed for ToMV (from 0.038 to 0.161), TSWV (from 0.047 to 0.075), and PMoV (from 0.039 to 0.059), whereas the positive control ranged from 0.903 to 2.076.

3.2. Effect of STV in Symptoms of Tomato Plants Mixed- Infected with PepMV and/or CMV

STV-infected tomato plants were single inoculated with PepMV or CMV, and with a combination of both viruses, to study the effect of STV in mixed infection. Leaf deformation and mosaic (severity scoring from 0 to 3, where 0 correspond to symptomless and 1, 2, and 3 to mild, moderate, and severe symptoms, respectively) were observed in infected tomato plants at different times (5, 10, 15, and 20 dpi) (**Figure 1**). Additionally, height and weight of tomato plants of different plant groups was taken at 20 dpi.

The leaf symptoms severity values (mean of plants symptoms on each group) in tomato plants infected with STV, CMV, and PepMV in single and/or mixed infections varied depending on the time and the virus combination (**Figure 2**). As expected, both STV single-infected and mock-inoculated plants remained symptomless for all the times [**5,6**]. At 5 dpi, only the STV + CMV co-infected plants showed mild symptoms (1.25), whereas no symptoms were observed in CMV-single infected plants. At 10 dpi, symptoms of STV + CMV co-infected plants were moderate (2.12), whereas those of STV + PepMV co-infected ones were mild (1.47). At this time, CMV and PepMV single-infected plants only showed mild symptoms (1.24 and 1.06, respectively). Regarding CMV + PepMV co-infection, these plants showed mild symptoms (1.00), whereas STV + CMV + PepMV triple-infected ones remained symptomless. At 15 dpi, the symptoms severity of STV + CMV co-infected plants decreased (from 2.12 to 1.55), whereas those of STV + PepMV double-infected ones increased (from 1.47 to 1.97). At this time, symptom severity of CMV single-infected plants (1.00) was lower than STV + CMV co-infected ones and higher in PepMV single-infected plants (2.63) than STV + PepMV co-infected ones. In contrast, the symptom severity of CMV + PepMV co- and STV + CMV + PepMV triple-infected plants was similar (1.43 and 1.37, respectively). Finally, at 20 dpi, symptom severity of STV + CMV and STV + PepMV co-infected plants decreased (from 1.55 to 1.26 and from 1.97 to 1.00, respectively) showing a remarkable difference respect to CMV single infection (2.62), but none with respect to PepMV-single infection (1.00). Regarding CMV + PepMV co-infected plants, symptom severity was slightly higher than STV + CMV + PepMV triple-infected ones (2.00 and 1.76, respectively).

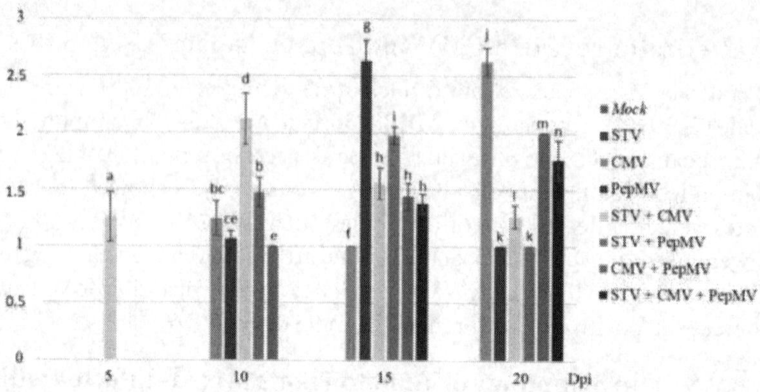

Figure 2. Graphic representation (mean values) of leaf symptoms severity (ordinate axis) of tomato plants infected with STV, PepMV, and CMV in single and mixed infections at 5, 10, 15, and 20 dpi (abscise axis). Leaf symptoms intensity was scored from 0 to 3, where 0 corresponds to symptomless, and 1, 2, and 3 to mild, moderate, and severe symptoms, respectively. Bars and letters up to the columns correspond to standard errors (from 0 to 0.23) and different plant groups (*p*-value ≤ 0.05), respectively.

Differences of height and weight (mean values) among groups of infected-tomato plants are shown in **Figure 3**. STV single-infected and mock-inoculated plants had similar height (64 and 65 cm, respectively). PepMV and CMV single-infected plants were significantly taller and smaller than the mock-inoculated ones (71.0 and 57.5 cm, respectively). The STV + PepMV co-infected plants (75.5 cm) were significantly taller than PepMV single-infected plants (71.0 cm), whereas height of STV + CMV co- and CMV single-infected plants was almost identical (57.7 and 57.5, respectively). The height of CMV + PepMV co-infected plants (64.0 cm) scored between STV + PepMV (75.5 cm) and STV + CMV (57.7 cm) co-infected plants, with significant differences with respect to both of them. Finally, STV + CMV + PepMV triple-infected plants were the smallest (50.0 cm), with significant differences with respect to the rest of virus-infected and mock-inoculated plants. With regard to the weight, mock-inoculated, STV, and CMV single-infected plants did not show significant differences (11.0, 11.9, and 12.3 g, respectively). PepMV single-infected plants reached the maximum value (32.7 g) of the assay, with significant differences with respect to the other plant groups, including STV + PepMV co-infection (21.7 g). The weight of STV + CMV co- and CMV single-infected plants was similar (13.2 and 12.3 g, respectively), whereas the weight of CMV + PepMV co-infected plants (15.5 g) scored between STV + PepMV and STV + CMV co-infected plants. Finally, the weight of STV + CMV + PepMV triple-infected plants (11.3 g) was significantly lower than that of CMV + PepMV co-infected ones.

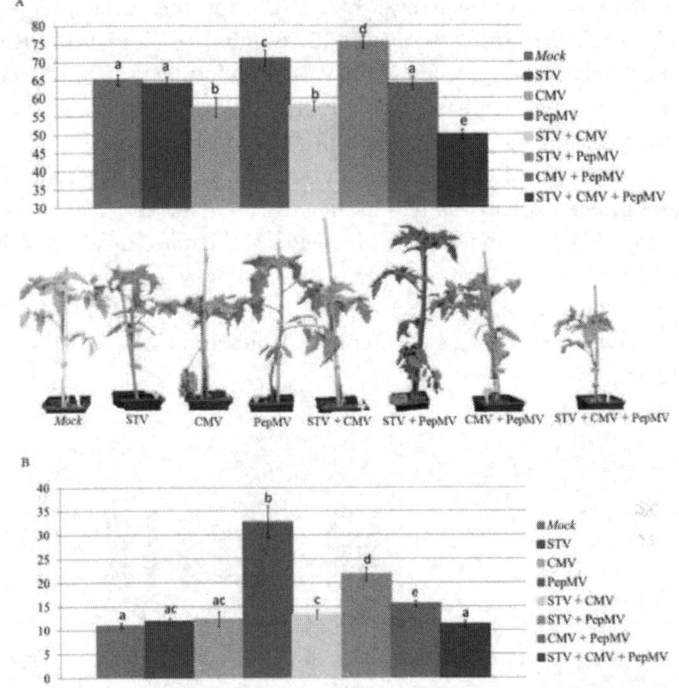

Figure 3. Graphical representation (mean values) of height (Panel A) and weight (Panel B) measured in cm and g, respectively, of tomato plants infected with STV, PepMV, and CMV in single and mixed infections at 20 dpi. Bars and letters up to the columns correspond to standard errors and different plant groups (p-value ≤ 0.05), respectively. At the bottom of (Panel A), we show the height of tomato plants infected with different virus combinations in comparison with mock-inoculated plants.

3.3. Effect of STV in Virus Accumulation of Tomato Plants Mixed-Infected with PepMV and/or CMV

Virus accumulation was studied at 5, 10, 15, and 20 dpi by RT-qPCR using specific primers and TaqMan probes for STV, PepMV and CMV (**Table S1**). The specificity assays of virus primer and probe sets showed no unspecific cross-amplifications. **Figure 4** shows the mean values of virus accumulation. STV titer remained almost constant ($10 \times 2.38^4 - 2.29$ $10 \times^5$ virus RNA copies/ng total RNA) overtime for single-, co-, and triple-infections with CMV and PepMV, so the other viruses did not affect STV accumulation (**Figure 4**, Panel A). PepMV accumulation pattern was quite similar in single-, co-, and triple- infections (**Figure 4**, Panel B): this pattern consisted in a decrease from 5 dpi ($2.70–5.28 \times 10^6$ virus RNA copies/ng total RNA) to 15 dpi ($10 \times 2.43–1.39^5$ virus RNA copies/ng total RNA) and an increase at 20 dpi ($10 \times 5.20–1.01^6$ virus RNA copies/ng total RNA). CMV showed different accumulation patterns, depending on the virus combination (**Figure 4**, Panel C): the CMV concentration showed a low variation at 5 and 10 dpi (from 5.21×10^3 to $\times$ 1.64 10^4 virus RNA copies/ng total RNA), but it increased strongly at 20 dpi (10×2.96^8 virus RNA copies/ng total RNA) in CMV single infected plants. However, co-infection with

STV produced a high increase of CMV at 10 and 15 dpi (10×8.30^5 and 10×4.46^7 virus RNA copies/ng total RNA, respectively), but, at 20 dpi, the CMV titer was similar in single- and STV + CMV co-infected plants (2.96 and 10×2.45^8 virus RNA copies/ng total RNA, respectively). The pattern of CMV accumulation changed when co-infected with PepMV: the CMV titer decreased at 15 dpi (from 10×8.93^3 to 10×1.05^2 virus RNA copies/ng total RNA) and increased slightly it at 20 dpi (10×7.47^3 virus RNA copies/ ng total RNA). Differences of CMV accumulation between CMV + PepMV co-infected and CMV-single infected plants were significant at 15 ,10, and 20 dpi. STV infection increased strongly CMV titer in STV + CMV + PepMV triple- infection at 15 ,10, and 20 dpi (10×5.95^3 10×8.76^6 and 10×4.16^7 virus RNA copies/ng total RNA, respectively) to be similar to those of CMV single-infected plants. The differences of CMV accumulation were significant between STV + CMV + PepMV triple- and CMV + PepMV co-infection at 15 ,10 and 20 dpi.

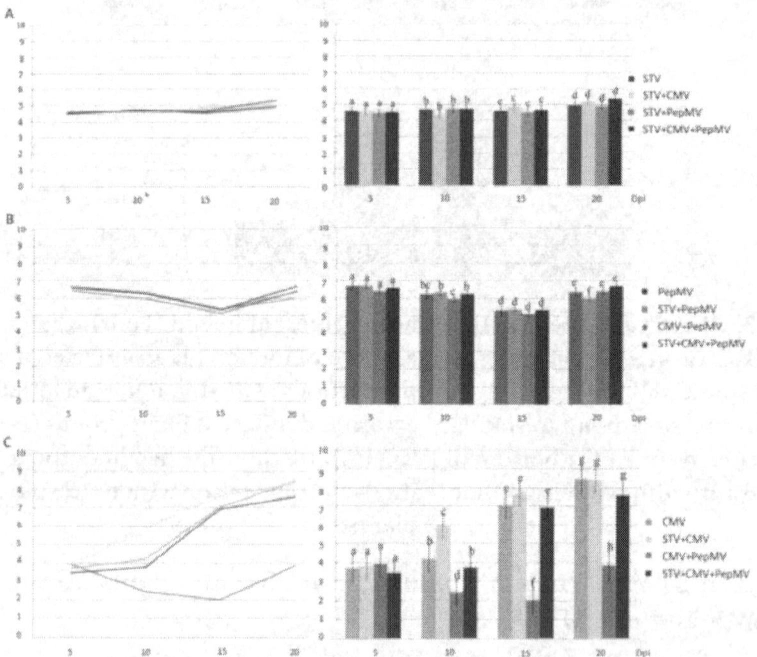

Figure 4. Virus accumulation (mean values) in tomato plants (ordinated axis) shown as log of concentration (no. RNA copies/ng of total RNA) of STV, PepMV, and CMV (Panel A, B, C, respectively) in single and mixed infections at 5, 10, 15, and 20 dpi (abscise axis). Bars and letters up to the columns correspond to standard errors and plant groups (in each dpi), respectively, showing differences (*p*-value ≤ 0.05). In each panel, virus accumulation is represented by columns (right) and in lineal representation (left).

3.4. Effect of STV in siRNA Accumulation of Tomato Plants Mixed-Infected with PepMV and/or CMV

The accumulation of siRNAs was determined by high throughput small RNA sequencing from total nucleic acids obtained at 15 dpi, since the greatest effect of STV in CMV accumulation was found between 10 and 15 dpi. Additionally, at 15 dpi, a strong effect of STV in CMV + PepMV co-infection was observed. Three biological replicates were sequenced for each group of tomato infected plants and mock-inoculated plants were used as the controls. Biological replicates considered to be outlayer by PCA analysis were excluded from further analysis and the total reads were filtered to obtain the useful reads of about 21–24 nts (**Table S2**). The percentages of useful reads with respect to the total ranged from 34% to 70%. The highest percentages of useful reads were found in the mock-inoculated and STV single-infected plants (59% and 70%, respectively), whereas, in the other virus-infected plants, they ranged from 34% to 48%. Expression profiling analysis of potential miRNAs performed with the OASIS 2 software showed a total of 78 siRNAs, which accumulated differentially in the plants that were infected with different virus combinations with respect to the control mock-inoculated plants (FDR < 0.05 and for log2FC > 0.56) (**Table 1**). Of those, 31 miRNAs were described previously in tomato and 47 corresponded to potential novel miRNAs described on other plant species, such as *Solanum tuberosum*, *Oryza sativa*, *Glycine max*, *Prunus persica*, or *Arabidopsis thaliana*. Three miRNAs with animal sequence homology, such as cow (*Bos taurus*) and mouse (*Mus musculus*), were also detected. It was found 5, 34, and 39 miRNAs with differential expression in STV, CMV, and PepMV single-infected plants, respectively. STV infection modified the number of miRNAs in STV + CMV and STV + PepMV co-infection with respect to CMV and PepMV single- infections (from 34 to 57 and from 39 to 37, respectively) (**Tables S3 and S4**). Slight changes in the number of miRNAs with differential expression were observed between the CMV + PepMV co-infected and the STV + CMV + PepMV triple-infected plants (from 24 to 25) (**Table S5**). Finally, in CMV + PepMV co-infected plants, less miRNAs expressed differentially were found than in CMV and PepMV single-infected plants (from 24 to 34 or 39, respectively) (**Table S6**). In addition to the change of the number of miRNAs expressed differentially, it was observed that STV infection significantly (FDR < 0.05 and fold-change was > 0.56) modified the accumulation of some miRNAs.

Table 1. Differential accumulation of miRNAs in tomato plants infected with different virus combinations with respect to the control mock-inoculated plants (FDR < 0.05 and log2FC > 0.56).

	miRNA with Differential Expression	
Sample	***Solanum lycopersicum***	**Novel *Solanum lycopersicum***
STV vs. Mock-inoculated	1	4
CMV vs. Mock-inoculated	14	20
PepMV vs. Mock-inoculated	14	25
STV + CMV vs. Mock-inoculated	26	31
STV + PepMV vs. Mock-inoculated	15	22
CMV + PepMV vs. Mock-inoculated	11	13
STV + CMV + PepMV vs. Mock-inoculated	10	15

The potential functions of 53 out of these 78 miRNAs were determined by searching on the bibliography or by analysis with the online psRNAtarget software and they were mainly related with fundamental plant process, such as cellular biotic and abiotic stress, metabolism, or plant development. For example, it was reported that sly-miR9470-5p was related to hydric and salt stress, and Potato virus Y (PVY) infection as well [**42,43**]. This miRNA was upregulated in plants that were infected with all of the virus combinations with respect to the control mock-inoculated plants. Additionally, mtr-miR172c-5p that was previously related to salt stress [**44**] was up-regulated in CMV-single, and STV +CMV co-infected plants, whereas it was down-regulated in STV + CMV + PepMV triple-infected plats with respect to the control mock-inoculated. Finally, psRNAtarget analysis showed that mmu-miR-466i-5p that was upregulated in STV + CMV co-infected plants with respect to the control mock- inoculated ones targeted a gene encoding for a thylakoidal chloroplastic protein.

To obtain the vsiRNAs populations, the useful reads were aligned with the complete nucleotide sequence of STV, CMV, and PepMV (KJ174690.1, AB188234, and KJ018164). For each virus combination the percentage of vsiRNAs with respect the useful reads were calculated using the mean values of the biological replicates (**Figure 5**). It was detected few STV derived vsiRNAs in STV single-infected tomato plant (23.56 useful reads, which corresponded to 0.0003% of vsiRNAs), but they increased with the presence of other viruses in mixed-infections: STV + CMV and STV + PepMV co-infected plants (839.88 and 329.39 useful reads, which corresponded to 0.0106% and 0.0028% of vsiRNAS, respectively) and STV + CMV + PepMV triple-infected plants (1559.32 useful reads that corresponded to 0.0081% of vsiRNAs).

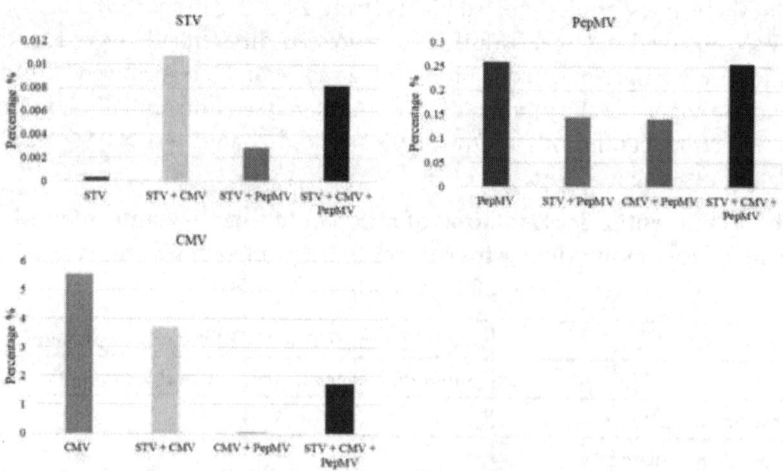

Figure 5. Graphic representation of the percentages (%) of vsiRNAs respect to the useful reads obtained by small RNA high throughput sequencing of the different STV, CMV, and PepMV virus combinations. The percentages of vsiRNAs were obtained from the useful reads mean values of the three biological replicates.

With regard to CMV derived vsiRNAs, high quantities were detected in CMV single- and STV + CMV co- infected plants (272,758.04 and 288,716.41 useful reads, which corresponded to 5.53% and 3.67% of vsiRNAs, respectively). However, in CMV + PepMV co-infected plants, CMV derived vsiRNAs were almost undetectable (7.82 useful reads, which corresponded to 0.000068% of vsiRNA), but they increased markedly with the STV presence in STV + CMV + PepMV triple-infected plants (330,588.95 useful reads which corresponded to 1.73% of vsiRNAs). Contrarily, PepMV derived vsiRNAs decreased in STV + PepMV co- and CMV + PepMV co-infected plants (8140.93 and 15,980.81 useful reads, which corresponded to 0.1446% and 0.1396% of vsiRNAs, respectively) with respect to PepMV-single infected ones (21,741.45 useful reads, which correspond to 0.2590% of vsiRNAs). However, in STV + CMV + PepMV triple infection (48,620.65 useful reads, which corresponded to 0.2551% of vsiRNAs), PepMV derived vsiRNAs increased with respect to PepMV single-infection.

The polarity of the vsiRNA plus (+) or minus (–) was also determined by aligning the useful reads with the positive and negative genomic virus strands (**Table 2**). For STV, in all virus combinations (STV single-, STV + CMV co-, STV + PepMV co-, and STV + CMV + PepMV triple-infected plants), more minus than plus vsiRNAs were detected (52.80–74.36% and 25.64–47.2%, respectively). For CMV, more minus than plus vsiRNAs were detected in CMV single-, STV + CMV co-, and STV + CMV + PepMV triple-infected tomato plants (67.39–72.37% and 27.35–32.61%, respectively) and less than CMV + PepMV co-infected ones (and 43.22 and 56.78%, respectively). For PepMV, similar amounts of plus and minus vsiRNAs were detected (50.35–53.39% and 47.25–50.41%, respectively).

Table 2. siRNAs polarity (plus or minus) in tomato plants infected with de different STV, CMV, and PepMV virus combinations. The numbers for each virus combination correspond to the mean of the biological replicates. The percentages of plus (+) and minus (–) vsiRNAs polarity base on the useful reads are in brackets.

	STV		CMV		PepMV	
	Plus (+)	Minus (–)	Plus (+)	Minus (–)	Plus (+)	Minus (–)
STV	11.12	12.44	-	-	-	-
	(47.2%)	(52.80%)	-	-	-	-
CMV	-	-	88,928.90	183,829.14	-	-
	-	-	(32.61%)	(67.39%)	-	-
PepMV	-	-	-	-	11,469.85	10,271.60
	-	-	-	-	(52.75%)	(47.25%)
STV + CMV	332.56	507.32	78,955.78	209,760.63	-	-
	(39.6%)	(60.40%)	(27.35%)	(72.65%)	-	-
STV + PepMV	84.43	244.96	-	-	4346.79	3794.14
	(25.64%)	(74.36%)	-	-	(53.39%)	(46.61%)
CMV + PepMV	-	-	4.44	3.38	8047.01	7933.80
	-	-	(56.78%)	(43.22%)	(50.35%)	(49.65%)
STV + CMV + PepMV	626.67	932.65	90,349.20	240,239.75	24,113.38	24,507.27
	(40.19%)	(59.81%)	(27.33%)	(72.67%)	(49.59%)	(50.41%)

Moreover, the distribution of the plus and minus vsiRNAs (average number of biological replicate) from each virus genome was determined by calculating the vsiRNAs frequency at each virus nucleotide position (**Figure 6**). STV and CMV vsiRNAs frequencies could not be represented in STV-single infected and CMV + PepMV co- infected tomato, since the amounts of vsiRNAs were so low. Both plus and minus vsiRNAs displayed a non-uniform distribution pattern along the virus genomes with hotspots (high accumulation of vsiRNAs) in specific genomic regions. These plus and minus vsiRNAs patterns were not symmetric for all viruses. For each virus, co-infection with the other virus did not produce remarkable variations of vsiRNAs patterns, but only changes in its accumulation level. Further estimations of vsiRNAs hotspots showed that minus STV vsiRNAs accumulated in the p42 (CP) coding region, which overlaps with the RdRp, meanwhile the plus vsiRNAs accumulated in the terminus part of the RdRp and the starting part of 3′ non-coding UTR. For CMV, plus and minus vsiRNAs accumulated more in the RNA3 (encoding for the MP and CP) than in the RNA2 (encoding for the RdRp and the 2b protein) and RNA1 (encoding for RdRP). In the RNA, mainly for the minus vsiRNAs, several hotspots were observed in the start and terminus parts of RdPp. In the RNA2 and 3, plus and minus hotspots were observed spread along genome, but in different positions, depending on the strand polarity. Some of the regions in the 2b, MP, and CP regions showed a high accumulation of vsiRNAs. Finally, for PepMV, plus and minus vsiRNAs hotspots localized along the virus genome but with some hotspots in the 5‘non-coding UTR, start part of the RdRp, TGB3, and CP regions.

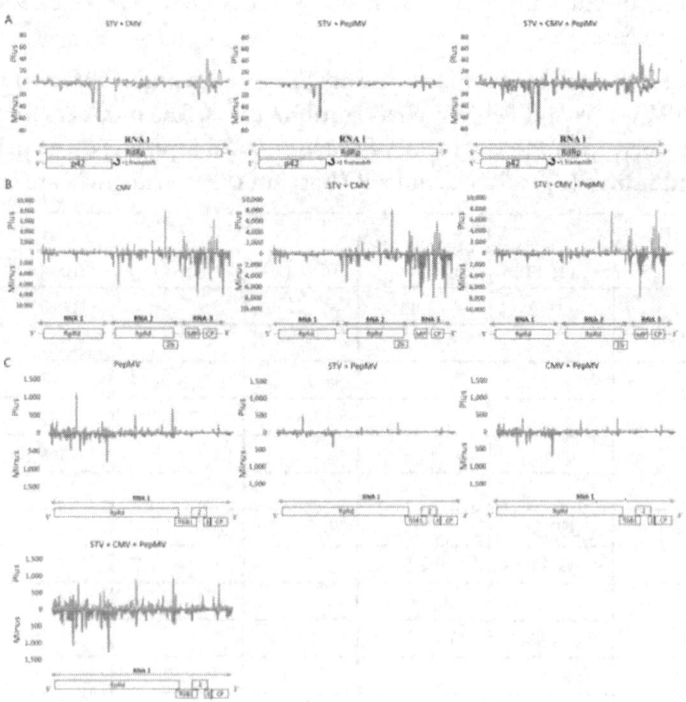

Figure 6. Graphical representation of plus (+) (blue) and minus (–) (orange) vsiRNAs frequencies along STV (Panel A), CMV (Panel B), and PepMV (Panel C) genomes for the different virus combinations. Virus genome organization is showed in the bottom of each graph. STV and CMV vsiRNAs frequencies were not represented, since amounts of vsiRNAs were so low in STV-single infected and CMV + PepMV double- infected tomato plants.

4. Discussion

STV is a persistent virus that is widespread, and high incidences have been reported in some Spanish tomato production areas, such as Valencian community and Canary Islands. Despite that STV was associated to some disorders, such as a lack of fruit maturation and coloration alteration [**3**], recent studies suggest that STV does not produce symptoms in in tomato STV single-infected plants [**5,6**]. However, STV is frequently detected in tomato fields in combination with other viruses, but, to date, the effect of STV in mixed-infections on plant symptom development is unknown. In this work, the interaction of STV with two important acute viruses infecting tomato crops, such as CMV and PepMV, was studied. For this purpose, an assay with tomato plants in virus single-, co- and triple-infections was performed. As expected, STV-single infected plants did not show any symptoms, corroborating the results that were obtained in previous research works [**5,6,7,45,46**]. In this assay, the STV titer remained constant over time (5–20 dpi) in single-infections, as reported previously [**5**], and the same occurred in in co- and triple-infections with CMV and PepMV The steady titer of STV during the infection contrasts with the majority of acute viruses, whose concentration varies, depending on the infection state [**38,39**].

STV and CMV in co-infections established a synergistic interaction that produced the earlier apparition of leaf symptoms, increasing their severity, and increasing CMV titer in the first stages of infection. STV also produced an increase of plant symptoms in STV + PepMV co-infected plants, but it did not produce changes in PepMVtiter. To our knowledge, this is the first report of a synergistic interaction between a persistent and two acute viruses. The best-known synergisms between acute viruses are those involving potyviruses. For example, infection with the potyvirus PVY and the potexvirus Potato virus X (PVX) increases the accumulation of PVX and the severity of symptoms [**47,48**]. It has been reported that potyviral VSR (HC-Pro) can suppress the defense mechanism that is based on the plant PTGS, favoring the replication and accumulation of the accompanying virus and enhancing the induced plant symptoms [**24,49**]. STV could codify for a VSR, but previous studies that were carried out in our lab showed that p42 had no VSR activity (unpublished data). Because STV only codifies for p42 and RpRd, further studies must be performed to confirm whether RdRp has VSR activity.

PepMV single-infected plants showed the maximum severity foliar symptoms (medium-severe) at 15 dpi and then decreases at 20 dpi. However, few changes of the viral titer were observed, with a slight decrease from 5 to 15 dpi and a recovery at 20 dpi. This accumulation pattern is not common in acute viruses, which normally increase the viral concentration at the beginning of the infection to reach a maximum that is followed by a stable or "plateau" stage, or sometimes with a slight decrease in the virus concentration [**38,39**]. Virus accumulation depends on many biotic and abiotic factors. For example, some Broad bean wilt virus 1 (BBWV-1) isolates showed abnormal accumulation patterns in pepper similar to that shown by PepMV in this study, whereas the same BBWV-1 isolates accumulated normally in tomato [**38**]. Furthermore, at 20 dpi, it was observed that PepMV infection induced an increase in the height and weight of the virus single-infected plants with respect to the control mock-inoculated ones. There are studies showing the beneficial effects of some acute viruses, as, for example, CMV that induces symptoms, but it is also able to increase the thermic resistance in beet *(Beta vulgaris)* infected plants [**50**]. The presence of STV in STV + PepMV co-infected plants increased their height respect to PepMV-single infected ones. However, in this plant group, the weight decreased as consequence of a stem slimming. It has been previously reported that the co-infection of PepMV and ToTV induces a slight increase in the height of infected tomato plants as compared to uninfected ones [**51**].

In CMV + PepMV co-infected plants, an antagonistic effect was observed with a decrease of CMV titer and symptoms were different to those that were induced by CMV or PepMV in single infections. Because CMV and PepMV are phylogenetically unrelated, this interaction cannot be explained as "cross protection" or "mutual exclusion", which are produced between closely related viruses. To date, a few antagonistic interactions between phylogenetically distant viruses have been reported, but the mechanisms underlying these interactions have not been determined. For example, the simultaneous infection of Cucumber green mottle mosaic virus (CGMMV) and Tomato leaf curl New Delhi virus (ToLCNDV) in squash plants *(Cucurbita maxima)* led to a reduction in ToLCNDVtiter, decreasing the virus-induced symptoms [**52**]. STV presence in plants STV + CMV + PepMV triple-infected plants suppressed the antagonistic effect between CMV and PepMV, restoring the CMV titer in CMV single-infected plants and modifying the symptom severity with respect to CMV + PepMV co-infection. To our knowledge, this is the first description of a virus modifying interaction being established between two other different viruses.

In this research work the effect of the interaction between STV, CMV, and PepMV on the populations of both plant miRNAs and viral vsiRNAs was also studied. The differential expression of 78 miRNAs was determined in tomato plants in single and mixed infection conditions with respect to the control mock-inoculated plants. Of all these miRNAs, 47 corresponded to novel miRNAs that were described for the first time in tomato. It was previously reported that plant infection by viruses, such as PVY and Papaya ringspot virus (PRSV), stimulated the synthesis of novel miRNAs [**43,53**]. miRNAs with differential expression that were found in this work were mainly involved in fundamental processes in the plant, such as development, metabolism, abiotic, and biotic stress. Thus, variations in accumulation of these miRNAs could lead to important changes in the plant. The number of miRNAs differentially expressed, and their level of accumulation, varied depending on the virus combination. Additionally, it was demonstrated that STV presence in the different groups of infected plants modified both the number and expression level of some miRNAs with respect to the CMV or PepMV single- and CMV + PepMV co-infections. Some examples of miRNAs with differential expression, depending on the virus combination, are: mtr-miR172c-5p was up-regulated in CMV-single, and STV +CMV co-infected plants, whereas it was down-regulated in STV + CMV + PepMV triple-infected plants with respect to the control mock-inoculated ones. This miRNA was previously related to salt stress [**44**]. sly-miR164b-3p had differential expression in STV + CMV co-infected plants, but not in CMV or STV single-infected ones (**Table S3**). The miRNA was related to saline and hydric stress as well as fruit maturation in tomato plants [**42,54**]. miRNA stu-miR398a-5p had differential expression in STV + PepMV double-infected plants, but not in PepMV or STV single- infected ones (Supplementary material S4). This miRNA is related to tolerance to the virus infection [**55**]. Finally, mmu-miR-466i-5p was upregulated in STV + CMV co-infected plants with respect to the control mock-inoculated ones (Supplementary material Table S3). This miRNA can target a gene encoding for a thylakoidal chloroplastic protein. In the last years, reports of changes in the miRNA expression as consequence of plant virus infection have been increasing. For example, miR159/319 and miR172 expression is modified by ToLCNDV infection in tomato [**56**], or miR163, miR164, and miR167 expression is modified by ToMV infection in *A. thaliana* [**57**].

STV, CMV, and PepMV plus and minus vsiRNAs were identified. The results obtained in this work showed that the vsiRNAs proportion varied, depending on the virus (single infection) and the combination whit other viruses (multiple infection). The amount of vsiRNAs generated in STV single-infected plants was very small, but it increased markedly in plants that were co-infected with PepMV or CMV (STV + PepMV and STV + CMV co-infections). These variations in the production of vsiRNAS from STV were not related to viral accumulation, since the concentration of STV did not change with the presence of other viruses. This is in concordance with the results reported by other authors that showed low vsiRNAs concentrations in STV single-infected tomato plants, but that increased in combination with other viruses in mixed-infections [**46,58,59**]. It was observed that the presence of STV varied the proportion of CMV and PepMVvsiRNAs in STV + CMV and STV + PepMV co-infections with respect to the CMV and PepMV-single infections, and that these changes were not related with the viral accumulation. The interaction of CMV and PepMV also influenced the formation from vsiRNAs in both viruses in co-infections, mainly with a strong reduction of CMV vsiRNAs, which were practically not detected. In this case, the low titer of CMV vsiRNAs correlated with the low titer of CMV RNA due to the antagonistic effect of PepMV. Differences in the vsiRNAs accumulation may be relevant in the development of plant symptoms due them having the ability to mimic the miRNAs by sequence homology. For example, there is experimental evidence that some vsiRNAs of the Sugarcane mosaic virus (SCMV) and Rice stripe virus (RSV) target genes in corn and rice, respectively, altering their development [**22**]. Additionally, vsiRNAs are generated by CMV satellite RNA Y in *N. tabacum* and by Tomato yellow leaf curl virus (TYLCV) in tomato target genes of these plants [**22,23,60**].

The study of the frequency of STV, CMV, and PepMV derived vsiRNAs per nucleotide site, in the positive and negative strands of the viral genomes, showed that the distribution of plus and minus vsiRNAs was not uniform in these viruses. It detected regions of accumulation with peaks (hotspots) that were usually different for the plus and minus vsiRNAs. However, the distribution patterns of vsiRNAs for each virus were not influenced by the presence of the other viruses in mixed infection. Differences were only observed in the vsiRNAs accumulation level, which was correlated with the number of total useful reads. This agrees with the non-uniform patterns of STV and PepMVvsiRNAs frequencies found by other authors [**59**]. Additionally, it has been reported that co-infections of PRSV and PapMV did not alter the its frequency patterns of vsiRNAs accumulation with respect to simple infections [**47**].

5. Conclusions

To date, the role played by STV in the development of some plant symptoms, such as disorders in tomato fruit coloration and maturation, was controversial. Despite recent studies showing that STV did not induce any plant symptoms in single-infections, the reported symptoms could be induced by the interaction of STV with other viruses. Here, we studied the effect of STV in co- and triple-infections with the widespread Cucumber mosaic virus (CMV) and Pepino mosaic virus (PepMV). The results showed that the persistent STV is relevant from a phytopathological point of view, since STV can interact with these viruses: (i) establishing a synergism with CMV or PepMV in which STV increased CMV titer and CMV induced symptoms at early infection stages, whereas PepMVtiter did not change in spite that PepMV induced symptoms exacerbated, (ii) suppressing the antagonism between CMV and PepMV, restoring the CMV titer, and modifying the plant symptom severity with respect to CMV + PepMV co-infection, and (iii) modifying the accumulation of both plant miRNAs and viral vsiRNAs with respect to PepMV and CMV in single- or co-infections. Most of these miRNAs are involved in essential plant process, Additionally, vsiRNAs could mimic the action of miRNAs targeting plant genes. Thus, it is important to establish control measures to avoid STV spread by preventing the commercialization of STV-infected seeds, since the virus is only horizontally transmitted.

Supplementary Materials

article/10.3390/microorganisms9040689/s1. Table S1: List of primers and probes used in this work. Table S2: Useful reads of each virus combination. Tables S3–S6: List of miRNAs expressed differentially in tomato plants infected with different virus combinations with respect to the control mock-inoculated ones.

References

1. Scholthof, K.G.; Adkins, S.; Czosnek, H.; Palukaitis, P.; Jacquot, E.; Hohn, T.; Hohn, B.; Saunders, K.; Candresse, T.; Ahlquist, P. Top 10 plant viruses in molecular plant pathology. *Mol. Plant Pathol.* **2011**, *12*, 938–954. [**Google Scholar**] [**CrossRef**]

2. Souiri, A.; Khataby, K.; Kasmi, Y.; Zemzami, M.; Amzazi, S.; Ennaji, M.M. Risk assessment and biosecurity considerations in control of emergent plant viruses. In *Emerging and Reemerging Viral Pathogens: Volume 2: Applied Virology Approaches Related to Human, Animal and Environmental Pathogens*; Elsevier: Amsterdam, The Netherlands, 2019; pp. 287–311. ISBN 9780128149669. [**Google Scholar**]

3. Sabanadzovic, S.; Valverde, R.A.; Brown, J.K.; Martin, R.R.; Tzanetakis, I.E. Southern tomato virus: The link between the families *Totiviridae* and *Partitiviridae*. *Virus Res.* **2009**, *140*, 130–137. [**Google Scholar**] [**CrossRef**]

4. Nibert, M.L.; Pyle, J.D.; Firth, A.E. A 1 ribosomal frameshifting motif prevalent among plant amalgaviruses. *Virology* **2016**, *498*, 201–208. [**Google Scholar**] [**CrossRef**] [**PubMed**] [**Green Version**]

5. Elvira-González, L.; Carpino, C.; Alfaro-Fernández, A.; Font-San Ambrosio, M.I.; Peiró, R.; Rubio, L.; Galipienso, L. A sensitive real-time RT-PCR reveals a high incidence of Southern tomato virus (STV) in Spanish tomato crops. *Spanish J. Agric. Res.* **2018**, *16*, 1008. [**Google Scholar**] [**CrossRef**] [**Green Version**]

6. Elvira-González, L.; Medina, V.; Rubio, L.; Galipienso, L. The persistent Southern tomato virus modifies miRNA expression without inducing symptoms and cell ultra-structural changes. *Eur. J. Plant Pathol.* **2019**, *156*, 615–622. [**Google Scholar**] [**CrossRef**]

7. Puchades, A.V.; Carpino, C.; Alfaro-Fernandez, A.; Font-San-Ambrosio, M.I.; Davino, S.; Guerri, J.; Rubio, L.; Galipienso, L. Detection of Southern tomato virus by molecular hybridisation. *Ann. Appl. Biol.* **2017**, *171*, 172–178. [**Google Scholar**] [**CrossRef**]

8. Padmanabhan, C.; Zheng, Y.; Li, R.; Sun, S.E.; Zhang, D.; Liu, Y.; Fei, Z.; Ling, K.S. Complete genome sequence of Southern tomato virus identified in China using next-generation sequencing. *Genome Announc.* **2015**, *3*, 10–1128. [**Google Scholar**] [**CrossRef**] [**PubMed**] [**Green Version**]

9. Verbeek, M.; Dullemans, A.M.; Espino, A.; Botella, M.; Alfaro-Fernández, A.; Font, M.I. First report of Southern tomato virus in tomato in the Canary Islands, Spain. *J. Plant Pathol.* **2015**, *97*. [**Google Scholar**] [**CrossRef**]

10. Cruz, S.S.; Roberts, A.G.; Prior, D.A.; Chapman, S.; Oparka, K.J. Cell-to-cell and phloem-mediated transport of Potato virus X. The role of virions. *Plant Cell* **1998**, *10*, 495–510. [**Google Scholar**] [**CrossRef**] [**PubMed**] [**Green Version**]

11. Aguilar, J.M.; Hernandez-Gallardo, M.D.; Cenis, J.L.; Lacasa, A.; Aranda, M.A. Complete sequence of the Pepino mosaic virus RNA genome. *Arch. Virol.* **2002**, *147*, 2009–2015. [**Google Scholar**] [**CrossRef**]

12. Morozov, S.Y.; Solovyev, A.G. Triple gene block: Modular design of a multifunctional machine for plant virus movement. *J. Gen. Virol.* **2003**, *84*, 1351–1366. [**Google Scholar**] [**CrossRef**] [**PubMed**]

13. Ling, K.-S. Pepino mosaic virus on tomato seed: Virus location and mechanical transmission. *Plant Dis.* **2008**, *92*, 1701–1705. [**Google Scholar**] [**CrossRef**] [**PubMed**] [**Green Version**]

14. Hanssen, I.M.; Mumford, R.; Blystad, D.-R.; Cortez, I.; Hasiów-Jaroszewska, B.; Hristova, D.; Pagán, I.; Pereira, A.-M.; Peters, J.; Pospieszny, H. Seed transmission of Pepino mosaic virus in tomato. *Eur. J. Plant Pathol.* **2010**, *126*, 145. [**Google Scholar**] [**CrossRef**]

15. Hanssen, I.M.; Gutiérrez-Aguirre, I.; Paeleman, A.; Goen, K.; Wittemans, L.; Lievens, B.; Vanachter, A.C.R.C.; Ravnikar, M.; Thomma, B. Cross protection or enhanced symptom display in greenhouse tomato co-infected with different Pepino mosaic virus isolates. *Plant Pathol.* **2010**, *59*, 13–21. [**Google Scholar**] [**CrossRef**] [**Green Version**]

16. Agüero, J.; Gómez-Aix, C.; Sempere, R.N.; García-Villalba, J.; García-Núñez, J.; Hernando, Y.; Aranda, M.A. Stable and broad spectrum cross-protection against Pepino mosaic virus attained by mixed infection. *Front. Plant Sci.* **2018**, *9*, 1810. [**Google Scholar**] [**CrossRef**] [**PubMed**]

17. Perry, K.L.; Zhang, L.; Palukaitis, P. Amino acid changes in the coat protein of Cucumber mosaic virus differentially affect transmission by the Aphids *Myzuspersicae* and Aphis *gossypii*. *Virology* **1998**, *242*, 204–210. [**Google Scholar**] [**CrossRef**] [**PubMed**]

18. Montasser, M.S.; Tousignant, M.E.; Kaper, J.M. Viral satellite RNAs for the prevention of Cucumber mosaic virus (CMV) disease in field-grown pepper and melon plants. *Plant Dis.* **1998**, *82*, 1298–1303. [**Google Scholar**] [**CrossRef**]

19. De la Torre-Almaráz, R.; Pallás, V.; Sánchez-Navarro, J.A. First report of Cucumber mosaic virus (CMV) and CARNA-5 in Carnation in Mexico. *Plant Dis.* **2016**, *100*, 1509. [**Google Scholar**] [**CrossRef**]

20. Moxon, S.; Jing, R.; Szittya, G.; Schwach, F.; Rusholme Pilcher, R.L.; Moulton, V.; Dalmay, T. Deep sequencing of tomato short RNAs identifies microRNAs targeting genes involved in fruit ripening. *Genome Res.* **2008**, *18*, 1602–1609. [**Google Scholar**] [**CrossRef**] [**Green Version**]

21. Zhai, J.; Jeong, D.H.; De Paoli, E.; Park, S.; Rosen, B.D.; Li, Y.; Gonzalez, A.J.; Yan, Z.; Kitto, S.L.; Grusak, M.A.; et al. MicroRNAs as master regulators of the plant NB-LRR defense gene family via the production of phased, trans-acting siRNAs. *Genes Dev.* **2011**, *25*, 2540–2553. [**Google Scholar**] [**CrossRef**] [**Green Version**]

22. Huang, J.; Yang, M.; Lu, L.; Zhang, X. Diverse functions of small RNAs in different plant–pathogen communications. *Front. Microbiol.* **2016**, *7*, 1552. [**Google Scholar**] [**CrossRef**] [**PubMed**] [**Green Version**]

23. Yang, Y.; Liu, T.; Shen, D.; Wang, J.; Ling, X.; Hu, Z.; Chen, T.; Hu, J.; Huang, J.; Yu, W. Tomato yellow leaf curl virus intergenic siRNAs target a host long noncoding RNA to modulate disease symptoms. *PLoSPathog.* **2019**, *15*, e1007534. [**Google Scholar**] [**CrossRef**] [**PubMed**]

24. Syller, J. Facilitative and antagonistic interactions between plant viruses in mixed infections. *Mol. Plant Pathol.* **2012**, *13*, 204–216. [**Google Scholar**] [**CrossRef**] [**PubMed**]

25. Pacheco, R.; García-Marcos, A.; Barajas, D.; Martiáñez, J.; Tenllado, F. PVX–potyvirus synergistic infections differentially alter microRNA accumulation in *Nicotiana benthamiana. Virus Res.* **2012**, *165*, 231–235. [**Google Scholar**] [**CrossRef**] [**PubMed**]

26. Mascia, T.; Gallitelli, D. Synergies and antagonisms in virus interactions. *Plant Sci.* **2016**, *252*, 176–192. [**Google Scholar**] [**CrossRef**]

27. Zhang, X.-F.; Zhang, S.; Guo, Q.; Sun, R.; Wei, T.; Qu, F. A new mechanistic model for viral cross protection and superinfection exclusion. *Front. Plant Sci.* **2018**, *9*, 40. [**Google Scholar**] [**CrossRef**] [**Green Version**]

28. Ratcliff, F.G.; MacFarlane, S.A.; Baulcombe, D.C. Gene silencing without DNA. rna-mediated cross-protection between viruses. *Plant Cell* **1999**, *11*, 1207–1216. [**Google Scholar**] [**CrossRef**] [**PubMed**] [**Green Version**]

29. Powell, C.A.; Pelosi, R.R.; Rundell, P.A.; Cohen, M. Breakdown of cross-protection of grapefruit from decline-inducing isolates of Citrus tristeza virus following introduction of the brown citrus aphid. *Plant Dis.* **2003**, *87*, 1116–1118. [**Google Scholar**] [**CrossRef**] [**Green Version**]

30. Elena, S.F. Evolutionary constraints on emergence of plant RNA viruses. *Recent Adv. Plant Virol.* **2011**, *14*, 283–300. [**Google Scholar**]

31. Aramburu, J.; Galipienso, L. First report in Spain of a variant of Tomato mosaic virus (ToMV) overcoming the Tm-22 resistance gene in tomato *(Lycopersicon esculentum). Plant Pathol.* **2005**, *54*, 566. [**Google Scholar**] [**CrossRef**]

32. Ling, K.-S.; Wechter, W.P.; Jordan, R. Development of a one-step immunocapture real-time TaqMan RT-PCR assay for the broad spectrum detection of Pepino mosaic virus. *J. Virol. Methods* **2007**, *144*, 65–72. [**Google Scholar**] [**CrossRef**] [**PubMed**]

33. Wijkamp, I.; van Lent, J.; Kormelink, R.; Goldbach, R.; Peters, D. Multiplication of Tomato spotted wilt virus in its insect vector, *Frankliniella occidentalis*. *J. Gen. Virol.* **1993**, *74*, 341–349. [**Google Scholar**] [**CrossRef**] [**PubMed**]

34. YaoLiang, Z.; ZhongJian, S.; Jiang, Z.; XiaoLi, L.; Gong, C.; RuDuo, L. Preliminary characterization of Pepino mosaic virus Shanghai isolate (PepMV-Sh) and its detection with ELISA. *Acta Agric. Shanghai* **2003**, *19*, 90–92. [**Google Scholar**]

35. Galipienso, L.; Herranz, M.C.; Pallás, V.; Aramburu, J. Detection of a tomato strain of Parietaria mottle virus (PMoV-T) by molecular hybridization and RT-PCR in field samples from north-eastern Spain. *Plant Pathol.* **2005**, *54*, 29–35. [**Google Scholar**] [**CrossRef**]

36. Kumar, S.; Prakash, H.S. Detection of tobacco mosaic virus and tomato mosaic virus in pepper seeds by enzyme linked immunosorbent assay (ELISA). *Arch. Phytopathol. Plant Prot.* **2016**, *49*, 59–63. [**Google Scholar**] [**CrossRef**]

37. Bald-Blume, N.; Bergervoet, J.H.W.; Maiss, E. Development of a molecular assay for the general detection of tospoviruses and the distinction between tospoviral species. *Arch. Virol.* **2017**, *162*, 1519–1528. [**Google Scholar**] [**CrossRef**]

38. Carpino, C.; Elvira-González, L.; Rubio, L.; Peri, E.; Davino, S.; Galipienso, L. A comparative study of viral infectivity, accumulation and symptoms induced by Broad bean wilt virus 1 isolates. *J. Plant Pathol.* **2019**, *101*, 275–281. [**Google Scholar**] [**CrossRef**]

39. Ferriol, I.; Ruiz-Ruiz, S.; Rubio, L. Detection and absolute quantitation of Broad bean wilt virus 1 (BBWV-1) and BBWV-2 by real time RT-PCR. *J. Virol. Methods* **2011**, *177*, 202–205. [**Google Scholar**] [**CrossRef**] [**PubMed**]

40. Falgueras, J.; Lara, A.J.; Fernández-Pozo, N.; Cantón, F.R.; Pérez-Trabado, G.; Claros, M.G. SeqTrim: A high-throughput pipeline for pre-processing any type of sequence read. *BMC Bioinform.* **2010**, *11*, 38. [**Google Scholar**] [**CrossRef**] [**PubMed**] [**Green Version**]

41. Rahman, R.-U.; Gautam, A.; Bethune, J.; Sattar, A.; Fiosins, M.; Magruder, D.S.; Capece, V.; Shomroni, O.; Bonn, S. Oasis 2: Improved online analysis of small RNA-seq data. *BMC Bioinform.* **2018**, *19*, 54. [**Google Scholar**] [**CrossRef**] [**Green Version**]

42. Zhao, G.; Yu, H.; Liu, M.; Lu, Y.; Ouyang, B. Identification of salt-stress responsive microRNAs from *Solanum lycopersicum* and *Solanum pimpinellifolium*. *Plant Growth Regul.* **2017**, *83*, 129–140. [**Google Scholar**] [**CrossRef**]

43. Prigigallo, M.I.; Križnik, M.; De Paola, D.; Catalano, D.; Gruden, K.; Finetti-Sialer, M.M.; Cillo, F. Potato virus Y infection alters small RNA metabolism and immune response in tomato. *Viruses* **2019**, *11*, 1100. [**Google Scholar**] [**CrossRef**] [**PubMed**] [**Green Version**]

44. Long, R.; Li, M.; Kang, J.; Zhang, T.; Sun, Y.; Yang, Q. Small RNA deep sequencing identifies novel and salt-stress-regulated microRNAs from roots of *Medicago sativa* and *Medicago truncatula*. In *The Model Legume Medicago truncatula*; Wiley: New York, NY, USA, 2020; pp. 963–976. [**Google Scholar**]

45. Alcala-Briseno, R.I.; Coskan, S.; Londono, M.A.; Polston, J.E. Genome sequence of Southern tomato virus in asymptomatic tomato "Sweet Hearts. " *Genome Announc.* **2017**, *5*, 10–1128. [**Google Scholar**] [**CrossRef**] [**PubMed**] [**Green Version**]

46. Fukuhara, T.; Tabara, M.; Koiwa, H.; Takahashi, H. Effect of asymptomatic infection with Southern tomato virus on tomato plants. *Arch. Virol.* **2019**, *165*, 11–20. [**Google Scholar**] [**CrossRef**]

47. Chávez-Calvillo, G.; Contreras-Paredes, C.A.; Mora-Macias, J.; Noa-Carrazana, J.C.; Serrano-Rubio, A.A.; Dinkova, T.D.; Carrillo-Tripp, M.; Silva-Rosales, L. Antagonism or synergism between Papaya ringspot virus and Papaya mosaic virus in *Carica papaya* is determined by their order of infection. *Virology* **2016**, *489*, 179–191. [**Google Scholar**] [**CrossRef**] [**Green Version**]

48. Landeo-Ríos, Y.; Navas-Castillo, J.; Moriones, E.; Cañizares, M.C. Supresión viral del silenciamientopor RNA enplantas. *Rev. Fitotec. Mex.* **2017**, *40*, 181–197. [**Google Scholar**] [**CrossRef**]

49. Moreno, A.B.; López-Moya, J.J. When viruses play team sports: Mixed infections in plants. *Phytopathology* **2020**, *110*, 29–48. [**Google Scholar**] [**CrossRef**] [**PubMed**]

50. Roossinck, M.J. The good viruses: Viral mutualistic symbioses. *Nat. Rev. Microbiol.* **2011**, *9*, 99. [**Google Scholar**] [**CrossRef**]

51. Gómez, P.; Sempere, R.N.; Amari, K.; Gómez-Aix, C.; Aranda, M.A. Epidemics of Tomato torrado virus, Pepino mosaic virus and Tomato chlorosis virus in tomato crops: Do mixed infections contribute to torrado disease epidemiology? *Ann. Appl. Biol.* **2010**, *156*, 401–410. [**Google Scholar**] [**CrossRef**]

52. Crespo, O.; Robles, C.; Ruiz, L.; Janssen, D. Antagonism of Cucumber green mottle mosaic virus against Tomato leaf curl New Delhi virus in zucchini and cucumber. *Ann. Appl. Biol.* **2020**, *176*, 147–157. [**Google Scholar**] [**CrossRef**]

53. Guo, W.; Wu, G.; Yan, F.; Lu, Y.; Zheng, H.; Lin, L.; Chen, H.; Chen, J. Identification of novel *Oryza sativa* miRNAs in deep sequencing-based small RNA libraries of rice infected with Rice stripe virus. *PLoS ONE* **2012**, *7*, e46443. [**Google Scholar**] [**CrossRef**] [**PubMed**]

54. Yin, J.; Liu, M.; Ma, D.; Wu, J.; Li, S.; Zhu, Y.; Han, B. Identification of circular RNAs and their targets during tomato fruit ripening. *Postharvest Biol. Technol.* **2018**, *136*, 90–98. [**Google Scholar**] [**CrossRef**]

55. Stare, T.; Ramšak, Ž.; Križnik, M.; Gruden, K. Multiomics analysis of tolerant interaction of potato with Potato virus Y. *Sci. Data* **2019**, *6*, 250. [**Google Scholar**] [**CrossRef**]

56. Naqvi, A.R.; Haq, Q.M.R.; Mukherjee, S.K. MicroRNA profiling of tomato leaf curl new delhi virus (TOLCNDV) infected tomato leaves indicates that deregulation of mir159/319 and mir172 might be linked with leaf curl disease. *Virol. J.* **2010**, *7*, 281. [**Google Scholar**] [**CrossRef**] [**PubMed**] [**Green Version**]

57. Tagami, Y.; Inaba, N.; Kutsuna, N.; Kurihara, Y.; Watanabe, Y. Specific enrichment of miRNAs in Arabidopsis thaliana infected with Tobacco mosaic virus. *DNA Res.* **2007**, *14*, 227–233. [**Google Scholar**] [**CrossRef**] [**Green Version**]

58. Niu, X.; Sun, Y.; Chen, Z.; Li, R.; Padmanabhan, C.; Ruan, J.; Kreuze, J.; Ling, K.; Fei, Z.; Gao, S. Using small RNA-seq data to detect siRNA duplexes induced by plant viruses. *Genes* **2017**, *8*, 163. [**Google Scholar**] [**CrossRef**] [**Green Version**]

59. Turco, S.; Golyaev, V.; Seguin, J.; Gilli, C.; Farinelli, L.; Boller, T.; Schumpp, O.; Pooggin, M.M. Small RNA-omics for virome reconstruction and antiviral defense characterization in mixed infections of cultivated Solanum plants. *Mol. Plant-Microbe Interact.* **2018**, *31*, 707–723. [**Google Scholar**] [**CrossRef**] [**PubMed**] [**Green Version**]

60. Smith, N.A.; Eamens, A.L.; Wang, M.-B. Viral small interfering RNAs target host genes to mediate disease symptoms in plants. *PLoSPathog.* **2011**, *7*, e1002022. [**Google Scholar**] [**CrossRef**] [**Green Version**]

Chapter-15

The Global Dimension of Tomato Yellow Leaf Curl Disease: Current Status and Breeding Perspectives

1. Tomato Yellow Leaf Curl Disease Causing Agents: Tomato Yellow Leaf Curl Virus (TYLCV) and TYLCV-Like Viruses

A large number of viruses can infect tomato (*Solanum lycopersicum* L.) [**1**]. These viruses directly or indirectly cause severe reductions in yield and fruit quality. Among them, tomato yellow leaf curl virus (TYLCV) threatens tomato production and currently ranks third after tobacco mosaic virus and tomato spotted wilt virus on the list of the most important plant viruses worldwide [**2,3**]. TYLCV and 12 TYLCV-like viruses belong to a complex of viruses causing tomato yellow leaf curl disease (TYLCD) [**4**]. The typical symptoms associated with TYLCD in tomato are leaf yellowing, curling and a marked stunting of plants (**Figure 1**). At the final stage of disease development, flowers and fruits are abscised followed by cessation of plant growth [**5**].

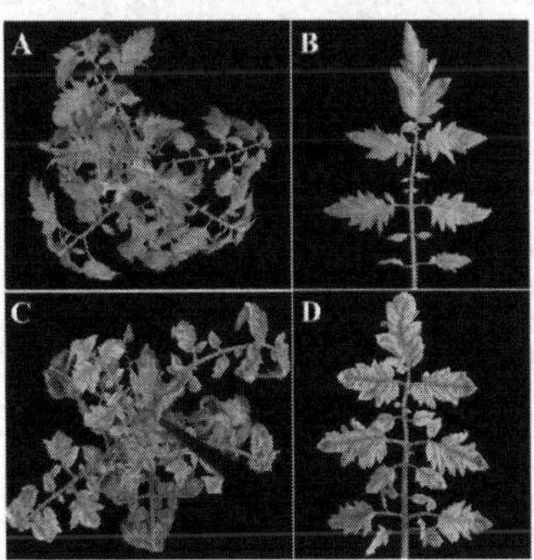

Figure 1. Typical symptoms associated with tomato yellow leaf curl disease. (A,B) Uninfected tomato plant of cv. Moneymaker. (C,D) Tomato plant infected with tomato yellow leaf curl virus (TYLCV). Photos were taken 37 days post inoculation using *Agrobacterium*-mediated inoculation of the infectious clone of TYLCV-Israel strain (TYLCV-IL).

TYLCD-causing viruses belong to the genus *Begomovirus* in the family *Geminiviridae* [**6**]. Begomoviruses possess one or two circular single-stranded DNA (ssDNA) genome(s) each of about 2.7–2.8 kb. TYLCV and most TYLCV-like begomoviruses have monopartite genomes consisting of one ssDNA molecule, except for tomato yellow leaf curl Kanchanaburi virus (TYLCKaV) and tomato yellow leaf curl Thailand virus (TYLCTHV). These two begomoviruses are bipartite, with a genome containing two ssDNA molecules, DNA-A and DNA-B (**Figure 2**) [**7**,**8**]. The monopartite TYLCV genome, equivalent to DNA-A of bipartite begomoviruses, contains six open reading frames (ORFs) organized in two transcriptional directions separated by an intergenic region (IR) (**Figure 2**) [**8**]. Based on the function, the proteins encoded by the six ORFs have been named: coat protein (CP/V1), virus movement protein (MP/V2), replication-associated protein (Rep/C1), transcriptional activation protein (TrAP/C2), replication enhancer protein (REn/C3) and a protein determining symptom expression and virus spreading (C4) [**9**]. Bipartite begomoviruses encode the nuclear shuttle protein (BV1/NSP) and movement protein (BC1/MP) on the DNA-B component [**10**]. All six proteins of monopartite begomoviruses/DNA-A of bipartite begomoviruses and both proteins encoded by the DNA-B component of bipartite begomoviruses are essential for successful systemic infection of host plants [**9**,**10**].

Many monopartite begomoviruses including two TYLCV-like viruses, tomato yellow leaf curl China virus (TYLCCNV) and tomato yellow leaf curl Yunnan virus (TYLCYnV), have been shown to associate with satellite DNA molecules, known as alphasatellites and betasatellites (**Figure 2**) [**11**,**12**,**13**,**14**]. They are small circular ssDNA molecules of approximately 1350 nucleotides in length. Betasatellites code for one single protein βC1, therefore they rely on helper viruses for replication, cell-to-cell and systemic movement, encapsidation, and insect vector transmission [**14**]. Emerging evidence shows that co-infection with betasatellites is essential for symptom induction by many monopartite begomoviruses such as TYLCCNV and TYLCYnV [**14**,**15**] and enhancing disease severity by a few bipartite begomoviruses, as is the case for TYLCTHV [**12**,**16**,**17**]. Alphasatellites are mainly present associated with monopartite begomoviruses, and are also frequently associated with betasatellites, although their role in infection is not yet fully understood [**14**,**18**].

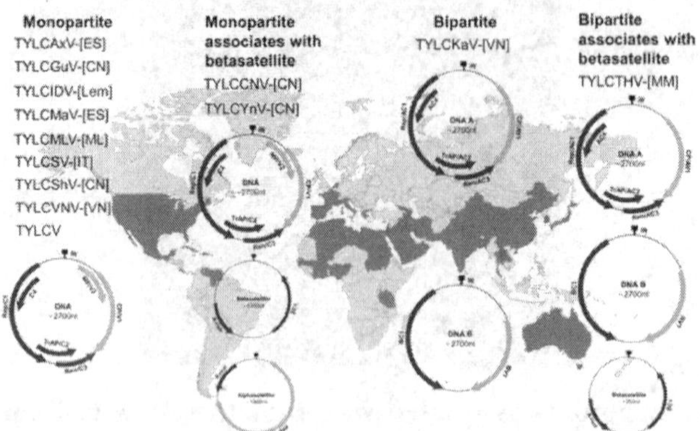

Figure 2. Thirteen virus species causing tomato yellow leaf curl disease (TYLCD) according to the International Committee on Taxonomy of Viruses list of 2019. These viruses are grouped into monopartite and bipartite viruses based on the

number of DNA genome components. Their association with alphasatellites and betasatellites are shown. Countries or territories where TYLCV has been officially reported are highlighted in orange-red in the background world map [19]. The complete names of the viruses are tomato yellow leaf curl (TYLC) Axarquia virus (TYLCAxV-[ES]), TYLC-Guangdong virus (TYLCGuV-[CN]), TYLC-Indonesia virus (TYLCIDV-[Lem]), TYLC-Malaga virus (TYLCMaV-[ES]), TYLC-Mali virus (TYLCMLV-[ML]), TYLC-Sardinia virus (TYLCSV-[IT]), TYLC-Shuangbai virus (TYLCShV-[CN]), TYLC-Vietnam virus (TYLCVNV-[VN]), TYLC-China virus (TYLCCNV-[CN]), TYLC-Yunnan virus (TYLCYnV-[CN]), TYLC-Kanchanaburi virus (TYLCKaV-[VN]), TYLC-Thailand virus (TYLCTHV-[MM]).

2. Global Spreading of TYLCD: Efficient Transmission of Whitefly Vector and Dynamic Nature of the Virus

TYLCD was first reported in the Jordan Valley, Israel, in the late 1930s, and it was not until the 1960s that TYLCV was officially identified as the causal virus of this disease [**20**]. Since then, the emergence of TYLCD and its subsequent spreading have been extremely rapid into the Mediterranean basin and most tropical and sub-tropical regions of the world (**Figure 2**) [**21,22**]. Nowadays, the disease is still spreading to new areas, with recent reported outbreaks in Costa Rica [**23**] and Trinidad and Tobago [**24**].

Although TYLCD can be found worldwide, only two strains, the Israel (TYLCV-IL) and Mild strains of TYLCV (TYLCV-Mld), are truly global TYLCD-causing agents [**25**]. Other begomoviruses associated with TYLCD have been found only in restricted regions, such as TYLCCNV and tomato yellow leaf curl Sardinia virus (TYLCSV) which have been limited to China and Mediterranean countries, respectively [**25,26**]. The global distribution of TYLCD is closely related to international trafficking/trading of planting material [**27**], and most importantly to a worldwide increase of the insect vector population and rapid evolution of virus variants [**20,25**].

2.1. An Efficient Transmission Vector: Whiteflies

Under natural conditions, TYLCV is transmitted exclusively by whiteflies (*Bemisiatabaci* Genn.) in either a persistent-circulative [**8**] or persistent-propagative manner [**28**]. A single whitefly is able to transmit TYLCV following an acquisition access period of 24 h. In order to reach up to 100% transmission efficiency, 5–15 whiteflies per tomato plant are needed [**29,30**].

B. tabaci is in fact a complex consisting of at least 24 cryptic species that differ in host range, virus-transmitting capacity, host plant adaptation, ability to induce physiological changes, and capacity of spreading and acquiring insecticide resistance [**31**]. Two species, Middle East–Asia Minor 1 (MEAM1, formerly known as B biotype) and Mediterranean (MED, formerly known as Q biotype), are considered as the most invasive and damaging species, which are also the predominant species that transmit TYLCV to tomato [**31,32**]. MED has a higher ability to develop insecticide resistance than other species, while MEAM1 is characterized by high fecundity and a wide host range [**29,33,34**]. Considering that begomoviruses are exclusively transmitted by *B. tabaci*, a change and/or increase of the vector population is one of the key factors associated with the high TYLCD prevalence [**7,25**]. Taking China as an example, the first invasion of *B. tabaci* MEAM1 appeared in the mid1990-s and was subsequently replaced by MED in 2003. Within a few years after its introduction, MED has become the predominant species in China, has invaded many areas and has been responsible for TYLCD outbreaks [**33**].

2.2. Driving Forces of Begomoviruses Evolution: Mutation and Recombination

TYLCV has a great potential to change due to factors including mutation and genetic recombination which enable rapid adaptation of the TYLCV complex to everchanging environmental conditions [**25**]. In general, mutation frequency should be lower for ssDNA viruses compared to RNA viruses due to the fact that they take advantage of host DNA polymerases for their replication [**27,35**]. However, the estimated substitution rates of TYLCV are approximately 10^{-4} to $^{5-}10$ nucleotide substitutions per site per year which is equivalent to that detected in RNA viruses [**27,35**]. Although mechanisms triggering high substitution rates found in ssDNA viruses have yet to be fully assessed, it appears that ssDNA viruses are able to escape host DNA polymerase proof-reading repair mechanisms of the replication errors [**7,27,35**]. By studying the TYLCCNV population [**36**], it was shown that mutations are not equally distributed along the genome, but are concentrated in the non-coding IR, Rep/C1 and C4 regions. The substitutions of guanine (G) to adenine (A) and cytosine (C) to thymine (T) are dominant for viral populations, whereas the reverse transitions (A to G and T to C) were not detected [**36**]. Considering that transcriptional gene silencing (TGS) provides a generic response to DNA viruses [**37**], the methylation directed by the TGS response takes place on cytosine/guanine. Reduced cytosine/guanine levels could, therefore, lead to lower efficiency in methylation.

Recombinant viruses frequently occur in nature. This has contributed greatly to the genetic diversification of TYLCV populations [**7,27,38**]. Well-documented examples of recombination having an association with (recent) outbreaks and/or epidemics in the Mediterranean Basin and the Middle East are shown in **Figure 3**. The analysis pinpointed two major groups of the TYLCV complex, one group with TYLCV backbone and the other group with TYLCSV backbone. Several new virus strains have been shown to be recombinants between TYLCV and TYLCSV [**7,39,40**]. The recombination sites are typically found in the regions of *Rep/C1* and *C4* genes, which are referred to as recombination hot spots (**Figure 3**). Rep/C1 and C4-encoded proteins play an important role in virulence [**41,42**]. The resulting recombinants are naturally selected for a better fitness through an efficient interaction with host factors [**7,27**].

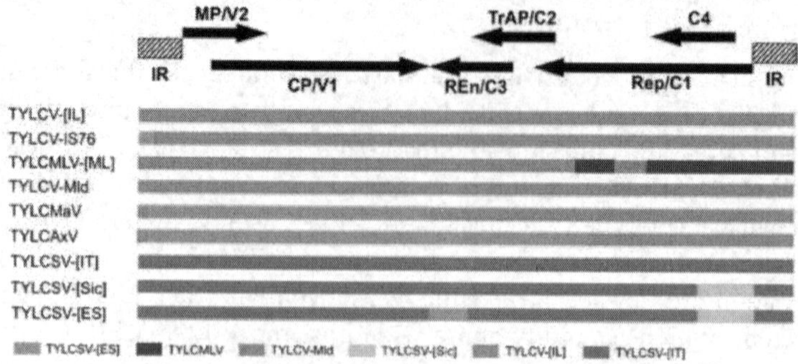

Figure 3. Diagram representing the recombination map of TYLCV complex. Blocks with same color shading represent regions of high identity. The positions of open

reading frames (horizontal arrows, CP/V1, MP/V2, Rep/C1, TrAP/C2, Ren/C3, C4) and intergenic regions (IR) are represented on the top of the graph. Tomato yellow leaf curl (TYLC) virus-Israel (TYLCV-[IL]), TYLC Mali virus—Mali (TYLCMLV-[ML]), TYLC virus—Mild [Israel] (TYLCV-Mld [IL]), TYLC Malaga virus-Spain (TYLCMaV-[ES]), TYLC Axarquia virus- Spain (TYLCAxV-[ES]), TYLC Sardinia virus—Italy (TYLCSV-[IT]), TYLC Sardinia virus—Sicily (TYLCSV-[Sic]), TYLC Sardinia virus—Spain (TYLCSV-[ES]).

3. Increasing Global Significance of TYLCD

In the past decades the occurrence of TYLCD has been reported in an increasing number of countries, showing that this major viral disease is still a spreading threat [**20**,**25**].

3.1. Mixed Infection: An Incubator of New Recombinant Viruses

During mixed infection, a high degree of intra- and inter-species recombination has been observed within the TYLCV complex or among begomoviruses [**7**]. For example, TYLCV-IL is the result of recombination between TYLCV-Mld and tomato leaf curl Karnataka virus (ToLCKV, a tomato-infecting begomovirus) which occurred in nature during mixed infection [**20**,**43**,**44**]. The Sardinia strain of TYLCSV (TYLCSV-Sar) likely emerged from a South African cassava mosaic virus (SACMV, a cassava begomovirus) ancestor by genetic exchange through recombination [**43**,**45**]. Co-infection of tomato plants with TYLCV and TYLCSV led to the emergence of two recombinant viruses associated with TYLCD, tomato yellow leaf curl Málaga virus (TYLCMaV) and tomato yellow leaf curl Axarquia virus (TYLCAxV) (**Figure 3**), which have acquired a broader host range than either of the parents [**25**]. A new virus strain (TYLCV-IS76) has arisen due to a recombination event between TYLCV-IL and the Spanish strain of TYLCSV (TYLCSV-ES) (**Figure 3**) [**39**,**46**]. Very recent natural recombinant strains, namely TYLCV-IL [IT:Sic23:16] [**47**] and TYLCV-IL-[IT:Sar IS141:16] [**48**], emerged by genetic exchange of parental strains TYLCV and TYLCSV. Both recombinants have been frequently detected in the field in Sicily (Italy) and Sardinia (Italy), respectively [**47**,**48**].

3.2. Alarming Scenario: Role of Betasatellites in TYLCV Epidemics

Association of TYLCV complex viruses with betasatellites is another factor linked with global TYLCD epidemics. The βC1 protein encoded by betasatellites has been shown to suppress the antiviral RNA interference (RNAi) pathways which provides benefit to the helper viruses for a successful infection [**14**]. βC1 protein can counteract the post-transcriptional gene silencing (PTGS) pathway by upregulating a Calmodulin-like protein (CaM), which in turn represses the expression of RNA-dependent RNA polymerase 6 (RDR6) by targeting the Suppressor of Gene Silencing 3 (SGS3, a co-factor of RDR6) for degradation [**49**,**50**]. βC1 is also able to mediate TGS suppression by physically interacting with and inhibiting S-adenosyl homocysteine hydrolase (SAHH) which is needed to maintain the methylation cycle [**51**]. In addition, begomovirus-betasatellite co-infection manipulates host insect defense to enhance whitefly behavior and performance. βC1 directly interacts with the transcription factor MYC2 to suppress plant terpene biosynthesis, thereby reducing whitefly resistance [**52**]. Compared to plants infected with TYLCCNV, plants co-infected with a betasatellite were shown to attract more whiteflies, with female whiteflies laying more eggs, which developed faster into adult whiteflies [**52**].

Further, betasatellites are capable of being trans-replicated by a wide range of helper begomoviruses in mixed infection. For example, monopartite begomoviruses including the Oman strain of TYLCV (TYLCV-Om) [**53**], tomato yellow leaf curl Mali virus (TYLCMLV) [**54**], and TYLCV-IL [**55**] can trans-replicate betasatellites associated with tomato leaf curl virus (ToLCV), cotton leaf curl Gezira virus (CLCuGV), and honeysuckle yellow vein mosaic virus (HYVMV), respectively. These observations pose an alarming threat that upon polyphagous feeding of whitefly, monopartite begomoviruses may form novel disease complexes by acquiring betasatellites from other begomoviruses [**14**].

Usually the resulting new disease complexes are characterized by more severe symptoms consisting of extremely stunted and distorted plants (**Figure 4**) [**33,55**]. In *N. benthamiana*, co-replication of TYLCV with ageratum yellow vein betasatellite (AYVB) increases the symptom severity level [**56,57**]. Co-infection of TYLCMLV and cotton leaf curl Gezira betallatellite (CLCuGB) by cross-feeding of whiteflies resulted in more severe symptoms [**33,54**]. In the presence of honeysuckle yellow vein mosaic betasatellite (HYVMVB), TYLCV-infected tomato plants developed more severe stunting symptoms [**55**]. A recent study showed that co-replication of TYLCV-IL with CLCuGB leads to a significant increase of TYLCV symptoms (**Figure 4**) [**58**].

Figure 4. New tomato disease complex with more severe symptoms caused by re-assortments of TYLCV-IL and a betasatellite. Leaf symptoms on tomato plants infected with TYLCV-IL alone or co-infection with CLCuGB (adapted from Conflon et al. [58] with permission). TYLCV-IL: tomato yellow leaf curl virus-Israel; CLCuGB: cotton leaf curl Gezira betasatellite.

In the Mediterranean basin and the Middle East, which are proposed to be centers of both TYLCV complex origin and diversification, the first betasatellite (i.e., CLCuGB) associated with either TYLCV-IL or TYLCV-Mld in tomato plants was identified very recently in Israel [**12,21,59**]. This is of great concern for tomato growers worldwide but especially in the Mediterranean region. So far, 61 betasatellite species have been officially reported [**4**]. China and the Indian subcontinent host more than 90% of betasatellite species. Preventing further spreading of betasatellites to the genetic pool of TYLCV complex can efficiently limit the appearance of new begomovirus-betasatellite disease complexes.

3.3. Emerging Problem: Plant Host Range Expansion

TYLCD infection has been detected in 49 plant species including economically important crops and weed species belonging to 16 families [**60**]. Alternate hosts that act as virus inoculum sources enable the persistence and spread of the virus especially in crop-free periods. Although TYLCV has a diverse host range in addition to tomato, its detection in alternate hosts is rare [**61**], and TYLCD global spread is associated primarily with tomato [**62**].

Another important and wide spreading leaf curl disease of tomato is tomato leaf curl disease (ToLCD) [**63**]. Fifty-five distinct viruses have been associated with ToLCD [**4**], with one of the most important being tomato leaf curl New Delhi virus (ToLCNDV) which was initially identified in the India sub-continent. ToLCNDV is a bipartite begomovirus species. It causes the most predominant disease affecting tomato and its epidemics were limited to Asian countries [**64**]. However, in recent years, ToLCNDV has been extending its host range to a broader spectrum, including Cucurbitaceae, Euphorbiaceae, Fabaceae, as well as Malvaceae, and is spreading rapidly to new geographical regions, including the Middle East and the Mediterranean basin [**64**]. Recent outbreaks of ToLCNDV in the Mediterranean basin have been associated with the emergence of a novel strain, ToLCNDV-ES, which affects mainly cucurbits including cucumber (*Cucumis sativus* L.), melon (*C. melo* L.) and zucchini (*Cucurbita pepo* L.). The isolates of the ToLCNDV-ES strain are well adapted to infect cucurbits, but have limited ability to infect tomato [**65,66**].

In different begomoviruses, betasatellites also affect the host ranges. For example, cassava, but not ageratum, is the host of the bipartite begomovirus Sri Lankan cassava mosaic virus (SLCMV). However, in the presence of a betasatellite associated with AYVV, SLCMV is also able to infect ageratum and induces typical yellow vein symptoms [**14,67**].

4. TYLCD Control: Mapped TYLCV Resistance Genes

In practice, preventing viruses from infecting the host mainly requires the control of virus vectors by the use of appropriate physical barriers (traps and screens) and chemical agents (insecticides). However, building physical barriers is not always feasible and the application of chemical compounds can result in the development of resistance against the used compound by whiteflies [**68,69**]. The best crop protection method is host resistance against viruses and/or whiteflies. In tomato breeding for TYLCV resistance, the most prominent approach is transferring virus resistance genes from wild tomato relatives into cultivated tomato.

So far, six resistance genes (*Ty-1*, *Ty-2*, *Ty-3*, *Ty-4*, *ty-5* and *Ty6-*) were identified from a few tomato wild species, including *S. habrochaites* and *S. chilense* (**Table 1**) [**70,71,72,73,74,75**]. Four of these TYLCV resistance genes (*Ty-1*/*Ty-3*, *Ty-2* and *ty5-*) have been cloned, representing three classes of antiviral defense mechanisms (**Table 1**) [**76,77,78,79**].

Table 1. Tomato (*Solanum lycopersicum*) wild relatives as resistance sources for tomato yellow leaf curl disease.

Resistance Gene [a]	Genetic Source		Chro-mosome	Inheritance Pattern	Gene Identity [c]	Reference
	Accession/Line [b]	Species				
Ty-1	LA1969	S. chilense	6	Dominant	RDR	[75,78]
Ty-2	B6013	S. habrochaites	11	Dominant	NLR	[71,77,79]
Ty-3	LA2779	S. chilense	6	Dominant	RDR	[73,78]
Ty-4	LA1932	S. chilense	3	Incomplete dominant		[74]
ty-5	Tyking	S. lycopersicum	4	Recessive	Pelota	[76]
Ty-6	LA2779	S. chilense	10	Incomplete dominant		[72]

[a] Bold font indicates cloned genes; [b] Tyking, the source of the ty-5 gene, is an old tomato cultivar; [c] RDR = RNA-dependent RNA polymerase; NLR = nucleotide binding leucine-rich repeat protein; Pelota = Message RNA Surveillance Factor Pelota.

Ty-1 and *Ty3-* originate from *S. chilense* accession LA1969 and LA2779, respectively, and are located on the long arm of tomato chromosome **73**] 6,**75,80**]. They are allelic to each other and encode an RNA-dependent RNA polymerase belonging to the RDRγ type with homology to *Arabidopsis* RDR4- ,3 and 5- (**Table 1**) [**78**]. *Ty-1* confers resistance to TYLCV based on enhanced TGS by increasing cytosine methylation of the viral genome [**81**]. *Ty-1*-mediated antiviral TGS response has been considered to be generic against geminiviruses as *Ty-1* was shown to also confer resistance to a bipartite begomovirus tomato severe rugose virus (ToSRV) [**81**] and the leafhopper-transmitted beet curly top virus (BCTV), a curtovirus [**57**]. However, co-inoculation with betasatellites expressing the βC1 protein compromises *Ty-1*-mediated resistance and induces disease symptoms [**57**]. Furthermore, mixed infection with the RNA virus cucumber mosaic virus (CMV) compromises the effectivity of resistance conferred by *Ty-1*, by interference of the CMV RNAi suppressor protein 2b on Argonaut 4 (AGO4) activity involved in TGS [**81,82**]. The emergence of a recombinant strain TYLCV-IS76 coincides with the increased use of *Ty-1* containing tomato varieties by farmers in Souss (Morocco) [**39,46**]. TYLCV-IS76 can accumulate better than its parental strains in *Ty-1*-carrying varieties that leads eventually to the displacement of both parental virus strains locally [**39,46**]. In Sicily (Italy), infection by recombinant variant TYLCV-IL-[IT:Sic23:16] resulted in TYLCD symptoms in tomato plants carrying the *Ty-1* gene. In the same geographical region, samples collected from plants without the *Ty1-* gene harbored a mixture of the recombinant TYLCV-IL-[IT:Sic23:16], and both parental begomoviruses [**47**]. In contrast, tomato plants with the *Ty-1* gene contain only recombinant genomes. In Sardinia (Italy), recombinant variant TYLCV-IL-[IT:Sar IS141:16] reduces the effectiveness of *Ty1-* mediated resistance as typical TYLCD symptoms were observed on *Ty1-* harboring tomato plants [**48**]. This variant was found to be positively selected in *Ty-1* resistant plants under field conditions [**83**]. Collectively, these examples show the limitation of *Ty-1*-mediated resistance. In many breeding programs worldwide, introgression of *Ty-1*/*Ty-3* into cultivated tomatoes has been predominant. Agricultural practices including monocultures (i.e., intensive utilization of *Ty-1*-carrying lines in the Mediterranean basin) are potentially harmful because they may facilitate the emergence of new begomoviruses/disease complexes and the spreading of epidemics [**39,46**].

Ty-2 was first reported in a tomato line H24 derived from *S. habrochaites* accession B6013 and was mapped on the long arm of chromosome 11 (**Table 1**) [**71,84**]. Fine-mapping of the *Ty-2* gene was a great challenge due to a chromosomal inversion present in *S. habrochaites* compared with *S. lycopersicum*, resulting in suppression of chromosome recombination [**85**]. By using intraspecific crosses between two *S. habrochaites* accessions, suppression of recombination in the *Ty2-* region was overcome, allowing the fine-mapping of the *Ty2-* gene [**85**]. *Ty-2* was shown to encode a nucleotide-binding leucine-rich repeat protein (NLR) [**77,79**]. The Rep/C1 protein of TYLCV presents the Avr determinant of *Ty-2*-based resistance [**77**].

The *ty-5* gene, a loss-of-function allele of the *Pelota* (*Pelo*) gene encoding a messenger RNA surveillance factor, hinders TYLCV multiplication, leading to resistance in tomato [**76**]. A tomato inbred line harboring *ty-5* displays resistance to monopartite begomoviruses associated with TYLCD [**86**] and the bipartite begomovirus tomato chlorotic mottle virus (ToCMoV) [**87**]. Pelo is involved in the ribosome recycling phase of protein synthesis [**76**], which is highly conserved among animals, plants and yeast [**88**]. pelo deficiency in *Drosophila* restricts replication of the RNA viruses cricket paralysis virus (CrPV), Drosophila C virus (DCV) and Drosophila X virus (DXV), and of the DNA virus invertebrate iridescent virus 6 (IIV**88**] (6]. The *Pelo* gene has also been shown in rice to be involved in resistance against bacterial blight disease caused by *Xanthomonas oryzae* pv. *oryzae* by elevating the salicylic acid pathway [**89,90**].

In addition to the previously mentioned *Ty*-genes, two other TYLCV resistance genes have been mapped, namely *Ty-4* and *Ty6-*. *Ty-4* originates from *S. chilense* accession LA1932 and has been mapped to the long arm of chromosome 3. *Ty4-* is reported to have a minor effect on TYLCV resistance, accounting only for %15.7 of the total variance [**74**]. *Ty-6*, originating from *S. chilense* accessions LA1938 or LA2779, is located on the long arm of chromosome **72**] 10,**91**]. *Ty-6* confers moderate resistance to TYLCV, but a high level of resistance to begomovirus tomato mottle virus (ToMoV). The most effective use of *Ty6-* is in combination with other *Ty*-genes such as *Ty-3* or *ty***92**] 5-].

5. Breeding Strategies: Mounting a Broad-Spectrum and Sustainable Begomovirus Resistance

At present, introgression of *Ty-1* or *Ty3-* into cultivated tomato has been the major focus in breeding programs worldwide. However, *Ty-1*-mediated resistance has been observed not to be effective in the field and during mixed infection [**39,46,47,48,57,81,93**]. Further, the breakdown of *Ty-2*-based resistance was reported by TYLCSV [**94**] and an isolate of the Mild strain of TYLCV (TYLCV-Mld) [**95**]. Therefore, efforts have been made to pyramid the *Ty*-genes. At the World Vegetable Centre in Taiwan, tomato lines carrying the *Ty-2* resistance gene in combination with other known *Ty*-genes (i.e., *Ty-1*/*Ty-3* and *ty5-*) have been generated, which are extensively used in breeding programs in many Asian countries and other regions of the world [**96,97**]. Additionally, there is an urgent need to further exploit wild tomato relatives for novel genes against TYLCD.

5.1. Fishing in the Gene Pool: Natural Variation of Wild Tomato Relatives

Germplasm screening for resistance to TYLCD has been performed by researchers worldwide ever since the mid-1980s, when TYLCD became a constraint of tomato production [**22,98,99**]. The emergence of resistance-breaking recombinant variants like TYLCV-IS76, TYLCV-IL-[IT:Sic23:16] and TYLCV-IL-[IT:Sar IS141:16] are very recent events (appearance in less than 5 years) [**39,46,47,48**]. As a result, breeders have to search constantly in the genetic pool for effective sources to tackle the rapid evolution of TYLCD-causing agents. The germplasm screened so far is rather extensive, representing a full range of genetic diversity of tomato. Highly resistant accessions exhibiting no TYLCD symptoms have been reported in a number of species (**Table 2**).

Table 2. Summary of previously identified symptomless, symptomatic and segregating accessions to tomato yellow leaf curl virus complex in wild tomato species.

Solanum spp. [a]	Number of Accessions [b]		
	Symptomless	Symptomatic	Segregating
S. arcanum	15	5	5
S. cheesmaniae/S. galapagense	0	9	0
S. chilense	54	6	4
S. chmielewskii	1	3	0
S. corneliomulleri	30	8	9
S. habrochaites	13	52	14
S. huaylasense	4	0	0
S. lycopersicoides	1	13	0
S. neorickii	2	5	2
S. pennellii	2	42	2
S. peruvianum	69	39	20
S. pimpinellifolium	9	455	0

aThe number of accessions was summarized from the following articles, [97, 98, 100,-124].

b For simplification, phenotypic responses were categorized into three groups, symptomless, symptomatic and segregating. Accessions belonging to the symptomatic category may contain a certain level of resistance/tolerance.

It is clear from these studies that the vast majority of accessions from *S. chilense* are resistant and many resistant accessions can also be found in *S. peruvianum* and other species of its complex (i.e., *S. arcanum, S. huaylasense* and *S. corneliomulleri*) (**Table 2**). In different independent screenings, the highest levels of resistance are found in accessions of these two species. Most of these accessions remain symptomless throughout the whole disease test [**97,111,123**]. Moreover, resistance in many of these accessions is characterized by reduction in viral accumulation [**123**]. Wild tomato species (*S. cheesmaniae, S. habrochaites, S. neorickii, S. pennelli*, and *S. pimpinellifolium*) belong to the "esculentum" complex which can be crossed with the cultivated tomato [**125**]. However, these species generally do not confer a high level of TYLCV resistance (**Table 2**) [**97,116,119**].

Next to having a list of symptomless accessions, it is of great importance to determine the virus accumulation levels of symptomless genotypes. This will help to clarify whether the symptomless accessions are also virus-free or not. In tomato varieties containing any of the mapped TYLCV resistance loci (*Ty*-loci), viral replication is not completely blocked since virus accumulation is still detected in systemic tissues [**77,91,114,126,127,128**]. Clearly, the TYLCD epidemic is extremely difficult to control, but utilization of symptomless virus-free tomato varieties, if possible, could minimize the chances of emergence of recombinant viruses during mixed infection.

In order to identify novel resistance loci, it is important to investigate whether the resistance reported in the symptomless accessions in **Table 2** are controlled by allelic variants of the known *Ty*-genes. This may be achieved by comparing the mapping positions of the resistance loci using existing (functional) molecular markers and/or to apply virus-induced gene silencing (VIGS) approaches in combination with allele mining [**97**].

5.2. A Challenging Task: Introgression Breeding for TYLCV Resistance

Many pre-breeding populations and breeding lines have been developed focusing on the introgression of resistance derived from accessions of *S. chilense*, *S. peruvianum*, *S. habrochaites* and to a lesser extent, *S. pimpinellifolium*. *S. pimpinellifolium* is a close wild relative of *S. lycopersicum* that is easily crossable with the cultivated tomato [**125**]. Therefore, it was the first wild species used to develop TYLCV-resistant lines which in many cases resulted in only partially resistant lines [**109,129,130**]. In other studies, breeding lines were developed from several *S. pimpinellifolium* accessions that show resistance to TYLCV with no viral symptoms, including: Hirsute-INRA and LA **99**] 1478]; LA1582 [**131**]; LA1921 [**98**]; PI 407543 and PI 407544 [**103**]; and G1.1554 (CGN15528) [**132**]. *S. habrochaites* accessions are also easily crossable with the cultivated tomato [**125**]. Promising TYLCV-resistant lines were developed using accessions LA1777 and LA0386 [**121**] and EELM388 and EELM-889 [**120**].

Despite the high incompatibility, accessions of *S. chilense* and *S. peruvianum* have been well utilized as the most resistant sources. Exploitation of TYLCV resistance in *S. peruvianum* began from the incorporation of the resistance from accession PI 126935 **133**]] and PI 126944 [**113,117,134**]. Resistant *S. chilense* accessions (LA1932, LA1938, LA1960, LA1969, LA1971, LA2779 and LA3473) have been intensively applied into breeding practices [**92,114,135,136,137**]. Previous studies indicated that resistance in these accessions is mediated by functional *Ty-1*/*Ty-3* alleles [**97,109,138**]. Therefore, attempts to explore and identify new resistance genes using highly resistant *S. chilense* accessions only resulted in the identification of additional *Ty1-*/*Ty-3* alleles. The analyzed functional *Ty1-*/ *Ty-3* alleles differ in only a few amino acids [**97,138**]. Whether each allele displays similar or different characteristics (e.g., durability) remains to be investigated.

5.3. Pyramiding Resistance Genes: Towards Durable and Broad-Spectrum Resistance

To produce durable and broad-spectrum resistance, an essential approach would be to use multiple genes, a process known as pyramiding/stacking. Desirably, the stacked genes should confer different types of resistance. The chance of viruses overcoming polygenic resistance is substantially reduced compared to monogenic resistance. It requires a virus to accumulate various mutations with a low incidence and probably a fitness cost to adapt to pyramided resistance genes which is not likely to occur [**139**]. In light of the effectivity of the *Ty-1* gene against a broad spectrum of DNA viruses, pyramiding distinct gene(s) including *Ty1-* offers one possibility to achieve the goal. Such efforts in tomato have been shown to lead to enhanced resistance relative to the level in the presence of single genes [**122,140**].

5.4. Additional Source for Resistance: Dysfunctional Susceptibility Genes

Susceptibility (*S*) genes encode host proteins exploited by the pathogen to facilitate infection and establish a compatible interaction [**141**]. Utilization of *S* genes in resistance breeding, therefore, implies impairing their function in disease susceptibility [**141**]. Impaired *S* genes are known to provide broad-spectrum resistance and are effective not only against many if not all strains or races of a given pathogen, but also against multiple pathogens [**141,142**]. In addition, the presence and function of numerous *S* genes was found to be conserved between plant species, providing the possibility to impair the *S* gene in different plant species [**143**]. The *ty-5* gene represents a loss of function allele of the *Pelota* gene [**76**], showing that the *Pelota* gene is a host susceptibility factor of TYLCV. Loss-of-function mutations at the *Pelo* homologous locus have been shown to provide resistance to a broad range of geminiviruses [**144**]. In *Capsicum annuum*, plants containing the mutated pepper *Pelo* gene confer resistance to monopartite begomovirus pepper leaf curl virus (PLCV) and bipartite begomovirus pepper yellow leaf curl Indonesian virus (PepYLCIV) [**144**].

Impaired *S* genes confer resistance in a recessively inherited mode. Recessive resistance against viruses is found with a higher frequency compared to resistance against other types of plant pathogens where most of the reported resistance sources, until now, are dominant [**145**]. Viruses require many host factors to complete their infection cycle [**10**]. Many of the naturally occurring recessive resistance genes code for the eukaryotic translation initiation factors (eIF) 4E and eIF4G, and their isoforms. These are effective against RNA viruses [**146**]. Research on recessive resistance to DNA viruses (geminiviruses) is lagging behind. So far, only a few host factors have been identified as naturally occurring recessive resistant alleles. For example, a recessive gene named *tgr-1* derived from a tomato breeding line has been demonstrated to confer a high level of resistance against ToLCV by impairing viral movement [**147**]. Resistance to bean golden yellow mosaic virus is controlled by a recessive locus *bgm-1* which reduces mosaic and yellowing symptoms of common bean [**148**].

Identification of recessive genes is not restricted by the naturally occurring traits only. In practice, if the naturally occurring recessive allele is absent in the gene pool, genetic variation can be created by artificial ways. Discovering host genes involved in viral infection processes can be facilitated by using forward (loss-of-susceptibility mutants) and reverse (candidate gene approach) genetic screens. Attempts to search for genes potentially involved in geminiviral infection can be achieved according to the following criteria: (1) host proteins interacting with geminiviral proteins; (2) host genes exclusively or preferentially expressed in phloem tissues, to which virus is restricted; (3) host genes involved in cellular processes required for geminivirus infection. Using a VIGS-based approach, 11 host genes were identified as involved in TYLCSV infection, with plants showing delayed, reduced or completely abolished infection after silencing [**149**]. Recent genetic editing techniques such as the clustered regularly interspaced palindromic repeats

(CRISPR)/CRISPR-associated genes (Cas) system are widely used to study gene functions [**150,151**]. This highly specific gene editing technique targeting host susceptibility genes offers plant breeders a unique opportunity to achieve durable resistance against TYLCD-associated viruses.

Forward genetic screening to identify *S* genes is based on artificial mutations. Both chemical and physical mutagenesis are used for this purpose. Among them, the use of the chemical mutagen ethyl methanesulfonate (EMS) is one of the most popular methods to induce large numbers of random point mutations across the whole genome [**152**]. Over the past years, several EMS tomato populations have been developed using different tomato cultivars [**153,154,155,156,157**]. The EMS-mutagenized populations can be subjected to phenotypical screening for resistance to TYLCD-causing agents. Cultivated tomatoes do not mount TYLCD defense responses. Once a plant shows no or reduced TYLCD symptoms (compared with the target tomato cultivar), this plant can be further characterized. To confirm the recessive nature of inheritance and find out the causal mutation for the altered phenotype, segregating populations can be developed.

5.5. Engineering Virus Resistance: Modification of Virus Genes

Up until now, host plants that are immune to TYLCV infection have never been reported. Breeding for TYLCV complex resistance remains challenging due to the emergence of resistance-breaking strains. To overcome these challenges, conventional transgenic approaches such as pathogen-derived resistance (PDR) has been utilized for improved geminiviruses resistance. This approach involves expression of truncated viral proteins [**158**] or viral sequences in an inverted-repeat format [**159**], leading to a resistant phenotype. Engineering *N. benthamiana* and tomato resistance to TYLCSV was accomplished by expression of a truncated TYLCSV-Rep/C1 protein. However, the transgenic plants did not protect against one of the closely related virus strains, TYLCSV-ES, a recombinant derived from TYLCSV which shares %93 amino acid sequence identity [**158**]. Similarly, expression of a truncated *Rep/C1* gene from TYLCV-Mld confers resistance in tomato but not to the TYLCV-IL strain [**160**]. All the examples demonstrate the limitation of PDR which shows strain specificity at least for Rep/C1 based resistance.

Recent research indicates that durable and broad-spectrum resistance can be achieved using the CRISPR/Cas9 system to target viral genes [**161,162,163,164**]. First, given the recombination ability of the TYLCV complex in the coding region, the CRISPR/Cas9 system targeting the non-coding intergenic region (IR) reduces the chance of non-homologous end-joining repair (NHEJ)-induced viral variants and enables durable virus interference [**161,162**]. Second, targeting a conserved sequence of the virus genome allows simultaneous interference with various TYLCD associated species/strains [**161,162,163**]. Single guide RNA (sgRNA) designed to target a conserved sequence (TAATATTAC) in the IR which serves as the origin of virion-strand DNA replication among geminiviruses and betasatellites of begomoviruses could be an effective approach to combat multiple viruses/virus complex with betasatellites under natural conditions, where mixed infection is commonly observed [**161,162,165**].

6. Conclusions and Prospects

Here, we have described how the ever-changing begomoviruses defeat widely adapted resistant varieties of tomato, spread rapidly throughout the world and expand their host ranges. All reported *Ty*-loci have been shown to allow virus replication though at different extents, defining them as symptomless carriers. The monoculture of resistant tomato varieties is potentially harmful considering that virus-carrying plants serve as reservoirs of new virus variants. Furthermore, the ability of betasatellites to indiscriminately recruit begomoviruses during mixed infections indicates that geographic regions not yet affected are at significant risk and efforts need to be made to control the spread of betasatellites. Sustainable tomato breeding programs can be achieved by pyramiding various genes that cover a diverse range of resistance mechanisms. Meanwhile, efforts to search for new resistance sources either in the large genetic diversity of tomato gene pool or through artificial approaches should be continued.

To summarize, some of the future studies aimed to increase success and durability of genetic resistance to TYLCV and related begomoviruses, in a scenario of globalization, climate change and viral disease emergence, should include:

- Screening of additional wild tomato accessions for natural resistance to TYLCV and related begomoviruses.
- Testing TYLCV-resistant tomato genotypes for resistance to TYLCV-betasatellite complexes.
- Testing the performance of tomato lines containing individual TYLCV-resistant loci (*Ty-1* to *Ty6-*) and their combinations for resistance to other globally emerging begomoviruses, e.g., ToLCNDV.
- Combining the individual *Ty*-genes with whitefly-resistance genes in order to study whether such a combination will prolong the effectiveness of the virus resistance genes.
- Identification of the specific interactions between the proteins encoded by TYLCV (wild or mutants) and the proteins encoded by the *Ty*-gene alleles, as was done for the *Ty-2* gene. This will allow us to forecast the effectiveness and durability of the *Ty*-genes in different tomato production areas by monitoring viral variants in the population.
- Screening for natural and/or induced mutations in *S* genes to get durable resistance to begomoviruses.
- Determining the role of mixed viral infections, more prevalent due to emergence of virus diseases and modifications of crops and vector geographical limits due to climate change, in modulating host resistance and durability.
- Altogether, these approaches should enable breeders to achieve durable resistance against TYLCD.

References

1. Hanssen, I.M.; Lapidot, M.; Thomma, B.P. Emerging viral diseases of tomato crops. *Mol. Plant Microbe Interact.* **2010**, *23*, 539–548. [**Google Scholar**] [**CrossRef**] [**Green Version**]

2. Rybicki, E.P. A Top Ten list for economically important plant viruses. *Arch. Virol.* **2015**, *160*, 17–20. [**Google Scholar**] [**CrossRef**]

3. Scholthof, K.B.G.; Adkins, S.; Czosnek, H.; Palukaitis, P.; Jacquot, E.; Hohn, T.; Hohn, B.; Saunders, K.; Candresse, T.; Ahlquist, P.; et al. Top 10 plant viruses in molecular plant pathology. *Mol. Plant Pathol.* **2011**, *12*, 938–954. [**Google Scholar**] [**CrossRef**]

4. International Committee on Taxonomy of Viruses. Available online: **https://talk.ictvonline.org/taxonomy/** (accessed on 11 January 2021).

5. Lapidot, M. Screening for TYLCV-resistance plants using whitefly-mediated inoculation. In *Tomato Yellow Leaf Curl Virus Disease: Management, Molecular Biology, Breeding for Resistance*; Czosnek, H., Ed.; Springer: Dordrecht, The Netherlands, 2007; pp. 329–342. [**Google Scholar**] [**CrossRef**]

6. Jeske, H. Barcoding of plant viruses with circular single-stranded DNA based on rolling circle amplification. *Viruses* **2018**, *10*, 469. [**Google Scholar**] [**CrossRef**] [**PubMed**] [**Green Version**]

7. Abhary, M.; Patil, B.L.; Fauquet, C.M. Molecular biodiversity, taxonomy, and nomenclature of *Tomato Yellow Leaf Curl*-like Viruses. In *Tomato Yellow Leaf Curl Virus Disease: Management, Molecular Biology, Breeding for Resistance*; Czosnek, H., Ed.; Springer: Dordrecht, The Netherlands, 2007; pp. 85–118. [**Google Scholar**] [**CrossRef**]

8. Gronenborn, B. The tomato yellow leaf curl virus genome and function of its proteins. In *Tomato Yellow Leaf Curl Virus Disease: Management, Molecular Biology, Breeding for Resistance*; Czosnek, H., Ed.; Springer: Dordrecht, The Netherlands, 2007; pp. 67–84. [**Google Scholar**] [**CrossRef**]

9. Castillo, A.G.; Morilla, G.; Lozano, R.; Collinet, D.; Perez-Luna, A.; Kashoggi, A.; Bejarano, E. Identification of plant genes involved in TYLCV replication. In *Tomato Yellow Leaf Curl Virus Disease: Management, Molecular Biology, Breeding for Resistance*; Czosnek, H., Ed.; Springer: Dordrecht, The Netherlands, 2007; pp. 207–221. [**Google Scholar**] [**CrossRef**]

10. Hanley-Bowdoin, L.; Bejarano, E.R.; Robertson, D.; Mansoor, S. Geminiviruses: Masters at redirecting and reprogramming plant processes. *Nat. Rev. Microbiol.* **2013**, *11*, 777. [**Google Scholar**] [**CrossRef**] [**PubMed**]

11. Ding, M.; Li, T.; Fang, Q.; Zhang, Z.; Zhou, X. *Tomato yellow leaf curl Yunnan virus*, a new begomovirus species associated with tomato yellow leaf curl disease in China. *J. Plant Pathol.* **2016**, *98*, 337–340. [**Google Scholar**] [**CrossRef**]

12. Nawaz-ul-Rehman, M.S.; Fauquet, C.M. Evolution of geminiviruses and their satellites. *FEBS Lett.* **2009**, *583*, 1825–1832. [**Google Scholar**] [**CrossRef**] [**Green Version**]

13. Xie, Y.; Wu, P.; Liu, P.; Gong, H.; Zhou, X. Characterization of alphasatellites associated with monopartite begomovirus/betasatellite complexes in Yunnan, China. *Virol. J.* **2010**, *7*, 178. [**Google Scholar**] [**CrossRef**] [**Green Version**]

14. Zhou, X. Advances in understanding begomovirus satellites. *Annu. Rev. Phytopathol.* **2013**, *51*, 357–381. [**Google Scholar**] [**CrossRef**] [**PubMed**]

15. Cui, X.; Tao, X.; Xie, Y.; Fauquet, C.M.; Zhou, X. A DNAβ associated with Tomato yellow leaf curl China virus is required for symptom induction. *J. Virol.* **2004**, *78*, 13966–13974. [**Google Scholar**] [**CrossRef**] [**Green Version**]

16. Guo, W.; Yang, X.; Xie, Y.; Cui, X.; Zhou, X. Tomato yellow leaf curl Thailand virus-[Y72] from Yunnan is a monopartite begomovirus associated with DNAβ. *Virus Genes* **2009**, *38*, 328–333. [**Google Scholar**] [**CrossRef**]

17. Sivalingam, P.N.; Varma, A. Role of betasatellite in the pathogenesis of a bipartite begomovirus affecting tomato in India. *Arch. Virol.* **2012**, *157*, 1081–1092. [**Google Scholar**] [**CrossRef**] [**PubMed**]

18. Yang, X.; Wang, B.; Li, F.; Yang, Q.; Zhou, X. Research advances in geminiviruses. In *Current Research Topics in Plant Virology*; Wang, A., Zhou, X., Eds.; Springer: Cham, Switzerland, 2016; pp. 251–269. [**Google Scholar**] [**CrossRef**]

19. European and Mediterranean Plant Protection Organization Global Database. 2021. Available online: **https://gd.eppo.int** (accessed on 23 February 2021).

20. Mabvakure, B.; Martin, D.P.; Kraberger, S.; Cloete, L.; van Brunschot, S.; Geering, A.D.W.; Thomas, J.E.; Bananej, K.; Lett, J.-M.; Lefeuvre, P.; et al. Ongoing geographical spread of Tomato yellow leaf curl virus. *Virology* **2016**, *498*, 257–264. [**Google Scholar**] [**CrossRef**]

21. Lefeuvre, P.; Martin, D.P.; Harkins, G.; Lemey, P.; Gray, A.J.A.; Meredith, S.; Lakay, F.; Monjane, A.; Lett, J.-M.; Varsani, A.; et al. The spread of Tomato yellow leaf curl virus from the Middle East to the world. *PLoSPathog.* **2010**, *6*, e1001164. [**Google Scholar**] [**CrossRef**] [**PubMed**] [**Green Version**]

22. Moriones, E.; Navas-Castillo, J. Tomato yellow leaf curl virus, an emerging virus complex causing epidemics worldwide. *Virus Res.* **2000**, *71*, 123–134. [**Google Scholar**] [**CrossRef**]

23. Barboza, N.; Blanco-Meneses, M.; Hallwass, M.; Moriones, E.; Inoue-Nagata, A.K. First report of Tomato yellow leaf curl virus in tomato in Costa Rica. *Plant Dis.* **2014**, *98*, 699. [**Google Scholar**] [**CrossRef**] [**PubMed**]

24. Chinnaraja, C.; Ramkissoon, A.; Ramsubhag, A.; Jayaraj, J. First report of Tomato yellow leaf curl virus infecting tomatoes in Trinidad. *Plant Dis.* **2016**, *100*, 1958. [**Google Scholar**] [**CrossRef**]

25. Navas-Castillo, J.; Fiallo-Olivé, E.; Sánchez-Campos, S. Emerging virus diseases transmitted by whiteflies. *Annu. Rev. Phytopathol.* **2011**, *49*, 219–248. [**Google Scholar**] [**CrossRef**]

26. Pan, H.; Chu, D.; Yan, W.; Su, Q.; Liu, B.; Wang, S.; Wu, Q.; Xie, W.; Jiao, X.; Li, R.; et al. Rapid spread of Tomato yellow leaf curl virus in China is aided differentially by

two invasive whiteflies. *PLoS ONE* **2012**, *7*, e34817. [**Google Scholar**] [**CrossRef**] [**Green Version**]

27. Seal, S.; van den Bosch, F.; Jeger, M. Factors influencing begomovirus evolution and their increasing global significance: Implications for sustainable control. *Crit. Rev. Plant Sci.* **2006**, *25*, 23–46. [**Google Scholar**] [**CrossRef**]

28. He, Y.Z.; Wang, Y.M.; Yin, T.Y.; Fiallo-Olivé, E.; Liu, Y.Q.; Hanley-Bowdoin, L.; Wang, X.W. A plant DNA virus replicates in the salivary glands of its insect vector via recruitment of host DNA synthesis machinery. *Proc. Natl. Acad. Sci. USA* **2020**, *117*, 16928–16937. [**Google Scholar**] [**CrossRef**]

29. Ghanim, M. A review of the mechanisms and components that determine the transmission efficiency of *Tomato yellow leaf curl virus* (Geminiviridae; *Begomovirus*) by its whitefly vector. *Virus Res.* **2014**, *186*, 47–54. [**Google Scholar**] [**CrossRef**]

30. Rosen, R.; Kanakala, S.; Kliot, A.; Pakkianathan, B.C.; Farich, B.A.; Santana-Magal, N.; Elimelech, M.; Kontsedalov, S.; Lebedev, G.; Cilia, M.; et al. Persistent, circulativetransmission of begomoviruses by whitefly vectors. *Curr. Opin. Virol.* **2015**, *15*, 1–8. [**Google Scholar**] [**CrossRef**]

31. De Barro, P.J.; Liu, S.S.; Boykin, L.M.; Dinsdale, A.B. *Bemisiatabaci*: A statement of species status. *Annu. Rev. Entomol.* **2011**, *56*, 1–19. [**Google Scholar**] [**CrossRef**]

32. Ning, W.; Shi, X.; Liu, B.; Pan, H.; Wei, W.; Zeng, Y.; Sun, X.; Xie, W.; Wang, S.; Wu, Q.; et al. Transmission of Tomato yellow leaf curl virus by *Bemisiatabaci* as affected by whitefly sex and biotype. *Sci. Rep.* **2015**, *5*, 10744. [**Google Scholar**] [**CrossRef**]

33. Gilbertson, R.L.; Batuman, O.; Webster, C.G.; Adkins, S. Role of the insect supervectors *Bemisiatabaci* and *Frankliniella occidentalis* in the emergence and global spread of plant viruses. *Annu. Rev. Virol.* **2015**, *2*, 67–93. [**Google Scholar**] [**CrossRef**] [**PubMed**]

34. Gottlieb, Y.; Zchori-Fein, E.; Mozes-Daube, N.; Kontsedalov, S.; Skaljac, M.; Brumin, M.; Sobol, I.; Czosnek, H.; Vavre, F.; Fleury, F.; et al. The transmission efficiency of Tomato yellow leaf curl virus by the whitefly *Bemisiatabaci* is correlated with the presence of a specific symbiotic bacterium species. *J. Virol.* **2010**, *84*, 9310–9317. [**Google Scholar**] [**CrossRef**] [**Green Version**]

35. Duffy, S.; Holmes, E.C. Phylogenetic evidence for rapid rates of molecular evolution in the single-stranded DNA begomovirus Tomato yellow leaf curl virus. *J. Virol.* **2008**, *82*, 957–965. [**Google Scholar**] [**CrossRef**] [**PubMed**] [**Green Version**]

36. Ge, L.; Zhang, J.; Zhou, X.; Li, H. Genetic structure and population variability of Tomato yellow leaf curl China virus. *J. Virol.* **2007**, *81*, 5902–5907. [**Google Scholar**] [**CrossRef**] [**PubMed**] [**Green Version**]

37. Pooggin, M. How can plant DNA viruses evade siRNA-directed DNA methylation and silencing? *Int. J. Mol. Sci.* **2013**, *14*, 15233–15259. [**Google Scholar**] [**CrossRef**] [**Green Version**]

38. Moriones, E.; Navas-Castillo, J. Rapid evolution of the population of begomoviruses associated with the tomato yellow leaf curl disease after invasion of a new ecological niche. *Span. J. Agric. Res.* **2008**, *6*, 147–159. [**Google Scholar**] [**CrossRef**]

39. Belabess, Z.; Dallot, S.; El-Montaser, S.; Granier, M.; Majde, M.; Tahiri, A.; Blenzar, A.; Urbino, C.; Peterschmitt, M. Monitoring the dynamics of emergence of a non-canonical recombinant of Tomato yellow leaf curl virus and displacement of its parental viruses in tomato. *Virology* **2015**, *486*, 291–306. [**Google Scholar**] [**CrossRef**] [**PubMed**] [**Green Version**]

40. Fiallo-Olivé, E.; Trenado, H.P.; Louro, D.; Navas-Castillo, J. Recurrent speciation of a Tomato yellow leaf curl geminivirus in Portugal by recombination. *Sci. Rep.* **2019**, *9*, 1–8. [**Google Scholar**] [**CrossRef**] [**Green Version**]

41. Castillo, A.G.; Collinet, D.; Deret, S.; Kashoggi, A.; Bejarano, E.R. Dual interaction of plant PCNA with geminivirus replication accessory protein (Ren) and viral replication protein (Rep). *Virology* **2003**, *312*, 381–394. [**Google Scholar**] [**CrossRef**]

42. Rodriguez-Negrete, E.; Lozano-Duran, R.; Piedra-Aguilera, A.; Cruzado, L.; Bejarano, E.R.; Castillo, A.G. Geminivirus Rep protein interferes with the plant DNA methylation machinery and suppresses transcriptional gene silencing. *New Phytol.* **2013**, *199*, 464–475. [**Google Scholar**] [**CrossRef**] [**Green Version**]

43. Moriones, E.; García-Andrés, S.; Navas-Castillo, J. Recombination in the TYLCV complex: A mechanism to increase genetic diversity. Implications for plant resistance development. In *Tomato Yellow Leaf Curl Virus Disease: Management, Molecular Biology, Breeding for Resistance*; Czosnek, H., Ed.; Springer: Dordrecht, The Netherlands, 2007; pp. 119–138. [**Google Scholar**] [**CrossRef**]

44. Navas-Castillo, J.; Sanchez-Campos, S.; Noris, E.; Louro, D.; Accotto, G.P.; Moriones, E. Natural recombination between Tomato yellow leaf curl virus-Is and Tomato leaf curl virus. *J. Gen. Virol.* **2000**, *81*, 2797–2801. [**Google Scholar**] [**CrossRef**] [**PubMed**]

45. Díaz-Pendón, J.A.; Sánchez-Campos, S.; Fortes, I.M.; Moriones, E. *Tomato yellow leaf curl sardinia virus*, a begomovirus species evolving by mutation and recombination: A challenge for virus control. *Viruses* **2019**, *11*, 45. [**Google Scholar**] [**CrossRef**] [**Green Version**]

46. Belabess, Z.; Peterschmitt, M.; Granier, M.; Tahiri, A.; Blenzar, A.; Urbino, C. The non-canonical Tomato yellow leaf curl virus recombinant that displaced its parental viruses in southern Morocco exhibits a high selective advantage in experimental conditions. *J. Gen. Virol.* **2016**, *97*, 3433–3445. [**Google Scholar**] [**CrossRef**] [**Green Version**]

47. Panno, S.; Caruso, A.G.; Davino, S. The nucleotide sequence of a recombinant Tomato yellow leaf curl virus strain frequently detected in Sicily isolated from tomato plants carrying the *Ty-1* resistance gene. *Arch. Virol.* **2018**, *163*, 795–797. [**Google Scholar**] [**CrossRef**]

48. Granier, M.; Tomassoli, L.; Manglli, A.; Nannini, M.; Peterschmitt, M.; Urbino, C. First report of TYLCV-IS141, a Tomato yellow leaf curl virus recombinant infecting tomato plants carrying the *Ty-1* resistance gene in Sardinia (Italy). *Plant Dis.* **2019**, *103*, 1437. [**Google Scholar**] [**CrossRef**]

49. Li, F.; Huang, C.; Li, Z.; Zhou, X. Suppression of RNA silencing by a plant DNA virus satellite requires a host calmodulin-like protein to repress *RDR6* expression. *PLoS Pathog.* **2014**, *10*, e1003921. [**Google Scholar**] [**CrossRef**] [**PubMed**]

50. Li, F.; Zhao, N.; Li, Z.; Xu, X.; Wang, Y.; Yang, X.; Liu, S.-S.; Wang, A.; Zhou, X. A calmodulin-like protein suppresses RNA silencing and promotes geminivirus infection by degrading SGS3 via the autophagy pathway in *Nicotiana benthamiana*. *PLoSPathog.* **2017**, *13*, e1006213. [**Google Scholar**] [**CrossRef**] [**PubMed**]

51. Yang, X.; Xie, Y.; Raja, P.; Li, S.; Wolf, J.N.; Shen, Q.; Bisaro, D.M.; Zhou, X. Suppression of methylation-mediated transcriptional gene silencing by βC1-SAHH protein interaction during geminivirus-betasatellite infection. *PLoSPathog.* **2011**, *7*, e1002329. [**Google Scholar**] [**CrossRef**] [**Green Version**]

52. Li, R.; Weldegergis, B.T.; Li, J.; Jung, C.; Qu, J.; Sun, Y.; Qian, H.; Tee, C.S.; van Loon, J.J.A.; Dicke, M.; et al. Virulence factors of geminivirus interact with MYC2 to subvert plant resistance and promote vector performance. *Plant Cell* **2014**, *26*, 4991–5008. [**Google Scholar**] [**CrossRef**] [**Green Version**]

53. Khan, A.J.; Idris, A.M.; Al-Saady, N.A.; Al-Mahruki, M.S.; Al-Subhi, A.M.; Brown, J.K. A divergent isolate of Tomato yellow leaf curl virus from Oman with an associated DNAβ satellite: An evolutionary link between Asian and the Middle Eastern virus–satellite complexes. *Virus Genes* **2008**, *36*, 169–176. [**Google Scholar**] [**CrossRef**]

54. Chen, L.F.; Rojas, M.; Kon, T.; Gamby, K.; Xoconostle-Cazares, B.; Gilbertson, R.L. A severe symptom phenotype in tomato in Mali is caused by a reassortant between a novel recombinant begomovirus (Tomato yellow leaf curl Mali virus) and a betasatellite. *Mol. Plant Pathol.* **2009**, *10*, 415–430. [**Google Scholar**] [**CrossRef**]

55. Ito, T.; Kimbara, J.; Sharma, P.; Ikegami, M. Interaction of Tomato yellow leaf curl virus with diverse betasatellites enhances symptom severity. *Arch. Virol.* **2009**, *154*, 1233–1239. [**Google Scholar**] [**CrossRef**] [**PubMed**]

56. Ueda, S.; Onuki, M.; Yamashita, M.; Yamato, Y. Pathogenicity and insect transmission of a begomovirus complex between Tomato yellow leaf curl virus and Ageratum yellow vein betasatellite. *Virus Genes* **2012**, *44*, 338–344. [**Google Scholar**] [**CrossRef**]

57. Voorburg, C.M.; Yan, Z.; Bergua-Vidal, M.; Wolters, A.M.A.; Bai, Y.; Kormelink, R. *Ty-1*, a universal resistance gene against geminiviruses that is compromised by co-replication of a betasatellite. *Mol. Plant Pathol.* **2020**, *21*, 160–172. [**Google Scholar**] [**CrossRef**] [**PubMed**] [**Green Version**]

58. Conflon, D.; Granier, M.; Tiendrébéogo, F.; Gentit, P.; Peterschmitt, M.; Urbino, C. Accumulation and transmission of alphasatellite, betasatellite and Tomato yellow leaf curl virus in susceptible and *Ty-1*-resistant tomato plants. *Virus Res.* **2018**, *253*, 124–134. [**Google Scholar**] [**CrossRef**] [**PubMed**] [**Green Version**]

59. Gelbart, D.; Chen, L.; Alon, T.; Dobrinin, S.; Levin, I.; Lapidot, M. The recent association of a DNA betasatellite with Tomato yellow leaf curl virus in Israel–A new threat to tomato production. *Crop Prot.* **2020**, *128*, 104995. [**Google Scholar**] [**CrossRef**]

60. Prasad, A.; Sharma, N.; Hari-Gowthem, G.; Muthamilarasan, M.; Prasad, M. Tomato yellow leaf curl virus: Impact, challenges, and management. *Trends Plant Sci.* **2020**, *25*, 897–911. [**Google Scholar**] [**CrossRef**] [**PubMed**]

61. Salati, R.; Nahkla, M.K.; Rojas, M.R.; Guzman, P.; Jaquez, J.; Maxwell, D.P.; Gilbertson, R.L. Tomato yellow leaf curl virus in the Dominican Republic: Characterization of

an infectious clone, virus monitoring in whiteflies, and identification of reservoir hosts. *Phytopathology* **2002**, *92*, 487–496. [**Google Scholar**] [**CrossRef**] [**Green Version**]

62. García-Arenal, F.; Zerbini, F.M. Life on the edge: Geminiviruses at the interface between crops and wild plant hosts. *Annu. Rev. Virol.* **2019**, *6*, 411–433. [**Google Scholar**] [**CrossRef**] [**Green Version**]

63. Zaidi, S.S.E.A.; Martin, D.P.; Amin, I.; Farooq, M.; Mansoor, S. *Tomato leaf curl New Delhi virus*: A widespread bipartite begomovirus in the territory of monopartite begomoviruses. *Mol. Plant Pathol.* **2017**, *18*, 901–911. [**Google Scholar**] [**CrossRef**] [**PubMed**]

64. Moriones, E.; Praveen, S.; Chakraborty, S. *Tomato leaf curl New Delhi virus*: An emerging virus complex threatening vegetable and fiber crops. *Viruses* **2017**, *9*, 264. [**Google Scholar**] [**CrossRef**] [**Green Version**]

65. Fortes, I.M.; Sánchez-Campos, S.; Fiallo-Olivé, E.; Díaz-Pendón, J.A.; Navas-Castillo, J.; Moriones, E. A novel strain of Tomato leaf curl New Delhi virus has spread to the Mediterranean basin. *Viruses* **2016**, *8*, 307. [**Google Scholar**] [**CrossRef**] [**Green Version**]

66. Juárez, M.; Rabadán, M.P.; Martínez, L.D.; Tayahi, M.; Grande-Pérez, A.; Gómez, P. Natural hosts and genetic diversity of the emerging Tomato leaf curl New Delhi virus in Spain. *Front. Microbiol.* **2019**, *10*, 140. [**Google Scholar**] [**CrossRef**]

67. Saunders, K.; Salim, N.; Mali, V.R.; Malathi, V.G.; Briddon, R.; Markham, P.G.; Stanley, J. Characterisation of Sri Lankan Cassava mosaic virus and Indian cassava mosaic virus: Evidence for acquisition of a DNA B component by a monopartite begomovirus. *Virology* **2002**, *293*, 63–74. [**Google Scholar**] [**CrossRef**] [**Green Version**]

68. Luo, C.; Jones, C.M.; Devine, G.; Zhang, F.; Denholm, I.; Gorman, K. Insecticide resistance in *Bemisiatabaci* biotype Q (Hemiptera: Aleyrodidae) from China. *Crop Prot.* **2010**, *29*, 429–434. [**Google Scholar**] [**CrossRef**]

69. Roditakis, E.; Grispou, M.; Morou, E.; Kristoffersen, J.B.; Roditakis, N.; Nauen, R.; Vontas, J.; Tsagkarakou, A. Current status of insecticide resistance in Q biotype *Bemisiatabaci* populations from Crete. *Pest Manag. Sci.* **2009**, *65*, 313–322. [**Google Scholar**] [**CrossRef**] [**PubMed**]

70. Anbinder, I.; Reuveni, M.; Azari, R.; Paran, I.; Nahon, S.; Shlomo, H.; Chen, L.; Lapidot, M.; Levin, I. Molecular dissection of Tomato leaf curl virus resistance in tomato line TY172 derived from *Solanum peruvianum*. *Theor. Appl. Genet.* **2009**, *119*, 519–530. [**Google Scholar**] [**CrossRef**]

71. Hanson, P.M.; Green, S.K.; Kuo, G. *Ty-2*, a gene on chromosome 11 conditioning geminivirus resistance in tomato. *Tomato Genet. Coop. Rep.* **2006**, *56*, 17–18. [**Google Scholar**]

72. Hutton, S.F.; Scott, J.W. *Ty-6*, a major begomovirus resistance gene located on chromosome 10. *Rept. Tomato Genet. Coop.* **2014**, *64*, 14–18. [**Google Scholar**]

73. Ji, Y.; Schuster, D.J.; Scott, J.W. *Ty-3*, a begomovirus resistance locus near the Tomato yellow leaf curl virus resistance locus *Ty-1* on chromosome 6 of tomato. *Mol. Breed.* **2007**, *20*, 271–284. [**Google Scholar**] [**CrossRef**]

74. Ji, Y.; Scott, J.W.; Schuster, D.J.; Maxwell, D.P. Molecular mapping of *Ty-4*, a new Tomato yellow leaf curl virus resistance locus on chromosome 3 of tomato. *J. Am. Soc. Hortic. Sci.* **2009**, *134*, 281–288. [**Google Scholar**] [**CrossRef**] [**Green Version**]

75. Zamir, D.; Ekstein-Michelson, I.; Zakay, Y.; Navot, N.; Zeidan, M.; Sarfatti, M.; Eshed, Y.; Harel, E.; Pleban, T.; van-Oss, H.; et al. Mapping and introgression of a Tomato yellow leaf curl virus tolerance gene, *Ty-1*. *Theor. Appl. Genet.* **1994**, *88*, 141–146. [**Google Scholar**] [**CrossRef**]

76. Lapidot, M.; Karniel, U.; Gelbart, D.; Fogel, D.; Evenor, D.; Kutsher, Y.; Makhbash, Z.; Nahon, S.; Shlomo, H.; Chen, L.; et al. A novel route controlling begomovirus resistance by the messenger RNA surveillance factor pelota. *PLoS Genet.* **2015**, *11*, e1005538. [**Google Scholar**] [**CrossRef**]

77. Shen, X.; Yan, Z.; Wang, X.; Wang, Y.; Arens, M.; Du, Y.; Visser, R.G.F.; Kormelink, R.; Wolters, A.M.A. The NLR protein encoded by the resistance gene *Ty-2* is triggered by the replication-associated protein Rep/C1 of Tomato yellow leaf curl virus. *Front. Plant Sci.* **2020**, *11*, 1384. [**Google Scholar**] [**CrossRef**]

78. Verlaan, M.G.; Hutton, S.F.; Ibrahem, R.M.; Kormelink, R.; Visser, R.G.F.; Scott, J.W.; Edwards, J.; Bai, Y. The Tomato yellow leaf curl virus resistance genes *Ty-1* and *Ty-3* are allelic and code for DFDGD-class RNA-dependent RNA polymerases. *PLoS Genet.* **2013**, *9*, e1003399. [**Google Scholar**] [**CrossRef**] [**Green Version**]

79. Yamaguchi, H.; Ohnishi, J.; Saito, A.; Ohyama, A.; Nunome, T.; Miyatake, K.; Fukuoka, H. An NB-LRR gene, *TYNBS1*, is responsible for resistance mediated by the *Ty-2* Begomovirus resistance locus of tomato. *Theor. Appl. Genet.* **2018**, *131*, 1345–1362. [**Google Scholar**] [**CrossRef**]

80. Verlaan, M.G.; Szinay, D.; Hutton, S.F.; de Jong, H.; Kormelink, R.; Visser, R.G.F.; Scott, J.W.; Bai, Y. Chromosomal rearrangements between tomato and *Solanum chilense* hamper mapping and breeding of the TYLCV resistance gene *Ty-1*. *Plant J.* **2011**, *68*, 1093–1103. [**Google Scholar**] [**CrossRef**]

81. Butterbach, P.; Verlaan, M.G.; Dullemans, A.; Lohuis, D.; Visser, R.G.F.; Bai, Y.; Kormelink, R. Tomato yellow leaf curl virus resistance by *Ty-1* involves increased cytosine methylation of viral genomes and is compromised by Cucumber mosaic virus infection. *Proc. Natl. Acad. Sci. USA* **2014**, *111*, 12942–12947. [**Google Scholar**] [**CrossRef**] [**Green Version**]

82. Hamera, S.; Song, X.; Su, L.; Chen, X.; Fang, R. Cucumber mosaic virus suppressor 2b binds to AGO4-related small RNAs and impairs AGO4 activities. *Plant J.* **2012**, *69*, 104–115. [**Google Scholar**] [**CrossRef**]

83. Belabess, Z.; Urbino, C.; Granier, M.; Tahiri, A.; Blenzar, A.; Peterschmitt, M. The typical RB76 recombination breakpoint of the invasive recombinant Tomato yellow leaf curl virus of Morocco can be generated experimentally but is not positively selected in tomato. *Virus Res.* **2018**, *243*, 44–51. [**Google Scholar**] [**CrossRef**]

84. Hanson, P.M.; Bernacchi, D.; Green, S.; Tanksley, S.D.; Muniyappa, V.; Padmaja, A.S.; Chen, H.-M.; Kuo, G.; Fang, D.; Chen, J.-T. Mapping a wild tomato introgression associated with Tomato yellow leaf curl virus resistance in a cultivated tomato line. *J. Am. Soc. Hortic. Sci.* **2000**, *15*, 15–20. [**Google Scholar**] [**CrossRef**] [**Green Version**]

85. Wolters, A.M.A.; Caro, M.; Dong, S.; Finkers, R.; Gao, J.; Visser, R.G.F.; Wang, X.; Du, Y.; Bai, Y. Detection of an inversion in the *Ty-2* region between *S. lycopersicum* and *S. habrochaites* by a combination of de novo genome assembly and BAC cloning. *Theor. Appl. Genet.* **2015**, *128*, 1987–1997. [**Google Scholar**] [**CrossRef**] [**PubMed**] [**Green Version**]

86. García-Cano, E.; Resende, R.O.; Boiteux, L.S.; Giordano, L.B.; Fernández-Muñoz, R.; Moriones, E. Phenotypic expression, stability, and inheritance of a recessive resistance to monopartite begomoviruses associated with tomato yellow leaf curl disease in tomato. *Phytopathology* **2008**, *98*, 618–627. [**Google Scholar**] [**CrossRef**] [**PubMed**] [**Green Version**]

87. Giordano, L.B.; Silva-Lobo, V.L.; Santana, F.M.; Fonseca, M.E.N.; Boiteux, L.S. Inheritance of resistance to the bipartite Tomato chlorotic mottle begomovirus derived from *Lycopersicon esculentum* cv. 'Tyking'. *Euphytica* **2005**, *143*, 27–33. [**Google Scholar**] [**CrossRef**]

88. Wu, X.; He, W.T.; Tian, S.; Meng, D.; Li, Y.; Chen, W.; Li, L.; Tian, L.; Zhong, C.; Han, F.; et al. *pelo* is required for high efficiency viral replication. *PLoSPathog.* **2014**, *10*, e1004034. [**Google Scholar**] [**CrossRef**] [**PubMed**] [**Green Version**]

89. Ding, W.; Wu, J.; Ye, J.; Zheng, W.; Wang, S.; Zhu, X.; Zhou, J.; Pan, Z.; Zhang, B.; Zhu, S. A *Pelota*-like gene regulates root development and defence responses in rice. *Ann. Bot.* **2018**, *122*, 359–371. [**Google Scholar**] [**CrossRef**] [**PubMed**]

90. Zhang, X.B.; Feng, B.H.; Wang, H.M.; Xu, X.; Shi, Y.F.; He, Y.; Chen, Z.; Sathe, A.P.; Wu, J.L. A substitution mutation in *OsPELOTA* confers bacterial blight resistance by activating the salicylic acid pathway. *J. Integr. Plant Biol.* **2018**, *60*, 160–172. [**Google Scholar**] [**CrossRef**] [**Green Version**]

91. Gill, U.; Scott, J.W.; Shekasteband, R.; Ogundiwin, E.; Schuit, C.; Francis, D.M.; Sim, S.-C.; Smith, H.; Hutton, S. *Ty-6*, a major begomovirus resistance gene on chromosome 10, is effective against Tomato yellow leaf curl virus and Tomato mottle virus. *Theor. Appl. Genet.* **2019**, *132*, 1543–1554. [**Google Scholar**] [**CrossRef**] [**Green Version**]

92. Scott, J.W.; Hutton, S.F.; Freeman, J.H. Fla. 8638B and Fla. 8624 tomato breeding lines with begomovirus resistance genes *ty-5* plus *Ty-6* and *Ty-6*, respectively. *HortScience* **2015**, *50*, 1405–1407. [**Google Scholar**] [**CrossRef**]

93. Michelson, I.; Zamir, D.; Czosnek, H. Accumulation and translocation of Tomato yellow leaf curl virus (TYLCV) in a *Lyco-persicon esculentum* breeding line containing the *L. chilense* TYLCV tolerance gene *Ty-1*. *Phytopathology* **1994**, *84*, 928–933. [**Google Scholar**] [**CrossRef**]

94. Barbieri, M.; Acciarri, N.; Sabatini, E.; Sardo, L.; Accotto, G.P.; Pecchioni, N. Introgression of resistance to two Mediterranean virus species causing tomato yellow leaf curl into a valuable traditional tomato variety. *J. Plant Pathol.* **2010**, *92*, 485–493. [**Google Scholar**] [**CrossRef**]

95. Ohnishi, J.; Yamaguchi, H.; Saito, A. Analysis of the Mild strain of Tomato yellow leaf curl virus, which overcomes *Ty-2*. *Arch. Virol.* **2016**, *161*, 2207–2217. [**Google Scholar**] [**CrossRef**] [**PubMed**]

96. Dhaliwal, M.S.; Jindal, S.K.; Sharma, A.; Prasanna, H.C. Tomato yellow leaf curl virus disease of tomato and its man-agement through resistance breeding: A review. *J. Hortic. Sci. Biotechnol.* **2020**, *95*, 425–444. [**Google Scholar**] [**CrossRef**]

97. Yan, Z.; Pérez de Castro, A.; Díez, M.J.; Hutton, S.F.; Visser, R.G.F.; Wolters, A.M.A.; Bai, Y.; Li, J. Resistance to Tomato yellow leaf curl virus in tomato germplasm. *Front. Plant Sci.* **2018**, *9*, 1198. [**Google Scholar**] [**CrossRef**] [**PubMed**]

98. Banerjee, M.K.; Kalloo, M.K. Sources and inheritance of resistance to leaf curl virus in *Lycopersicon*. *Theor. Appl. Genet.* **1987**, *73*, 707–710. [**Google Scholar**] [**CrossRef**]

99. Kasrani, M.A. Inheritance of resistance to Tomato yellow leaf curl virus (TYLCV) in *Lycopersicon pimpinellifolium*. *Plant Dis.* **1989**, *73*, 435–437. [**Google Scholar**] [**CrossRef**]

100. Azizi, A.; Mozafari, J.; Shams-bakhsh, M. Phenotypic and molecular screening of tomato germplasm for resistance to Tomato yellow leaf curl virus. *Iran. J. Biotechnol.* **2008**, *6*, 199–206. [**Google Scholar**]

101. De la Peña, R.; Kadirvel, P.; Venkatesan, S.; Kenyon, L.; Hughes, J. Integrated approaches to manage Tomato yellow leaf curl viruses. In *Biocatalysis and Biomolecular Engineering*; Hou, C.T., Shaw, J.-F., Eds.; John Wiley & Sons: Hoboken, NJ, USA, 2010; pp. 105–132. [**Google Scholar**] [**CrossRef**]

102. El-Dougdoug, N.K.; Mahfouze, S.A.; Ahmed, S.A.; Othman, B.A.; Hazaa, M.M. Identification of biochemical and molecular markers in Tomato yellow leaf curl virus resistant tomato species. *Sci. Agric.* **2013**, *2*, 46–53. [**Google Scholar**]

103. Ji, Y.; Scott, J.W.; Hanson, P.; Graham, E.; Maxwell, D.P. Sources of resistance, inheritance, and location of genetic loci conferring resistance to members of the tomato-infecting begomoviruses. In *Tomato Yellow Leaf Curl Virus Disease: Management, Molecular Biology, Breeding for Resistance*; Czosnek, H., Ed.; Springer: Dordrecht, The Netherlands, 2007; pp. 343–362. [**Google Scholar**] [**CrossRef**]

104. Jordá, C.; Picó, B.; Díez, M.J.; Nuez, F. Cribado de germoplasmaresistente a TYLCV. Desarrollo de un método de diagnósticoadecuado. In Proceedings of the VIII Congreso Nacional de la Sociedad Española de Fitopatología, Córdoba, Spain, 23–27 September 1996; p. 218. [**Google Scholar**]

105. Kasrawi, M.A.; Suwwan, M.A.; Mansour, A. Sources of resistance to Tomato yellow leaf curl virus (TYLCV) in *Lycopersicon* species. *Euphytica* **1988**, *37*, 61–64. [**Google Scholar**] [**CrossRef**]

106. Pereira-Carvalho, R.C.; Boiteux, L.S.; Fonseca, M.E.N.; Díaz-Pendón, J.A.; Moriones, E.; Fernández-Muñoz, R.; Charchar, J.M.; Resende, R.O. Multiple resistance to *Meloidogyne* spp. and bipartite and monopartite *Begomovirus* spp. in wild *Solanum* (*Lycopersicon*) accessions. *Plant Dis.* **2010**, *94*, 179–185. [**Google Scholar**] [**CrossRef**] [**PubMed**] [**Green Version**]

107. Pérez de Castro, A.; Díez, M.J.; Nuez, F. Identificación de nuevasfuentes de resistencia al virus del rizadoamarillo del tomate (TYLCV). *ActasHorticultura* **2004**, *41*, 119–122. [**Google Scholar**]

108. Pérez de Castro, A.; Díez, M.J.; Nuez, F. Caracterización de entradas de *Lycopersicon peruvianum* y *L. chilense* porsuresistencia al Tomato yellow leaf curl virus (TYLCV). *ActasPortuguesasHorticultura* **2005**, *8*, 48–54. [**Google Scholar**]

109. Pérez de Castro, A.; Díez, M.J.; Nuez, F. Exploiting partial resistance to Tomato yellow leaf curl virus derived from *Solanum pimpinellifolium* UPV16991. *Plant Dis.* **2008**, *92*, 1083–1090. [**Google Scholar**] [**CrossRef**] [**Green Version**]

110. Pérez de Castro, A.; Díez, M.J.; Nuez, F. Resistencia a la enfermedad del rizadoamarillo del tomateen la especiesilvestre *Solanum lycopersicoides*. *ActasHorticultura* **2010**, *55*, 169–170. [**Google Scholar**]

111. Picó, B.; Díez, M.J. Screening *Lycopersicon* spp. for resistance to TYLCV. In Proceedings of the 2nd International Workshop on Bemisia and Geminiviral Disease, USDA-ARS, San Juan, Puerto Rico, 7–12 June 1998; p. 43. [**Google Scholar**]

112. Picó, B.; Díez, M.J.; Nuez, F. Viral diseases causing the greatest economic losses to the tomato crop. II. The *Tomato yellow leaf curl virus*—A review. *Sci. Hortic.* **1996**, *67*, 151–196. [**Google Scholar**] [**CrossRef**]

113. Picó, B.; Díez, M.J.; Nuez, F. Evaluation of whitefly-mediated inoculation techniques to screen *Lycopersicon esculentum* and wild relatives for resistance to Tomato yellow leaf curl virus. *Euphytica* **1998**, *101*, 259–271. [**Google Scholar**] [**CrossRef**]

114. Picó, B.; Ferriol, M.; Diez, M.J.; Nuez, F. Developing tomato breeding lines resistant to Tomato yellow leaf curl virus. *Plant Breed.* **1999**, *118*, 537–542. [**Google Scholar**] [**CrossRef**]

115. Picó, B.; Ferriol, M.; Diez, M.; Nuez, F. Cribado de fuentes de resistencia de *Lycopersicon* spp. al Tomato yellow leaf curl virus medianteagroinoculaciónen disco de hoja. *ActasHorticultura* **1999**, *24*, 105–112. [**Google Scholar**]

116. Picó, B.; Sifres, A.; Elía, M.; Díez, M.J.; Nuez, F. Searching for new resistance sources to Tomato yellow leaf curl virus within a highly variable wild *Lycopersicon* genetic pool. *Acta Physiol. Plant.* **2000**, *22*, 344–350. [**Google Scholar**] [**CrossRef**]

117. Picó, B.; Herraiz, J.; Ruiz, J.; Nuez, F. Widening the genetic basis of virus resistance in tomato. *Sci. Hortic.* **2002**, *94*, 73–89. [**Google Scholar**] [**CrossRef**]

118. Pilowsky, M.; Cohen, S. Screening additional wild tomatoes for resistance to the whitefly borne Tomato yellow leaf curl virus. *Acta Physiol. Plant* **2000**, *22*, 351–353. [**Google Scholar**] [**CrossRef**]

119. Soler, S.; Pico, B.; Sifres, A.; Diez, M.; De Frutos, R.; Nuez, F. Multiple virus resistance in a collection of *Lycopersicon* spp. In Proceedings of the Fifth Congress of the European Foundation for Plant Pathology, Taormina, Italy, 18–22 September 2000; European Foundation for Plant Pathology: Taormina, Italy, 2000; p. 17. [**Google Scholar**]

120. Tomás, D.M.; Cañizares, M.C.; Abad, J.; Fernández-Muñoz, R.; Moriones, E. Resistance to Tomato yellow leaf curl virus accumulation in the tomato wild relative *Solanum habrochaites* associated with the C4 viral protein. *Mol. Plant Microbe Interact.* **2010**, *24*,

849–861. [**Google Scholar**] [**CrossRef**] [**Green Version**]

121. Vidavsky, F.; Czosnek, H. Tomato breeding lines resistant and tolerant to Tomato yellow leaf curl virus issued from *Lycopersicon hirsutum*. *Phytopathology* **1998**, *88*, 910–914. [**Google Scholar**] [**CrossRef**] [**Green Version**]

122. Vidavski, F.; Czosnek, H.; Gazit, S.; Levy, D.; Lapidot, M. Pyramiding of genes conferring resistance to Tomato yellow leaf curl virus from different wild tomato species. *Plant Breed.* **2008**, *127*, 625–631. [**Google Scholar**] [**CrossRef**]

123. Vidavsky, F.; Leviatov, S.; Milo, J.; Rabinowitch, H.; Kedar, N.; Czosnek, H. Response of tolerant breeding lines of tomato, *Lycopersicon esculentum*, originating from three different sources (*L. peruvianum*, *L. pimpinellifolium* and *L. chilense*) to early controlled inoculation by Tomato yellow leaf curl virus (TYLCV). *Plant Breed.* **1998**, *117*, 165–169. [**Google Scholar**] [**CrossRef**]

124. Zakay, Y.; Navot, N.; Zeidan, M.; Kedar, N.; Rabinowitch, H.; Czosnek, H.; Zamir, D. Screening *Lycopersicon* accessions for resistance to Tomato yellow leaf curl virus: Presence of viral DNA and symptom development. *Plant Dis.* **1991**, *75*, 279–281. [**Google Scholar**] [**CrossRef**]

125. Díez, M.J.; Nuez, F. Tomato. In *Vegetables II. Handbook of Plant Breeding*; Prohens, J., Nuez, F., Eds.; Springer: New York, NY, USA, 2008; pp. 249–323. [**Google Scholar**] [**CrossRef**]

126. Maruthi, M.N.; Czosnek, H.; Vidavski, F.; Tarba, S.Y.; Milo, J.; Leviatov, S.; Venkatesh, H.M.; Padmaja, A.S.; Kulkarni, R.S.; Muniyappa, V. Comparison of resistance to Tomato leaf curl virus (India) and Tomato yellow leaf curl virus (Israel) among *Lycopersicon* wild species, breeding lines and hybrids. *Eur. J. Plant Pathol.* **2003**, *109*, 1–11. [**Google Scholar**] [**CrossRef**]

127. Pereira-Carvalho, R.; Díaz-Pendón, J.; Fonseca, M.; Boiteux, L.; Fernández-Muñoz, R.; Moriones, E.; Resende, R.O. Recessive resistance derived from tomato cv. Tyking-limits drastically the spread of Tomato yellow leaf curl virus. *Viruses* **2015**, *7*, 2518–2533. [**Google Scholar**] [**CrossRef**] [**PubMed**] [**Green Version**]

128. Pérez de Castro, A.; Díez, M.J.; Nuez, F. Evaluation of breeding tomato lines partially resistant to Tomato yellow leaf curl Sardinia virus and Tomato yellow leaf curl virus derived from *Lycopersicon chilense*. *Can. J. Plant Pathol.* **2005**, *27*, 268–275. [**Google Scholar**] [**CrossRef**]

129. Pérez de Castro, A.; Díez, M.J.; Nuez, F. Inheritance of Tomato yellow leaf curl virus resistance derived from *Solanum pimpinellifolium* UPV16991. *Plant Dis.* **2007**, *91*, 879–885. [**Google Scholar**] [**CrossRef**] [**Green Version**]

130. Pilowsky, M.; Cohen, S. Inheritance of resistance to Tomato yellow leaf curl virus in tomatoes. *Phytopathology* **1974**, *64*, 632–635. [**Google Scholar**] [**CrossRef**]

131. Geneif, A.A. Breeding for resistance to Tomato leaf curl virus in tomatoes in the Sudan. In Proceedings of the VIII African Symposium on Horticultural Crops, Wad Medani, Sudan, 20–24 March 1983; Volume 143, pp. 469–484. [**Google Scholar**]

132. Víquez-Zamora, M.; Caro, M.; Finkers, R.; Tikunov, Y.; Bovy, A.; Visser, R.G.F.; Bai, Y.; van Heusden, S. Mapping in the era of sequencing: High density genotyping and its application for mapping TYLCV resistance in *Solanum pimpinellifolium. BMC Genom.* **2014**, *15*, 1152. [**Google Scholar**] [**CrossRef**] [**PubMed**] [**Green Version**]

133. Pilowsky, M.; Cohen, S. Tolerance to Tomato yellow leaf curl virus derived from *Lycopersicon peruvianum. Plant Dis.* **1990**, *74*, 248–250. [**Google Scholar**] [**CrossRef**]

134. Julián, O.; Herráiz, J.; Corella, S.; di-Lolli, I.; Soler, S.; Díez, M.J.; Pérez de Castro, A. Initial development of a set of introgression lines from *Solanum peruvianum* PI 126944 into tomato: Exploitation of resistance to viruses. *Euphytica* **2013**, *193*, 183–196. [**Google Scholar**] [**CrossRef**]

135. Hutton, S.F.; Scott, J.W. Fla. 7907C: A Fla. 7907 near-isogenic tomato inbred line containing the begomovirus resistance gene, *Ty-1. HortScience* **2017**, *52*, 658–660. [**Google Scholar**] [**CrossRef**] [**Green Version**]

136. Hutton, S.F.; Ji, Y.; Scott, J.W. Fla. 8923: A tomato breeding line with begomovirus resistance gene *Ty-3* in a 70-kb *Solanum chilense* introgression. *HortScience* **2015**, *50*, 1257–1259. [**Google Scholar**] [**CrossRef**] [**Green Version**]

137. Pérez de Castro, A.; Julián, O.; Díez, M.J. Genetic control and mapping of *Solanum chilense* LA1932, LA1960 and LA1971-derived resistance to Tomato yellow leaf curl disease. *Euphytica* **2013**, *190*, 203–214. [**Google Scholar**] [**CrossRef**]

138. Caro, M.; Verlaan, M.G.; Julián, O.; Finkers, R.; Wolters, A.M.A.; Hutton, S.F.; Scott, J.W.; Kormelink, R.; Visser, R.G.F.; Díez, M.J.; et al. Assessing the genetic variation of *Ty-1* and *Ty-3* alleles conferring resistance to Tomato yellow leaf curl virus in a broad tomato germplasm. *Mol. Breed.* **2015**, *35*, 132. [**Google Scholar**] [**CrossRef**] [**PubMed**] [**Green Version**]

139. Fuchs, M. Pyramiding resistance-conferring gene sequences in crops. *Curr. Opin. Virol.* **2017**, *26*, 36–42. [**Google Scholar**] [**CrossRef**] [**PubMed**]

140. Prasanna, H.C.; Sinha, D.P.; Rai, G.K.; Krishna, R.; Kashyap, S.P.; Singh, N.K.; Singh, M.; Malathi, V.G. Pyramiding *Ty-2* and *Ty-3* genes for resistance to monopartite and bipartite Tomato leaf curl viruses of India. *Plant Pathol.* **2015**, *64*, 256–264. [**Google Scholar**] [**CrossRef**]

141. Pavan, S.; Jacobsen, E.; Visser, R.G.F.; Bai, Y. Loss of susceptibility as a novel breeding strategy for durable and broad-spectrum resistance. *Mol. Breed.* **2010**, *25*, 1–12. [**Google Scholar**] [**CrossRef**] [**PubMed**] [**Green Version**]

142. Hashimoto, M.; Neriya, Y.; Yamaji, Y.; Namba, S. Recessive resistance to plant viruses: Potential resistance genes beyond translation initiation factors. *Front. Microbiol.* **2016**, *7*, 1695. [**Google Scholar**] [**CrossRef**] [**Green Version**]

143. van Schie, C.C.; Takken, F.L. Susceptibility genes 101: How to be a good host. *Annu. Rev. Phytopathol.* **2014**, *52*, 551–581. [**Google Scholar**] [**CrossRef**]

144. Prins, M.W.; Van Enckevort, L.J.G.; Versluis, H.P. Geminivirus Resistant Plants. U.S. Patent No. 2020/0392530 A1, 27 June 2019. [**Google Scholar**]

145. Kang, B.C.; Yeam, I.; Jahn, M.M. Genetics of plant virus resistance. *Annu. Rev. Phytopathol.* **2005**, *43*, 581–621. [**Google Scholar**] [**CrossRef**] [**Green Version**]

146. Bastet, A.; Robaglia, C.; Gallois, J.L. eIF4E resistance: Natural variation should guide gene editing. *Trends Plant Sci.* **2017**, *22*, 411–419. [**Google Scholar**] [**CrossRef**] [**PubMed**]

147. Bian, X.Y.; Thomas, M.R.; Rasheed, M.S.; Saeed, M.; Hanson, P.; De Barro, P.J.; Rezaian, M.A. A recessive allele (*tgr-1*) conditioning tomato resistance to geminivirus infection is associated with impaired viral movement. *Phytopathology* **2007**, *97*, 930–937. [**Google Scholar**] [**CrossRef**] [**Green Version**]

148. Blair, M.W.; Rodriguez, L.M.; Pedraza, F.; Morales, F.; Beebe, S. Genetic mapping of the Bean golden yellow mosaic geminivirus resistance gene *bgm-1* and linkage with potyvirus resistance in common bean (*Phaseolus vulgaris* L.). *Theor. Appl. Genet.* **2007**, *114*, 261–271. [**Google Scholar**] [**CrossRef**]

149. Czosnek, H.; Eybishtz, A.; Sade, D.; Gorovits, R.; Sobol, I.; Bejarano, E.; Rosas-Díaz, T.; Lozano-Durán, R. Discovering host genes involved in the infection by the Tomato yellow leaf curl virus complex and in the establishment of resistance to the virus using Tobacco Rattle Virus-based post transcriptional gene silencing. *Viruses* **2013**, *5*, 998–1022. [**Google Scholar**] [**CrossRef**] [**Green Version**]

150. Kalinina, N.O.; Khromov, A.; Love, A.J.; Taliansky, M.E. CRISPR Applications in plant virology: Virus resistance and beyond. *Phytopathology* **2020**, *110*, 18–28. [**Google Scholar**] [**CrossRef**]

151. Tomlinson, L.; Yang, Y.; Emenecker, R.; Smoker, M.; Taylor, J.; Perkins, S.; Smith, J.; MacLean, D.; Olszewski, N.E.; Jones, J.D.G. Using CRISPR/Cas9 genome editing in tomato to create a gibberellin-responsive dominant dwarf DELLA allele. *Plant Biotechnol. J.* **2018**, *17*, 132–140. [**Google Scholar**] [**CrossRef**] [**PubMed**] [**Green Version**]

152. Oladosu, Y.; Rafii, M.Y.; Abdullah, N.; Hussin, G.; Ramli, A.; Rahim, H.A.; Miah, G.; Usman, M. Principle and application of plant mutagenesis in crop improvement: A review. *Biotechnol. Biotechnol. Equip.* **2016**, *30*, 1–16. [**Google Scholar**] [**CrossRef**] [**Green Version**]

153. Gady, A.L.; Hermans, F.W.; Van de Wal, M.H.; van Loo, E.N.; Visser, R.G.F.; Bachem, C.W. Implementation of two high through-put techniques in a novel application: Detecting point mutations in large EMS mutated plant populations. *Plant Methods* **2009**, *5*, 13. [**Google Scholar**] [**CrossRef**] [**PubMed**] [**Green Version**]

154. Meissner, R.; Jacobson, Y.; Melamed, S.; Levyatuv, S.; Shalev, G.; Ashri, A.; Elkind, Y.; Levy, A. A new model system for tomato genetics. *Plant J.* **1997**, *12*, 1465–1472. [**Google Scholar**] [**CrossRef**]

155. Menda, N.; Semel, Y.; Peled, D.; Eshed, Y.; Zamir, D. In silico screening of a saturated mutation library of tomato. *Plant J.* **2004**, *38*, 861–872. [**Google Scholar**] [**CrossRef**]

156. Minoia, S.; Petrozza, A.; D'Onofrio, O.; Piron, F.; Mosca, G.; Sozio, G.; Cellini, F.; Bendahmane, A.; Carriero, F. A new mutant genetic resource for tomato crop improvement by TILLING technology. *BMC Res. Notes* **2010**, *3*, 69. [**Google Scholar**] [**CrossRef**] [**PubMed**] [**Green Version**]

157. Saito, T.; Asamizu, E.; Mizoguchi, T.; Fukuda, N.; Matsukura, C.; Ezura, H. Mutant resources for the miniature tomato (*Solanum lycopersicum* L.) 'Micro-Tom'. *Hortic. J.* **2009**, *78*, 6–13. [**Google Scholar**] [**CrossRef**]

158. Shepherd, D.N.; Martin, D.P.; Thomson, J.A. Transgenic strategies for developing crops resistant to geminiviruses. *Plant Sci.* **2009**, *176*, 1–11. [**Google Scholar**] [**CrossRef**]

159. Ammara, U.E.; Mansoor, S.; Saeed, M.; Amin, I.; Briddon, R.W.; Al-Sadi, A.M. RNA interference-based resistance in transgenic tomato plants against Tomato yellow leaf curl virus-Oman (TYLCV-OM) and its associated betasatellite. *Virol. J.* **2015**, *12*, 38. [**Google Scholar**] [**CrossRef**] [**PubMed**] [**Green Version**]

160. Antignus, Y.; Vunsh, R.; Lachman, O.; Pearlsman, M.; Maslenin, L.; Hananya, U.; Rosner, A. Truncated *Rep* gene originated from Tomato yellow leaf curl virus-Israel [Mild] confers strain-specific resistance in transgenic tomato. *Ann. Appl. Biol.* **2004**, *144*, 39–44. [**Google Scholar**] [**CrossRef**]

161. Ali, Z.; Abulfaraj, A.; Idris, A.; Ali, S.; Tashkandi, M.; Mahfouz, M.M. CRISPR/Cas9-mediated viral interference in plants. *Genome Biol.* **2015**, *16*, 238. [**Google Scholar**] [**CrossRef**] [**Green Version**]

162. Ali, Z.; Ali, S.; Tashkandi, M.; Zaidi, S.S.E.A.; Mahfouz, M.M. CRISPR/Cas9-mediated immunity to geminiviruses: Differential interference and evasion. *Sci. Rep.* **2016**, *6*, 26912. [**Google Scholar**] [**CrossRef**] [**Green Version**]

163. Chaparro-Garcia, A.; Kamoun, S.; Nekrasov, V. Boosting plant immunity with CRISPR/Cas. *Genome Biol.* **2015**, *16*, 254. [**Google Scholar**] [**CrossRef**] [**Green Version**]

164. Ji, X.; Si, X.; Zhang, Y.; Zhang, H.; Zhang, F.; Gao, C. Conferring DNA virus resistance with high specificity in plants using virus-inducible genome-editing system. *Genome Biol.* **2018**, *19*, 197. [**Google Scholar**] [**CrossRef**] [**PubMed**] [**Green Version**]

165. Zaidi, S.S.E.A.; Mansoor, S.; Ali, Z.; Tashkandi, M.; Mahfouz, M.M. Engineering plants for geminivirus resistance with CRISPR/Cas9 system. *Trends Plant Sci.* **2016**, *21*, 279–281. [**Google Scholar**] [**CrossRef**] [**PubMed**]

Chapter-16

Resistant Sources and Genetic Control of Resistance to TOLCNDV in Cucumber

1. Introduction

Cucurbits are cultivated in tropical, subtropical, and temperate regions of the New and Old world and supply essential vitamins and minerals to current diets in countries around the world, being a major source of food for humans. Crops belonging to the three most economically important genera, *Cucumis* (melon and cucumber), *Citrullus* (watermelon), and *Cucurbita* (zucchini, pumpkin, squash and gourd), rank in the first positions in global vegetable and fruit production. Spain is one of the main world producers of cucurbits [**1**], and the first exporting country in Europe. However, the production of these crops has been severely affected by diseases, in particular those caused by viruses [**2**,**3**] that have a high economic impact. Among them, *Tomato leaf curl New Delhi virus* (ToLCNDV), a member of the genus *Begomovirus*, family *Geminiviridae*, has spread rapidly in southern Spain since the first detection in 2012 and represents a major risk in the production of zucchini, melon, and cucumber.

ToLCNDV was first detected in tomato (*Solanum lycopersicum* L.) in India in **4**] 1995] and, later, it was found in other south and southeast Asian countries in several hosts, particularly species of the *Solanaceae* and *Cucurbitaceae* families [**5**,**6**]. ToLCNDV was limited to Asian countries until 2012, when it was reported affecting cucurbits (mainly zucchini (*Cucurbita pepo* L.), melon (*Cucumis melo* L.), and cucumber (*Cucumis sativus* L.)) in different Mediterranean countries, first in Spain and later in Tunisia, Italy, Morocco, Greece, and Algeria [**7**,**8**,**9**,**10**,**11**,**12**]. More recently, the virus has been identified in cucurbit plants in Portugal and Estonia [**13**], and in species of the *Solanaceae* family in Italy [**14**], so ToLCNDV is rapidly spreading through Europe.

ToLCNDV consists of two circular single-stranded DNA molecules of approximately 2.7 kb each (designated as DNA-A and DNA-B) [**15**]. The symptoms caused by ToLCNDV depend on the species and the time of infection, but it usually induces curling, leaf mottling and mosaic of young leaves, short internodes, and fruit skin roughness [7], often resulting in a significant yield reduction. ToLCNDV is naturally transmitted by the whitefly *Bemisiatabaci* (Gennadius) byotipes MED and MEAM1 in a persistent manner [**15**,**16**,**17**] although some isolates are also mechanically sap-transmitted to different hosts [**16**,**18**,**19**]. Recently, seed-transmissible strains of ToLCNDV have been described infecting chayote (*Sechium edule* (Jacq) Sw) in India [**20**], and zucchini squash in Italy [**21**]. Against this background, the European and Mediterranean Plant Protection Organization (EPPO) included this virus in the EPPO Alert List [**22**].

ToLCNDV is currently managed using cultural practices and chemical treatment against its vector. However, these control methods have limited effectiveness and can be expensive. Therefore, the development of resistant varieties through conventional breeding provides an effective and sustainable solution for reducing the impact of the disease caused by this virus. In cucurbits, monogenic resistance to ToLCNDV has been described in sponge gourd (*Luffa cylindrica* M. Roem.) [**17,23**]. In melon, resistance has been identified in five Indian melon genotypes belonging to subsp. *agrestis* (Naudin) Pangalo (three accessions of the *momordica* horticultural group and two wild *agrestis*) [**19**]. A major QTL in chromosome 11 was found controlling the resistance to ToLCNDV in one of the wild *agrestis* accessions, with epistatic interactions of two additional regions in chromosomes 2 and **24**] 12]. Finally, resistance has also been identified in pumpkin (*Cucurbita moschata* L.) accessions from diverse origins [**25**]. A major recessive gene located in chromosome 8, in a region syntenic to the candidate region in chromosome 11 of melon, was found controlling the resistance to ToLCNDV in this species [**26**].

The first step for breeding resistant cucumber cultivars is the search for resistant sources. Cucumber germplasm has been screened for resistance to different viral diseases, but to our knowledge, no resistance has been described for ToLCNDV in cucumber [**3**]. In this report, we evaluated the response to ToLCNDV of a cucumber germplasm collection by mechanical inoculation. The identification of three Indian *C. sativus* accessions highly resistant to the virus, which remained symptomless and showed a reduced viral accumulation, provides the first sources for breeding ToLCNDV-resistant cucumber cultivars. Moreover, we have identified one QTL controlling the resistance to ToLCNDV in *C. sativus* using segregating populations derived from one of these resistant sources and a susceptible accession.

2. Materials and Methods

2.1. Plant Material

A nuclear collection of 40 Spanish landraces of *C. sativus* var. *sativus* (**Table 1**), held at the genebank of the Institute for the Conservation and Breeding of Agricultural Biodiversity at the Polytechnic University of Valencia (COMAV-UPV), was first screened in a climatic chamber against ToLCNDV by mechanical inoculation. These accessions represent the variability of the full COMAV collection, consisting of 217 accessions collected from diverse Spanish origins and multiplied by COMAV [**27,28**]. This collection includes accessions belonging to the typical "short" (20) "long" (16), and "French" (4) cucumber types (**Table 1**), which are highly appreciated on national and international markets because of their quality. Additionally, 23 *C. sativus* var. *sativus* accessions from different geographical origins (**Table 2**) of the "short" (12), "medium" (5), and "long" (5) cucumber types, and one unknown type, were also tested. Seeds of these accessions were firstly provided by the Centre for Genetic Resources (CGN germplasm collection, the Netherlands), and then multiplied at COMAV.

Table 2. Response of *C. sativus* accessions from different origins to the mechanical inoculation with ToLCNDV. Mean and range of symptoms scored in plants per genotype (at 15 and 30 dpi) according to the scale: 0, absence of symptoms; 1, mild symptoms; 2, moderate symptoms; 3, severe symptoms; 4, very severe symptoms or plant death. Mean score of viral load detected in each plant of the assayed accessions by tissue printing at 15 dpi according to the scale of high (+++), intermediate (++), low (+), or absent (-) viral accumulation. Data not available are shown as n/a.

Table 1. Response of Spanish landraces of *C. sativus* to the mechanical inoculation with ToLCNDV. Mean and range of symptoms scored in plants per genotype (at 15 and 30 dpi) according to the scale: 0, absence of symptoms; 1, mild symptoms; 2, moderate symptoms; 3, severe symptoms; 4, very severe symptoms or plant death. Mean score of viral load detected in each plant of the assayed accessions by tissue printing at 15 dpi according to the scale of high (+++), intermediate (++), low (+), or absent (-) viral accumulation. Data not available are shown as n/a.

			Symptoms at 15 Dpi		Symptoms at 30 Dpi		Viral Load
Type	Genebank Code	Spanish Province	Mean	Range	Mean	Range	Tissue Printing
Short	BGV000047	Zaragoza	1.2	(1–2)	1.8	(1–2)	+++
	BGV000408	Cádiz	1.6	(1–2)	1.6	(1–3)	+++
	BGV000437	Jaén	1.2	(0–3)	1.8	(1–2)	++
	BGV000467	Jaén	1.0	(0–2)	1.4	(0–3)	+++
	BGV000479	Córdoba	1.0	(1)	1.0	(0–2)	+++
	BGV000512	Huelva	1.0	(0–2)	1.4	(1–2)	+++
	BGV002495	Tenerife	n/a		0.6	(0–2)	++
	BGV003714	Cuenca	2.0	(1–3)	1.7	(1–2)	+++
	BGV004026	Cáceres	1.4	(1–2)	2.5	(1–3)	+++
	BGV004304	Murcia	n/a		2.0	(1–3)	+++
	BGV008299	Valencia	2.0	(1–3)	2.4	(2–3)	++
	BGV010301	Guadalajara	1.4	(0–3)	1.6	(0–3)	+++
	BGV010314	Guadalajara	1.7	(1–2)	1.2	(0–2)	+
	BGV010636	Soria	2.4	(1–4)	3.4	(2–4)	++
	BGV011582	Teruel	0.8	(0–2)	1.8	(1–2)	++
	BGV011734	Valladolid	2.0	(0–4)	3.2	(3–4)	++
	BGV011736	Ávila	2.6	(0–4)	2.8	(2–4)	++
	BGV011742	Albacete	3.4	(3–4)	3.8	(3–4)	++
	BGV014959	Huesca	3.8	(3–4)	3.6	(3–4)	++
	BGV015469	Cáceres	1.6	(0–2)	2.8	(2–4)	+

			Symptoms at 15 Dpi		Symptoms at 30 Dpi		Viral Load
Type	Genebank Code	Spanish Province	Mean	Range	Mean	Range	Tissue Printing
Long	BGV000372	Granada	0.6	(0–1)	1.2	(0–2)	++
	BGV000381	Málaga	0.6	(0–1)	2.4	(2–3)	+++
	BGV000416	Cádiz	1.4	(0–2)	2.0	(1–3)	++
	BGV001310	Asturias	1.0	(0–2)	1.6	(1–2)	+++
	BGV002494	Tenerife	2.0	(2)	1.0	(0–2)	+++
	BGV004305	Murcia	1.6	(1–2)	1.2	(1–2)	+++
	BGV004309	Murcia	1.4	(1–2)	3.0	(3)	+++
	BGV004851	Castellón	0.0	(0)	1.0	(0–2)	++
	BGV004926	Valencia	2.0	(2)	2.0	(2)	+++
	BGV004936	Valencia	1.4	(1–2)	1.3	(1–2)	++
	BGV011586	Orense	0.6	(0–3)	0.6	(0–3)	+
	BGV011724	Teruel	1.8	(0–4)	1.8	(0–4)	+
	BGV014967	Guadalajara	2.2	(0–4)	2.4	(1–4)	++
	BGV015229	Vizcaya	2.5	(1–3)	2.5	(2–3)	++
	BGV015696	Alicante	2.4	(0–4)	2.6	(0–4)	++
	BGV015700	Girona	2.8	(1–4)	3.4	(2–4)	++
French	BGV010290	Granada	2.8	(0–4)	3.8	(3–4)	+++
	BGV011735	Zaragoza	2.3	(0–3)	2.3	(0–4)	++
	BGV014961	Castellón	3.0	(0–4)	2.6	(0–4)	+++
	BGV014969	Cantabria	1.6	(0–3)	2.4	(0–4)	+++

2.2. Virus Source, Mechanical Inoculation, and Symptom Evaluation

As an inoculum source, zucchini plants of the MU-CU-16 accession were agroinoculated by injection into petioles with an infectious clone of ToLCNDV [**25**]. ToLCNDV transmission to cucumber plants was performed by mechanical inoculation at the stage of one true leaf, as described by López et al. [**19**]. Briefly, inoculum was prepared by grinding 1 g of symptomatic leaf tissue from agroinfiltrated plants in the presence of inoculation buffer in a 1:4 (*w*:*v*) proportion. The expanded true leaf and one cotyledon of each plant were dusted with carborundum (600 mesh) and then inoculated by rubbing with a cotton-bud stick, gently soaked in the crude homogenized inoculum.

For the mechanical inoculation, seeds were disinfected in a 10% solution of sodium hypochlorite for 3 min and washed for 5 min in distilled water. Germination was performed in Petri plates with moistened cotton at 37 °C for 48 h. Seedlings were transplanted to pots in a growth chamber under a photoperiod of 16 h day at 25 °C and 8 h night at 18 °C and 70% relative humidity. Seedlings at the one true leaf stage were mechanically inoculated, leaving two uninoculated plants per genotype as controls. Inoculated plants were individually evaluated at 15 and 30 days post inoculation (dpi) for the presence and severity of virus symptoms. Symptoms on upper leaves were recorded by visual evaluation using the following scale: 0, no symptoms; 1, mild symptoms; 2, moderate symptoms; 3, severe symptoms; 4, very severe symptoms or dead plant (**Figure 1**). Additionally, every plant was assayed for the presence of virus using the tissue printing technique and conventional PCR with the protocols described below. Additionally, the viral load of ToLCNDV was determined by qPCR in a selected number of accessions with the best resistance response (CGN22297, CGN22986, CGN23089, CGN23423, and CGN23633). The number of plants tested of each accession varied between 3 and 6 due to seed availability and germination. The most resistant accessions were selected for further analysis with additional plants.

Figure 1. Symptom scoring in cucumber plants showing ToLCNDV symptoms corresponding to the scale: 0 absence of symptoms, 1: mild symptoms, 2: moderate symptoms, 3: severe symptoms, and 4: very severe symptoms or dead plant.

2.3. ToLCNDV Detection by Tissue Printing

For detection of ToLCNDV in tissue prints, plant petioles of the upper leaves at 15 dpi were cut with a razor blade and cross-sections were blotted onto positively charged nylon membranes (Hybond-N, Amersham) immediately after cutting. Membranes were air dried, fixed by UV irradiation (700 × 100 mJ/cm^2), and hybridized with a digoxigenin-labeled RNA probe. The riboprobe was generated by transcription with T7 RNA polymerase from a recombinant pTZ57R plasmid (Fermentas) with an insert corresponding to the complete CP gene of ToLCNDV in a negative orientation, following the manufacturer's instructions (Roche Diagnostics). Prehybridization, hybridization, and washing of the membranes were performed as previously reported [**29**], except that the hybridization was conducted at 60 °C. Chemiluminiscent detection using CSPD reagent as substrate was performed as recommended by the manufacturer (Roche Diagnostics). Films were exposed to the membranes at room temperature for 30–60 min.

2.4. ToLCNDV Detection by PCR and qPCR

To confirm the presence of the virus by PCR, total DNA from apical leaves of inoculated and control plants was extracted at 30 dpi using the CTAB method [**30**]. DNA was quantified using a NanoDrop 1000 spectrophotometer (Thermo Scientific) and diluted with sterile deionized water to a final concentration of 50 ng μL^{-1}. One-microliter aliquots of total DNA (50 ng) were used as templates in PCR reactions with the ToLCNDV-specific primer pairs To-A1F and To-A1R from DNA-A [**25**]. The resulting PCR products of 504 bp in length were analyzed by electrophoresis in 1.0% agarose gels in TAE buffer.

The relative ToLCNDV accumulation in individual plants of the most promising accessions was determined at 30 dpi by qPCR, and susceptible plants were used as controls. DNA was diluted to a final concentration of 5 ng μL^{-1} and all plants were analyzed in two technical replicates using a *LightCycler*® 480 System (Roche). In each qPCR reaction, 15 ng of genomic DNA were used as templates, in a final volume of 10 μL. We used 2.5 μL of MasterMix qPCR No-ROX PyroTaqEvaGreen 5x (Cmbbioline) and 0.35 μL (10 μM) of each primer and 3.8 μL of H_2O. Primers ToLCNDVF1 (5′-AATGCCGACTACACCAAGCAT-3′, positions 1145–1169) and ToLCNDVR1 (5′-GGATCGAGCAGAGAGTGGCG-3′, positions 1399–1418), derived from the Spanish isolate Murcia 11.1 (segment DNA-A, accession number KF749225), were used for the amplification of a 273 bp fragment of viral DNA-A. The *β-actin* of *C. sativus* gene was amplified in all samples as reference control using an adapted design of primers used in previous works on watermelon [**31**], ClACT-F (5′-CCATGTATGTTGCCATTCAG-3′) and ClACT-R (5′-GGATAGCATGGGGAAGAGCA-3′). Cycling conditions consisted of incubation at 95 °C for 15 min, 45 cycles of 95 °C for 5 s, 60 °C for 30 s, and 72 °C for 30 s. Relative ToLCNDV levels were calculated using the $2^{-\Delta\Delta Ct}$ expression of the Livak method [**32**], where ΔΔCt is the difference between the ΔCt of each sample and the ΔCt of the calibrator sample.

2.5. Validation of Response to the Viral Infection and Generation of F_1, F_2, and BC Populations

Plants of each of the three resistant accessions were transplanted, grown, and selfed in a whitefly-proof greenhouse and the plant with the best resistant behavior, CGN23089-2, was crossed with plants of the accession BGV011742, highly susceptible to ToLCNDV. Seventeen seeds of the F_1 hybrid and from 15 to 20 seeds of the selfing offspring of each parent were disinfected and seedlings were transplanted to pots and grown in a climatic chamber under controlled conditions. All plants were mechanically inoculated with ToLCNDV and phenotyped according to symptomatology and viral accumulation determined by qPCR, at 15 and 30 dpi, following the procedure described above. Three plants of the genotype BGV011742 were also included as susceptible controls in the validation assay. Means of $2^{(-\Delta\Delta Ct)}$ values of each genotype were analyzed by analysis of variance (ANOVA) and least significance difference (LSD) multiple range tests using STATGRAPHIC 18 ™ (Statgraphics Technologies, Inc., The Plains, VA, USA), to evaluate statistically significant differences between them, with a level of confidence of %95.

In the following growing season, three plants of the F_1 progeny were cultivated in a greenhouse to generate F_2 segregating populations by selfing, and $BC_{CGN23089}$ and $BC_{BGV011742}$ by backcrossing to plants of CGN23089 and BGV011742, respectively. All plants of these populations were screened against ToLCNDV with the same inoculation protocol and disease assessment procedure described above. The chi-squared (χ^2) test ($p < 0.05$) was used to determine the goodness of fit between the expected and observed ratios of resistant:susceptible segregation in the three populations.

2.6. Genotyping of Segregating Populations

To genotype the segregating populations, single nucleotide polymorphism (SNP) markers were selected from two sources: a genotyping-by-sequencing (GBS) assay, including the BGV011742 accession, performed in previous studies by our group, and the available data of a GBS assay used to characterize the United States National Plant Germplasm System (NPGS) collection of cucumber [**33**]. Among the analyzed accessions, the Indian genotype PI 197087 is the same as the accession CGN23089 kept in the CGN germplasm collection, as described in the passport data from both genebanks (**https://cgngenis.wur.nl** (accessed on 28 February 2021); **https://www.ars-grin.gov/** (accessed on 14 February 2021)). Both sequences were aligned against the cucumber genome Gy14 v.2 available in the Cucurbit Genomics Database (**http://cucurbitgenomics.org**; (accessed on 15 March 2021)) using the Bowtie2 tool [**34**], and SNP variants were found by Freebayes version 1.0.2 [**35**]. A panel of 47 SNPs was designed to cover the seven chromosomes of the cucumber genome and used to genotype parents, F_2, and BC segregating populations by an Agena Bioscience iPLEX® Gold MassARRAY (Agena Biosciences, CA, USA) system at the Epigenetic and Genotyping Unit of the University of Valencia (Unitat Central d'InvestigacióenMedicina (UCIM), Faculty of Medicine, Malaga, Spain). F_2 genotyping results were used to construct a genetic map using the Kosambi map function in MAPMAKER **36**] 3.0], and a QTL analysis was performed applying the composite interval mapping approach (CIM) in Qgene 4.0 [**37**]. Symptom score, ToLCNDV relative accumulation at 30 dpi, and a qualitative trait of resistance, assigning to each plant a category of 0 if the phenotype was susceptible and 1 if it was resistant, were used to identify markers linked to the resistance to ToLCNDV. LOD threshold was estimated performing 1000 permutation tests per trait, with $p < 0.05$. The proportion of phenotypic variance explained (R^2), the additive and dominance effects, and the interval position of the QTL, according to a LOD drop of up to the significant LOD threshold level, were estimated for the peak LOD of each significant QTL. Since these traits were not normally distributed, we used a Kruskal–Wallis non-parametric test to support CIM QTL detection, using MapQTL version 4.1 software [**38**], considering associations significant at $p \leq 0.05$.

Means of symptom scores and $2^{(-\Delta\Delta Ct)}$ at 30 dpi of $BC_{CGN23089}$ plants were calculated for the closest SNPs to the QTL peak LOD, according to each genotypic class (b and h). To determine statistically significant differences between means ($p \leq 0.05$), ANOVA and LSD multiple range tests were performed using STATGRAPHIC 18™ (Statgraphics Technologies, Inc., The Plains, VA, USA) statistical software.

3. Results

3.1. Response of the Spanish Landraces of C. sativus to the Mechanical Transmission of ToLCNDV

A core collection of 40 accessions from different Spanish provinces held at the COMAV Genebank was assayed. Most of the 40 tested cucumber accessions were highly susceptible to the mechanical transmission of ToLCNDV, showing moderate symptoms of mottling to severe symptoms characterized by mosaic and yellowing of young leaves (**Figure 1**). Symptoms started to appear at different days after mechanical inoculation. On average, symptom scores in plants of the "short", "long", and "French" types of *C. sativus* increased from 15 to 30 dpi, with average scores from 0.19 ± 1.78 to 0.19 ± 1.52 ,0.2 ± 2.1 to ± 1.9 0.19, and 0.31 ± 2.43 to 0.31 ± 2.8 in each group, respectively, in a range of 0 to 4 (**Table 1**). As had already been observed in the cucumber-growing areas of the southeast of Spain, plants of the "French" type were the most susceptible [**39**].

To further characterize the response to ToLCNDV, the viral load of all plants at 15 dpi was evaluated by molecular hybridization by tissue printing. To carry out a more precise confirmation, viral accumulation was determined by semi-quantitative PCR at 30 dpi. Viral load ranged from intermediate to high (**Table 1**) in most of the genotypes at 15 dpi. In addition, similar high ToLCNDVtiters were detected by PCR at 30 dpi in all genotypes (data not shown). Only accessions BGV000479, BGV002495, BGV002494, and BGV004851 developed mild symptoms at the end of the trial, although with moderate or high viral accumulation. On average, most of the plants of the BGV011586 accession displayed low symptomatology and accumulated low viral titers, but some plants developed high symptoms from the beginning of the assay, suggesting variability in the response to ToLCNDV within this genotype.

3.2. Response of the *C. sativus* Accessions from Different Origins to the Mechanical Transmission of ToLCNDV

The cucumber accessions from different countries showed variable responses to ToLCNDV infection. Susceptible accessions behaved similarly to the Spanish landraces, displaying moderate to severe yellowing and mottling that in most cases increased from 15 to 30 dpi. Accessions belonging to the "long" type, all originating from China, had on average higher symptom scores (**Table 2**) at 15 dpi, with a mean of 1.86 ± 0.2. Similarly, the only accession of unknown type (CGN19655, originating from the U.S.A.) was highly susceptible, with a symptom score of 1.4 at 15 dpi (**Table 2**). Viral titers detected with PCR were high or very high in all these accessions (data not shown).

Interestingly, lower severity of the ToLCNDV infection was observed in some Indian genotypes of the "medium" and "short" types. Accession CGN22297 was symptomless or had very mild symptoms at 15 dpi (mean symptom score of 0.4, in a range between 0 and 1), although some plants developed moderate symptomatology at the end of the assay (plants with symptom scores ranging from 0 to 2) (**Table 2**). All plants of accessions CGN22297, CGN22986, CGN23089, CGN23423, and CGN23633 remained symptomless, or had symptom scores lower than one, throughout the screening assay. On these five accessions, ToLCNDVtiters were low or not detected by probe hybridization at 15 dpi (**Table 2**), although in all of them the virus was detected at 30 dpi after PCR analysis.

Among the remaining "medium" and "short" type assayed accessions from different countries, some initially had a promising behavior (mean symptom score 1.14 and 1.36, respectively, ranging from 0.3 to 3.5), but typical severe ToLCNDV symptomatology and high or very high viral titers were identified in all of them at different stages of the disease (**Table 2**).

3.3. ToLCNDV Quantification in Resistant Genotypes

Individual plants of the five Indian accessions with better response after infection with ToLCNDV were tested by qPCR to further determine the viral accumulation at 30 dpi. One or two plants of the susceptible accessions BGV002494, BGV010301, BGV011742, and BGV014959 were used as controls, with all of them showing the highest level of relative viral titers, but the resistant genotypes presented variability between their relative ToLCNDV accumulations (**Figure 2**A). Plants of the CGN23089, CGN23423, and CGN23633 accessions had uniformly low viral loads, with $2^{(-\Delta\Delta Ct)}$ values 10 × 1.9^3 times lower than the levels accumulated by the susceptible plants, on average. Instead, in both CGN22297 and CGN22986 accessions, some plants were identified with low $2^{(-\Delta\Delta Ct)}$ values and some with high viral load, similar to that detected in one of the susceptible genotypes (**Figure 2**A). After this further characterization, the accessions CGN23089, CGN23423, and CGN23623, which were those with the lowest symptoms scores at 15 and 30 dpi and with the lowest viral titers estimated with different methods, were selected for further characterization.

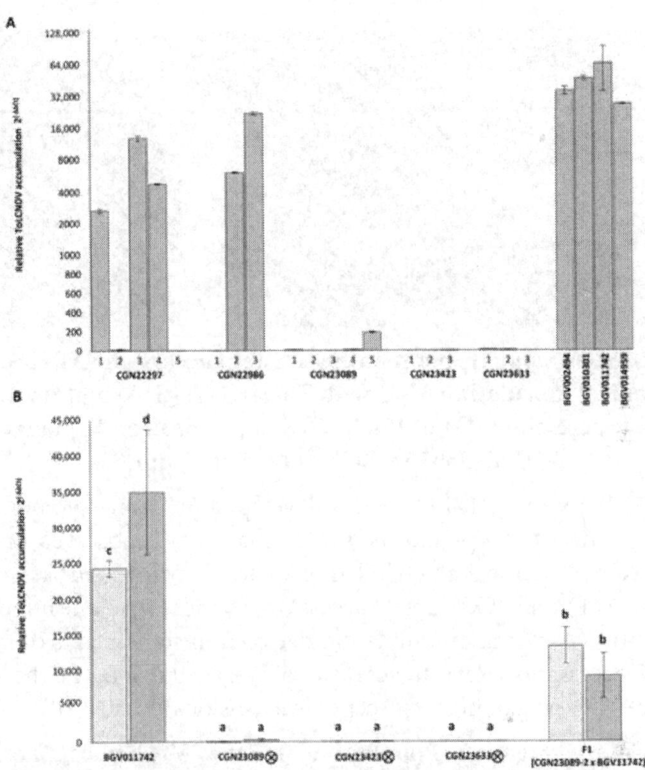

Figure 2. (A) Relative ToLC NDV accumulation ($2^{(-\Delta\Delta Ct)}$) at 30 days after mechanical inoculation (dpi) with ToLCNDV in the five asymptomatic Indian accessions (CGN22297, CGN22986, CGN23089, CGN23423, and CGN23633) and in four susceptible controls (BGV002494, BGV010301, BGV011742, and BGV014959). (B) Relative ToLCNDV accumulation ($2^{(-\Delta\Delta Ct)}$) at 15 and 30 dpi (light and dark bars, respectively) of plants obtained by selfing the CGN23089-3, CGN23423-2, and CGN23623-2 genotypes and of the F1 (CGN23089-2 x BGV011742) hybrids. On the *x* axis, accessions and number of plants of each accession are indicated. Bars sharing the same letter are not significantly different, according to ANOVA and LSD tests ($p \leq 0.05$).

3.4. Response of Self-Pollinated and F_1 Progenies to the Mechanical Transmission of ToLCNDV

The selfing offspring of the plants CGN23089-3, CGN23423-2, and CGN23623-2, and the F_1 hybrid derived from the cross CGN2-23089 x BGV011742 (one of the most susceptible Spanish landraces selected as a susceptible parent for this cross), were mechanically inoculated with ToLCNDV in a second assay to confirm the resistance. As expected, plants of the susceptible BGV011742 parent showed severe symptoms at 30 dpi (**Figure 3**B), while all plants of the self-pollinated offspring had a similar behavior to that observed in the resistant plants of the first assay, displaying mild to no symptoms at 15 and 30 dpi (**Figure 3**A).

Figure 3. (A) Asymptomatic plant of the resistant accession CGN23089 30 days after mechanical inoculation (dpi) with ToLCNDV. (B) Symptomatic plant of the susceptible accession BGV011742. (C) Symptoms in an F_1 plant of the cross CGN23089 x BGV011742 at 30 dpi.

The F_1 (CGN2-23089 x BGV011742) plants developed moderate symptomatology (two on the symptom scale) at 15 dpi and the same behavior was observed up to the end of the assay (**Figure 3**C). On average, viral titer in the F_1 hybrid surpassed those of the CGN23089, CGN23423, and CGN23633 accessions by more than one hundred times, but it was similar to the high viral accumulation detected in some plants of the CGN22297 and CGN22986 accessions. Nevertheless, the average viral load in the F_1 hybrid was almost three times lower than in the susceptible accessions (**Figure 2**B).

3.5. Response of Segregating Populations to the Mechanical Transmission of ToLCNDV

After ToLCNDV mechanical inoculation, both F_2 and $BC_{CGN23089}$ segregated for symptom development and viral load, while all assayed plants of $BC_{BGV011742}$ developed severe symptoms and high viral accumulation. The number of resistant and susceptible plants found in each segregating population is shown in **Table 3**, according to symptomatology and viral load at 30 dpi. At the end of the assay, 31 plants of F_2 remained symptomless or had slight symptoms (scores 0 to 1), and 65 showed moderate to severe symptomatology (scores 2 to 4). This segregation fit an expected ratio of 1:3 (resistant:susceptible), compatible with a single recessive gene controlling the resistance ($p = 0.099$) (**Table 3**). On average, viral accumulation correlated to symptom severity following an exponential model ($y = 44.594e^{1.564x}$, $R^2 = 0.8512$), with $2^{(-\Delta\Delta Ct)}$ viral load values of up to 10^4 times higher in susceptible plants than in resistant plants (**Figure 4**). In $BC_{CGN23089}$, 21 plants were resistant (scores 0 to 1) and 33 were susceptible (scores 2 to 3), fitting a 1:1 expected segregation for recessive monogenic control ($p = 0.1025$) (**Table 3**). Within each symptom score category, plants of both BC populations accumulated similar viral titers (**Figure 4**).

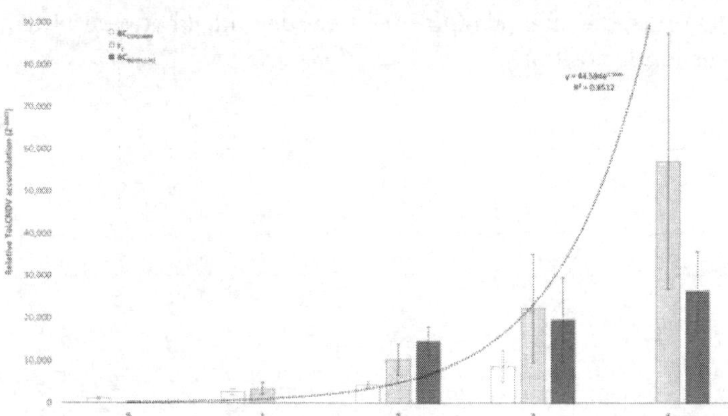

Figure 4. Mean of relative viral accumulation ($2^{-\Delta\Delta Ct}$) at 30 days after mechanical inoculation in plants of F_2 (CGN2-23089 x BGV011742) (light gray bars), $BC_{CGN23089}$ (white bars), and $BC_{BGV011742}$ (dark gray bars) in each symptom score category. Dotted line shows the tendency of the variable adjusted to an exponential model.

Table 3. Number of resistant and susceptible plants in each segregating population according to symptom development and ToLCNDVtiters. The probability of X^2 value was calculated for the expected ratio of one recessive gene controlling the resistance.

Populations	Resistant	Susceptible	Expected Frequencies	X^2
F_2	31	65	1:3	2.722 (p = 0.0990)
$BC_{CGN23089}$	21	33	1:1	2.667 (p = 0.1025)
$BC_{BGV011742}$	0	11	-	-

3.6. Genotyping and Linkage Analysis in Segregating Populations

After genotyping the F_2 and BC populations with the 47 SNP markers evenly distributed throughout the *C. sativus* genome, only 18 SNPs were polymorphic between the CGN23089 and BGV011742 accessions. Genotypic results of F_2 were used to construct a linkage map of the seven chromosomes, spanning a total of 554 cM of genetic distance with an average of 34.67 cM between markers (**Table S1**).

To identify genomic regions linked to the resistance to ToLCNDV in cucumber, a QTL analysis was performed. Symptoms at 30 dpi, viral accumulation at 30 dpi determined by probe hybridization, and the qualitative trait of resistance showed significant association with three overlapping QTLs in chromosome 2, explaining between 15.1 and 17.3% of the observed phenotypic variance (**Table 4**). A fourth QTL was linked to viral accumulation determined by qPCR (ΔΔCt), but the LOD peak obtained (2.54) was slightly under the LOD threshold (2.75) (**Figure 5**). The closest marker to all significant QTLs (ToLCNDVCs_Sy30-2, ToLCNDVCs_VT30-2, and ToLCNDVCs_Re-2) was SNPCS2_3 (physical position 12,760,375 pb), with LOD peaks between 3.07 and 3.93. All these QTLs were statistically validated by a Kruskal–Wallis test, with $p \leq 0.005$. Two additional QTLs were identified in chromosome 1 (**Figure 5**), but their effects were not significant on all the traits. Thus, they were excluded from the analysis. According to the regions with a significant LOD value, the interval of the QTL was delimited between 11,657,498 pb and 21,993,369 pb genomic positions.

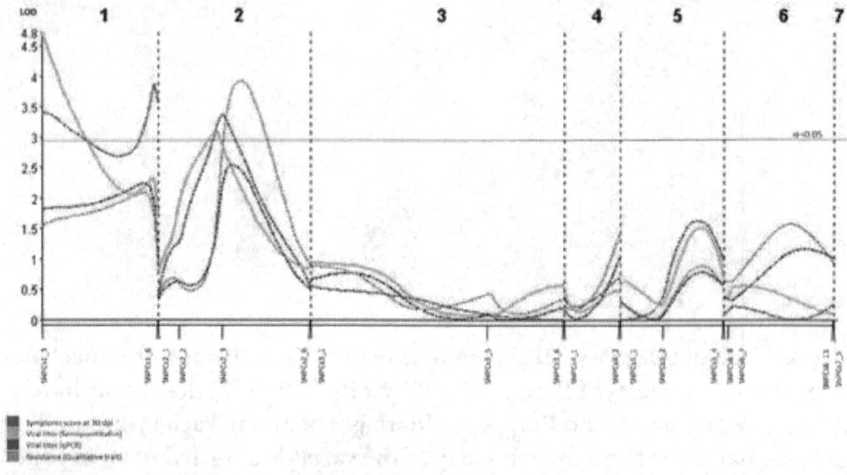

Figure 5. QTL analysis of F_2 (CGN2-23089 x BGV011742) using symptom score at 30 days after mechanical inoculation (dpi), viral titers of ToLCNDV at 30 dpi (semiquantitative and quantitative detection) and qualitative resistance as traits.

Table 4. Quantitative trait loci (QTLs) identified in the F_2 segregating population using composite interval mapping (CIM) and Kruskal–Wallis tests.

Trait	Chr [a]	Nearest Marker [b]	CIM					Kruskal–Wallis	
			Interval [c] (cM)	Add Effect [d]	Dom Effect [e]	LOD [f]	R^{2g}	K^{*h}	Significance [i]
Symptoms 30 dpi	2	SNPCs2_3	28–40	0.46	0.73	3.38	0.15	16.94	******
Viral load (Semiquantitative)	2	SNPCs2_3	24–32	0.18	0.79	3.07	0.14	13.01	****
Viral load (Quantitative, ΔΔCt)	2	SNPCs2_3	-	−1.45	−4.61	2.54	0.12	14.05	*****
Resistance (Qualitative trait)	2	SNPCs2_3	34–54	−0.23	−0.57	3.93	0.17	13.02	****

[a] Chromosome; [b] the closest marker to the LOD peak, [c] interval position of the putative QTL, identified in the F_2 (CGN23089-2 x BGV011742) by CIM, in cM on the genetic map; [d] Add effect: additive effect of the BGV011742 allele; [e] Dom effect: dominant effect of the BGV011742 allele; [f] LOD: higher logarithm of the odds score; [g] R^2: percentage of phenotypic variance explained by the QTL; [h] K*: the Kruskal–Wallis test statistic; [i] Significance level in the Kruskal–Wallis test ****: 0.005, *****: 0.001, ******: 0.0005.

To validate the effect of the chromosome 2 region in the $BC_{CGN23089}$ population, the mean of symptom scores and relative viral accumulation at 30 dpi were calculated for each genotypic class of the two closest SNPs to the identified QTL interval (SNPCs2_2 and SNPCs3_2). The lowest level of symptoms and viral load was observed in plants with a homozygous genotype (b) for both markers, while heterozygous (h) plants for any of these markers showed more severe symptomatology and accumulated more ToLCNDV particles (**Figure 6**).

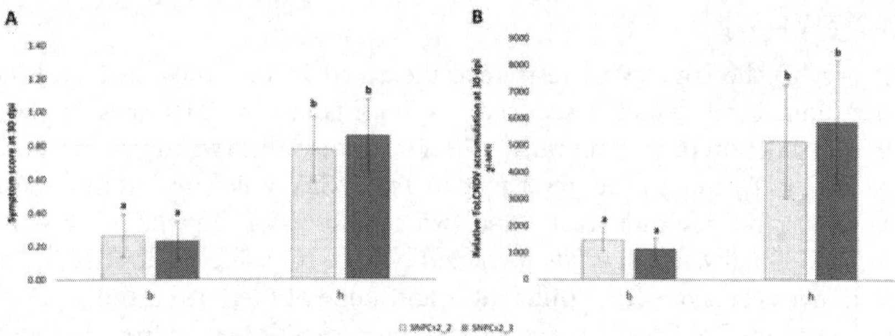

Figure 6. Mean of symptom score (A) and relative viral accumulation (B) at 30 days after mechanical inoculation in $BC_{BGV011742}$ according to each genotypic class of SNPCs2_2 and SNPCs3_2 markers (chromosome 2). On the *x* axis, homozygous genotype of CGN23089 allele is represented as "b", heterozygous genotype is represented as "h". Bars with same letter are not significantly different at $p \leq 0.05$.

4. Discussion

Forty accessions of cucumber collected from different provinces of Spain were screened, in order to find sources of resistance against ToLCNDV, but none of the accessions showed immunity or high resistance to the virus. Most accessions were highly susceptible after ToLCNDV mechanical inoculation, and only five showed intermediate-level symptoms and less viral load. The high susceptibility, observed across this collection representative of the cucumber Spanish diversity, reveals that ToLCNDV represents a major threat to cucumber cultivation.

The cucumber accessions of other origins showed variable results. All accessions from China, and the single accessions from Japan, Sri Lanka, Iran, the United States, and D.R. Congo used in this study were susceptible to ToLCNDV. Interestingly, we have identified resistance in Indian accessions. CGN22297 and CGN22986 showed variable responses in symptom development and in viral load, suggesting that the resistance was not fixed in these accessions. The accessions CGN23089, CGN23423, and CGN23633 were uniformly resistant, symptomless, and had very low ToLCNDV accumulation compared to susceptible controls.

Finding virus resistance in *C. sativus* is not unexpected as this species has often been used as a source of resistance for different cucurbit viruses. For example, resistance genes to different potyviruses have been identified mainly in three cucumber accessions: 'Suriman', 'Taichung Mou Gua' (TMG1-), and 'Dina40] '1-]. In the inbred cucumber line '02245', one *locus* controlling resistance to papaya ring spot virus (PRSV) and another controlling resistance to *Watermelon mosaic virus* (WMV), both recessives, were found by Tian et al. [**41,42**]. In the same line, resistance to the cucumovirus *Cucumber mosaic virus* (CMV) is quantitatively inherited [**43**] and in *C. sativus* var. *hardwickii*, Munshi et al. [**44**] identified CMV resistance controlled by a single recessive gene. Additionally, resistance to *Cucumber vein yellowing virus* (CVYV) has been reported in the Spanish landrace C.sat**45**] 10-], to *Cucurbit yellow stunting disorder virus* (CYSDV) controlled by more than one recessive gene [**46**], and in two Indian accessions of *C. sativus* to *Cucumber green mottle mosaic virus* (CGMMV) [**47**].

To date, most of the sources of resistance identified in cucurbits against ToLCNDV come from India. For instance, resistance to sponge gourd was identified in germplasm collected from different regions in India [**17**]. A dominant allele was found controlling the resistance [**23**]. In *Cucumis melo*, resistance to ToLCNDV was found in three accessions of the *momordica* horticultural group and two accessions of the wild *agrestis* group, all from India [**19**]. Finally, in *Cucurbita moschata*, genetic resistance to ToLCNDV has been identified in five accessions from different origins, one of them from India [**25,48**]. The fact that most of the ToLCNDV-resistant cucurbit accessions come from India could be related to the co-evolution of host and pathogen in this part of the world where ToLCNDV was detected infecting cucurbits many years ago [**49**].

The analysis of the F_1 generation derived from the resistant accession CGN23089 suggests that the resistance to ToLCNDV found in cucumber is recessive. It is interesting to note that recessive control of resistance is frequent in several virus resistance systems. Recessive resistance genes interfere the viral life cycle at different levels: single cells, cell-to-cell movement, long-distance transport through the plant, and/or preventing high levels of virus accumulation [**50**]. In cucumber, the mechanism of resistance to ToLCNDV is characterized by a drastic and significant reduction of virus titer and infected plants are asymptomatic or exhibit mild disease symptoms. This type of resistance is similar to that observed in the rest of the resistances identified in the pathosystem of ToLCNDV–host. In cucurbits, the high level of ToLCNDV DNA accumulation in plant tissue results in the development of severe symptoms and leads to a major reduction in yield in the case of susceptible cultivars, but this does not happen for the cultivars showing resistance. The virus DNA level remains low and approximately constant and has minimal effects on the yield and health of plants [**17,19,23,24,25,26**]. In tomato, ToLCNDV viral DNA also determined the level of resistance and yield loss in test varieties of tomato under the same environmental conditions. Resistant cultivars showing a low level of viral DNA in their tissue when compared to other susceptible cultivars have been reported previously [**51**]. This also happens in the case of CGMMV in cucumber [**47**], CYSDV and WMV [**52,53**], and *Papaya ringspot virus* (PRSV) in squash and watermelon [**54**]. Further studies will be needed to establish the mechanism that limits ToLCNDV accumulation in resistant plants.

The accessions identified in this study are good candidates for breeding programs to avoid damage caused by ToLCNDV in *C. sativus*. Given the importance of ToLCNDV and the scarcity of sources of resistance to ToLCNDV in cucumber, the virus resistance found in accessions CGN23089, CGN23423, and CGN23633 should be introgressed into commercial cultivars. Our inheritance analyses indicate that the resistance to ToLCNDV in the CGN23089 accession is mainly controlled by one recessive gene, and this was supported by the detection of one QTL in chromosome 2 of the *C. sativus* genome. Despite the fact that this region was significantly linked to symptom development and viral load of ToLCNDV in cucumber, the percentage of phenotypic variance explained by the QTL (R^2) is moderate. A higher density of the SNP panel covering the whole genome, along with a finer mapping of the candidate region, might likely increase this percentage. Nevertheless, the results obtained here, even with a small number of markers, contribute significantly to obtaining preliminary information about the *locus* implicated in ToLCNDV resistance in cucumber, and are in accordance with previous studies of genetic control of resistance to ToLCNDV in cucurbits. In melon, a major *locus* in chromosome 11 and two additional regions in chromosomes 2 and 12 controlling the resistance of the wild *agrestis* accession WM7- were found [**24**]. In a recent publication, Romay et al. [**55**] identified in the same Indian accession WM-7 one recessive (*bgm-1*) and two dominant (*Bgm-2* and *Tolcndv*) genes controlling the resistance to ToLCNDV. In *Cucurbita moschata*, a major recessive gene located in chromosome 8 was found controlling the resistance in an Indian accession. This candidate region of *C. moschata* is syntenic to the region responsible for ToLCNDV resistance in chromosome 11 of melon [**26**]. Since both *loci* for resistance to ToLCNDV are syntenic and share a common cluster of genes, we looked for this cluster in synteny with the cucumber genome, which was located in chromosome 6 (from 6,527,862 pb to 6,756,572 pb genomic positions) (**Table S2**). Two SNPs used in this work are close to this region (SNPCs6_8 and SNPCs6_7, at 6,705,461pb and 7,276,564 pb, respectively), but none of them was significantly associated with the resistance to ToLCNDV. Thus, the candidate region identified here may be in a different region, associated with different resistance genes. Among the list of annotated genes in the candidate region of cucumber chromosome 2 (**Table S3**), there are three LRR receptor-like serine/threonine-protein kinases (CsGy2G012160, CsGy2G015920, and CsGy2G016150) implicated in resistance to ToLCNDV and other geminiviruses [**56,57,58,59**], four NAC domain transcription factors (CsGy2G015830, CsGy2G016100, CsGy2G016110, and CsGy2G016220), gene family associated with an increase in tomato plant susceptibility during ToLCNDV infection and resistance to a begomovirus in common bean (*Phaseolus vulgaris* L.) [**60,61**], and an RNA-directed DNA methylation protein (CsGy2G016290.1), one of the components of the RNA silencing pathway used against plant viruses in the defense response [**62**]. More interestingly, a 26S proteasome non-ATPase regulatory subunit 12 (CsGy2G015260) is included in this region. In tomato, a 26S proteasomal subunit RPT4a (SlRPT4) interferes with the genome transcription of ToLCNDV and induces the hypersensitive response [**63**]. Although SlRPT4 protein has an active ATPase activity, a possible effect of CsGy2G015260 against ToLCNDV infection must be further explored.

Our first approximation of candidate genes for resistance to ToLCNDV is being broadened with new sequencing assays, which will provide new molecular markers to finely map the identified QTL and facilitate marker-assisted breeding for ToLCNDV resistance in cucumber.

5. Conclusions

In this paper, germplasm accessions of cucumber (*Cucumis sativus*) from different geographical origins were screened for resistance to ToLCNDV. Three Indian accessions (CGN23089, CGN23423, and CGN23633), as well as all plants of their progenies obtained by selfing, were highly resistant to the mechanical inoculation, and remained symptomless and showed a reduced viral accumulation. Plants of the CGN23089 accession were crossed with plants of the susceptible accession BGV011742, and F_1 hybrids were used to construct segregating populations (F_2 and backcrosses), which were genotyped with SNPs distributed along the *C. sativus* genome. The results suggest a monogenic recessive genetic control, and a QTL in chromosome 2 of cucumber was identified controlling the resistance. The described SNPs linked to the resistance can be used in breeding programs to obtain cucumber cultivars with tolerance to ToLCNDV.

Supplementary Materials

Table S1: List of polymorphic SNPs in the F_2 (CGN2-23089 x BGV011742) population. Their position in the genome is according to Version 2 of the cucumber genome Gy14 (**http://cucurbitgenomics.org**). The positions in the genetic map were calculated using Kosambi's function and used for QTL analysis. Table S2: Syntenic genes between candidate region for ToLCNDV resistance in chromosome 8 of *C. moschata*, chromosome 11 of *C. melo*, and chromosome 6 of *C. sativus* genomes (v1.0, v3.6.1, and Gy-14v2, respectively), determined with Tripal "SyntenyViewer", available at cucurbitgenomics.com. Table S3: Annotated genes in the candidate region of chromosome 6 of *C. sativus* genome (Gy-14v2) available at cucurbitgenomics.com.

References

1. FAO. Food and Agriculture Organization of the United Nations. Food and Agriculture Data. Available online: **http://www.fao.org/faostat/en/#data/QC** (accessed on 5 March 2021).
2. Lecoq, H.; Katis, N. Control of cucurbit viruses. *Adv.Virus Res.* 2014, *90*, 255–296. [**Google Scholar**] [**CrossRef**] [**PubMed**]
3. Martín-Hernández, A.M.; Picó, B. Natural resistances to viruses in cucurbits. *Agronomy* 2021, *11*, 23. [**Google Scholar**] [**CrossRef**]
4. Srivastava, K.M.; Hallan, V.; Raizada, R.K.; Chandra, G.; Singh, B.P.; Sane, P.V. Molecular cloning of Indian tomato leaf curl virus genome following a simple method of concentrating the supercoiled replicative form of viral DNA. *J. Virol. Methods* 1995, *51*, 297–304. [**Google Scholar**] [**CrossRef**]
5. Moriones, E.; Praveen, S.; Chakraborty, S. Tomato leaf curl New Delhi virus: An emerging virus complex threatening vegetable and fiber crops. *Viruses* 2017, *9*, 264. [**Google Scholar**] [**CrossRef**] [**PubMed**] [**Green Version**]
6. Zaidi, S.S.; Martin, D.P.; Amin, I.; Farooq, M.; Mansoor, S. Tomato leaf curl New Delhi virus: A widespread bipartite begomovirus in the territory of monopartite begomoviruses. *Mol. Plant Pathol.* 2017, *18*, 901–911. [**Google Scholar**] [**CrossRef**]
7. Juárez, M.; Tovar, R.; Fiallo-Olivé, E.; Aranda, M.A.; Gosálvez, B.; Castillo, P.; Moriones, E.; Navas-Castillo, J. First detection of tomato leaf curl New Delhi virus infecting zucchini in Spain. *Plant Dis.* 2014, *98*, 857–858. [**Google Scholar**] [**CrossRef**]
8. Mnari-Hattab, M.; Zammouri, S.; Belkadhi, M.S.; Bellon Doña, D.; ben Nahia, E.; Hajlaoui, M.R. First report of tomato leaf curl New Delhi virus infecting cucurbits in Tunisia. *New Dis. Rep.* 2015, *31*, 21. [**Google Scholar**] [**CrossRef**] [**Green Version**]
9. Panno, S.; Iacono, G.; Davino, M.; Marchione, S.; Zappardo, V.; Bella, P.; Tomassoli, L.; Accotto, G.P.; Davino, S. First report of tomato leaf curl New Delhi virus affecting zucchini squash in an important horticultural area of southern Italy. *New Dis. Rep.* 2016, *33*, 6. [**Google Scholar**] [**CrossRef**] [**Green Version**]

10. Sifres, A.; Sáez, C.; Ferriol, M.; Selmani, E.A.; Riado, J.; Picó, B.; López, C. First report of tomato leaf curl New Delhi virus infecting zucchini in Morocco. *Plant Dis.* 2018, *102*, 1045. [**Google Scholar**] [**CrossRef**]

11. Orfanidou, C.G.; Malandraki, I.; Beris, D.; Keksidou, O.; Vassilakos, N.; Varveri, C.; Katis, N.; Maliogka, N.I. First report of tomato leaf curl New Delhi virus in zucchini crops in Greece. *J. Plant Pathol.* 2019, *101*, 799. [**Google Scholar**] [**CrossRef**] [**Green Version**]

12. Kheireddine, A.; Sifres, A.; Sáez, C.; Picó, B.; López, C. First report of tomato leaf curl New Delhi virus infecting cucurbit plants in Algeria. *Plant Dis.* 2019, *103*, 3291. [**Google Scholar**] [**CrossRef**]

13. EPPO. European and Mediterranean Plant Protection Organization. Available online: // **gd.eppo.int/taxon/TOLCND/distribution** (accessed on 20 November 2019).

14. Luigi, M.; Bertin, S.; Manglli, A.; Troiano, E.; Davino, S.; Tomassoli, L.; Parrella, G. First report of tomato leaf curl New Delhi virus causing yellow leaf curl of pepper in Europe. *Plant Dis.* 2019, *103*, 2970. [**Google Scholar**] [**CrossRef**]

15. Jyothsna, P.; Haq, Q.M.I.; Singh, P.; Sumiya, K.V.; Praveen, S.; Rawat, R.; Briddon, R.W.; Malathi, V.G. Infection of tomato leaf curl New Delhi virus (ToLCNDV), a bipartite begomovirus with betasatellites, results in enhanced level of helper virus components and antagonistic interaction between DNA B and betasatellites. *Appl. Microbiol. Biotechnol.* 2013, *97*, 5457–5471. [**Google Scholar**] [**CrossRef**] [**PubMed**]

16. Chang, H.H.; Ku, H.M.; Tsai, W.S.; Chien, R.C.; Jan, F.J. Identification and characterization of a mechanical transmissible begomovirus causing leaf curl on oriental melon. *Eur. J. Plant Pathol.* 2010, *127*, 219–228. [**Google Scholar**] [**CrossRef**]

17. Islam, S.; Munshi, A.D.; Verma, M.; Arya, L.; Mandal, B.; Behera, T.K.; Kumar, R.; Lal, S.K. Genetics of resistance in Luffa cylindrica Roem. against tomato leaf curl New Delhi virus. *Euphytica* 2010, *174*, 83–89. [**Google Scholar**] [**CrossRef**]

18. Sohrab, S.S.; Karim, S.; Varma, A.; Abuzenadah, A.M.; Chaudhary, A.G.; Damanhouri, G.A.; Mandal, B. Characterization of tomato leaf curl New Delhi virus infecting cucurbits: Evidence for sap transmission in a host specific manner. *Afr. J. Biotechnol.* 2013, *12*, 5000–5009. [**Google Scholar**] [**CrossRef**] [**Green Version**]

19. López, C.; Ferriol, M.; Picó, M.B. Mechanical transmission of tomato leaf curl New Delhi virus to cucurbit germplasm: Selection of tolerance sources in *Cucumis melo*. *Euphytica* 2015, *204*, 279–691. [**Google Scholar**] [**CrossRef**]

20. Sangeetha, B.; Malathi, V.G.; Alice, D.; Suganthy, M.; Renukadevi, P. A distinct seed-transmissible strain of tomato leaf curl New Delhi virus infecting chayote in India. *Virus Res.* 2018, *251*, 81–91. [**Google Scholar**] [**CrossRef**]

21. Kil, E.J.; Vo, T.T.B.; Fadhila, C.; Ho, P.T.; Lal, A.; Troiano, E.; Parrella, G.; Lee, S.K. Seed transmission of tomato leaf curl New Delhi virus from zucchini squash in Italy. *Plants* 2020, *9*, 563. [**Google Scholar**] [**CrossRef**]

22. EPPO. European and Mediterranean Plant Protection Organization. Available online: **https://www.eppo.int/QUARANTINE/Alert_List/viruses/ToLCNDV.htm** (accessed on 28 February 2017).

23. Islam, S.; Munshi, A.D.; Verma, M.; Arya, L.; Mandal, B.; Behera, T.K.; Kumar, R.; Lal, S.K. Screening of *Luffa cylindrica* Roem. for resistance against tomato leaf curl New Delhi virus, inheritance of resistance, and identification of SRAP markers linked to the single dominant resistance gene. *J. Hortic. Sci. Biotechnol.* 2011, *86*, 661–667. [**Google Scholar**] [**CrossRef**]

24. Sáez, C.; Esteras, C.; Martínez, C.; Ferriol, M.; Narinder, P.S.D.; López, C.; Picó, B. Resistance to tomato leaf curl New Delhi virus in melon is controlled by a major QTL located in chromosome 11. *Plant Cell Rep.* 2017, *36*, 1571–1584. [**Google Scholar**] [**CrossRef**]

25. Sáez, C.; Martínez, C.; Ferriol, M.; Manzano, S.; Velasco, L.; Jamilena, M.; López, C.; Picó, B. Resistance to tomato leaf curl New Delhi virus in Cucurbita spp. *Ann. Appl. Biol.* 2016, *169*, 91–105. [**Google Scholar**] [**CrossRef**]

26. Sáez, C.; Martínez, C.; Montero-Pau, J.; Esteras, C.; Blanca, J.; Sifres, A.; Ferriol, M.; López, C.; Picó, B. A major QTL located in chromosome 8 of *Cucurbita moschata* is responsible for resistance to tomato leaf curl New Delhi virus (ToLCNDV). *Front. Plant Sci.* 2020, *11*, 207. [**Google Scholar**] [**CrossRef**] [**PubMed**] [**Green Version**]

27. Valcárcel, J.V.; Peiró, R.M.; Pérez-de-Castro, A.; Díez, M.J. Morphological characterization of the cucumber (*Cucumis sativus* L.) collection of the COMAV's Genebank. *Genet. Resour.CropEvol.* 2018, *65*, 1293–1306. [**Google Scholar**] [**CrossRef**]

28. Valcárcel, J.V.; Pérez-de-Castro, A.; Díez, M.J.; Peiró, R.M. Molecular characterization of the cucumber (*Cucumis sativus* L.) accessions of the COMAV's genebank. *Span. J. Agric. Res.* 2018, *16*, e0701. [**Google Scholar**] [**CrossRef**]

29. Aparicio, F.; Soler, S.; Aramburu, J.; Galipienso, L.; Nuez, F.; Pallás, V.; López, C. Simultaneous detection of six RNA plant viruses affecting tomato crops using a single digoxigenin-labelled polyprobe. *Eur.J. Plant Pathol.* 2009, *123*, 117–123. [**Google Scholar**] [**CrossRef**]

30. Doyle, J.J.; Doyle, J.L. Isolation of plant DNA from fresh tissue. *Focus* 1990, *12*, 13–15. [**Google Scholar**]

31. Kong, Q.; Yuan, J.; Gao, L.; Zhao, L.; Cheng, F.; Huang, Y.; Bie, Z. Evaluation of appropriate reference genes for gene expression normalization during watermelon fruit development. *PLoS ONE* 2015, *10*, e0130865. [**Google Scholar**] [**CrossRef**]

32. Livak, K.J.; Schmittgen, T.D. Analysis of relative gene expression data using real-time quantitative PCR and the 2(-Delta Delta Ct) method. *Methods* 2001, *25*, 402–408. [**Google Scholar**] [**CrossRef**] [**PubMed**]

33. Wang, X.; Bao, K.; Reddy, U.K.; Bai, Y.; Hammar, S.A.; Jiao, C.; Wehner, T.C.; Ramírez-Madera, A.O.; Weng, Y.; Grumet, R.; et al. The USDA cucumber (*Cucumis sativus* L.) collection: Genetic diversity, population structure, genome-wide association studies, and core collection development. *Hortic. Res.* 2018, *5*, 64. [**Google Scholar**] [**CrossRef**]

34. Langmead, B.; Salzberg, S. Fast gapped-read alignment with Bowtie 2. *Nature Methods.* 2012, *9*, 357–359. [**Google Scholar**] [**CrossRef**] [**PubMed**] [**Green Version**]

35. Garrison, E.; Marth, G. Haplotype-based variant detection from short-read sequencing. *arXiv* 2012, arXiv:1207.3907. [**Google Scholar**]

36. Lincoln, S.; Daly, M.; Lander, E. *Constructing Genetic Maps with MAPMAKER/EXP 3.0. Whitehead Institute Technical Report*, 3rd ed.; Whitehead Institute: Cambridge, MA, USA, 1992. [**Google Scholar**]

37. Joehanes, R.; Nelson, J.C. QGene 4.0, an extensible Java QTL-analysis platform. *Bioinformatics* 2008, *24*, 2788–2789. [**Google Scholar**] [**CrossRef**] [**PubMed**] [**Green Version**]

38. Van Ooijen, J.W. *MapQTL® 6 Software for the Mapping of Quantitative Trait Loci in Experimental Population of Diploid Species*; Kayzma BV: Wageningen, The Netherlands, 2009. [**Google Scholar**]

39. RAIF, Red de Alerta e InformaciónFitosanitaria de Andalucía. Situación Actual del Virus del Rizado de Nueva Delhi (ToLCNDV) Bajo InvernaderoenCultivo de PEPINO en la Provincia de Almería. 2015. Available online: **//www.juntadeandalucia.es/agriculturapescaydesarrollorural/raif/ca/novedades/-/asset_publisher/4rpcT3lrh8uL/content/situacion-actual-del-virus-del-rizado-de-nueva-delhi-tolcndv-bajo-invernadero-en-cultivo-de-pepino-en-la-provincia-de-almeri-1?inheritRedirect=false** (accessed on 25 March 2021).

40. Weng, Y.; Wehner, T.C. Cucumber gene catalog 2017. *Rep. Cucurbit Genet. Coop.* 2017, *39–40*, 17–54. [**Google Scholar**]

41. Tian, G.L.; Yang, Y.H.; Zhang, S.P.; Miao, H.; Lu, H.W.; Wang, Y.; Xie, B.Y.; Gu, X.F. Genetic analysis and gene mapping of papaya ring spot virus resistance in cucumber. *Mol. Breed.* 2015, *35*, 110. [**Google Scholar**] [**CrossRef**]

42. Tian, G.L.; Miao, H.; Yang, Y.H.; Zhou, J.; Lu, H.W.; Wang, Y.; Xie, B.Y.; Zhang, S.P.; Gu, X.F. Genetic analysis and fine mapping of watermelon mosaic virus resistance gene in cucumber. *Mol. Breed.* 2016, *36*, 131. [**Google Scholar**] [**CrossRef**]

43. Shi, L.; Yang, Y.; Xie, Q.; Miao, H.; Bo, K.; Song, Z.; Wang, Y.; Xie, B.; Zhang, S.; Gu, X. Inheritance and QTL mapping of cucumber mosaic virus resistance in cucumber (*Cucumis sativus* L.). *PLoS ONE* 2018, *13*, e0200571. [**Google Scholar**] [**CrossRef**]

44. Munshi, A.D.; Panda, B.; Mandal, B.; Bisht, I.S.; Rao, E.S.; Kumar, R. Genetics of resistance to cucumber mosaic virus in *Cucumis sativus* var. *hardwickii R. Alef. Euphytica* 2008, *164*, 501–507. [**Google Scholar**] [**CrossRef**]

45. Picó, B.; Villar, C.; Nuez, F.; Weber, W.E. Screening *Cucumis sativus* landraces for resistance to cucumber vein yellowing virus. *Plant Breed.* 2003, *122*, 426–430. [**Google Scholar**] [**CrossRef**]

46. Aguilar, J.M.; Abad, J.; Aranda, M.A. Resistance to cucurbit yellow stunting disorder virus in cucumber. *Plant Dis.* 2006, *90*, 583–586. [**Google Scholar**] [**CrossRef**]

47. Crespo, O.; Janssen, D.; Robles, C.; Ruiz, L. Resistance to cucumber green mottle mosaic virus in *Cucumis sativus*. *Euphytica* 2018, *214*, 201. [**Google Scholar**] [**CrossRef**]

48. Romero-Masegosa, J.; Martínez, C.; Aguado, E.; García, A.; Cebrián, G.; Iglesias-Moya, J.; Paris, H.S.; Jamilena, M. Response of *Cucurbita* spp. to tomato leaf curl New Delhi virus inoculation and identification of a dominant source of resistance in *Cucurbita moschata*. *Plant Pathol.* 2021, *70*, 206–218. [**Google Scholar**] [**CrossRef**]

49. Dhillon, N.P.S.; Monforte, A.J.; Pitrat, M.; Pandey, S.; Singh, P.K.; Reitsma, K.R.; García-Mas, J.; Sharma, A.; McCreight, J.D. Melon landraces of India: Contributions and importance. *Plant Breed. Rev.* 2012, *35*, 85–150. [**Google Scholar**] [**CrossRef**]

50. Díaz-Pendón, J.A.; Truniger, V.; Nieto, C.; García-Mas, J.; Bendahmane, A.; Aranda, M.A. Advances in understanding recessive resistance to plant viruses. *Mol. Plant Pathol.* 2004, *5*, 223–233. [**Google Scholar**] [**CrossRef**] [**PubMed**]

51. Wege, C. *Movement and localization of Tomato yellow leaf curl viruses in the infected plant*; In Tomato Yellow Leaf Curl Virus Disease; Springer: Dordrecht, The Netherlands, 2007; pp. 185–206. [**Google Scholar**] [**CrossRef**]

52. Marco, C.F.; Aguilar, J.M.; Abad, J.; Gómez-Guillamón, M.L.; Aranda, M.A. Melon resistance to cucurbit yellow stunting disorder virus in characterized by reduced virus accumulation. *Phytopathology* 2003, *93*, 844–852. [**Google Scholar**] [**CrossRef**] [**PubMed**]

53. Díaz-Pendón, J.A.; Fernández-Muñoz, R.; Gómez-Guillamón, M.L.; Moriones, E. Inheritance of resistance to watermelon mosaic virus in *Cucumis melo* that impairs virus accumulation, symptom expression, and aphid transmission. *Phytopathology* 2005, *95*, 840–846. [**Google Scholar**] [**CrossRef**] [**PubMed**] [**Green Version**]

54. Pacheco, D.A.; Rezende, J.A.M.; Piedade, S.M.dS. Biomass, virus concentration, and symptomatology of cucurbits infected by mild and severe strains of papaya ringspot virus. *Sci. Agric.* 2003, *60*, 691–698. [**Google Scholar**] [**CrossRef**] [**Green Version**]

55. Romay, G.; Pitrat, M.; Lecoq, H.; Wipf-Scheibel, C.; Millot, P.; Girardot, G.; Desbiez, C. Resistance against melon chlorotic mosaic virus and tomato leaf curl New Delhi virus in melon. *Plant Dis.* 2019, *103*, 2913–2919. [**Google Scholar**] [**CrossRef**] [**PubMed**] [**Green Version**]

56. Kundu, A.; Pal, A. Identification and characterization of elite inbred lines with MYMIV-resistance in Vigna mungo. *Field Crops Res.* 2012, *135*, 116–125. [**Google Scholar**] [**CrossRef**]

57. Kushwaha, N.; Singh, A.K.; Basu, S.; Chakraborty, S. Differential response of diverse solanaceous hosts to tomato leaf curl New Delhi virus infection indicates coordinated action of NBS-LRR and RNAi-mediated host defense. *Arch. Virol.* 2015, *160*, 1499–1509. [**Google Scholar**] [**CrossRef**]

58. Jeevalatha, A.; Siddappa, S.; Kumar, A.; Kaundal, P.; Guleria, A.; Sharma, S.; Nagesh, M.; Singh, B.P. An insight into differentially regulated genes in resistant and susceptible genotypes of potato in response to tomato leaf curl New Delhi virus-[potato] infection. *Virus Res.* 2017, *232*, 22–33. [**Google Scholar**] [**CrossRef**]

59. Yamaguchi, H.; Ohnishi, J.; Saito, A.; Ohyama, A.; Nunome, T.; Miyatake, K.; Fukuoka, H. An NB-LRR gene, TYNBS1, is responsible for resistance mediated by the Ty-2 Begomovirus resistance locus of tomato. *Theor. Appl. Genet.* 2018, *131*, 1345–1362. [**Google Scholar**] [**CrossRef**]

60. Bhattacharjee, P.; Das, R.; Mandal, A.; Kundu, P. Functional characterization of tomato membrane-bound NAC transcription factors. *Plant Mol. Biol.* 2017, *93*, 511–532. [**Google Scholar**] [**CrossRef**]

61. Soler-Garzón, A.; Oladzad, A.; Beaver, J.; Beebe, S.; Lee, R.; Lobaton, J.D.; Macea, E.; McClean, P.; Raatz, B.; Rosas, J.C.; et al. NAC Candidate gene marker for bgm-1 and interaction with QTL for resistance to bean golden yellow mosaic virus in common bean. *Front. Plant Sci.* 2021, *12*. [**Google Scholar**] [**CrossRef**] [**PubMed**]

62. Erdmann, R.M.; Picard, C.L. RNA-Directed DNA Methylation. *PLoS Genet.* 2020, *16*, e1009034. [**Google Scholar**] [**CrossRef**] [**PubMed**]

63. Sahu, P.P.; Sharma, N.; Puranik, S.; Chakraborty, S.; Prasad, M. Tomato 26S proteasome subunit RPT4a regulates ToLCNDV transcription and activates hypersensitive response in tomato. *Sci. Rep.* 2016, *6*, 27078. [**Google Scholar**] [**CrossRef**] [**PubMed**] [**Green Version**]

Index